The Organic Chemistry of Iron

Volume 2

ORGANOMETALLIC CHEMISTRY

A Series of Monographs

BRIAN G. RAMSEY: Electronic Transitions in Organometalloids, 1969.
R. C. POLLER: The Chemistry of Organotin Compounds, 1970.
RUSSELL N. GRIMES: Carboranes, 1970.
PETER M. MAITLIS: The Organic Chemistry of Palladium, Volumes I and II, 1971.
DONALD S. MATTESON: Organometallic Reaction Mechanisms of the Nontransition Elements, 1974.
RICHARD F. HECK: Organotransition Metal Chemistry: A Mechanistic Approach, 1974.
P. W. JOLLY AND G. WILKE: The Organic Chemistry of Nickel, Volume I, Organonickel Complexes, 1974. Volume II, Organic Synthesis, 1975.
P. C. WAILES, R. S. P. COUTTS, AND H. WEIGOLD: Organometallic Chemistry of Titanium, Zirconium, and Hafnium, 1974.
U. BELLUCO: Organometallic and Coordination Chemistry of Platinum, 1974.
P. S. BRATERMAN: Metal Carbonyl Spectra, 1974.
L. MALATESTA AND S. CENINI: Zerovalent Compounds of Metals, 1974.
THOMAS ONAK: Organoborane Chemistry, 1975.
R. P. A. SNEEDEN: Organochromium Compounds, 1975.
A. G. SHARPE: The Chemistry of Cyano Complexes of the Transition Metals, 1976.
ERNST A. KOERNER VON GUSTORF, FRIEDRICH-WILHELM GREVELS, AND INGRID FISCHLER: The Organic Chemistry of Iron, Volume 1, 1978. Volume 2, 1981.
G. DEGANELLO: Transition Metal Complexes of Cyclic Polyolefins, 1978.
BRIAN E. MANN AND BRIAN F. TAYLOR: The ^{13}C NMR Data for Organometallic Compounds, 1981.

The Organic Chemistry of Iron

Volume 2

edited by

ERNST A. KOERNER VON GUSTORF

FRIEDRICH-WILHELM GREVELS

INGRID FISCHLER

Institut für Strahlenchemie
Max-Planck-Institut für Kohlenforschung
Mülheim a. d. Ruhr, Germany

Academic Press **1981**

A Subsidiary of Harcourt Brace Jovanovich, Publishers

New York London Toronto Sydney San Francisco

Academic Press Rapid Manuscript Reproduction

ACADEMIC PRESS, INC.
111 Fifth Avenue, New York, New York 10003

United Kingdom Edition published by
ACADEMIC PRESS, INC. (LONDON) LTD.
24/28 Oval Road, London NW1 7DX

Library of Congress Cataloging in Publication Data
Main entry under title:

The Organic chemistry of iron.

(Organometallic chemistry)
Includes bibliographical references and index.
1. Organoiron compounds. I. Koerner von Gustorf, Ernst A., Date. II. Grevels, Friedrich Wilhelm. III. Fischer, Ingrid.
QD412.F4073 547'.05621 77-16071
ISBN 0-12-417102-8 (v. 2) AACR1

PRINTED IN THE UNITED STATES OF AMERICA

81 82 83 84 9 8 7 6 5 4 3 2 1

Contents

Contributors

Numbers in parentheses indicate the pages on which the authors' contributions begin.

P. Chini* (189), Istituto di Chimica Generale ed Inorganica, dell'Universitá, Via G. Venezian 21, Milano, 20133, Italy

Robert C. Kerber (1), Department of Chemistry, State University of New York at Stony Brook, Stony Brook, New York 11794

R. B. King (155), Department of Chemistry, University of Georgia, Athens, Georgia 30602

László Markó (283), Department of Organic Chemistry, University of Chemical Engineering, Veszprem, Hungary

Bernadett Markó-Monostory (283), Hungarian Oil and Gas Research Institute, Veszprem, Hungary

*Deceased.

Foreword

This book is the second and final volume of "The Organic Chemistry of Iron," edited by Dr. F. W. Grevels and Dr. I. Fischler in memory of Dr. Ernst A. Koerner Von Gustorf, who originally conceived the project.

It covers a series of selected topics in organo–iron chemistry, including complexes with poly-olefins, arenes, and sulphur-containing ligands, as well as an account of iron-metal bonds by the late Paolo Chini. Although these articles were for the most part commissioned and written some years ago, the authors have kindly added new material and have revised them to include later work.

We should like again to express our thanks to Dr. Grevels and Dr. Fischler and all the authors for their efforts, help, and patience in the face of the many difficulties that the monograph encountered due to the untimely death of Dr. Koerner Von Gustorf in 1975.

P. M. MAITLIS
F. G. A. STONE
R. WEST
August 1981

IRON COMPLEXES OF TRIENES, TETRAENES, AND POLYENES

ROBERT C. KERBER

Department of Chemistry
State University of New York at Stony Brook
Stony Brook, New York 11794, U.S.A.

This chapter is dedicated to the memory of

Ernst A. Koerner von Gustorf

outstanding chemist, colleague, and friend

ISBN 0-12-417102-8

TABLE OF CONTENTS

I. INTRODUCTION

This chapter will discuss the synthesis and properties of compounds in which iron is coordinated to an organic ligand which contains (in the free state) three or more double bonds, and of compounds derived from such organic ligands, even when the ligand has undergone change during the coordination process. Emphasis will be on structure and chemical properties, particularly those chemical properties (*e.g.*, fluxionality, nucleophilicity) which are especially pronounced in these compounds compared to ordinary diene complexes.

The number of electrons and orbitals available in a polyene for coordination to iron are, of course, greater than in simple alkenes and dienes. As a consequence, the stoichiometries and structural types of complexes which result from interaction of polyenes with iron-containing reagents are much more diverse than those from monoenes and dienes. I have chosen, therefore, to organize the material in this chapter primarily according to the numbers of iron atoms and ancillary ligands L (usually CO, less frequently phosphines or arsines) in the complexes, and only secondarily according to the nature of the free polyene ligand. I hope that this deviation from custom will result in some new insights into the properties of these organometallic species.

No previous review has dealt with precisely this subject. A recent review *(165)* has extensively discussed iron, ruthenium and osmium carbonyl complexes of cyclic polyenes; complexes of transition metals in general with such ligands had earlier been reviewed *(40,202)*. Fluxionality, a property especially pronounced in many of these complexes, has been reviewed by Cotton and coworkers *(59,120,122)*. Transition metal complexes of special subclasses of polyenes have also been reviewed: azulenes *(105)*, fulvenes *(268)*, and trienes *(372)*.

Literature coverage of this chapter is intended to be complete to mid-1976; some references to the end of 1976 are also included. Nomenclature is informal; I have opted for simplicity and clarity rather than for total adherence to a formal system.

II. COMPLEXES OF $CpFe(CO)_2$ (Cp = η^5-C_5H_5)

The cyclopentadienyldicarbonyliron group readily forms compounds with σ-bonds to organic groups, and these compounds have an extensively explored chemistry. The only σ-bonded polyene complexes of iron, 1 and 2 *(118)*, contain this group. It was suggested on the basis of IR that 1, described as a

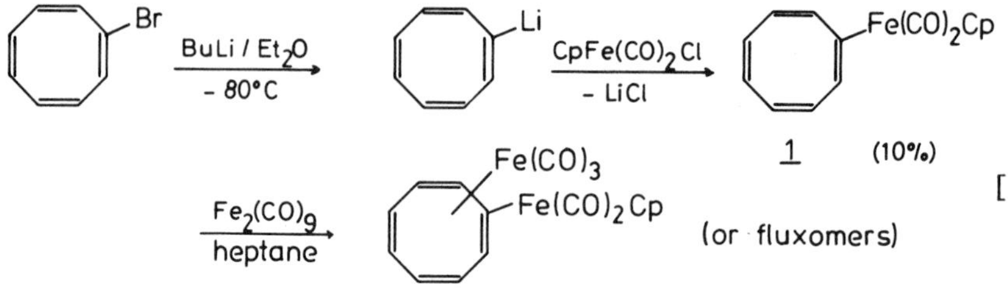

1 (10%) (or fluxomers)

2 (7%)

[1]

"surprisingly robust" compound, may exist in two distinct conformations, presumably rotamers about the C-Fe bond, in solution. 2, with its eight-membered ring presumably partially flattened by coordination to the $Fe(CO)_3$ group *(173)*, did not show this effect. The ^{1}H-NMR spectrum of 2 showed two-fold symmetry of the eight-membered ring, suggesting fluxionality of the $Fe(CO)_3$ group. Coalescence was not observed at -100°C.

Attempts to introduce an (η^1-cyclopentadienyl)dicarbonyliron group at the 7-position of free or coordinated cycloheptatrienes have so far failed, η^3-coordinated products being obtained instead (eq. [2] *(115,116)* and [3] *(287)*). The product 3 was fluxional, with only one resonance in the room temperature NMR and a coalescence temperature of about -10°C.

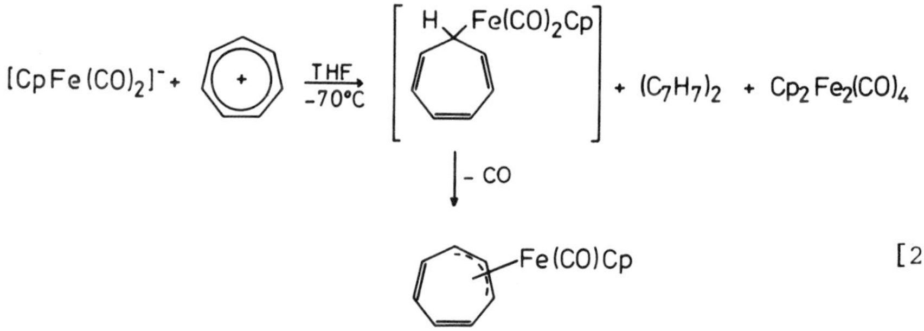

[2]

3 (4%)

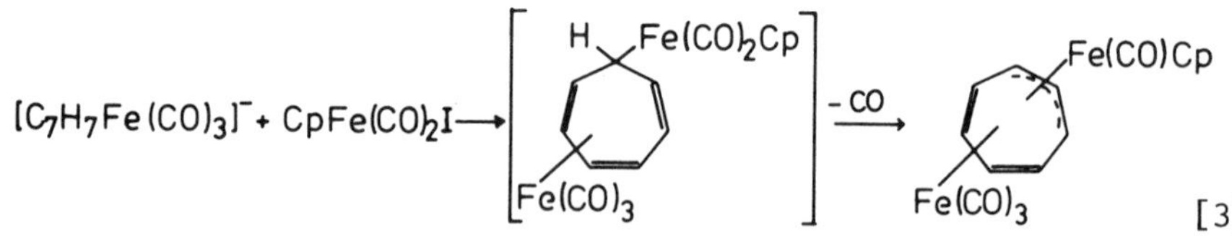

[3]

4

Formation of compounds analogous to 4 has also been reported for carbonyl halides of Mn, Mo, Re, and Rh *(41,141)*.

The apparent instability of (η^1-cycloheptatrien-7-yl)-metal complexes stands in stark contrast to the numerous (η^1-cyclopentadien-5-yl)metal compounds known, and is probably related to the antiaromatic character of cycloheptatrienyl anion, of which they are formally derivatives.

Another stable (η^1-cyclopentadienyl)dicarbonyliron compound, 5, the first complex of heptafulvene, was prepared as shown in equation [4] *(176,269)*. 5 may be viewed as a tro-

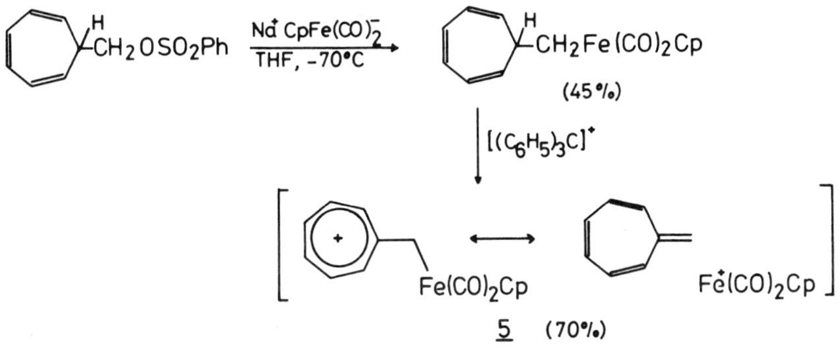

[4]

pylium ion stabilized by carbon-metal hyperconjugation. Its assumption of this η^1- rather than η^2-coordination was confirmed by X-ray crystallography *(109)*, which showed the iron atom to be 2.16(5) Å from C(8) of the essentially planar heptafulvene ligand, and 3.00 Å from C(7), well beyond normal bonding distance. The iron lay on the C_2 axis of the heptafulvene ligand as expected for optimum hyperconjugation involving the C-Fe bond.

III. COMPLEXES OF $[CpFe(CO)_2]^+$ AND $Fe(CO)_4$

Each of these groups, having 16 electrons in the valence shell of iron, can achieve an inert gas configuration by η^2-coordination to a double bond. An example of such a complex, isomeric to 5 but much less stable, is the cyclooctatetraene complex 6 *(148)*. 6 is unstable above 0°C in nitromethane solution and, unlike most cyclooctatetraene complexes, is not fluxional up to that temperature. Presumably the eight-mem-

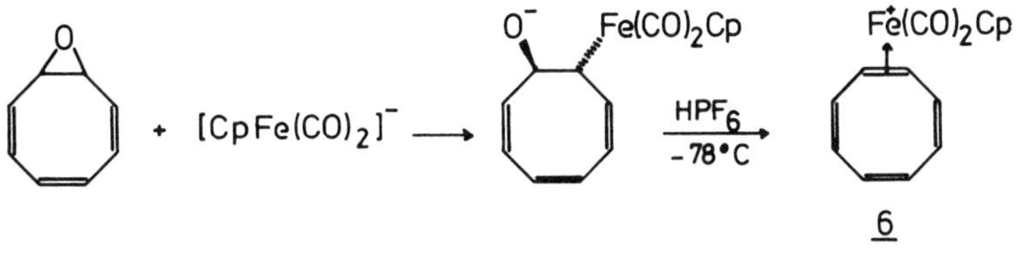

[5]

bered ring retains its tub conformation in complex 6, with the iron on the *exo* side. The reason for the low stability of 6 compared to most η^2-alkene complexes of $[CpFe(CO)_2]^+$ is not clear.

A related complex also of low stability is the cyclopentadienyl-dicarbonyliron complex of benzocyclobutadiene *(357a)*, prepared by an unusual oxidative demetallation (eq. [6]).

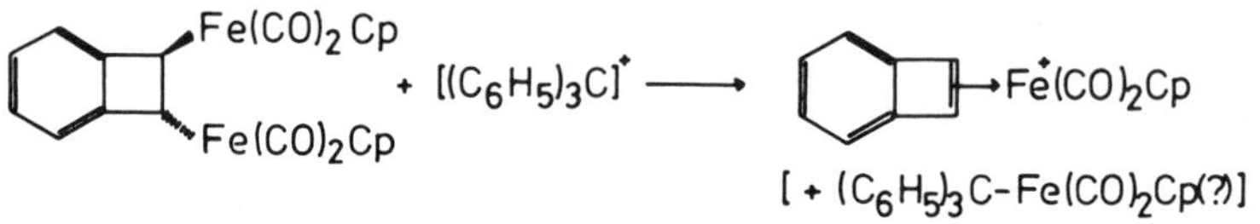

[6]

Two cyclopentadienyldicarbonyliron complexes of butatriene, 7 and 8, have recently been reported *(34)* (eq. [7]).

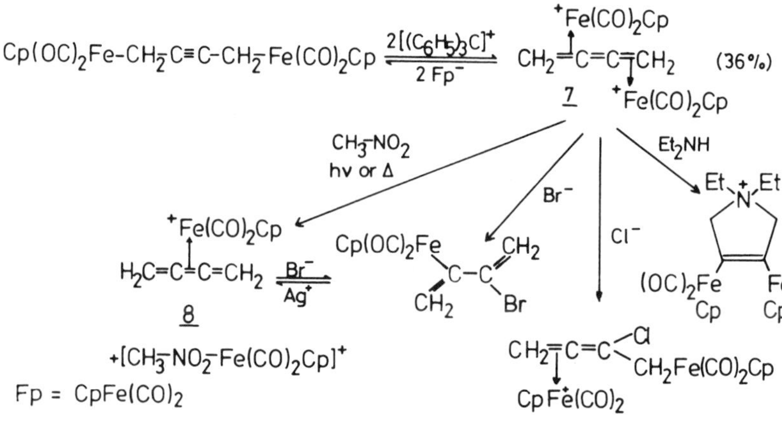

[7]

The reactions of 7 with nucleophiles are obviously complex and deserving of further study. The migration of the remaining iron group upon loss of the first when 7 → 8 is clearly indicated by the ^{1}H-NMR spectrum, and shows greater thermodynamic stability of the centrally-coordinated butatriene.

A much earlier example of a centrally-coordinated butatriene was obtained by reaction of $Fe_2(CO)_9$ with tetraphenylbutatriene *(260)* (eq. [8]). When a larger amount of $Fe_2(CO)_9$

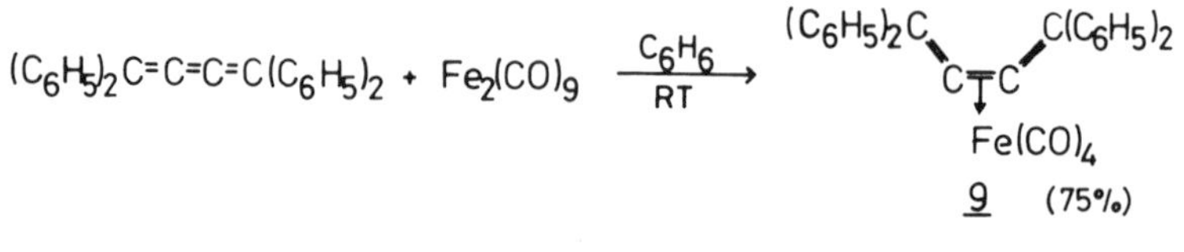

[8]

was used, the product was a $Fe_2(CO)_6$ complex (Sect. IX). 9, unlike many $Fe(CO)_4$ complexes, was quite stable and gave the butatriene back on oxidative decomposition with $FeCl_3$. An X-ray crystallographic structure determination *(62)* of 9 not only confirmed the coordination of the central double bond, but also showed a bending of the carbon backbone away from the iron (C=C=C angle 151°). Reaction of tetraphenylhexapentaene with iron carbonyls gave a mixture of complexes from which pure products were not isolated *(260,321)*.

However, reaction of tetrakis(*t*-butyl)hexapentaene with either $Fe_2(CO)_9$ in ether or with $Fe_3(CO)_{12}$ in refluxing hexane gave modest yields of a single $Fe(CO)_4$ complex, in addition to an $Fe_2(CO)_6$ complex *(279)*. Again, coordination of the central double bond is clearly indicated by both ^{1}H- and ^{13}C-NMR results.

In contrast to these cumulene-$Fe(CO)_4$ complexes, most $Fe(CO)_4$ complexes are rather unstable compounds. They are seldom isolated from reactions of polyenes with iron carbonyls, although they are often postulated as intermediates on the way to more stable products, since free or solvated $Fe(CO)_4$ is generally assumed to be the common reactive intermediate in thermolysis (≥ 80°C) or photolysis of $Fe(CO)_5$ and thermolysis of $Fe_2(CO)_9$ (≥ 20°C). Consistent with this notion, isolation of $Fe(CO)_4$ complexes usually results only from reactions run at about room temperature and preferably with short reaction times.

Thus, Murdoch and Weiss *(320)* in 1963 isolated an $Fe(CO)_4$ complex *inter alia* from reaction of $Fe_2(CO)_9$ and 1,3,5-hexatriene in a 1:1 ratio in petroleum ether at 40°C. The material was unstable and underwent disproportionation to free hexatriene and an $Fe_2(CO)_8$ complex and conversion to an $Fe(CO)_3$ complex upon attempted purification. The site of coordination of the $Fe(CO)_4$ unit thus remains uncertain. Use of greater quantities of $Fe_2(CO)_9$ in the synthesis gave only the $Fe(CO)_3$, $Fe_2(CO)_8$, and $Fe_2(CO)_7$ complexes of hexatriene.

Reaction of styrene and a number of substituted styrenes with $Fe_2(CO)_9$ gave only complexes of the double bond with an

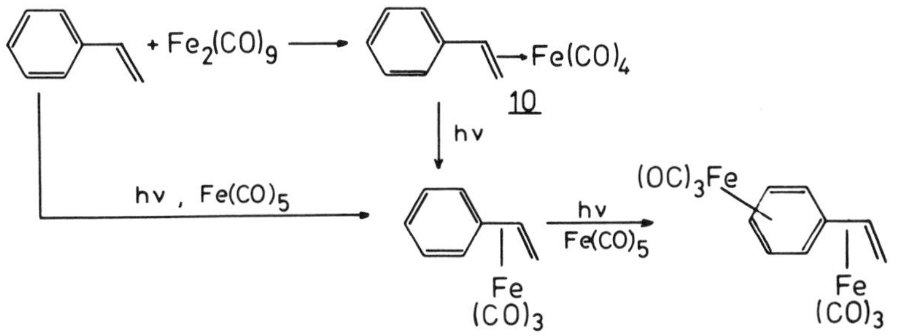

[9]

$Fe(CO)_4$ unit, 10 *(385,386,387)*. Photolysis of 10 or photochemical reaction of the free styrene with $Fe(CO)_5$ gave further products containing one or two $Fe(CO)_3$ groups (eq. [9]). These results support the notion of transient $Fe(CO)_4$ complexes in reactions of polyenes with iron carbonyls; in the cases of dienes or polyenes, however, intramolecular displacement of a CO by a double bond appears to be more facile than displacement by the phenyl ring in the styrene complex 10.

$Fe(CO)_4$ complexes of alkenes are often more stable when the alkene is strained. Thus, semibullvalene yields an isolable $Fe(CO)_4$ complex on photochemical reaction with $Fe(CO)_5$, whereas bullvalene does not *(22)*. Products which appear to derive from a transient $Fe(CO)_4$ complex are, however, obtained (eq. [10], see Section IX).

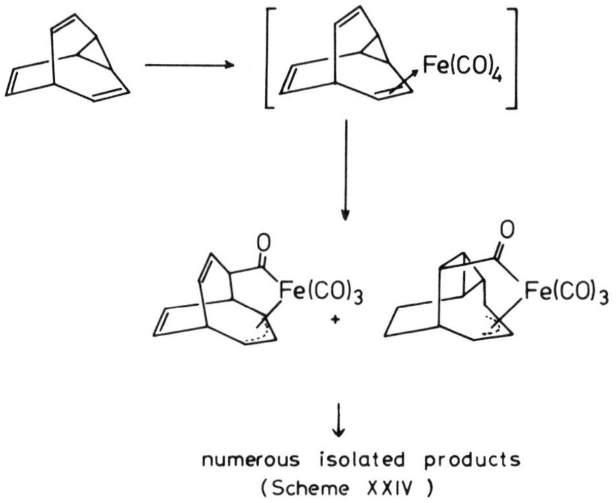

[10]

The still more highly strained 1,2-dimethylenecyclobutene gave two $Fe(CO)_4$ complexes on 16 h reaction with $Fe_2(CO)_9$ in pentane *(280)* (eq. [11]). Interestingly, reaction in THF is

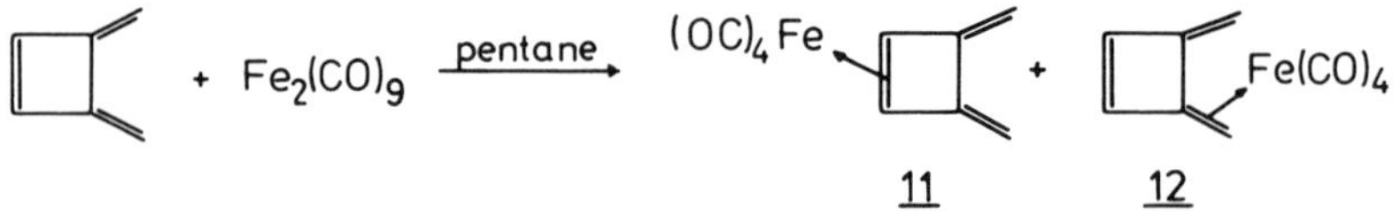

[11]

11 12

complete in only five hours, and results in formation of 11 only (31 % yield). 11 slowly isomerizes to 12 on standing in solution, presumably by dissociation of $Fe(CO)_4$ then recombination. Thus, formation of 11 is the result of kinetic control, but 12 is thermodynamically more stable. Stable $Fe(CO)_4$ complexes were not formed from 1,2-dimethylenecyclo-

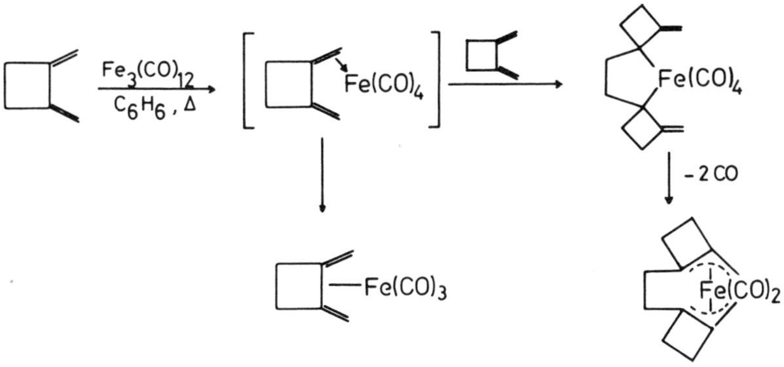

13

[12]

butane, for which the scheme (eq. [12]) was proposed *(280)*. Formation of the unusual product 13, first reported in 1966 *(274,275)*, was attributed to the large spread imposed upon the 1,3-diene unit by the fused cyclobutane ring, which inhibits formation of the tricarbonyliron complex. This is also consistent with the failure of 12 to form the diene-tricarbonyliron complex.

An $Fe(CO)_4$ complex has been reported by one group of workers studying the reaction of bicyclo[6.1.0]nonatriene with $Fe_2(CO)_9$ in ether *(163)*, along with a number of other products. This material may be formed from the cyclobutene valence tautomer of the starting hydrocarbon (eq. [13]).

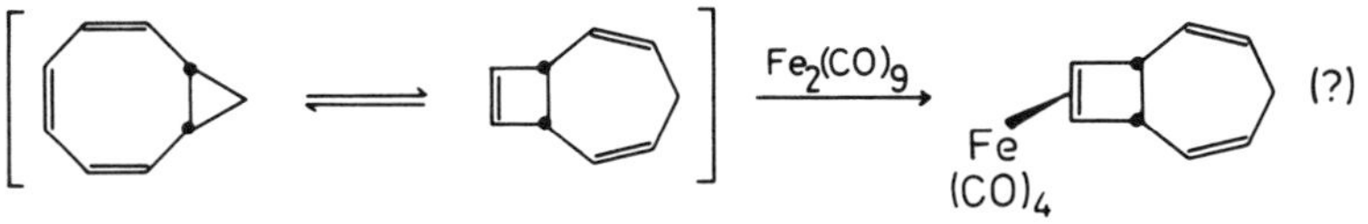

[13]

IV. COMPLEXES OF $Fe(CO)_3$ AND $Fe(CO)_2L$

An $Fe(CO)_3$ moiety is four electrons short of the inert gas configuration, which can therefore be achieved by coordination to two double bonds. Indeed, the diene-tricarbonyliron complexes constitute a very large class of organoiron compounds, described in a chapter by R.B. King in volume 1 of this book. In the case of polyene ligands, coordination of only two double bonds can lead to possibilities of coordination isomerism and altered reactivity in the free double bond(s), which will be discussed in this section.

A. $Fe(CO)_3$ COMPLEXES OF ACYCLIC POLYENES

Organometallic complexes of acyclic polyenes have been much less extensively studied than those of cyclic polyenes, probably because of lower availability and/or stability of the polyenes.

The complexes have most frequently been obtained by reaction of the free polyene with an iron carbonyl. Thus, reaction of hexatriene with $Fe_2(CO)_9$ in a 1:1 ratio gave, after distillation, the tricarbonyliron complex 14 in 34 % yield *(320)*. Use of larger amounts of $Fe_2(CO)_9$ gave essentially the same amount of 14, along with increased amounts of $Fe_2(CO)_7$ and $Fe_2(CO)_8$ complexes of hexatriene. An uncharacterized material, presumably at least partly 14, was also obtained from hexatriene and $Fe_3(CO)_{12}$ *(407)*. 1,6-Diphenylhexatriene gave a 79 % yield of the $Fe(CO)_3$ complex on reaction with $Fe(CO)_5$ in decalin at 150°C *(400)*. The use of pentacarbonyliron in refluxing butyl ether (142°C) or of dodecacarbonyltriiron in refluxing benzene (80°C) have been recommended over nonacarbonyldiiron for complexation of trienes *(30,168)*.

Unsymmetrically substituted trienes can give coordination isomers upon complexation with an $Fe(CO)_3$ group. If the triene bears an electron-withdrawing group at one end, the iron preferentially or exclusively coordinates the diene moiety at that end (see Table 1). Coordination of diene units with *syn* substituents (such as C in Table I) tends to be avoided if possible in these systems. Thus, a degree of selectivity can be obtained in direct complexation reactions. The selectivity largely results from thermodynamic rather than kinetic factors, at least when reactions are run at 80 - 140°C, as for the entries in Table I (see discussion below).

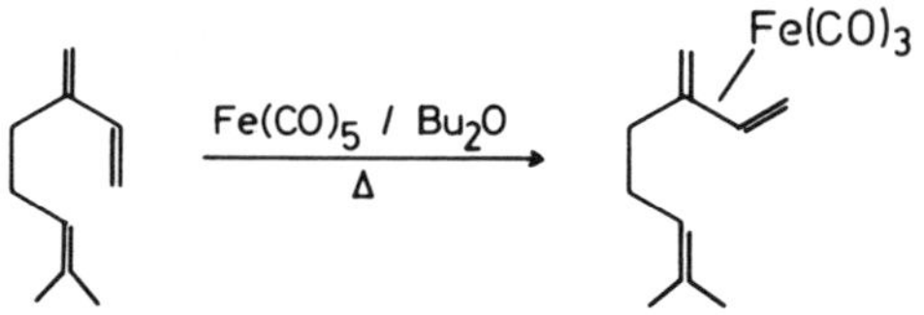

myrcene 15 (80%) [14]

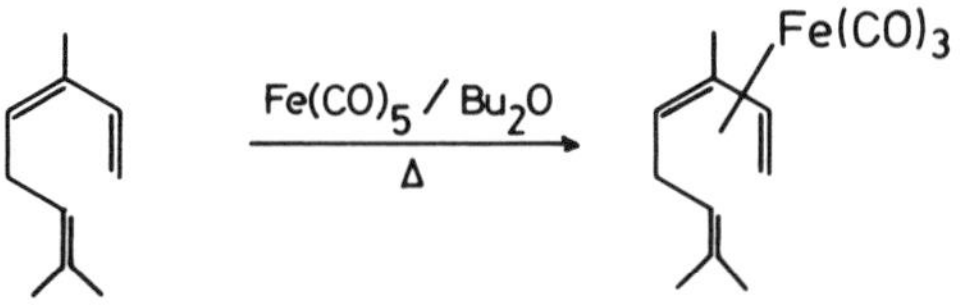

cis-ocimene 17 (83%) [15]

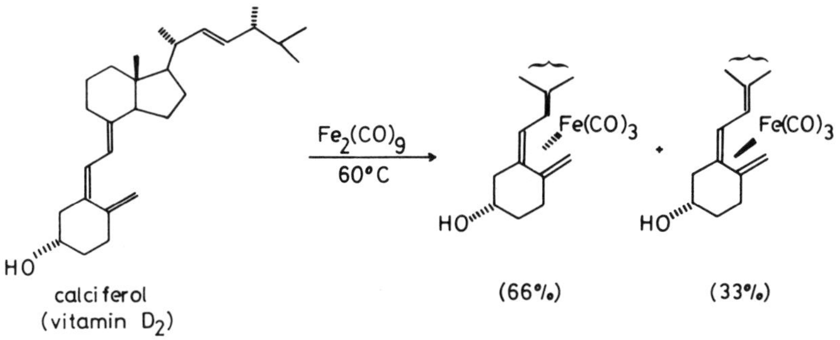

[16]

Not surprisingly, trienes in which one double bond is out of conjugation with the other two give complexes of the conjugated diene moiety, a result which has proven useful in a number of studies involving terpenoids (eqs. [14] *(30,168)*, [15] *(30)*, [16] *(32,330)*).

In a number of cases, some of which appear as "other products" in Table 1, coordination is accompanied by H migration or other rearrangement. This happens most commonly when the original polyene is unconjugated or, if conjugated, is unable to adopt the *s-cis* conformation needed for a diene-tricarbonyliron moiety, *e.g.* eq. [17] *(82)*. Likewise, reac-

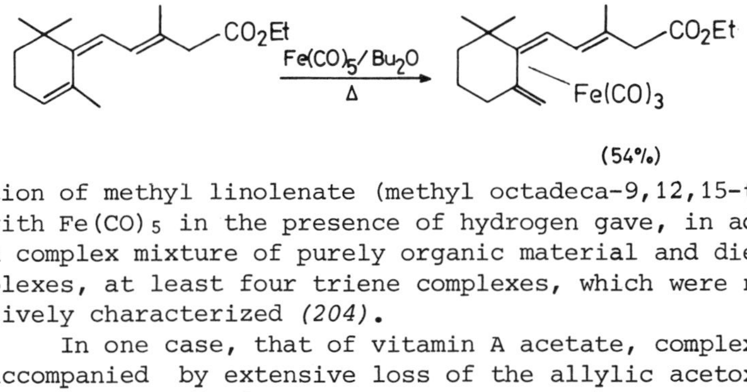

[17]

tion of methyl linolenate (methyl octadeca-9,12,15-trienoate) with $Fe(CO)_5$ in the presence of hydrogen gave, in addition to a complex mixture of purely organic material and diene complexes, at least four triene complexes, which were not extensively characterized *(204)*.

In one case, that of vitamin A acetate, complexation was accompanied by extensive loss of the allylic acetoxy group *(331,48)* (eq. [18]). A non-allylic acetoxy group poses, in

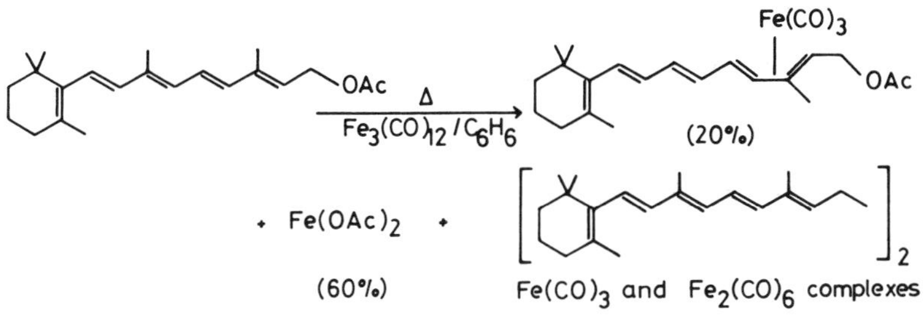

[18]

Table 1: Results of Complexation of Trienes 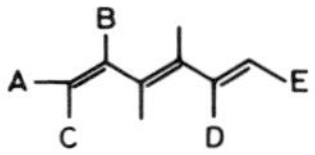

A	B	C	D	E	Iron Carbonyl	t [°C]	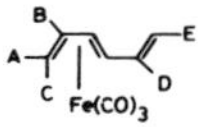	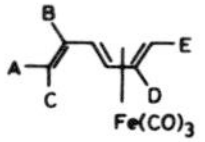	Other Products	*Ref.*
C_6H_5–CH=CH	H	H	H	C_6H_5	$Fe_3(CO)_{12}$	80	3 %	18 %	---	*403*
p–CH_3–C_6H_4	H	H	H	CHO	$Fe_3(CO)_{12}$	75	13 %	16 %	3 %[a]	*403*
C_6H_5	H	H	H	CHO	$Fe_3(CO)_{12}$	75	15 %	55 %	1 %[a]	*403*
C_6H_5	H	H	H	CHO	$Fe_3(CO)_{12}$	80	7 %	27 %	4 %[a]	*64*
C_6H_5–CH=CH	H	H	H	CHO	$Fe_3(CO)_{12}$	80	–	10 %	traces	*64*
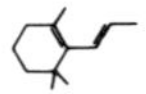	H	CH_3	CH_3	CHO	$Fe_3(CO)_{12}$	80	–	45 %	---	*48,49,313*
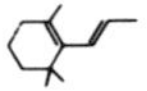	H	CH_3	CH_3	CH_2OAc	$Fe_3(CO)_{12}$ $Fe(CO)_5$	80,140	–	20 %	60 %	*331,48*
		CH_3	H	CO_2CH_3	$Fe(CO)_5$	142	–	46 %	---	*82*
CH_3	H	CH_3	CH_3	CH_3[b]	$Fe_3(CO)_{12}$	80	–	6 %[c]	---	*281*
CH_3	H	CH_3	CH_3	CH_3[b]	$Fe(CO)_5$	142	–	64 %	16 %	*168*
CH_3	H	CH_3	CH_3	CH_3[b]	$Fe_2(CO)_9$	80	–	8 %		*168*
CH_3	H	CH_3	CH_3	CH_3[b]	$Fe_3(CO)_{12}$	80	–	65 %	---	*30*
CH_3	H	CH_3	CH_3	CH_3[b]	$Fe_3(CO)_{12}$	142[d]	–	15 %[c]	55 %	*30*

[a] $Fe(CO)_3$ complex of α,β-unsaturated aldehyde group; [b] *allo*-ocimene: mixture of *cis* and *trans* isomers; [c] CH_3 substituent at E *syn* rather than *anti*; [d] Prolonged refluxing: 150 h.

contrast, no difficulty, as shown by the successful complexation of calciferol acetate *(32)*.

The nature of some of the rearrangements accompanying complexation has been explored in some detail in the case of *allo*-ocimene. This material is obtained as a mixture of (4-*E*, 6-*E*) and (4-*E*, 6-*Z*) isomers of 2,6-dimethyl-2,4,6-octatriene. Banthorpe, Fitton and Lewis *(30)* found that reaction with $Fe_3(CO)_{12}$ in refluxing benzene for twelve hours gave a single complex, 16, in 65 % yield. An isomeric complex, 17, was obtained from *cis*-ocimene, (5-*Z*)-2,6-dimethyl-2,5,7-octatriene, in 83 % yield. Refluxing either 16 or 17 for 150 h in butyl ether gave two new isomers, 18 and 19, in 55 % and 15 % yields, respectively (Scheme I). The latter was also slowly produced from 16 on alumina chromatography. Previous workers had obtained 19 only *(281)* or 18 and 19 *(168)* directly from *allo*-ocime. Treatment of either 16 or 17 with triphenylcarbenium fluoroborate then sodium borohydride gave a mixture of 16 - 19 in the same ratio, presumably *via* the common cation 20.

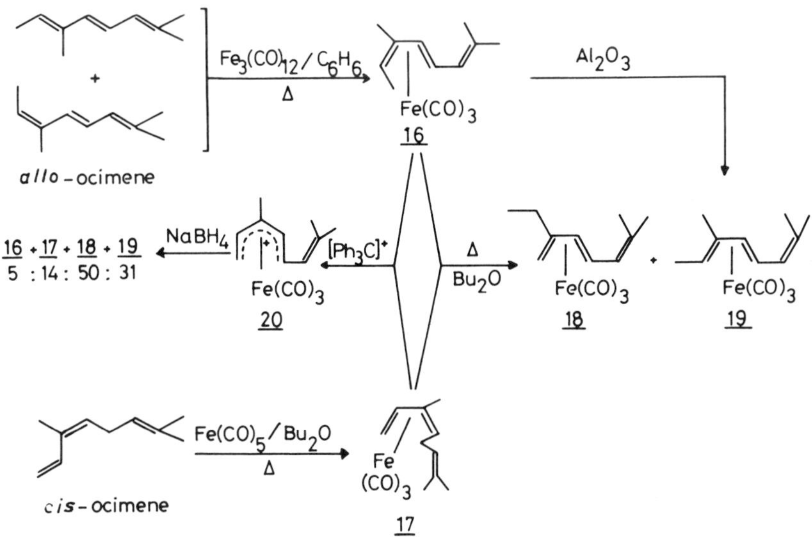

Scheme I: Reactions of *allo*-ocimene with iron carbonyls *(30)*.

The structures of 16 - 19 were based on NMR spectroscopy and decomplexation with ceric ammonium nitrate, which gave the free trienes cleanly in all cases. Use of MnO_2, $FeCl_3$ or triphenylphosphine gave isomerization to mixtures accompanying decomplexation.

The second general route to triene-$Fe(CO)_3$ complexes is the introduction of a free double bond into a diene-$Fe(CO)_3$ complex. This has commonly been done by either a Wittig reaction or by deprotonation of a dienyl cation formed from the diene complex. The Wittig reaction has proven particularly useful for synthesizing $Fe(CO)_3$ complexes of unsymmetrical trienes, in which the position of the iron is unequivocally defined *(314,401,403)*. Yields obtained were in the range 44 - 98 % (eq. [19]).

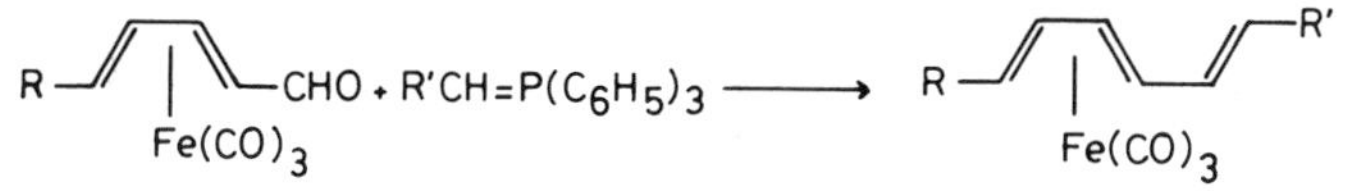

[19]

R = H ; C_6H_5 ; *p*-CH_3O-C_6H_4 ; *p*-CH_3-C_6H_4 ; 3,4-$Cl_2C_6H_3$; CH_3 ;
R' = H ; C_6H_5 ; *p*-CH_3-C_6H_4.

Protonation of dienol complexes quantitatively gives dienyl cation complexes 20 (Scheme II). These, on treatment with bases, can give either substituted diene complexes or, by proton loss, triene complexes 14. The latter may be further attacked by dienyl cation complexes 20, giving dimeric complexes 21 *(10,298,299)*.

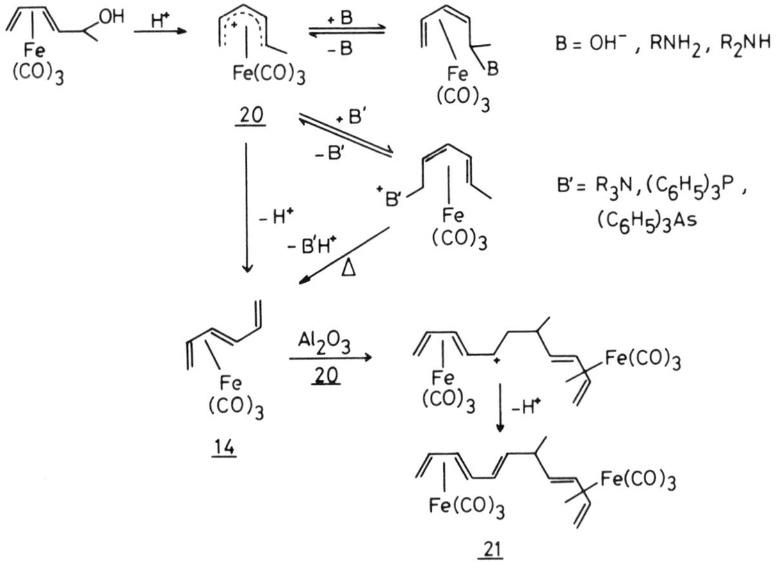

Scheme II: Reactions of Dienyl Cation Complexes with Bases.

Dimers 21 (two stereoisomers) were obtained and characterized from chromatography of cation 20 with basic alumina *(10,298)*. The hexatriene complex 14 can be obtained in "almost quantitative" yield by slow heating of the *syn* triethylamine adduct *(299)*.

A rare example of a $Fe(CO)_3$ complex of a cross-conjugated triene was also obtained by deprotonation *(57)* (eq. [20]).

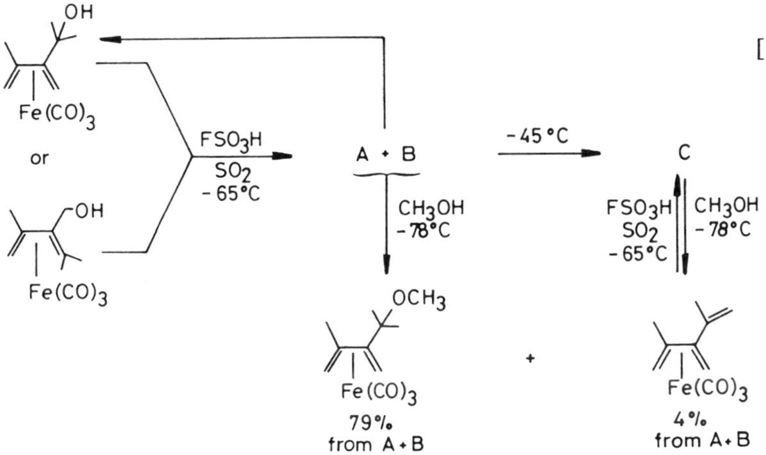

[20]

The intermediates A, B, and C were detected by ^{1}H-NMR at -65°C: all were indicated to be $[(C_8H_{13})Fe(CO)_8]^+$ cations, but specific structures were not assigned. Possibilities include:

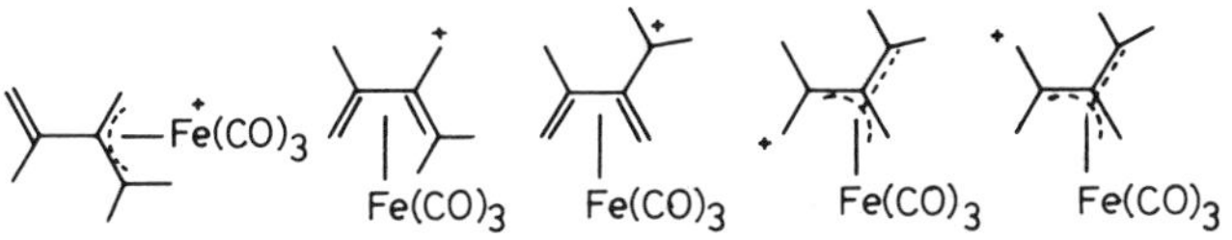

The only other $Fe(CO)_3$ complexes of cross-conjugated trienes were also obtained indirectly (eqs. [21] *(189)* and [22] *(20)*. The latter two products in eq. [22] are not reported to be formed from quadricyclane and $Fe_2(CO)_9$, only the

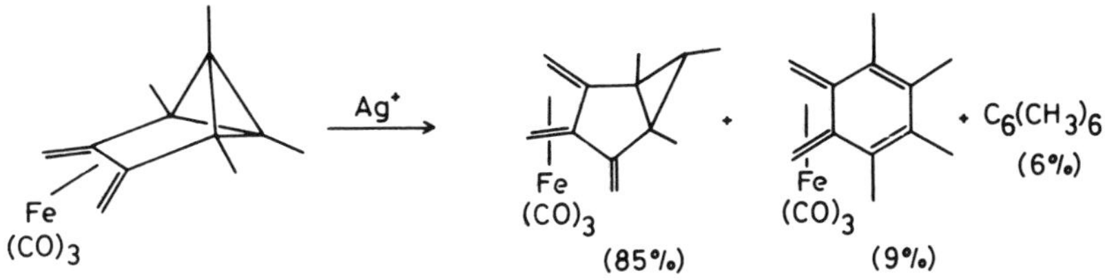

[21]

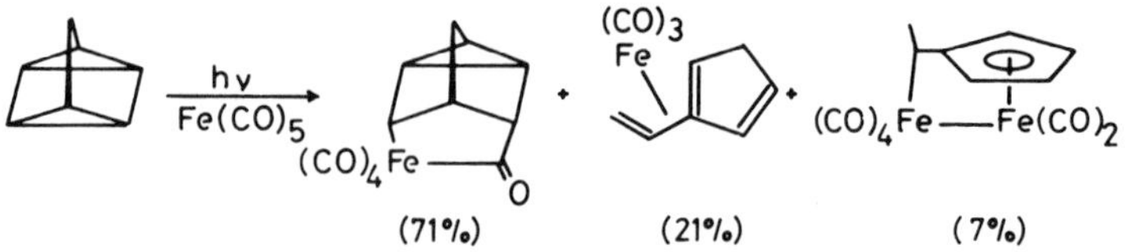

[22]

former being reported *(256)*.

No isomerization studies on these cross-conjugated diene complexes have yet appeared. However, extensive studies on linear triene and tetraene complexes have been carried out by Whitlock and coworkers. For example, at 120°C 22 and 23 interconvert on heating to form a 50:50 mixture of the two isomers *(401)* (eq. [23]). The Arrhenius activation energy for

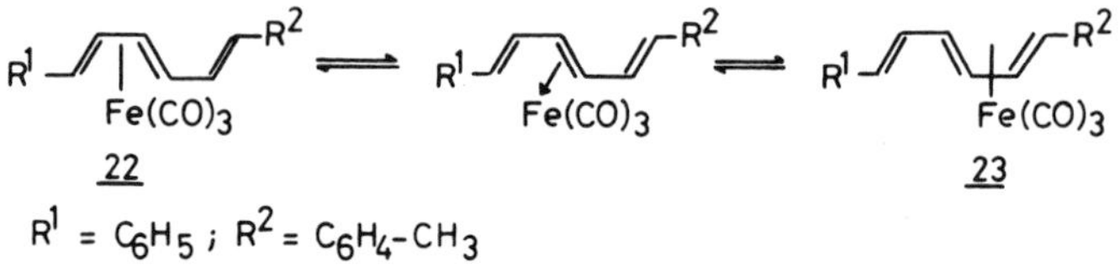

[23]

the first-order interconversion was 33 kcal mol^{-1}, and the entropy of activation + 16 cal $mol^{-1}K^{-1}$. These data suggested a dissociative mechanism to give the transient η^2-intermediate shown. This intermediate was inferred to have a very short lifetime with respect to recollapse to 22 or 23, since it was not effectively trapped by added triphenylphosphine - bis-(triphenylphosphine)tricarbonyliron formed at a rate much slower than equilibration of 22 and 23.

Similar results were also found for a variety of triene and tetraene $Fe(CO)_3$ complexes *(403)* (Table 2). The $Fe(CO)_3$ group showed a slight thermodynamic preference for coordination at the more electron-deficient end of the polyene, and the terminal tetraene complexes were more stable than the internal. The isomer favoured in each case was the one whose four-electron localization energy was smaller *(334)*. Rates of interconversion were somewhat more substituent-sensitive than equilibria, but were not significantly solvent-dependent.

Curiously, the two end-coordinated tetraene complexes 24 and 25 were found to interconvert at a rate too fast to allow the internally-coordinated 26 to be an intermediate (eq. [24]). This was interpreted in terms of migration of the η^2-coordinated $Fe(CO)_3$ group along the tetraene chain and rotation around formal single bonds both being much faster than reclosure to the η^4-diene complex of intermolecular trapping by phosphines.

Using a chiral triene complex, Whitlock and Markezich *(402)* found back-and-forth movement of the $Fe(CO)_3$ group to be

Table 2: Isomerization of Triene- and Tetraene-tricarbonyl-iron complexes *(401-403)*.

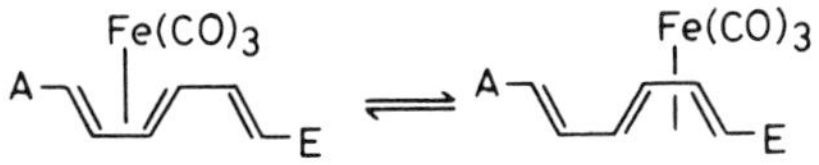

A	E	t[°C]	K	k • 10^3 [min^{-1}]
C_6H_5	CHO	64.9	2.88	1.20(6)
p-CH_3-C_6H_4	H	119.0	14.2	5.3(3) [a]
C_6H_5-CH=CH	C_6H_5	99.4	2.96	2.7(2)
p-CH_3-C_6H_4-CH=CH	*p*-CH_3-C_6H_4	99.4	3.2	3.7(1)
p-CH_3O-C_6H_4	3,4-$Cl_2C_6H_3$	100.1	1.26	4.4(3)
p-CH_3O-C_6H_4	C_6H_5	99.4	1.21	1.4(6)
p-CH_3-C_6H_4	C_6H_5	120.0	1.0	-
CO_2-CD_3	CO_2-CH_3	119.4	1.07	0.74
CH_3-O_2C-CH=CH	CO_2-CH_3	119.4	18.6	9.5

[a] E_a = 32(2) kcal mol^{-1}; $\Delta S^{\neq}$ = +16 cal $mol^{-1}K^{-1}$.

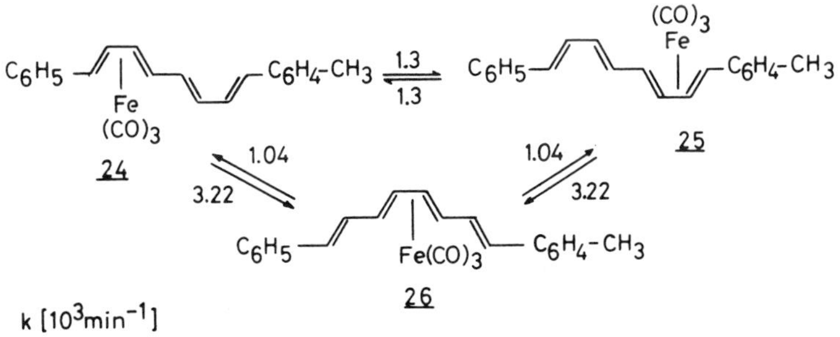

[24]

2.6 times faster than racemization, and hence the iron moves preferably suprafacially across *s-cis* diene units (racemization requires, in effect, transfer across an *s-trans* diene unit).

A puzzling last example of an $Fe(CO)_3$ complex is the product $[C_3(C_6H_5)_4]Fe(CO)_3$ reported by Nakamura *et al. (327, 328)* from tetraphenylallene and $Fe(CO)_5$ in refluxing iso-

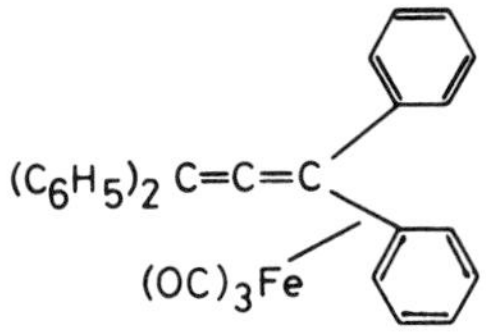

octane. Warming with triphenylphosphine gave the free allene back; pyrolysis also gave 1,3,3-triphenylindene. These results and the infrared spectrum are consistent with the product being a styrene-type complex. A number of examples of such complexes are now known *(385-387)*; some have been characterized by X-ray crystallography *(153)*.

Chemical properties of acyclic polyene-tricarbonyliron complexes have not been systematically investigated. One reaction reported is further complexation of uncoordinated multiple bonds. Thus, (hexatriene)tricarbonyliron, 14, gave a fully-coordinated $Fe_2(CO)_7$ complex on reaction with excess nonacarbonyldiiron *(320)*, and the tetraene 27 likewise underwent further complexation *(312)* (eq. [25]).

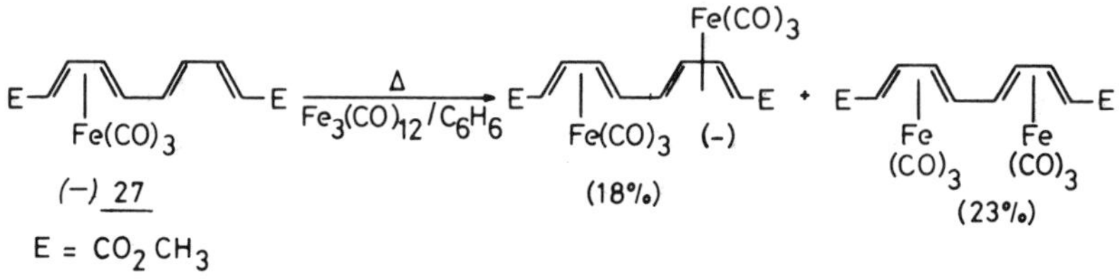

[25]

Some reactions which suggest a useful role for the $Fe(CO)_3$ group in protecting diene units within polyenes have been reported. For example, the secondary alcohol group in (α-calciferol)tricarbonyliron could be oxidized to a ketone using *N*-chlorosuccinimide/dimethylsulfide. The reaction was reversible using lithium tri(*t*-butoxy)aluminium hydride (eq. [26]).

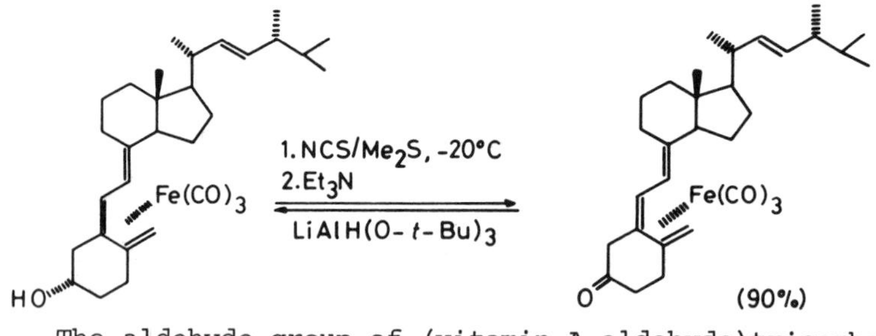

[26]

The aldehyde group of (vitamin-A aldehyde)tricarbonyliron could similarly be reduced using sodium borohydride; the alcohol product was unstable, however *(48)*.

The free double bond of the myrcene complex 15, which could not be hydrogenated catalytically, was successfully reduced with diimine (23 %) or, better, with diborane then acetic acid *(30)*. Free double bonds of 1,3,5-triene-tricarbonyliron complexes are also effectively hydroborated *(50,314)* (eqs. [27]-[29]). Electrophilic additions of acetic acid and

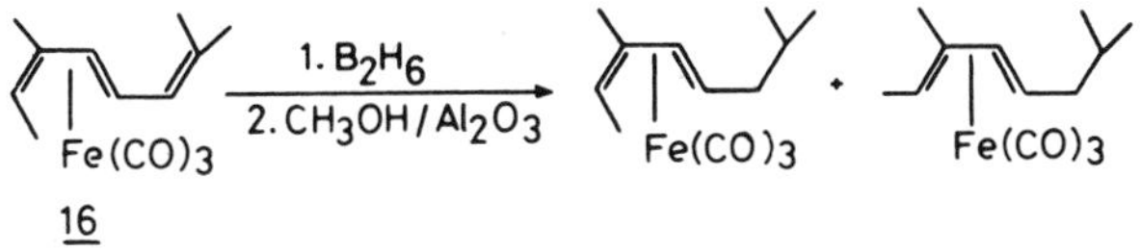

[27]

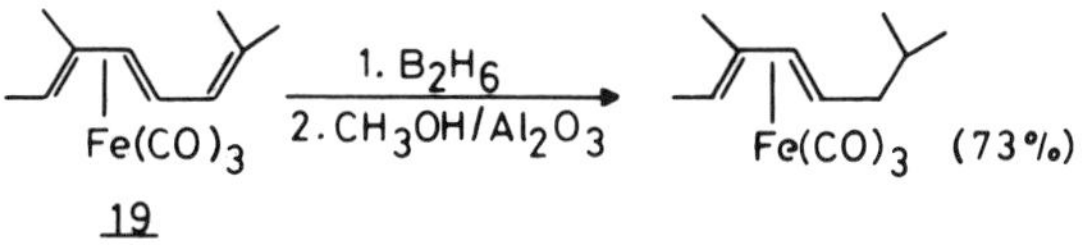

[28]

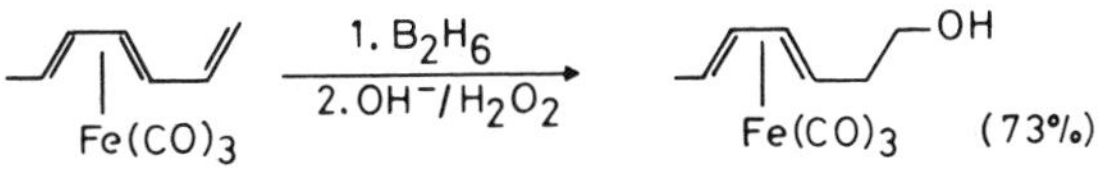

[29]

water to the free double bond of 15 were also successful *(30)*. Acetylation with acetyl chloride/aluminium chloride at -78°C proceeded at the free double bond *(50)* (eq. [30]). Use of oxalylchloride resulted in intramolecular acylation with ring closure *(50)* (eq. [31]).

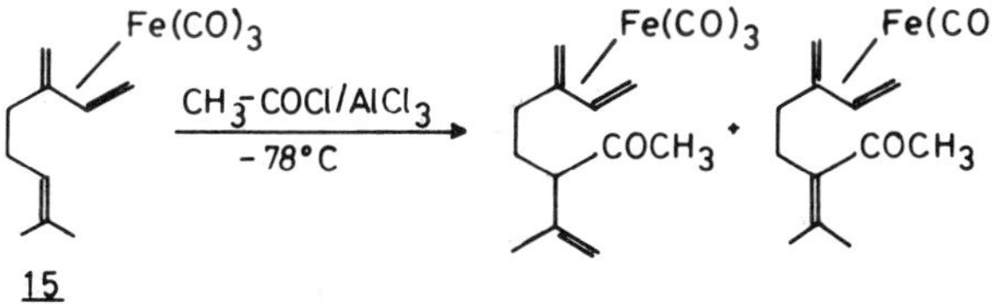

[30]

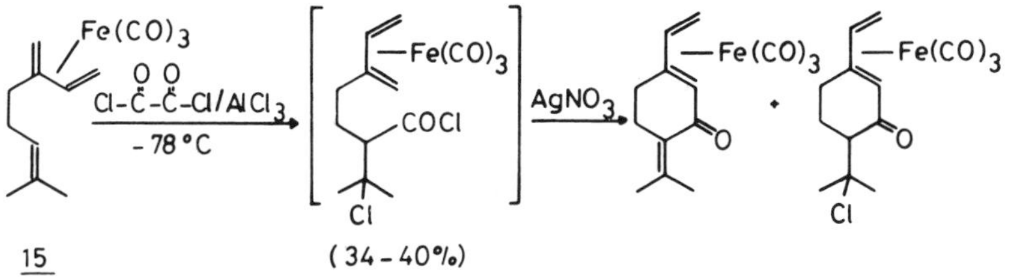

[31]

B. *$Fe(CO)_3$ COMPLEXES OF POLYENES WITH FOUR- TO SIX-MEMBERED RINGS.*

As previously mentioned, reaction of 1,2-dimethylene-cyclobutene with $Fe_3(CO)_{12}$ in boiling benzene gives only $Fe(CO)_4$ complexes *(280)*. In contrast, reaction of bicyclo-[3.2.0]hepta-1,4,6-triene occurs with hydrogen migration to give a cyclobutadiene-tricarbonyliron derivative *(280)* (eq. [32]).

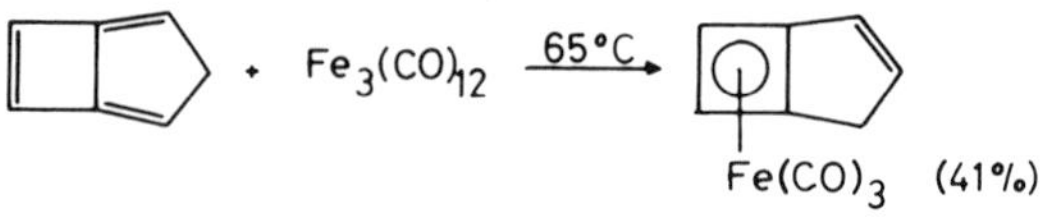

[32]

A number of other examples of cyclobutadiene complexes having additional unsaturation have been reported. These include the vinylcyclobutadiene complex 28, which forms spontaneously on passing (α-hydroxyethylcyclobutadiene)tricarbonyliron through acidic alumina *(314)* (eq. [33]). This reaction

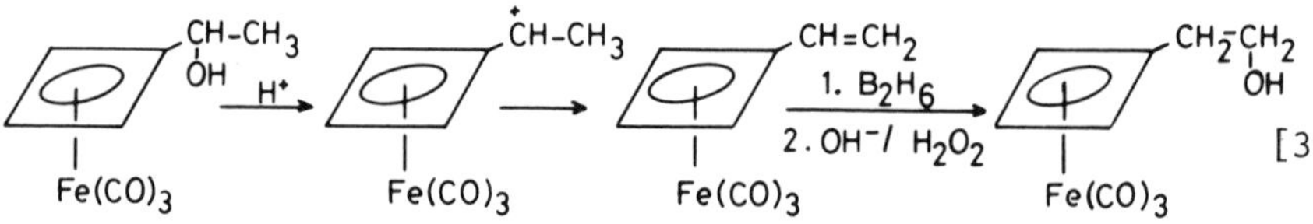

28

[33]

is a manifestation of the stability of the iron-stabilized cation. 28 undergoes hydroboration/oxidation to give solely the primary alcohol *(314)*. (1,2-Divinylcyclobutadiene)tricarbonyliron, reportedly an unstable liquid, was prepared by Wittig reaction of the diformylcyclobutadiene derivative *(42)* (eq. [34]). The diphenyl- and tetraphenyl-derivatives were

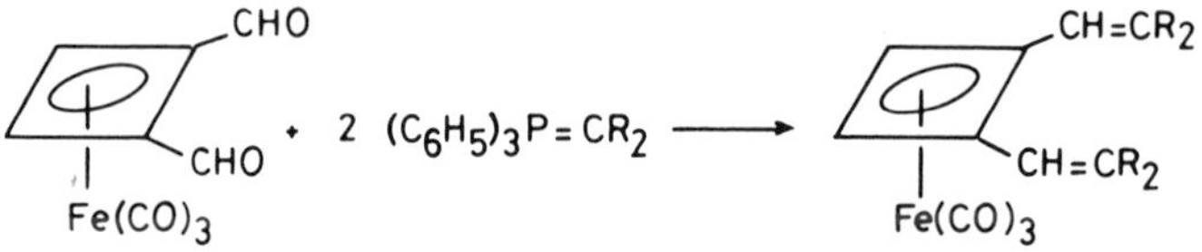

[34]

$R = H, C_6H_5$

more stable. Acid-catalysed intermolecular aldol condensation of the analogous diacetylcyclobutadiene complex, 29, has also been reported *(352)* (eq. [35]). The complex 29 undergoes an

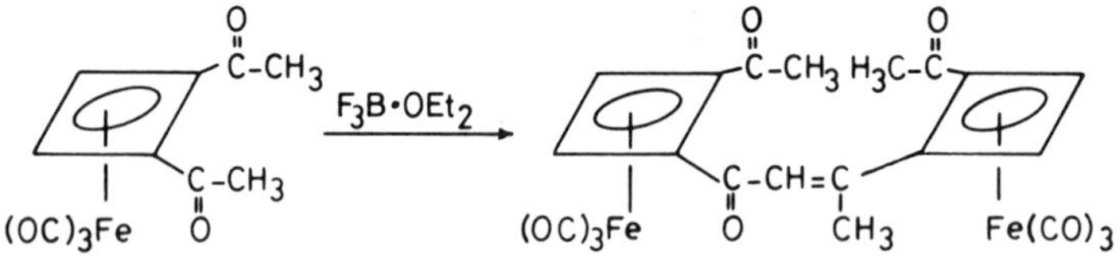

29 (2 stereoisomers)

[35]

intramolecular condensation with triethyl orthoformate in 95 % sulfuric acid (eq. [36]). These reactions presumably are all assisted by the cation-stabilizing effect of the cyclobutadiene-tricarbonyliron moiety.

Complexes of benzocyclobutadiene are also known. A remarkable synthesis of the parent such complex from 1,4-

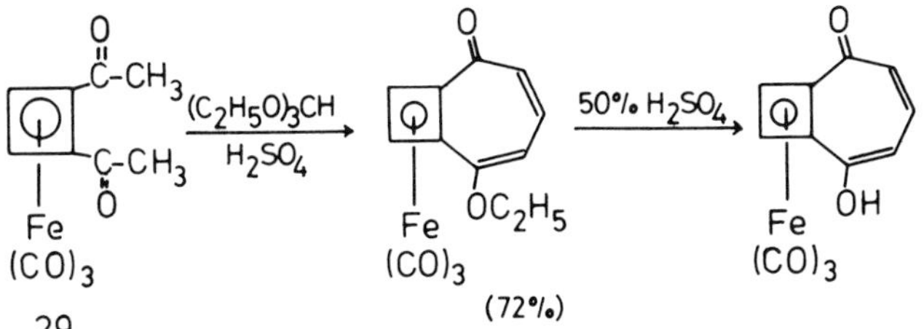

[36]

29

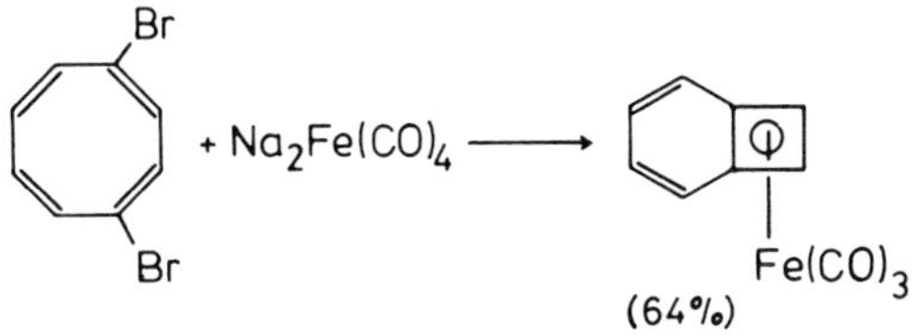

[37]

dibromocyclooctatetraene has recently been reported *(225)* (eq. [37]). The same product was obtained, in 33 % yield, using $Fe_2(CO)_9$. An X-ray crystallographic study *(153)* of a benzocyclobutadiene-tricarbonyliron derivative revealed substantial bond alternation in the six-membered ring, as indicated in the structure shown. Reaction of (benzocyclobutadiene)tricarbonyliron with $Fe_3(CO)_{12}$ *(151,225)* or with $Fe(CO)_5$ under irradiation *(383)* gave $Fe_2(CO)_6$ complexes described in Section IX.

A well-studied class of trienes having five-membered rings are the pentafulvenes. Reactions with iron carbonyls give complex mixtures of products, depending on conditions of reaction and substituents *(268)*. However, only in the case of 6,6-diarylpentafulvenes are the expected tricarbonyliron complexes obtained in significant yields *(395,397)* (eq. [38]),

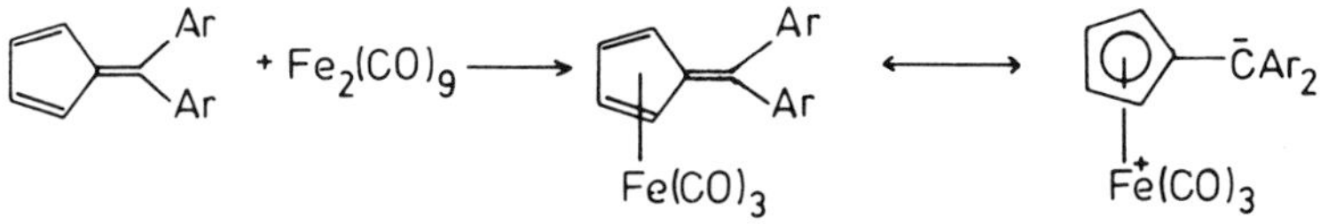

[38]

though they are thought to be intermediates in reactions leading to more complex products in other cases *(39,175a,268)*. Their instability in the absence of aryl substituents may result from interaction of the tricarbonyliron group with C(5), giving an incipient η^5-cyclopentadienyl-tricarbonyliron grouping. Consistent with this notion is the ready protonation of these complexes to form benzhydryl-substituted η^5-cyclopentadienyl-tricarbonyliron cations *(395,397)*.

A compound originally reported *(167)* to be the tricarbonyliron complex of spiro[4.2]hepta-4,6-diene, 30, has recently been shown to be in fact the acyliron complex, 31 *(180)*. 30 may well be an intermediate in formation of 31 (eq. [39]). Reactions analogous to formation of 31 were also ob-

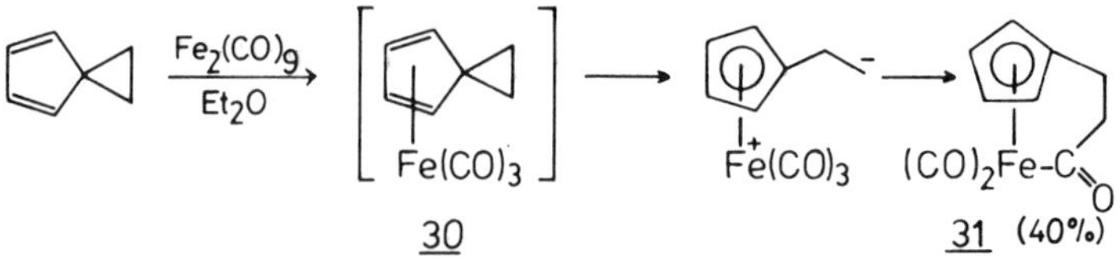

[39]

served in reaction of spiro[2,4-cyclopentadiene-1,7'-norcara-2',4'-diene] with $Fe_2(CO)_9$ *(317)*. The driving force for these rearrangements is presumably the formation of the η^5-cyclopentadienyl-iron moiety.

A number of steroidal trienes have been reported to undergo complexation with iron carbonyls, acetylergosterol being most often studied *(31,191,330)* (eq. [40]). The yield

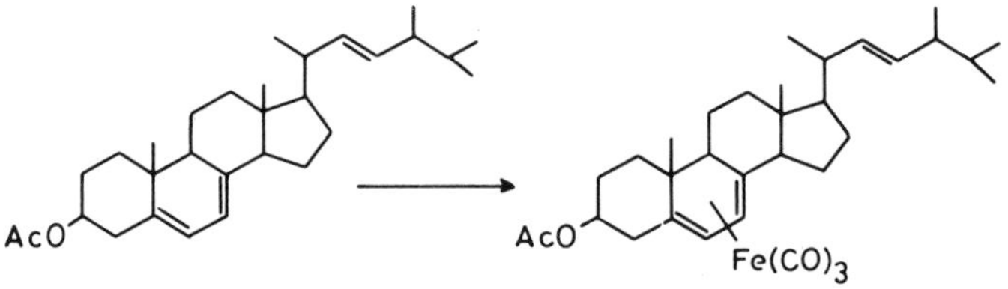

[40]

of tricarbonyliron complex was 16 % using $Fe_3(CO)_{12}$ in refluxing benzene *(330)*, and 70 % using (benzylideneacetone)-tricarbonyliron *(191)*. Benzoylergosterol gave a 70 % yield using $Fe(CO)_5$ in refluxing dibutylether (144°C) and an 80 % yield using (*p*-methoxybenzylidene)tricarbonyliron and $Fe_2(CO)_9$ together *(31)*. In contrast to the acylergosterols, ergosterol itself gave no $Fe(CO)_3$ complex with $Fe_3(CO)_{12}$ *(330)*.

The $Fe(CO)_3$ group in acetylergosterol complexes successfully protects the conjugated diene unit, allowing reactions at the free double bond. Hydroxylation (OsO_4, pyridine) and hydrogenation (H_2, PtO_2) have been successfully carried out, with the $Fe(CO)_3$ group being then removed using $FeCl_3$ *(31, 191)*.

In the eliminative complexation reaction, the elements of a small molecule (*e.g.* Br_2, HCl, H_2O) are eliminated from the organic species undergoing complexation. This reaction is of great importance for preparing complexes of unstable polyenes, (see, for example, Chapter by J.M. Landesberg in Vol. I of this book). A relevant example is the reaction of α,α'-dibromo-*o*-xylene with various carbonyliron reagents to give the $Fe(CO)_3$ complex of *o*-xylylene (eq. [41]). The yield is 6 % using $Fe_2(CO)_9$ in ether *(356)*, 4 % using $Fe(CO)_5$/trimethylamine oxide in benzene *(368)*, and 35 % using disodium tetracarbonylferrate *(255)*. The complexation of the exocyclic diene unit must result from a mechanism involving initial oxidative addition of a low-valent iron species to the C-Br

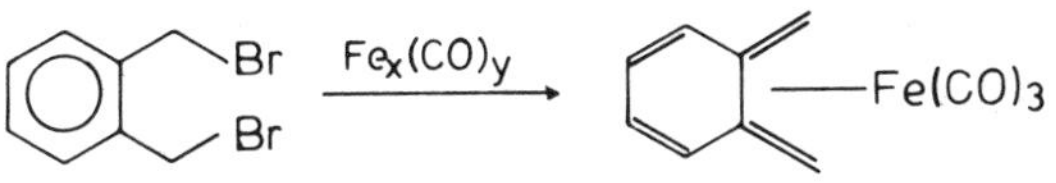

32

[41]

bond(s). The product 32 is very stable, undergoing pyrolysis to benzocyclobutene at 500°C *(356)*. Some chemical properties of 32 have been reported, including formation of three $Fe_2(CO)_6$ complexes upon irradiation with $Fe(CO)_5$ *(384)*, and acetylation *(255)* (eq. [42]). Reaction of 32 with a slight

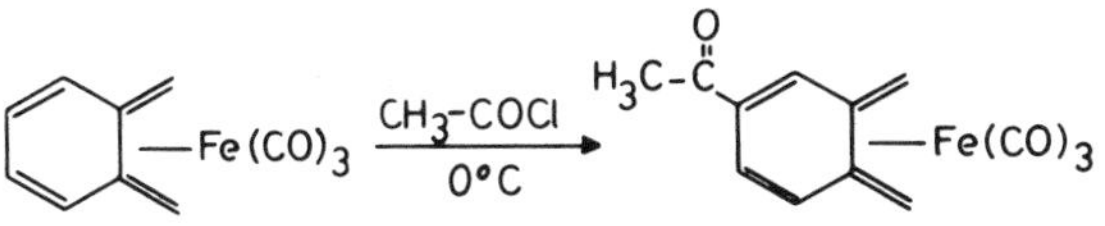

32

[42]

excess of aluminium chloride in benzene gives 2-indanone (48 %) *(255)*. A complex of an isoindene has also been prepared by eliminative complexation *(255,356)* (eq. [43]). An

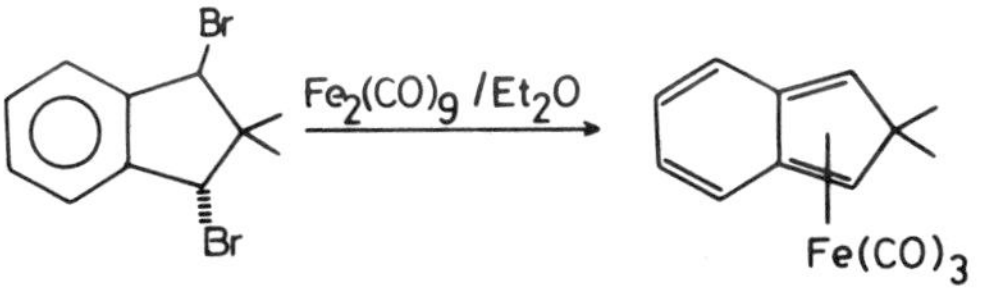

33 (40%)

[43]

X-ray crystallographic study of 33 might be of interest, since NMR coupling constants suggested bond localization in the erstwhile benzene ring *(356)*, and since the *gem*-dimethyl group should make the normal conformation of the $Fe(CO)_3$ group relative to the diene highly hindered.

The coordination of the exocyclic diene unit of these *p*-xylylenes rather than the endocyclic one is consistent with predictions of Nicholson *(334)* based on localization energies. However, the stabilized *p*-xylylene below gives both possible $Fe(CO)_3$ complexes (eq. [44]) *(231)*. This may be due to steric

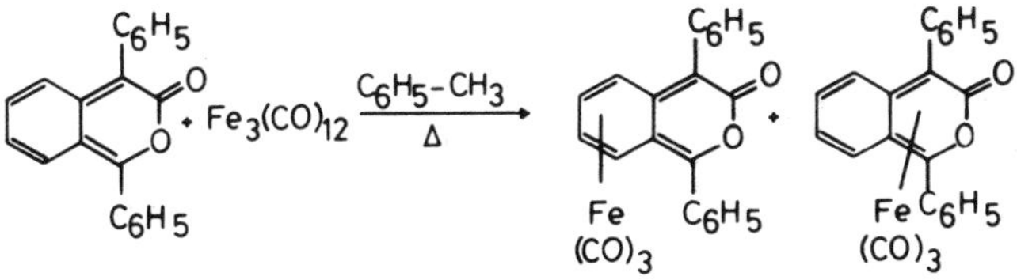

[44]

hindrance involving the two phenyl substituents in the latter, expected product. Interestingly, the two isomers interconvert on heating (temperature unspecified).

Ginsburg and coworkers have extensively studied the

chemistry of the tricyclic tetraenes (X = various three-atom

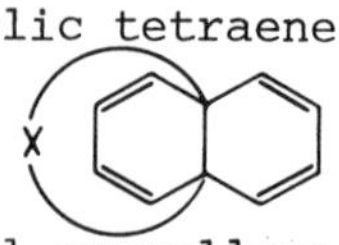

chains), which they call propellanes *(210)*. Reactions of propellanes with $Fe(CO)_5$ or $Fe_2(CO)_9$ give predominantly $Fe_2(CO)_6$ complexes, and hence propellane complexes are discussed in Section IX of this chapter.

1,4,5,6,9,10-Hexahydroanthracene reacted with $Fe_3(CO)_{12}$ with hydrogen migration *(47)* (eq. [45]). 34 was the only

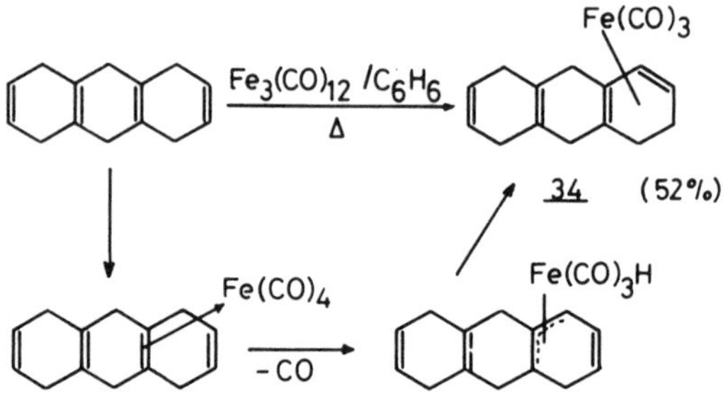

[45]

reported product; its formation suggests initial complexation of an internal double bond rather than the expected external one.

Formation of the $Fe(CO)_3$ complex of a cyclohexa-1,3-diene unit appears to be strongly favoured, and results when hydrogen migration will allow *(43)* (eq. [46]). Use of a

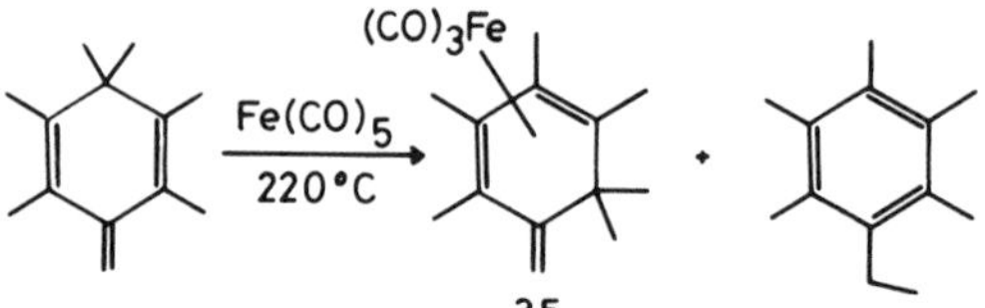

35

[46]

labeled methylene group confirmed that the above migration occurred by intramolecular hydrogen migration *(43)* (eq. [47]). In bridged cyclohexa-1,4-diene derivatives, the hydrogen migration is of course impossible. In these cases one does obtain $Fe(CO)_3$ complexes of an unconjugated diene unit (eqs. [48] *(271)*, [49] *(185)*, and [50] *(185)*). Both 36 and 37 give Wagner-Meerwein rearrangements upon protonation, giving rearranged iron-stabilized cations from the initially-formed allylic cations *(181,185)*.

The ability of the iron in the diene-tricarbonyliron group to expand its coordination so as to stabilize a car-

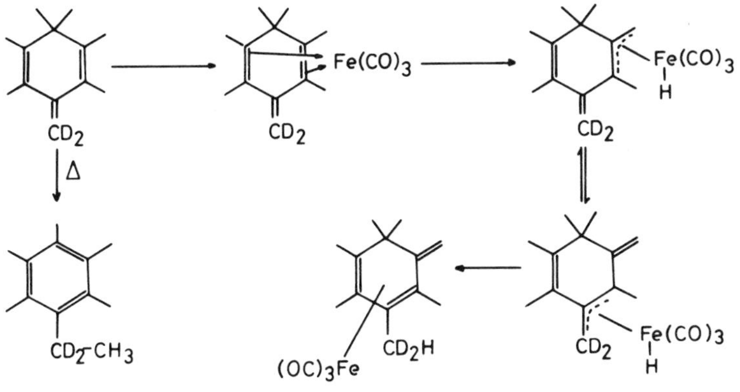

[47]

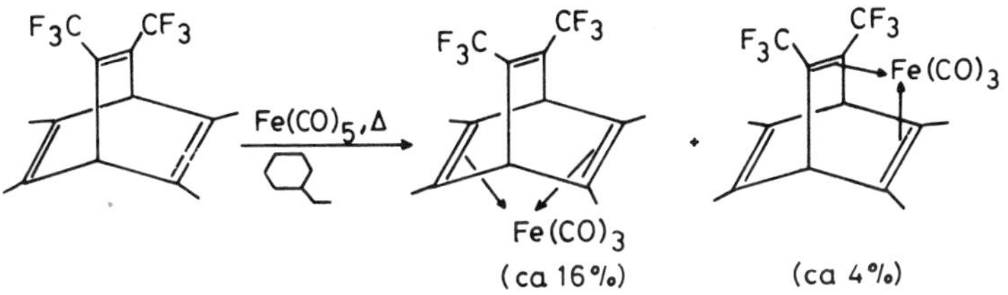

[48]

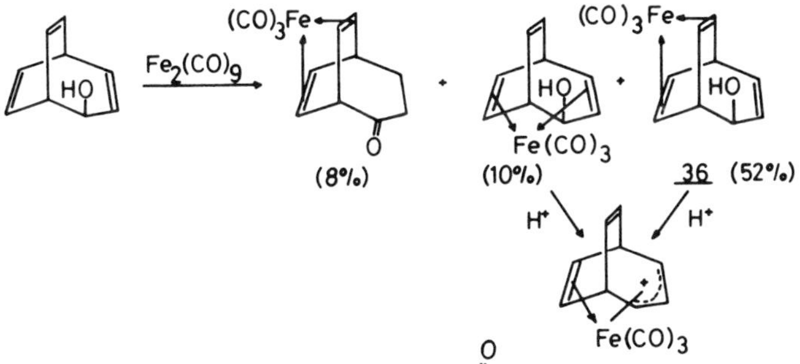

[49]

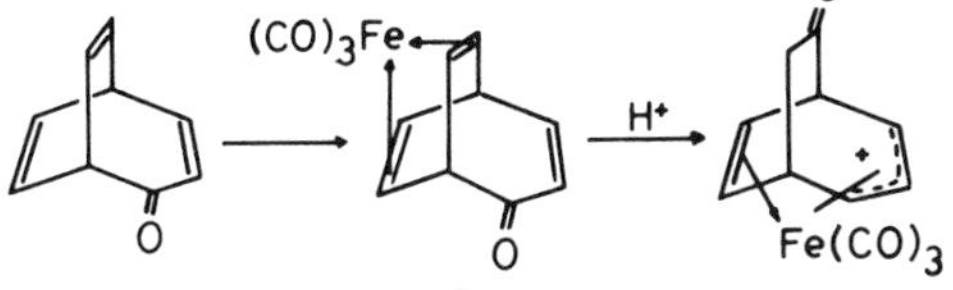

[50]

37

benium ion as a pentadienyl-tricarbonyliron cation is particularly pronounced in cyclic cases. This propensity is useful for introducing functional groups or for extending a diene into a triene, as already discussed for acyclic complexes. In the steroid series, for example, a complex of 1,3-cholestadiene was converted to a complex of 1,3,5-cholestatriene by this means *(5)* (eq. [51]). Similar reactions have been studied in detail in a simpler system *(147)* (eq. [52]).

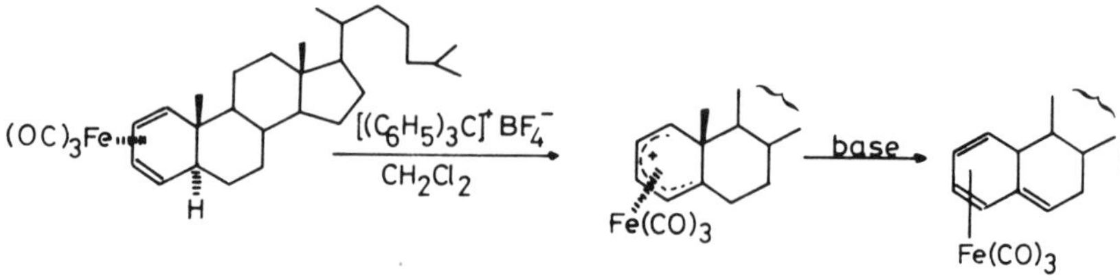

[51]

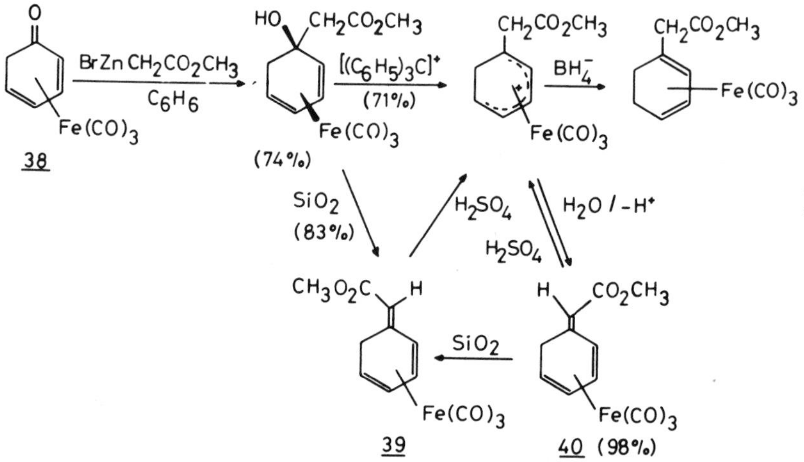

[52]

Attempts to form C-C bonds at the carbonyl group of 38 using alkyllithium, Wittig, and Grignard reagents failed, possibly due to their high basicity, causing enolate (*i.e.* phenoxide) formation followed by loss of the iron from the aromatic ring. The structures of the *anti* and *syn* triene complexes 39 and 40 (which were originally *(147a)* assigned cycloheptatriene structures) were confirmed by X-ray crystal-

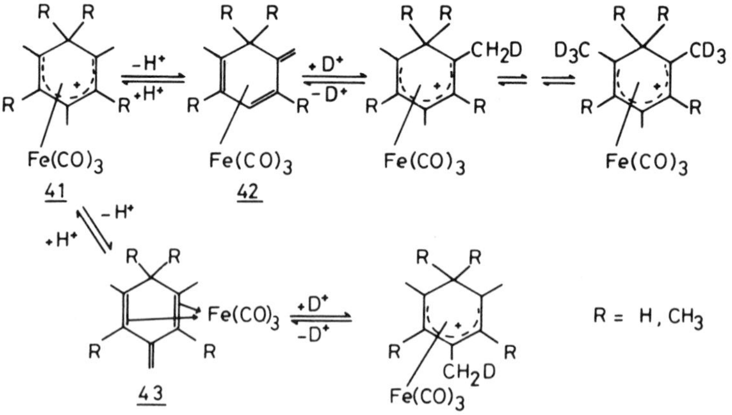

[53]

lography *(147)*. The *anti* isomer 39 is evidently more thermodynamically stable, but the *syn* is formed more rapidly in basic media.

In another study of equilibria between cyclohexadienyltricarbonyliron cations and triene complexes (eq. [53]), the ion 41 was found to undergo exchange with deuterium from the acidic medium almost exclusively at the 2- and 6-methyl groups *via* the 1-4-η-coordinated triene 42. Exchange at the 4-methyl requires the 1,2,4,5-η^4-coordinated cross-conjugated triene 43, which is evidently less stable. The d_6-product was formed with less than 5 % d_7 *(366)*.

Cyclohexadienyl complexes have also been converted to triene complexes by introducing the third double bond as a nucleophile (eqs. [54] *(343)*, [55] *(45)*, [56] *(50)*, and [57] *(46)*).

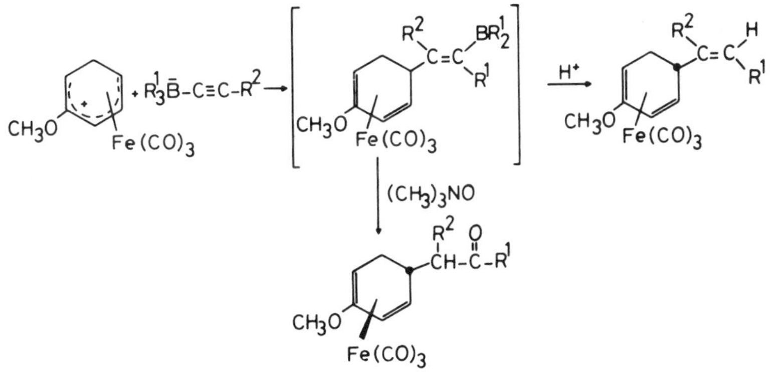

[54]

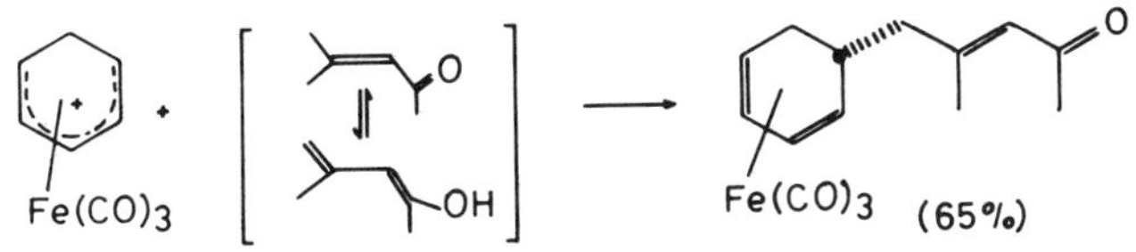

[55]

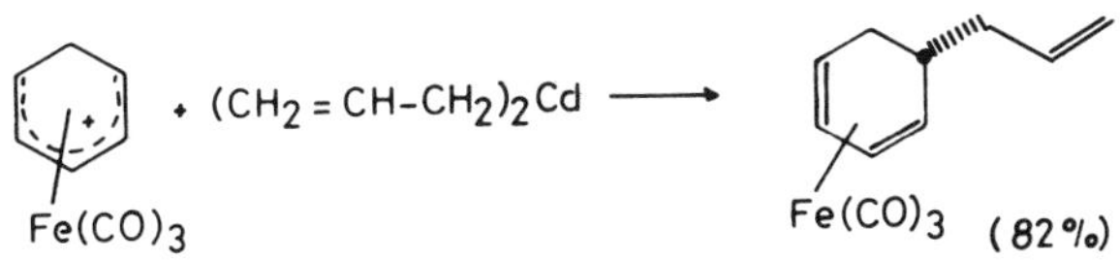

[56]

One last, extraordinary $Fe(CO)_3$ complex of a cyclic triene is the product obtained by photolysing (cyclohexa-1,3-diene)tricarbonyliron in the presence of excess hexafluorobut-2-yne *(58)* (eq. [58]). The structure was assigned on spectroscopic grounds, and by comparison to the analogous

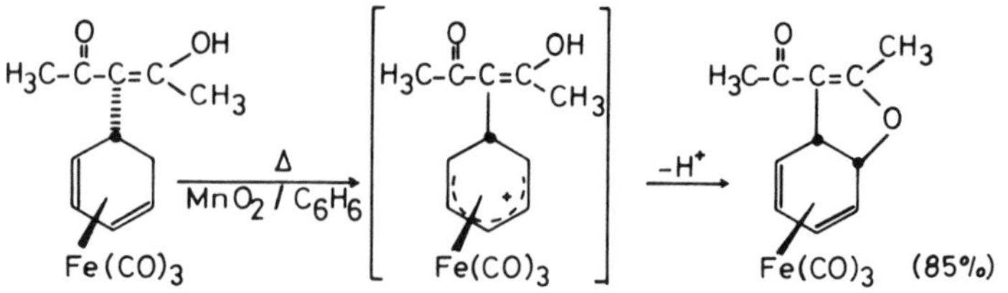

[57]

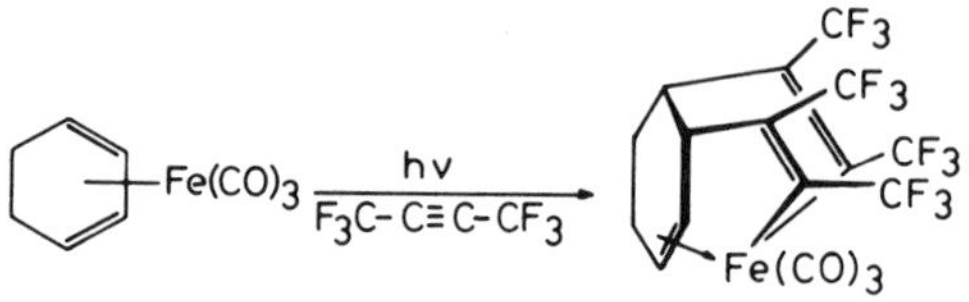

[58]

ruthenium complex, whose structure was confirmed by X-ray crystallography.

By analogy with the photoinsertion reactions of haloalkenes into diene-tricarbonyliron complexes *(269a)* the mechanism in Scheme III may be suggested.

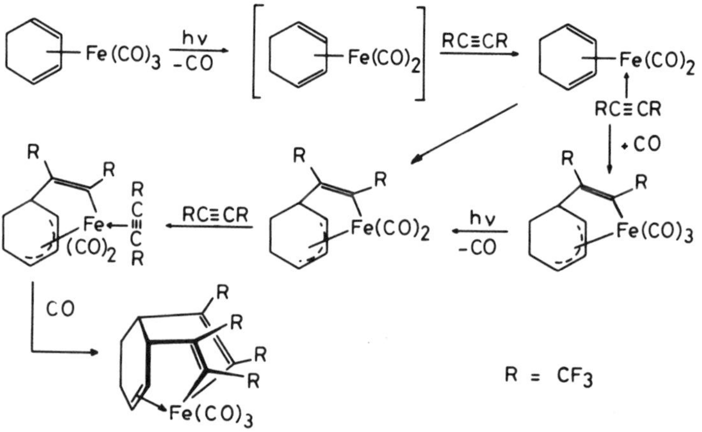

Scheme III: Possible Mechanism of Photoinsertion of Hexafluorobut-2-yne.

C. $Fe(CO)_3$ COMPLEXES OF CYCLOHEPTATRIENES

(Cycloheptatriene)tricarbonyliron, 44, has usually been prepared by prolonged refluxing of cycloheptatriene with $Fe(CO)_5$ at 110°C. The reaction is remarkable for its slowness: the yield is 20 % after 3 days *(178)* and 48 % after 7 days *(80,310)*. Use of higher temperature (135°C) gave byproducts including (cycloheptadiene)tricarbonyliron *(80,150)*.

The successful use of $Fe_3(CO)_{12}$ was also mentioned *(78,125)*, but $Fe_2(CO)_9$ gives primarily the $Fe_2(CO)_6$ complex *(190)*. The apparent best synthesis involves UV irradiation of a refluxing benzene solution of cycloheptatriene and $Fe(CO)_5$, which gives 64 % 44 in two days *(249)*.

Substituted cycloheptatrienes have been less extensively studied. Methyl thujate (3-carbomethoxy-7,7-dimethylcycloheptatriene) gave a polymer on heating with $Fe(CO)_5$ *(80)*. 7-Methoxycycloheptatriene gave an $Fe(CO)_3$ complex in unspecified yield with $Fe_2(CO)_9$ *(305)*; (1-3-η^3-cycloheptatrienyl)(η^5-cyclopentadienyl)dicarbonylmolybdenum gave 5 % of an $Fe(CO)_3$ complex with $Fe_2(CO)_9$ and 55 % on irradiation with $Fe(CO)_5$ *(141)*. Reaction of either isomer of benzocycloheptatriene with $Fe(CO)_5$ at 120°C gave high yields of 45 *(44)* (eq. [59]).

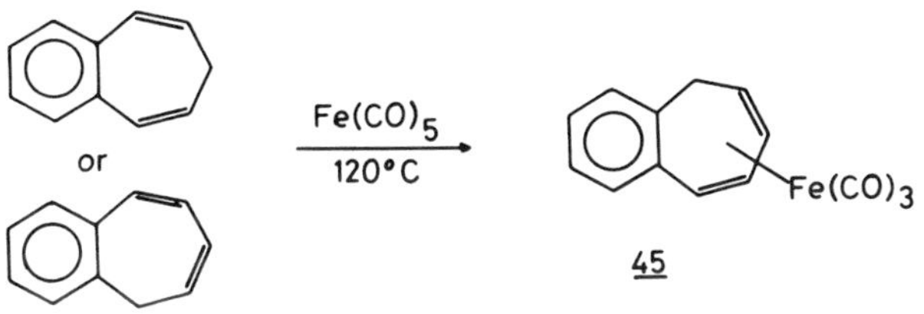

[59]

The formyl derivative, 46, of (cycloheptatriene)tricarbonyliron was obtained (19 %) on reaction of cyclooctatetraene oxide with $Fe_3(CO)_{12}$ in benzene *(306)* (eq. [60]).

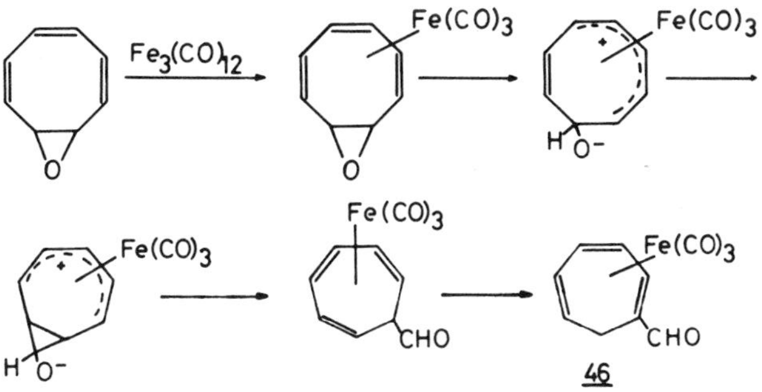

[60]

Although 44, a low-melting solid, has not been studied by X-ray crystallography, there can be little doubt that it possesses the biplanar structure found for the tropone and azepine complexes described below, the four carbons of the planar coordinated diene moiety intersecting the plane of the uncoordinated carbons at C(1) and C(4). ^{13}C-NMR indicates rapid (pseudo) rotation of the $Fe(CO)_3$ group relative to the triene at room temperature, with decoalescence at -73°C and an activation energy at that temperature of 11.6 kcal *(288)*.

The small "bite" of the diene ligand (a typical double

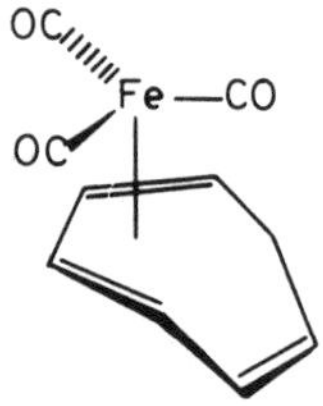

44

bond - iron - double bond angle being 62° *(288)*) in these pentacoordinate compounds is responsible for their adoption of the square pyramidal geometry invariably found in crystallographic studies *(124,146)*. Rotation of the $Fe(CO)_3$ group by 60° relative to the diene (with attendant changes in bond angles) converts the geometry essentially to that of the trigonal bipyramid, with one diene "double bond" equatorial and

one axial. Continued rotation restores the square pyramid, with basal and equatorial CO's exchanged. The high rotational barrier in diene-tricarbonyliron complexes compared to most pentacoordinate species reflects the difficulty of accommodating the diene ligand with its small "bite" within the trigonal bipyramid *(288)*.

The $Fe(CO)_3$ group does not move as a whole with respect to the cycloheptatriene ring at temperatures up to 120°C, as shown by NMR studies. That is, the iron remains bound to the same diene unit on the normal NMR time scale. However, Brookhart *(71a)* has shown by the spin saturation transfer technique that motion does occur at 80-85°C, with a $\Delta G^{\ddagger}$ of 23 kcal. The $Fe(CO)_3$ complex of 7-carbomethoxy-7-phenylcycloheptatriene (which forms substantial norcaradiene) shows a very similar $\Delta G^{\ddagger}$ of 24 kcal. Such degenerate isomerizations can occur *via* a number of possible intermediates (Scheme IV). Of these pathways, (a) is difficult to reconcile with the failure of the 7-carbomethoxy-7-phenyl derivative to rearrange more readily *(71a)*. Route (b) is the one shown to account for $Fe(CO)_3$ group shifts in acyclic trienes, but the energy barrier in the latter cases is somewhat higher than in 44. Route (c) may be expected in photochemical reactions rather than in thermal reactions below 100°C.

Cycloheptatriene $Fe(CO)_3$ complexes bearing organometallic substituents $Mo(CO)_2Cp$ *(141)*, $Mn(CO)_3$ *(41)*, $Rh(CO)_2$ *(41)*, and

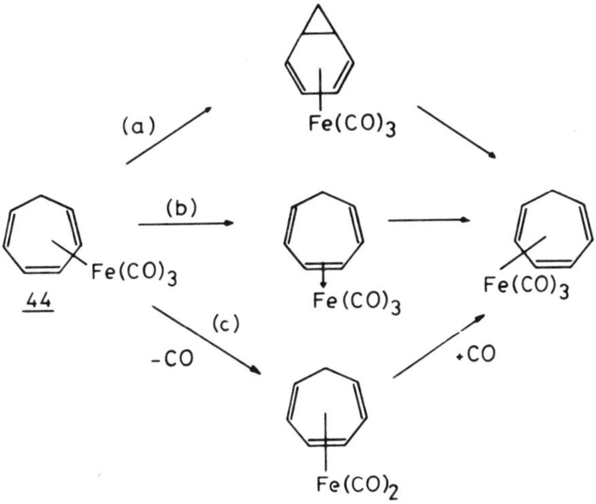

Scheme IV: Possible Intermediates in $Fe(CO)_3$ Shift in 44.

$Re(CO)_3$ *(41)* are fluxional at room temperature. In the molybdenum complex, the two metals are presumably *trans* and not directly bonded; the C_7H_7 ring can be thought of as rotating by 1,2-shifts between them, with decoalescence at 10°C and an activation energy of 13 kcal *(141)*. In the other bimetallic complexes, the metals appear to be directly bonded, and the rotational barrier of the C_7H_7 ring is much lower - NMR line broadening is observed only at -150°C, but limiting low temperature spectra have not been obtained *(41,287)*. The reason for the greater ease of $Fe(CO)_3$ group shifts in the 7-trimethylgermyl and trimethylsilyl derivatives of 44 as compared to 44 itself (T_c 100°C and 115°C, respectively, for the Ge and Si compounds *(292)*) is less clear. Proton or $(CH_3)_3M$ group shifts are excluded by the spectra, and route (c) of Scheme IV is excluded by absence of ^{13}CO exchange. Stabilization of the norcaradiene complex of route (a) by the substituents is possible, but hard to reconcile with the failure of the phenyl and carbomethoxy substituents to provide analogous stabilization.

The ions $[(C_7H_7)Fe(CO)_3]^+$, 47, and $[(C_7H_7)Fe(CO)_3]^-$, 48, (see below) are both fluxional at room temperature, the preferred 1,2-shift route being unhindered by the presence of an insulating saturated carbon in the ring as in 44. A comparison of activation energies for these two species would be most enlightening with respect to electronic effects on fluxionality. A limiting spectrum of the (tropylium)tricarbonyliron cation, 47, is observed at -80°C *(399a)*. 47 is made by removal of methoxide from the 7-methoxycycloheptatriene complex *(305)* (eq. [61]). In contrast to the cases of (benzocyclo-

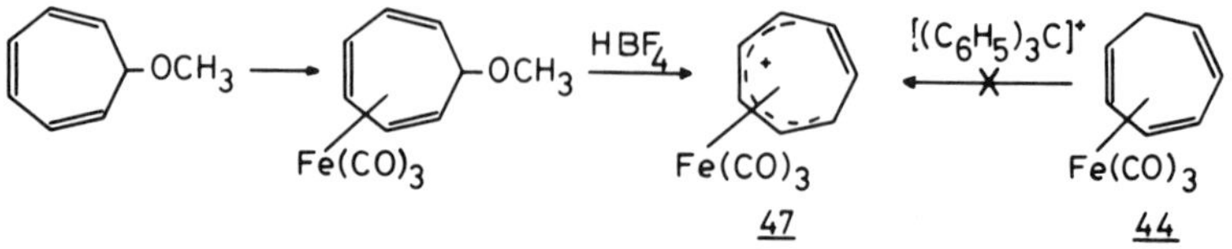

[61]

heptatriene)tricarbonyliron *(44)*, 45, and (cycloheptadiene)-tricarbonyliron *(150)*, triphenylcarbenium fluoroborate does not remove a hydride ion from 44 to generate 47 *(150,101)*. This is not due to thermodynamic factors, since all three cations in question (and free tropylium ion itself) have very similar pK_{R^+} values of 4.5 ± 0.2. Instead, it reflects the great reactivity of the free double bond in 44 toward attack by electrophiles, which is impossible in the other cases.

That 47 is best viewed as a fluxional 1-5-η^5-cycloheptatrienyl complex is indicated by an IR band at 731 cm^{-1} due to a free double bond *(305)*, and by its complex C-H absorption in the IR. Its carbonyl stretching frequencies (2120, 2070 cm^{-1}) are essentially the same as those of other dienyl-tricarbonyliron cations.

The cycloheptatriene complex 44 undergoes facile proton exchange with CH_3OD containing 10 % sodium methoxide *(307)*, under which conditions free cycloheptatriene does not exchange. Use of stronger base, butyl- or phenyllithium in THF at -78°C or potassium *t*-butoxide in THF *(161)*, gives the metal salt of 48, which is fluxional even at -65°C. The carbonyl stretching frequencies (1942 and 1868 cm^{-1} *(307)*) suggest that 48 is best considered a fluxional 1-3-η^3-complex.

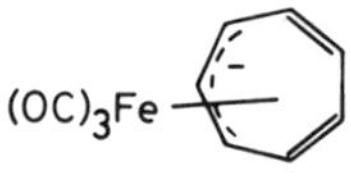

48

The thermodynamic stability of 48 (*i.e.* pk_A of 44) has not been reported, but should be accessible to electrochemical measurement. However, 48 is very readily oxidized, the dimer (bitropyl)hexacarbonyldiiron being formed on reaction of 48 with low-valent carbonyl halides *(41)* and even allyl halides *(161)*. 48 is unreactive toward alkyl halides and reacts as a nucleophile toward chlorosilanes and acyl chlorides *(161)*.

As alluded to previously in the case of triphenylcarbenium ion, the most general and extensively studied reaction of 44 is its reaction with electrophiles to form substituted cycloheptadienyl-tricarbonyliron cations (eq. [62]). The reactivity of the double bond of 44 toward electrophiles is

44 [62]

remarkably high, comparable to that of an enamine, though there appear to be no quantitative data. The facile protonation of 44 was recognized already in 1961 *(150,155)*, as was addition of triphenylcarbenium ion *(150)*. Protonation occurs on the face *anti* to the iron *(249)*. The cycloheptadienium ions thus formed react with nucleophiles by a variety of modes: I^- attacks the iron to form (cycloheptadienyl)dicarbonyliron iodide *(150)*, tertiary amines remove a proton to give 44 back *(101)*; methanol attacks the coordinated cycloheptadienyl ligand to give the (*anti*-5-methoxycycloheptadiene)tricarbonyliron under conditions of kinetic control and the *syn* isomer under thermodynamic control *(230)*, and borohydride reduces the ligand in both the 1- and 2-positions *(27)* (eq. [63]). Reaction with sodium hydroxide in a solvent

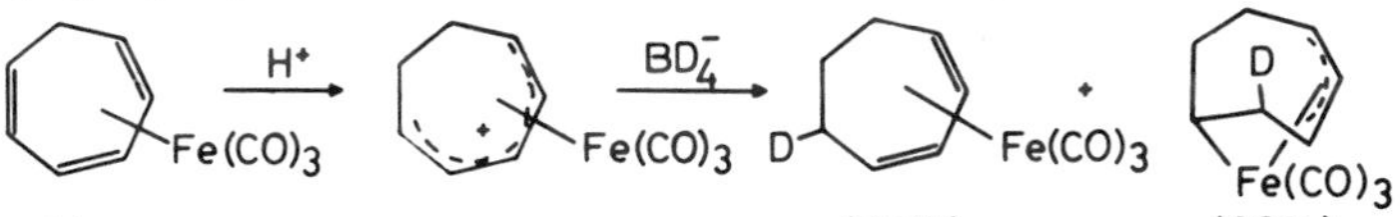

44 (30%) (60%) [63]

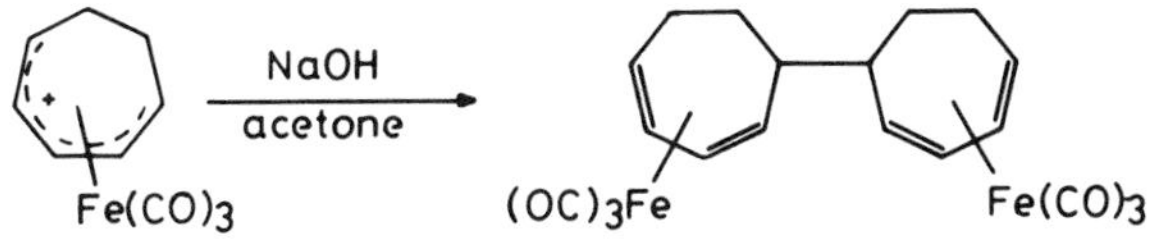

[64]

mixture containing acetone gave a dimer *(304)* (eq. [64]).

Even carbenium ions less reactive than triphenylcarbenium ion are capable of adding to 44, including tropylium ion *(295,300)*, (heptadienyl)tricarbonyliron cation *(300)*, and (slowly) the (cycloheptadienyl)tricarbonyliron cation *(300)*. The latter leads to slow electrophile-initiated polymerization of 44 *(300)* (eq. [65]).

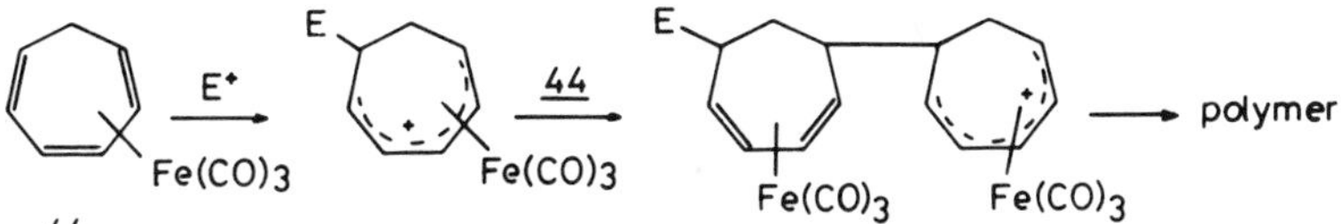

44 [65]

Johnson and coworkers have extensively explored the acylation of 44 (Scheme V) *(249,253)* and reactions of the acylated products. The aldehyde 46 is the only one isolated from the

Vilsmeier-type formylation of 44, although other positional isomers may be present before sublimation. Spectroscopic data on 46 suggest interaction between the aldehyde group (ν(CO) 1684 cm^{-1}) and the diene-tricarbonyliron group. It is

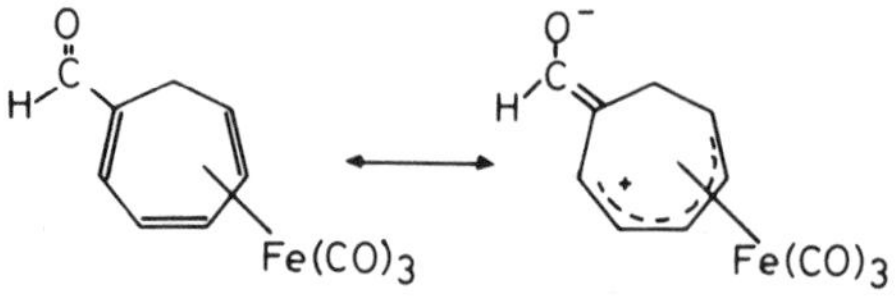

46

highly probable, therefore, that 46 is obtained from this reaction as a consequence of its thermodynamic stability *(306)*.

The addition and removal of electrophiles are highly stereoselective and occur at the face *anti* to the $Fe(CO)_3$ group, as shown in several cases in Scheme V.

A related electrophilic substitution of a derivative of 44 is the aminomethylation of (cyclohepta-2,4-dienone)tricarbonyliron *(147)*, which must occur *via* the enol (eq. [66]). (Cyclohexadienone)tricarbonyliron, which does not enolize, does not undergo this reaction.

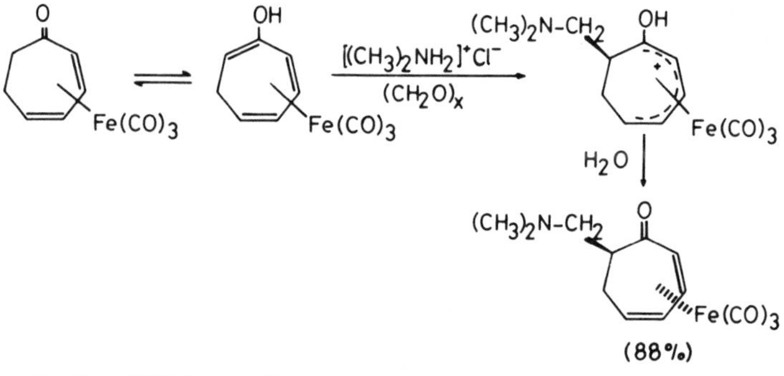

[66]

Cycloaddition of tetracyanoethylene (TCNE) to 44 also proceeds by electrophilic attack at the free double bond *(178,217)*, leading to a novel 1,3-adduct *(218,393)* (eq. [67]). Other highly electrophilic dienophiles, including hexafluoroacetone, react similarly *(217,218)*. The bitropyl-tricarbonyl-

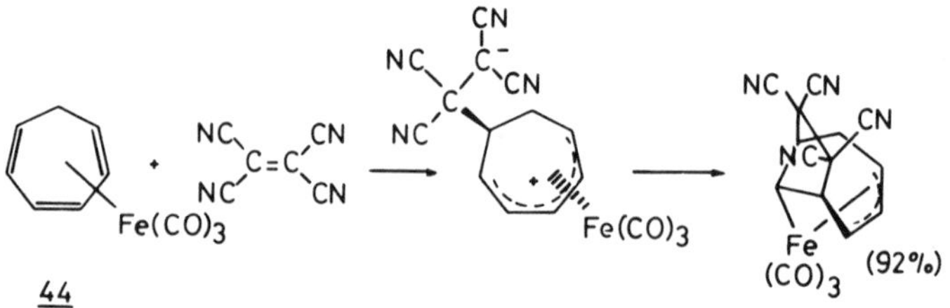

44

[67]

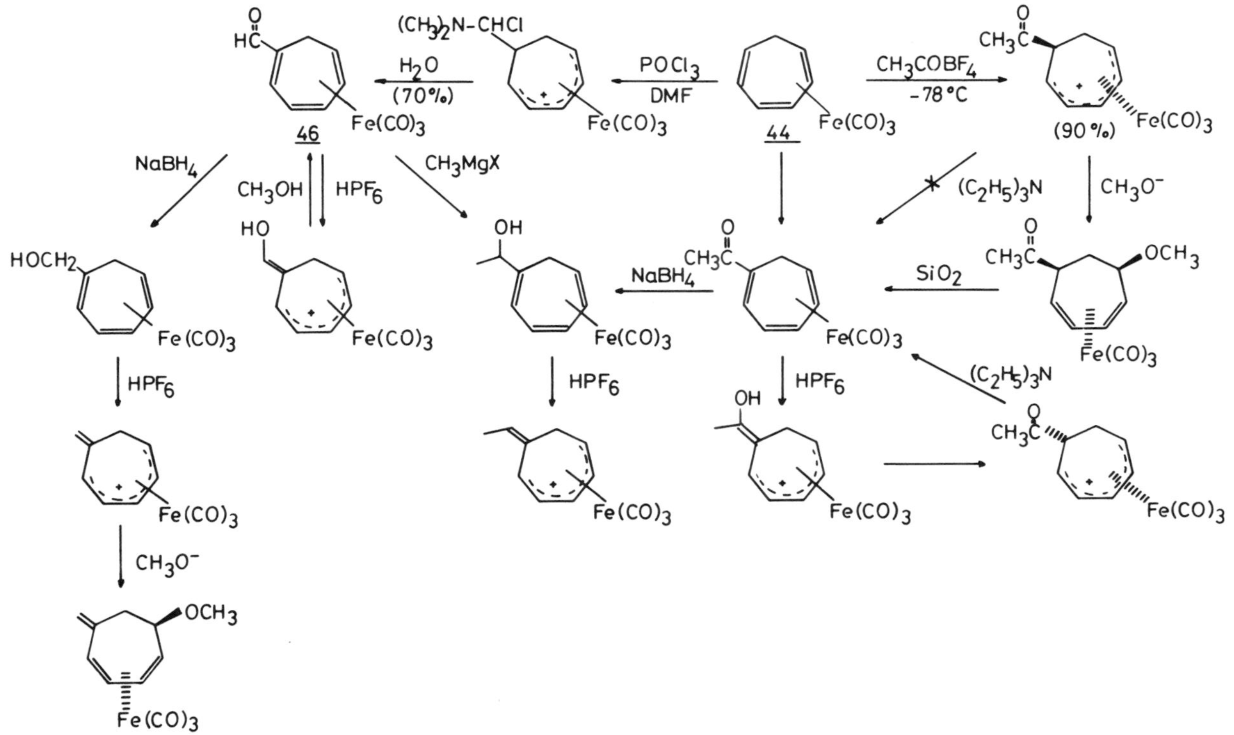

Scheme V: Acylation of <u>44</u> and Reactions of Acyl Products *(249)*.

iron complex 49 reacts with TCNE at the uncomplexed ring, presumably by a concerted mechanism, faster than at the complexed ring *(295)* (eq. [68]).

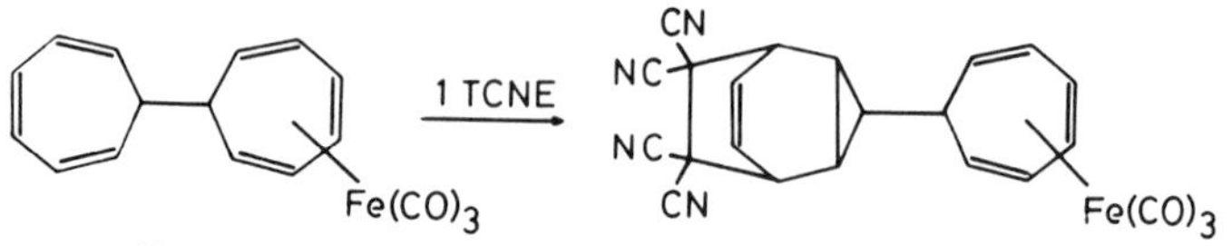

49

[68]

The free double bond of 44 could be hydrogenated in low yield by use of hydrogen and Raney nickel at room temperature *(80)*. It also underwent hydroboration/oxidation to give (cyclohepta-3,5-dienol)tricarbonyliron *(314)*, of undefined stereochemistry. Simmons-Smith methylenation gave the *anti* adduct in 23 % yield *(349)*, although in other hands *(249)* (bitropyl)hexacarbonyldiiron had been obtained.

44 undergoes attack by phosphorus nucleophiles at iron. Either displacement of a carbonyl ligand or of the triene, giving *trans*- $(R_3P)_2Fe(CO)_3$, can occur. The reactive nucleophile trimethylphosphite (TMP) reacts cleanly at temperatures below 60°C to give (1-4-η-cycloheptatriene)dicarbonyl(trimethylphosphite)iron, 50 (R = OMe), the product of carbonyl displacement *(348)* (eq. [69]). With excess TMP at 100°C,

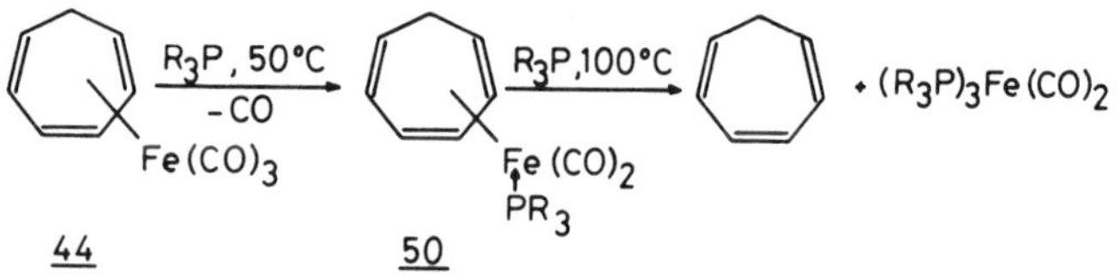

44 50

[69]

either 44 or 50 quantitatively yields dicarbonyl-tris(trimethylphosphite)iron, indicating displacement of cycloheptatriene from 50. Triphenyl- and triethylphosphine react at moderate temperatures to give predominantly carbonyl displacement, but formation of some tricarbonyl-bis(phosphine)iron resulting from triene displacement was also indicated by IR *(348)*. Other workers isolated the tricarbonyl-bis(phosphine)-iron in 29 % yield and dicarbonyl-tris(phosphine)iron in 13 % yield from reaction at 125°C *(310)*. The kinetics of reaction of 44 with triphenylphosphine at 154°C indicates the reaction to be first order in 44, and zero order in phosphine *(257)*. A carbonyl dissociation mechanism was indicated by this kinetic result and by the product analysis, which showed only (cycloheptatriene)dicarbonyl(triphenylphosphine)iron. This is consistent with the tendency of 44 to decompose at temperatures above 130°C, a process presumably also initiated by CO dissociation. This process can also account for formation of (cyclooctatetraene)tricarbonyliron from 44 on heating with cyclooctatetraene at 125°C (25 %) *(310)*.

Photolysis of 44 at room temperature in the presence of excess triphenylphosphine reportedly gave *only* 50 ($R=C_6H_5$), in contrast to other diene complexes, which also gave $(R_3P)_2Fe(CO)_3$ *(101)*. It is possible that CO dissociation in 44 is favoured by formation of a transient 1-6-η-$Fe(CO)_2$ complex (eq. [70]). Attempts to detect such an intermediate

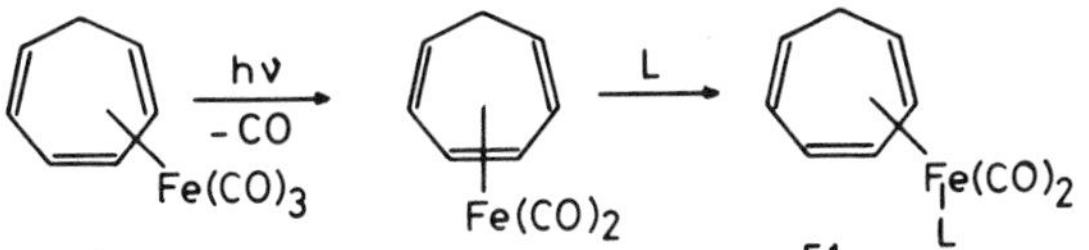

44 51 [70]

directly upon low temperature (-50°C) photolysis have not been successful. Studies underway involving photoracemization of 44 may provide better evidence.

Photolysis of 44 (and derivatives) in the presence of acetylenic dienophiles leads to interesting [2+6] adducts *(152)*. The dienophiles studied have included dimethyl acetylenedicarboxylate and diphenylacetylene. The same adducts are obtained by photolyzing 44 at low temperature in THF, followed by addition of the dienophile after shutting the light off. The mechanism ought to be related to that of photodisplacement of phosphines described above; one may therefore suggest the formation of 51 (with L = dienophile) followed by intramolecular cycloaddition within the iron's coordination sphere (eq. [71]). The dienophile in all cases is

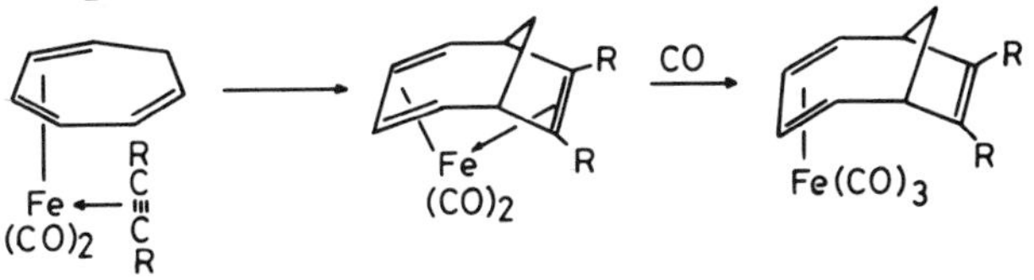

51 (L = RC≡CR) [71]

syn to the iron with respect to the cycloheptatriene ligand, consistent with the mechanism. An alternative mechanism, proposed by the original authors *(152)*, postulates formation of a THF-solvated η^2-cycloheptatriene complex.

Use of hexafluorobut-2-yne as dienophile gives rise to a still more complicated reaction with a 2:1 adduct as ultimate product *(58)* (eq. [72]), which may arise from further reaction of a 1:1 adduct like those described above.

Use of dimethyl maleate, a non-acetylenic dienophile, resulted in formation of a different sort of 1:1 adduct from those given by acetylenes *(152)* (eq. [73]).

The $Fe(CO)_3$ complex of tropone, 52, (and those of substituted tropones) are readily prepared by direct complexation.

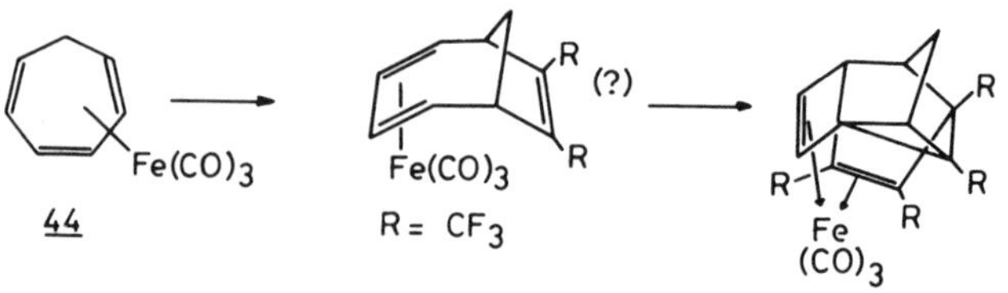

[72]

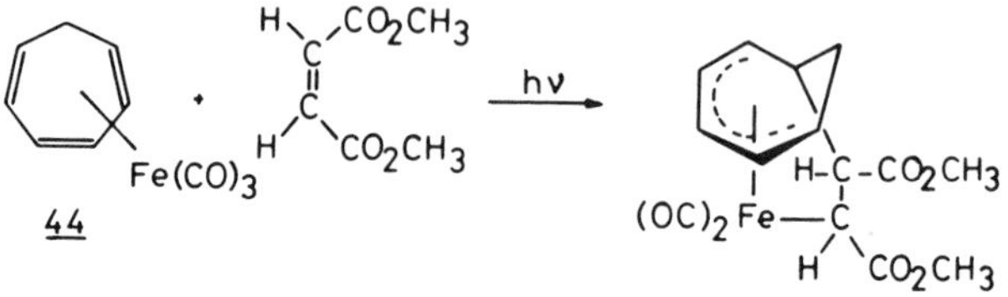

[73]

Use of $Fe(CO)_5$ at high temperature or with irradiation has not been reported. $Fe_3(CO)_{12}$ gives the product in low (7 %) yield *(238,272)*, but much better yields *(186,238)* result from use of $Fe_2(CO)_9$ (62 % *(239)*, 65 % *(203)*), or $Fe(CO)_5$ and trimethylamine oxide (62 % *(368)*). Use of $Fe_2(CO)_9$ is also quite successful with a number of 2-substituted tropones *(182)*, which give complexes with the substituent on the uncoordinated double bond, in 60-80 % yield.

52 and some derivatives can also be successfully prepared by reaction of terminal acetylenes with iron carbonyls. 52 itself is obtained in highest yield using $Fe_2(CO)_9$ and acetylene at 20-24 atm and 20-25°C *(396)*. A mixture of two triphenyltropone-tricarbonyliron complexes, 53, and 54, is similarly obtained from phenylacetylene and $Fe_3(CO)_{12}$ in 46 % combined yield *(61)*. The mechanism of forming tropone complexes under these conditions must be quite complicated. The immediate precursor of the tropone complexes, however, appears to be a substance composed of three erstwhile acetylene moieties, two irons, and six carbonyls which can be isolated using reaction temperatures below 60°C. An X-ray structure of such a material revealed the structure 55 for the product from phenylacetylene *(270)*. On treatment with triphenylphosphine or on refluxing in benzene, 55 forms the two isomers of (triphenyltropone)tricarbonyliron *(61,234)*.

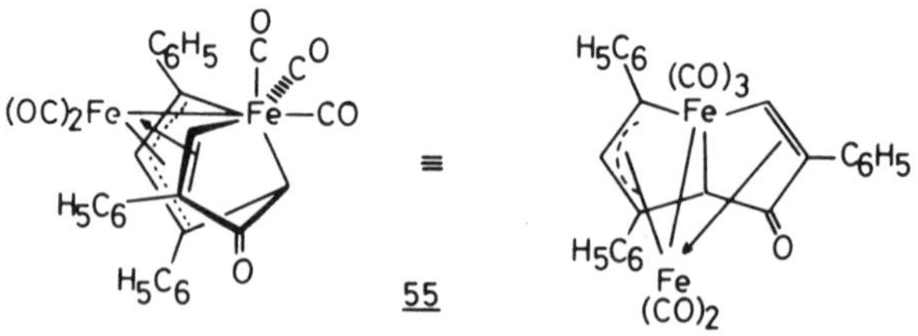

55

X-ray structures of both 52 *(174)* and the more stable triphenyl derivative, 53 *(371)*, have been reported. In contrast to early predictions *(74)*, both show a biplanar structure, with the plane of the three uncoordinated carbons bent away from the iron side of the coordinated diene plane. The iron is somewhat closer to the carbonyl carbon (2.99 Å) than to C(6) (3.11 Å) at the other end of the diene unit, suggesting some direct interaction as shown below. This is supported

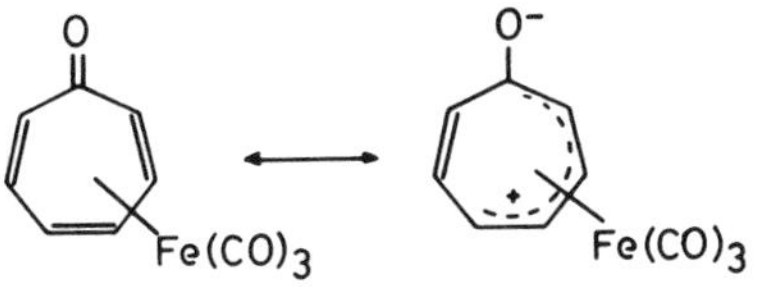

52a 52b

by the low ketone carbonyl stretching frequency (1637 cm^{-1} compared to 1712 cm^{-1} in (cyclohepta-3,5-dienone)tricarbonyl-iron) and the high dipole moment of 52, 4.25 D *(396)*. Structures like 52b should also stabilize transition states in shifts of the $Fe(CO)_3$ group about the seven-membered ring, in which event the iron should more readily move in 52 than in 44. In actuality, 52 is rigid on the NMR time-scale as is 44, the NMR spectrum being invariant up to 120°C. However, the two triphenyltropone complexes 53 and 54 slowly interconvert in solution (eq. [74]), even at room temperature. The ratio

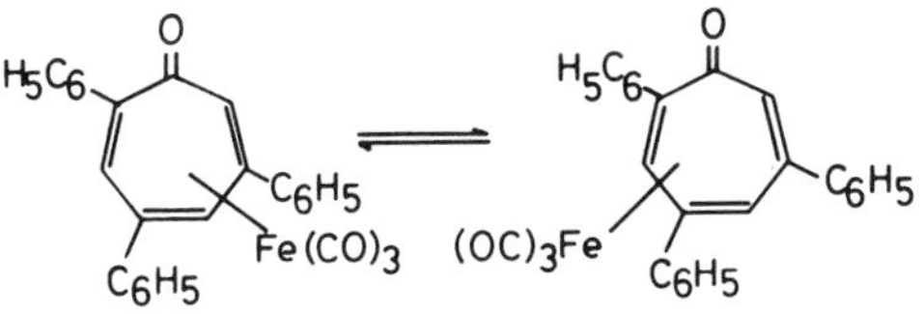

53 54 [74]

of 53/54 at equilibrium at 80°C was 2 *(61)*. The obtention of 4-7-η-$Fe(CO)_3$ complexes as the sole products when 2-substituted tropones undergo complexation with $Fe_2(CO)_9$ at 60°C (eq. [75]), conditions where isomer equilibration is probable,

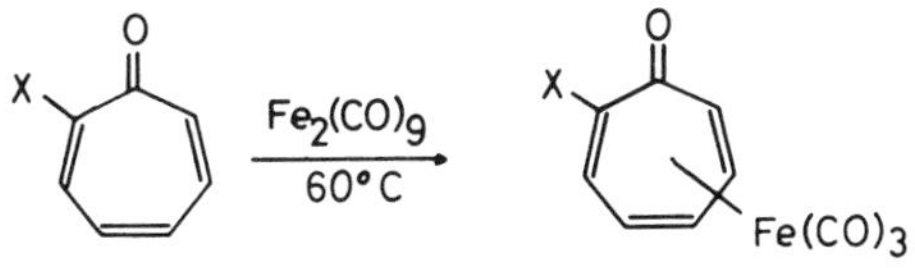

[75]

X = CH_3, C_6H_5, Cl

suggests that these are the thermodynamically more stable isomers, presumably for steric reasons *(182,184)*.

Despite the above qualitative evidence for $Fe(CO)_3$ group shifts in tropone-$Fe(CO)_3$ complexes, no quantitative data comparable to those on 44 have appeared as yet; a study by spin saturation transfer would be of interest.

The ketone group of 52 shows qualitatively normal carbonyl group reactions, giving a hydrazone and 2,4-dinitrophenylhydrazone, for example *(396)*; 53, in contrast, is said to give no 2,4-DNP derivative *(61)*. It undergoes attack by carbanionoid reagents, including CH_3Li *(184)*, $(CH_3)_2CHMgBr$ *(250)*, and 2-lithio-2-trimethylsilyl-1,3-dithiane *(214)*, at the carbonyl group. The 2-methyl derivative likewise undergoes carbonyl group reduction by lithium aluminium hydride *(184)*.

The uncoordinated double bond of 52 also gives some seemingly characteristic reactions, such as hydrogenation over palladium *(396)*. However, in addition to the expected $Fe(CO)_3$ complex of cyclohepta-2,4-dienone, which is obtained in 61 % yield, a 20 % yield of (cyclohepta-3,5-dienone)tricarbonyliron was also obtained when the hydrogenation was carried out in petroleum ether. In ethanol, dimeric and trimeric products were obtained.

The double bond functions as a dipolarophile toward diazomethane and diazoethane *(203,257a)* (eq. [76]). The

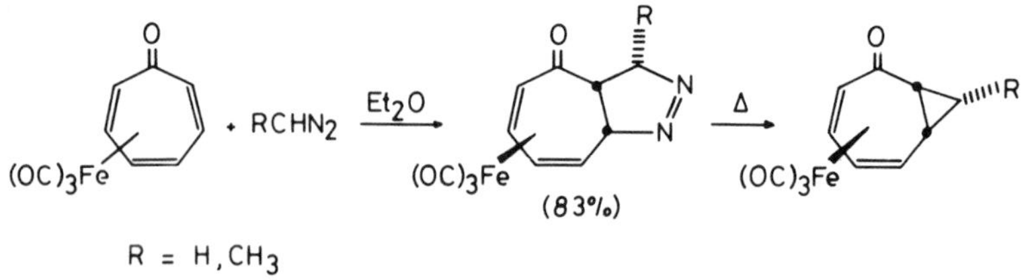

[76]

adducts lose N_2 at 80-110°C, giving homotropone complexes in quantitative yield.

The 2,5-diphenyl and 3,5,7-triphenyl derivatives of 52 undergo reversible electrochemical reduction, with two waves at *ca* 1.6 and 1.9 volts *(169)*. ESR spectroscopy revealed a radical species with g = 2.025; no splittings by protons were detected. The high g-value suggests that reduction occurs into an orbital with substantial iron character.

Tropone complexes, like cycloheptatriene complexes, undergo facile protonation, and this reaction has been extensively studied, even in the gas phase by chemical ionization mass spectrometry *(243)*.

Protonation of either the oxygen or the carbon(7) of 52 could lead to a stabilized η^5-pentadienyl-tricarbonyliron cation, and it is difficult to predict *a priori* which possibility should be preferred. In the event, extraction of a solution of 52 from methylene chloride into sulfuric acid or protonation of 52 with HBF_4 in acetic anhydride gave the salt resulting from C-protonation, 56 *(186,239)*. This straight-

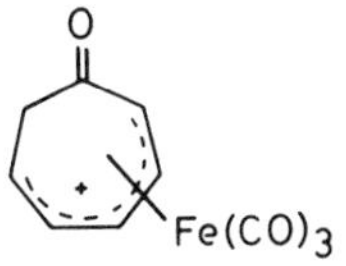

56

forward result was, however, beclouded by the observation that, using D_2SO_4, *three* deuterium atoms became incorporated into the cation, at the 2- *and* 7-positions *(186)*. With the weaker CF_3-CO_2D, D^+ added only to the *anti* side at C(7) of 52 *(239)*. The exchange of three hydrogens in sulfuric acid could be accounted for by either of two processes (Scheme VI) or by more complex schemes involving dications.

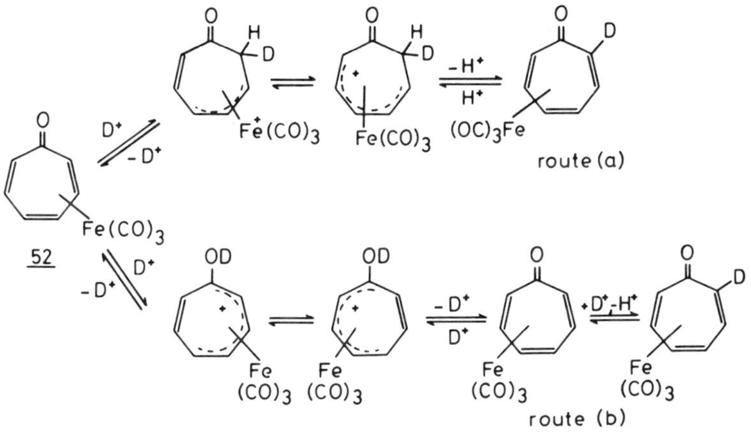

Scheme VI: Possible Routes for H(7) Exchange in 52.

Route (a) is inconsistent with the behaviour of 44 and of (cyclohepta-2,4-dienone)tricarbonyliron, which protonates only on the ketone oxygen *(239)*. Route (b), which provides a pathway for interconversion of C(2) and C(1) *via* the O-protonated intermediate, is consistent with the known fluxional behaviour of (tropylium)tricarbonyliron and with the behaviour of the dienone complex. Experiments with 7-substituted derivatives of 52 *(182)* support route (b) strongly. Thus, protonation of 52 at oxygen is kinetically favoured, but protonation at C(7) is thermodynamically preferable. Protonation at a coordinated carbon need not be invoked.

The stabilized cation 56 reacts with most nucleophiles (CH_3OH, N_3^-, $C_6H_5NH_2$, $(CH_3)_3CNH_2$, H-SiR_3) by covalency formation at the *anti*-6-position *(183,239)*. Phenyllithium removes the proton to regenerate 52 *(183)*. Cyanide and borohydride ions attack both the 2- and 6-positions from the side *anti* to the tricarbonyliron group *(183)*.

The reactions of 52 with electrophiles except the proton have not been reported. The result of attack of triphenylphosphine on 52 depends on conditions. Reaction at 100°C in a closed system gives predominantly (69 %) free tropone and only 11 % of (tropone)dicarbonyl(triphenylphosphine)iron *(396)*. At 80°C (open system) a greater quantity of the latter results. Carbonyl displacement is the predominant result with (triphenyltropone)tricarbonyliron, 53 *(61)*, and (cycloheptadienone)tricarbonyliron *(396)* in open systems, but again 53 and 54 undergo predominant (at least 82 %) organic ligand displacement in closed systems. Evidently the ability of CO to escape is an important factor in controlling these reactions (eq. [77]). Disengagement of the organic ligand from

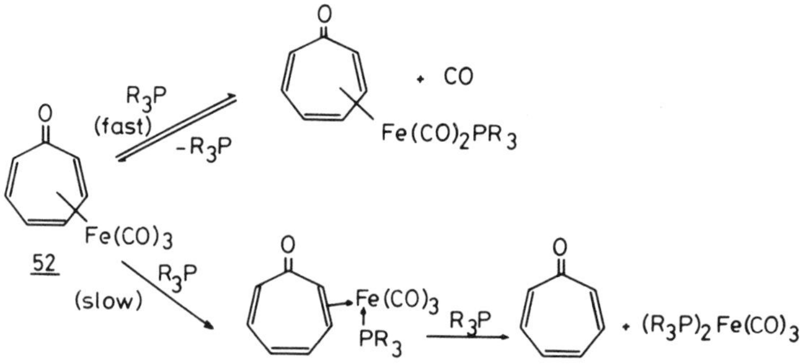

[77]

52 has been successfully accomplished (71 %) using trimethylamine oxide *(367)*. The more commonly used ceric ammonium nitrate has not evidently been applied to 52.

$Fe(CO)_3$ complexes of heptafulvene might be expected greatly to resemble those of tropone. In fact, the resemblances are much greater for complexes of 8-substituted heptafulvenes than for those of heptafulvene itself.

Heptafulvene complexes have seldom been obtained by direct complexation of heptafulvenes, because of the sensitivity of the latter to heat and light. 8,8-Diphenylheptafulvene alone has been directly converted to its $Fe(CO)_3$ complex, 57 ($R = C_6H_5$), by reaction with (benzylideneacetone)tricarbonyliron (70 %) *(233)*. Reaction of 2-(7-cycloheptatrienyl)-2-propanol with the same reagent, (BDA)$Fe(CO)_3$, gives the complex 57 ($R = CH_3$) directly. Reaction of 7-cycloheptatrienylmethanol with (BDA)$Fe(CO)_3$ *(233)*, or with $Fe_2(CO)_9$ *(177,269)*,

58

[78]

gives, however, not 57 (R = H) but rather the isomeric 58 (eq. [78]). The structure of 58 has been confirmed by X-ray crystallography *(108)*. Other derivatives of 57 have been obtained by a number of indirect routes from 44 and 52 (Scheme VII) (R = C_6H_5, CH_3, $C_6H_5CH_2$).

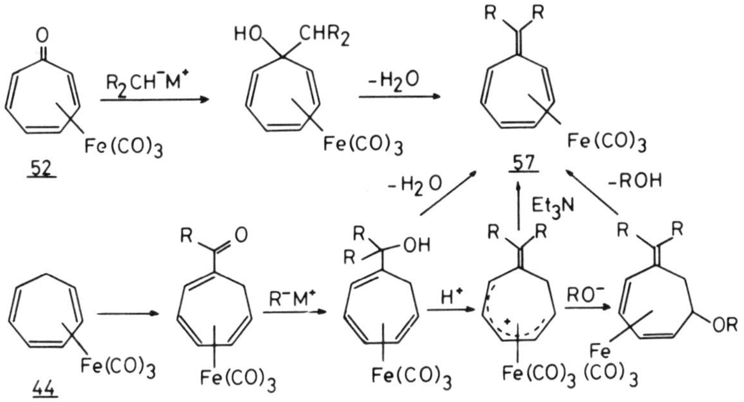

Scheme VII: Indirect Routes to Substituted Heptafulvene Complexes *(184,214,248-250,253)*.

Attempts to apply the reactions of Scheme VII to synthesis of the parent (heptafulvene)tricarbonyliron 57 (R = H), however, failed to give either 57 or the stable 58. Instead, a dimer was obtained *(184,250)* (eq. [79]). The monomeric 57

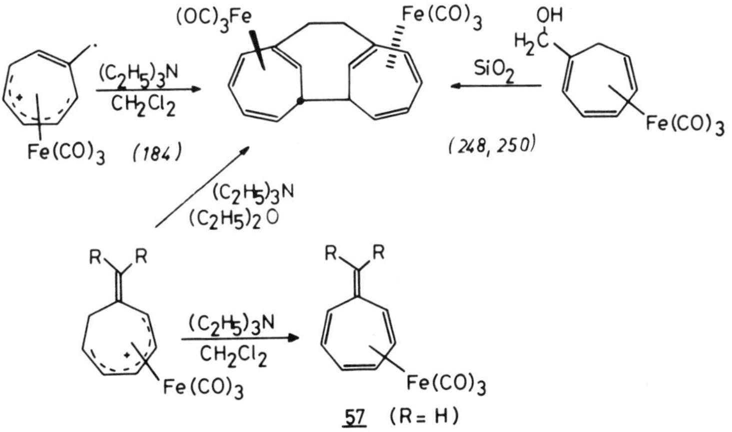

[79]

[80]

(R = H) was obtained by deprotonation of the exocyclic methylene cation *(354)* (eq. [80]). It proved to be unstable, undergoing slow dimerization at room temperature *(354)*. The structure of the dimer, shown above, was established by extensive

NMR studies and then X-ray crystallography *(184)*. The more stable 58 was not produced in any of these deprotonation reactions, presumably because shifts of the tricarbonyliron group from 1-4-η-coordination to 1,6,7,8-η-coordination cannot be achieved *via* a sequence of low-energy intermediates.

Only limited evidence on fluxionality of 1-4-η-bound heptafulvenes has appeared. Both 1-4-η isomers of complexes of 8-monosubstituted heptafulvenes were obtained using reactions in Scheme VII *(250,354)* from isomerically pure starting materials, indicating that the $Fe(CO)_3$ group does shift back and forth at room temperature, though not at a rate observable by NMR even at 100°C *(250)*.

Like the closely related 44 and 52, complexes 57 undergo facile electrophilic substitution at the uncoordinated endocyclic double bond *(250)*. Attack at the exocyclic double bond, to give derivatives of (tropylium)tricarbonyliron, 45, does not appear to occur. Electrophiles investigated include the proton and the $POCl_3$-DMF Vilsmeier reagent.

In contrast to 57, the more stable isomer 58 is less chemically reactive. It resists hydrogenation or reaction with tetracyanoethylene *(269)*. It does react with excess $Fe_2(CO)_9$ with complexation of the free diene unit. At 140°C it apparently decomposes (at least in part) to free heptafulvene, judging from the obtention of the heptafulvene-dimethyl acetylenedicarboxylate adduct when the reaction is run in the presence of that dienophile. 58 decomposes at once on treatment with fluoroboric acid.

In contrast to cycloheptatriene, tropone and heptafulvenes, oxepins exist at room temperature in significant amount as the bicyclo[4.1.0]hexadiene ("benzene oxide") tautomers. Perhaps for this reason or because of their relative inaccessibility, complexation of oxepins has not been extensively studied. Photochemical reaction of oxepin itself with $Fe(CO)_5$ at -60°C gave as the main products benzene (60 %) and phenol (30 %) *(26,195)*. A very low yield (3 %) of an $Fe(CO)_3$ complex, 59, (R = H), was also obtained. Similar results were

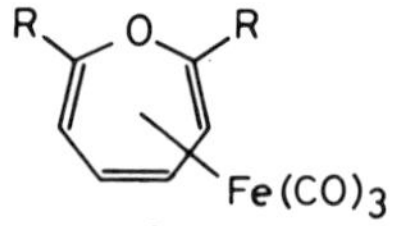

59

obtained with 2,6-dimethyloxepin, which gave 5 % of 59 ($R = CH_3$). 2,3-Benzoxepin, which shows no valence tautomerism, gave 22 % of the tricarbonyliron complex *(195)*. The substantial formation of benzene and phenol in the reactions of oxepin and dimethyloxepin was thought to arise *via* an oxepin- or (benzeneoxide)tetracarbonyliron complex (eq. [81]).

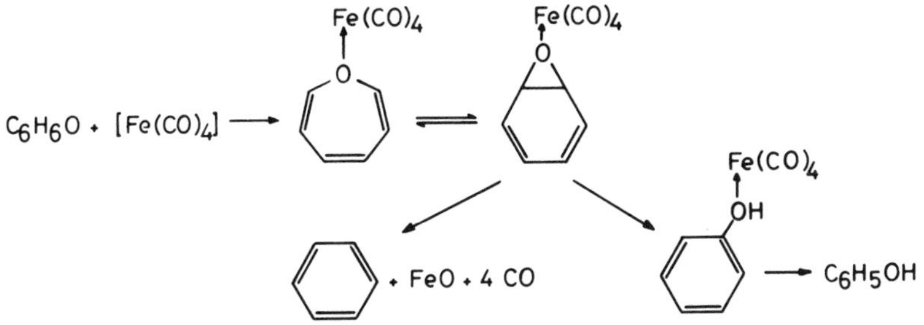

[81]

The complexes 59, unlike 44, 52, and 57 show fluxionality in the NMR. The dimethyl complex shows a coalescence temperature of 46°C, and a $\Delta G^{\ddagger}$ of 15.8 kcal *(26)*. The barrier to $Fe(CO)_3$ group motion is thus about 7 kcal less in 59 than in the analogous cycloheptatriene complex, 44. This has been attributed to facilitation of a 1,2-shift by the unpaired electrons on oxygen, which can stabilize the intermediate.

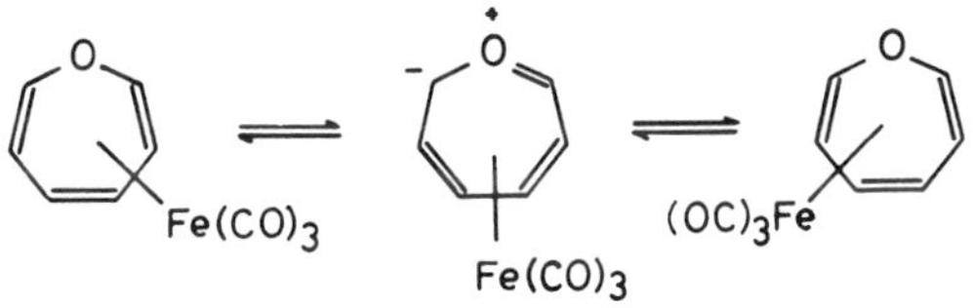

59 59'

Trapping of this carbonyl ylide by an appropriate dipolarophile should be possible.

No chemical reactions of oxepin complexes have been reported, but one would certainly expect a high reactivity toward electrophiles.

Direct complexation of *N*-carbalkoxyazepines, which have no detectable bicyclic tautomers present at room temperature, proceeds very effectively. Thus, *N*-carbethoxyazepine gives a 77 % yield of the $Fe(CO)_3$ complex, 60, on irradiation with $Fe(CO)_5$ *(200)* and 69 % on reaction with $Fe_2(CO)_9$ *(338)*. The $Fe_2(CO)_9$ method is said to be simpler and cleaner, but the $Fe(CO)_5$ method has been more extensively used to prepare *N*-carbalkoxyazepine-tricarbonyliron complexes, and also an *N*-methanesulfonyl analog *(338)*, as well as a complex of 2,4,6-trimethyl-*N*-carbethoxyazepine *(200)*.

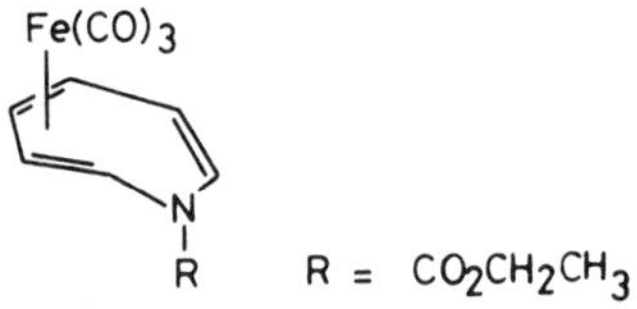

60

X-ray crystallographic investigations of 60 *(258,342)* and of (azepine)tricarbonyliron itself *(208)*, formed by saponification of 60 *(200)* have revealed for both a biplanar structure with the uncoordinated enamine moiety bent away from the $Fe(CO)_3$ group. The bond lengths suggest enamine delocalization within the former.

Dynamic NMR studies of 60 at 74°C reveal rapid motion of the $Fe(CO)_3$ group relative to the azepine ligand, with a decoalescence temperature of 34°C, and a $\Delta G^{\ddagger}$ of 15.5 kcal, essentially the same as 59 *(223,338)*. On further cooling, hindered rotation of the carbethoxy group is also revealed, with a decoalescence temperature of -28°C and a $\Delta G^{\ddagger}$ of 13.1 kcal. The fluxionality of 60, like that of 59, is apparently due to the influence of the electrons of the heteroatom; a study of (azepine)tricarbonyliron should therefore reveal a still lower barrier to $Fe(CO)_3$ group motion.

As previously indicated, alkoxide nucleophiles attack 60 at the urethan carbonyl group, leading to loss of the carbalkoxy group *(200)*. Electrophiles react readily at the uncoordinated double bond, but in a different manner from the foregoing examples *(209)*. As suggested by the X-ray and NMR results, the double bond interacts more strongly with the nitrogen than with the diene-tricarbonyliron moiety, and thus 60 reacts essentially as an enamine (eq. [82]). Both the

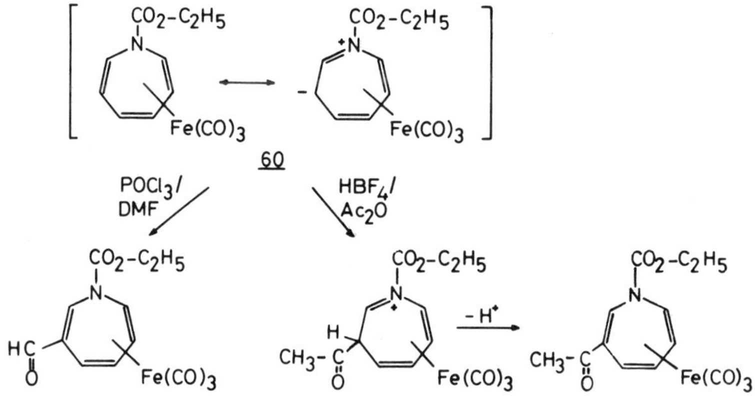

[82]

formyl *(406)* and acetyl *(389)* derivatives have been characterized by X-ray crystallography. The structures are biplanar, like 60, and show strong interaction between the nitrogens and the acyl groups *via* the free double bond.

Reaction of 60 (methyl ester) with the electrophile tetracyanoethylene *(218)* does not proceed as would be expected from the reactions with other nucleophiles. The products obtained (Scheme VIII) are not consistent with attack at the 6-carbon. This may reflect the inability of the 6-adduct

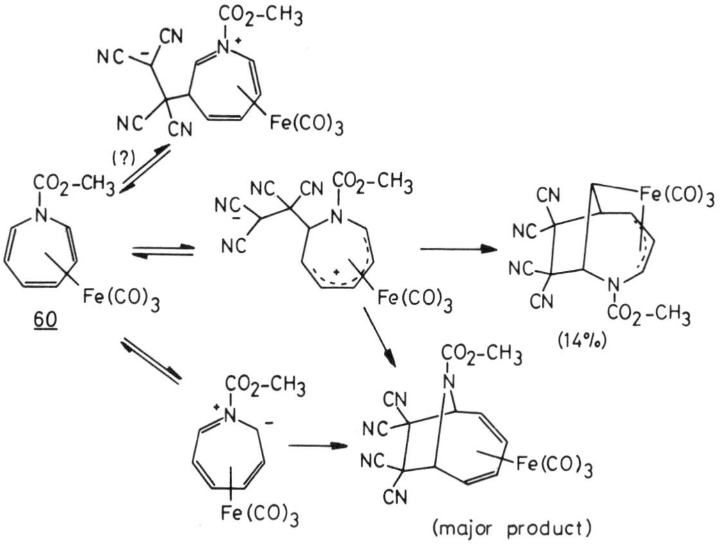

Scheme VIII: Possible Route of Reaction of 60 with Tetracyanoethylene.

to give any product except a strained cyclobutane. The minor product is precisely analogous to that formed by 44 and TCNE, and must be formed by the same mechanism, as shown. The major product, a [6+2] cycloadduct, may arise *via* the same dipolar intermediate, although no analogous product forms in other cases involving such dipolar intermediates. More interestingly, it may arise *via* trapping of the 1,3-dipolar intermediate previously invoked to explain the fluxional character of 60. The latter proposal should be verifiable by means of kinetic measurements.

A number of tricarbonyliron complexes of 1H-1,2-diazepines, 61 have been prepared by direct complexation with $Fe_2(CO)_9$ *(97,373,374,375)* (eq. [83]). High yields (84-95 %)

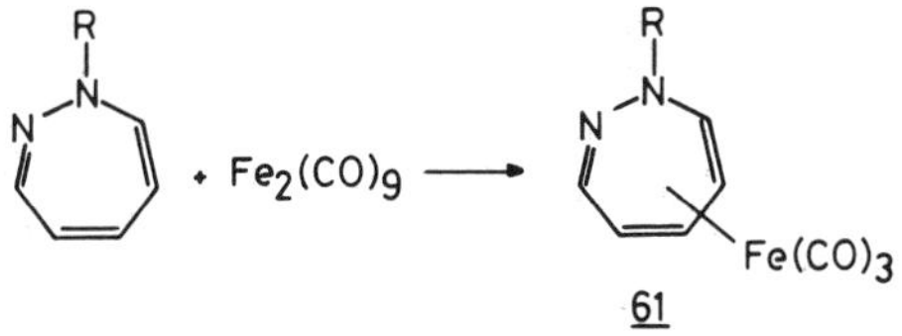

[83]

R = CO_2-CH_2-CH_3 , CO_2-$CH(CH_3)_2$, COC_6H_5 , $COCH_3$
SO_2-C_6H_5 , SO_2-C_6H_4-CH_3 , CH_3

were obtained with *N*-acyldiazepines in benzene solution, lower (16-36 %) with *N*-sulfonyldiazepines in acetonitrile *(375)*.

Some *C*-substituted azepines have also undergone complexation with $Fe_2(CO)_9$ *(97)*.

X-ray structures of two derivatives of 61, having R = $CO_2CH(CH_3)_2$ *(4)* and $CO_2CH_2CH_3$ *(158)*, have indicated the expected biplanar structure of the seven-membered rings.

Ester groups on 61 undergo ready saponification with sodium ethoxide at 0°C, giving 61 (R = H) *(95)*. This in turn can be acetylated or alkylated on N using sodium hydride / acetyl chloride and butyl lithium / benzyl bromide. 61 (R = H) can also be methylated directly with methyl fluorosulfonate *(96)*. The *N*-alkyl and -acyl derivatives of 61 show no fluxional character even at 100°C, but the parent 61 (R = H) is fluxional *(95)*. An acid-catalysed process formally involving simultaneous shifts of the hydrogen and the $Fe(CO)_3$ group, with $\Delta G^{\ddagger}$ = 13.7 kcal, is indicated (eq. [84]). The 3-methyl

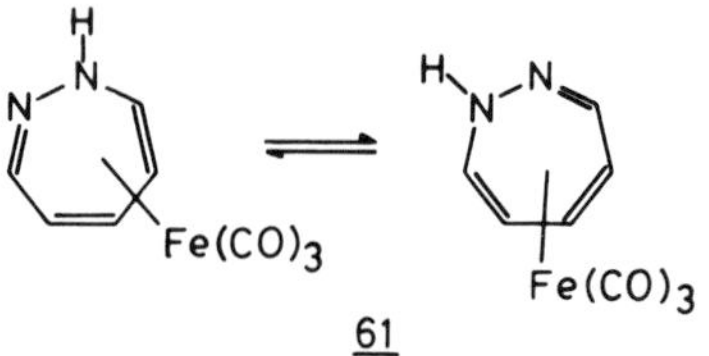

[84]

61

analogue is rigid, with the $Fe(CO)_3$ group bonded to carbon atoms C(4)-C(7), but the 5-methyl compound is again fluxional, $\Delta G^{\ddagger}$ = 15.0 kcal.

61 (R = H) undergoes N(2)-protonation, giving a fluxional cation *(96)*. X-ray crystallography of the trifluoroacetate salt indicated an η^4-structure *(99)*. The shift of the $Fe(CO)_3$ group presumably occurs *via* the alternative η^5-structure (eq. [85]).

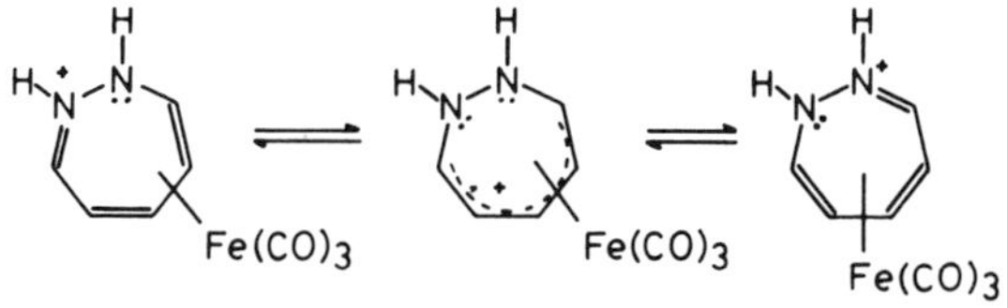

[85]

D. *$Fe(CO)_3$ COMPLEXES OF CYCLOOCTATRIENES AND THEIR VALENCE TAUTOMERS*

Reaction of 1,3,5-cyclooctatriene, pure or mixed with the 1,3,6-isomer, with $Fe(CO)_5$ at temperatures above 120°C gives, in addition to a small amount of an $Fe_2(CO)_6$ complex, only one $Fe(CO)_3$ complex, isolated in 20-60 % yield *(80,199,301, 310,311,325)*. This product, originally thought to be a complex, 62, of 1,3,6-cyclooctatriene *(199)*, was soon shown in fact to be (bicyclo[4.2.0]octadiene)tricarbonyliron, 63 *(301, 311,325)*. The dipole moment of 63 was found to be 2.37 D

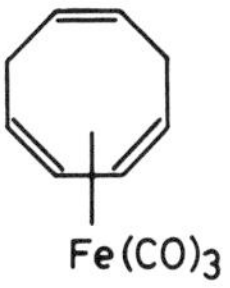

62

63

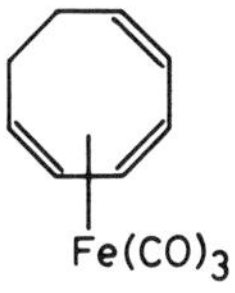

64

(199). Reaction of the cyclooctatriene mixture with $Fe_3(CO)_{12}$ at 80–110°C gave both 63 and the 1,3,5-cyclooctatriene complex 64, the latter in 5–20 % yield *(272,301,311)*. Use of still lower temperature conditions gave 64 as the isolated product – 10–15 % using $Fe_2(CO)_9$ *(325)*, 56 % using $Fe(CO)_5$ with irradiation *(70)*, the latter being clearly the synthetic route of choice. From this information, it was recognized in 1963 *(301)* that 63 was the more thermodynamically stable isomer. The product yields in fact depend little on the carbonyliron reagent used or on the ratio of cyclooctatriene isomers in the starting material, temperature being the key variable. It is curious that 62 has never been obtained under any conditions. Complexation of 5,8-bis(trimethylsilyl)cycloocta-1,3,6-triene with $Fe_2(CO)_9$ at 101°C gives the analog of 63 in 8 % yield, which is consistent with the results from the unsubstituted triene *(154)*.

Brookhart has studied the isomerization of 64 to 63, which occurs quantitatively (K_{eq} ⩾100) at 102°C. The $\Delta G^{\ddagger}$ was found to be 29.3 kcal for the conversion *(70)*. 64 was found not to be fluxional on the NMR time scale; indeed, the 1,2-shift of the $Fe(CO)_3$ group which might initiate interconversion of the enantiomers of 64 is the same as that required for conversion to 63 (*cf.* path (a) of Scheme IV). Thus, interconversion may be frustated by the availability of the latter alternative.

It is interesting to note that coordination of 1,3,5-cyclooctatriene with the $Fe(CO)_3$ group greatly stabilizes the bicyclic tautomer, but reduces the rate of conversion of the monocycle to the tautomer. The relative stabilization of 63 was attributed to ring strain in the presumed biplanar conformation of 64 *(70,215)*.

Reaction of 1,3,5-cyclooctatriene with (benzylideneacetone)tricarbonyliron (BDA)$Fe(CO)_3$ *(233)*, proceeds differently from the reactions with iron carbonyls themselves in that only 63 is obtained (83 %) *(215)*, even at 61°C. Kinetic studies have shown that this reaction involves highly selective reaction of (BDA)$Fe(CO)_3$ with the bicyclic tautomer *(215)* (Scheme IX). The selectivity may result from steric preference for coordination to a double bond of the planar cyclohexadiene rather than the tub-shaped cyclooctatriene, or, more

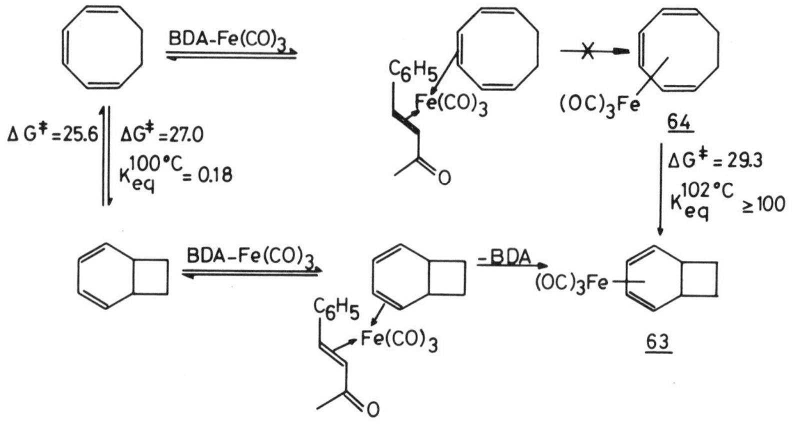

Scheme IX: Rationale for Selectivity of (BDA)Fe(CO)$_3$ *(215)*.

plausibly, reversible formation of an η^2-coordinated cyclooctatriene complex which does not readily convert to the η^4-coordinated product because of the large conformational change from tub to biplanar structure required. Kinetic studies on (BDA)Fe(CO)$_3$ and cyclooctadiene might be useful in connection with this proposal.

Complexes 63 and 64 have also been obtained by indirect routes. Refluxing (tricyclooctene)tetracarbonyliron in hexane gave 63 by metal-assisted disrotatory ring opening *(370)* (eq. [86]). A number of preparations of 64 from (cyclooctatetra-

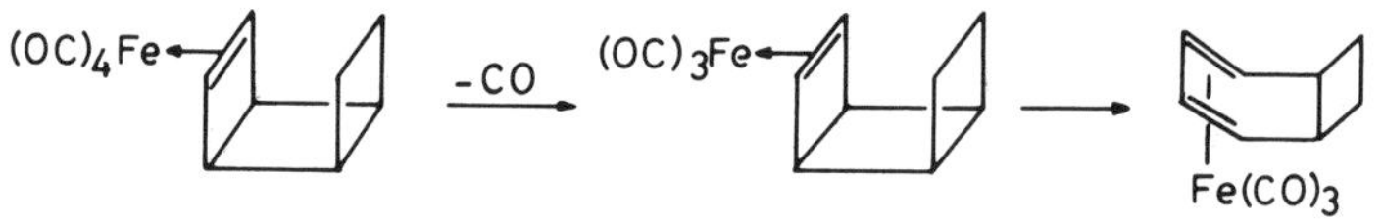

[86]

ene)tricarbonyliron, 65, have been reported. The obvious route, hydrogenation of a double bond, does not readily occur *(157)*. However, 65 has been reduced electrochemically in the presence of trimethylammonium bromide as proton source to give 64 quantitatively *(188)*. Indirect reduction of 65 by consecutive transfer of a proton and a hydride has been extensively studied. A variety of products has been obtained, depending on conditions (eq. [87]). The chief product, and the one originally reported to result from use of $NaBH_4$ in tetrahydrofuran, is 67 *(157,301)*. "Under different conditions" (presumably involving at least localized heating), a mixture containing 63 was formed *(157)*. With aqueous sodium borohydride at 0°C, 67 forms (75 %) along with 66 (15 %) or its

$(C_8H_8)Fe(CO)_3 \xrightarrow{H^+} [(C_8H_9)Fe(CO)_3]^+ \xrightarrow{H^-}$ 63 + 64 +

65

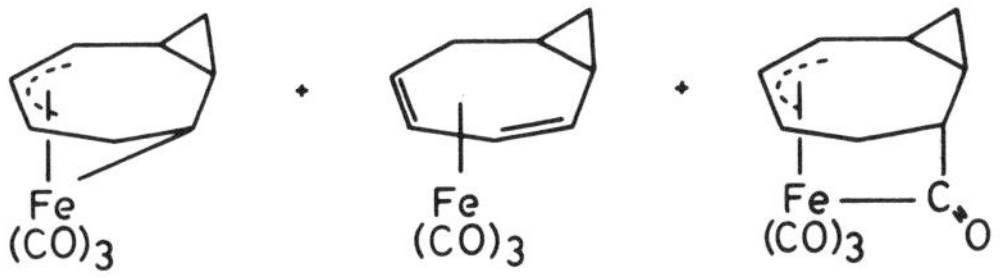

66 67

[87]

carbon monoxide insertion product *(19,27)*.

Both 66 and 67 undergo enantiomerization on heating *(24,68)*, as detected by averaging of NMR signals. At 40°C, 66 interconverts with its enantiomer 66', in effect by simultaneous shift of the $Fe(CO)_3$ group and the cyclopropane ring. A mechanism involving (1,2,6,7-η-cycloocta-1,3,6-triene)tricarbonyliron was proposed *(24)*, although a concerted process is not excluded (eq. [88]). At slightly higher temperature

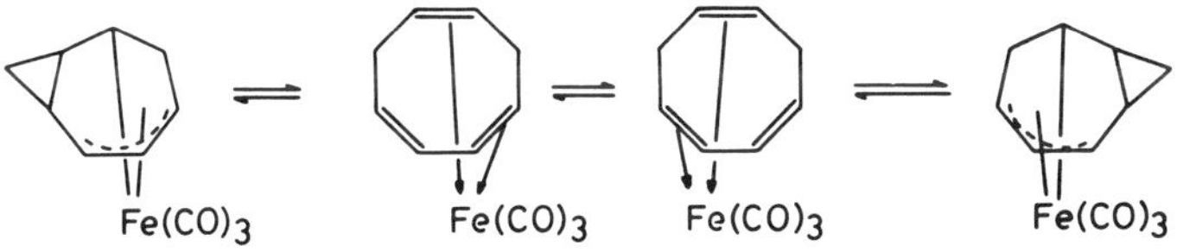

66 66' [88]

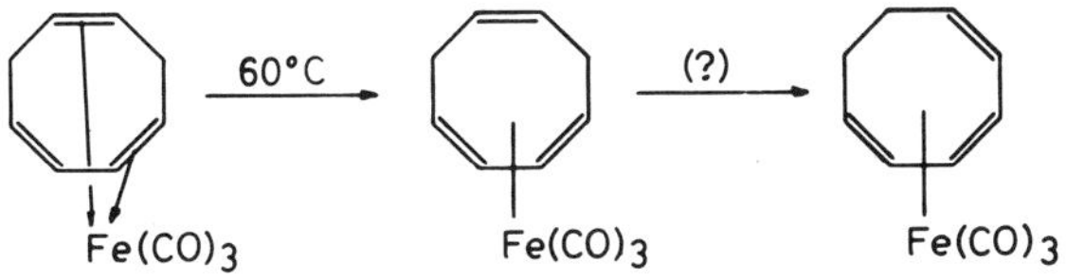

62 64 [89]

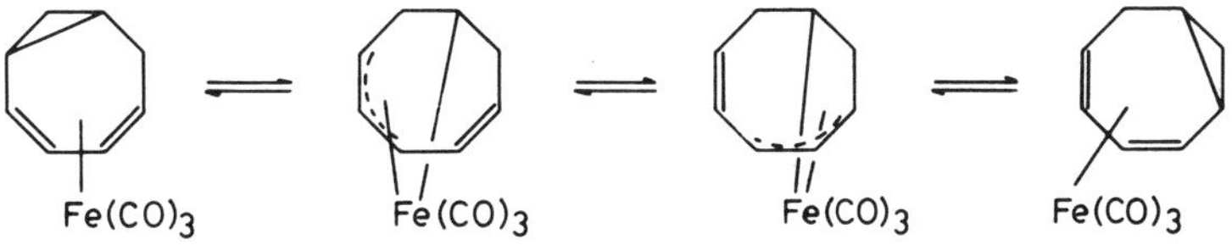

67 67' [90]

(60–70°C), rearrangement to 64 occurs quantitatively, a process which may occur through the unknown 62 *(24)* (eq. [89]). The more stable 67 enantiomerizes at 75°C *(24,68)* (eq. [90]). At 120°C, 67 rearranges *via* 64 (detectable by gas chromatography *(68)*) to 63 (eq. [91]). Derivatives of 64 have also been obtained by addition of nucleophiles to the cation

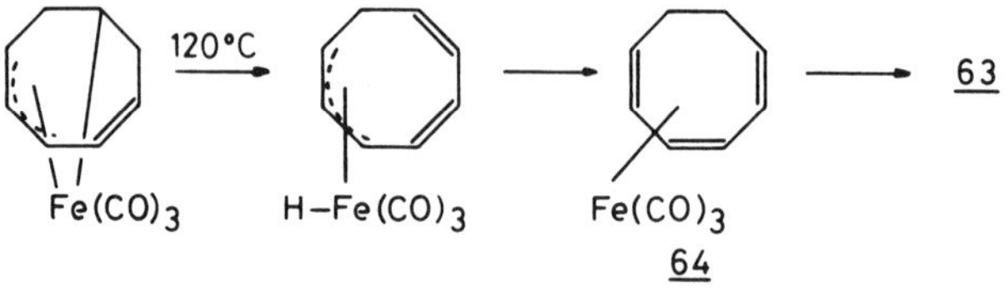

[91]

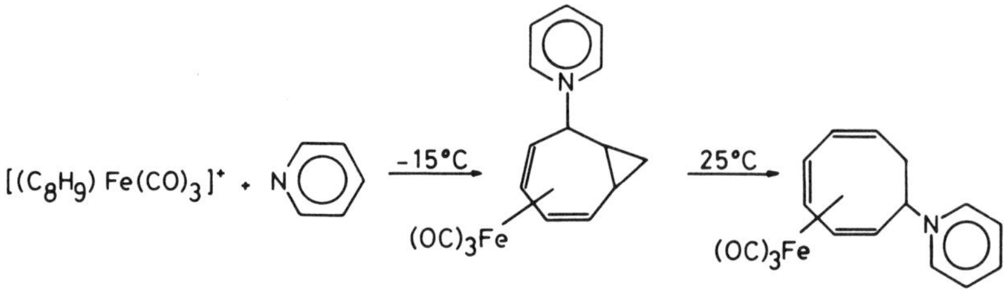

[92]

$[(C_8H_9)Fe(CO)_3]^+$ *(21)* (eq. [92]). Beyond these very interesting rearrangement reactions, little chemistry of 63 and 64 has been reported. 64 undergoes protonation as expected to give (cyclooctadienyl)tricarbonyliron cation *(301)*. Either 63 (or 64, which rearranges to 63) reacts with triphenylphosphine at 165°C to give cyclooctatriene and with maleic anhydride to give the Diels-Alder adduct *(311)*. An interesting reaction occurs upon treatment of 64 with aluminium chloride (eq.

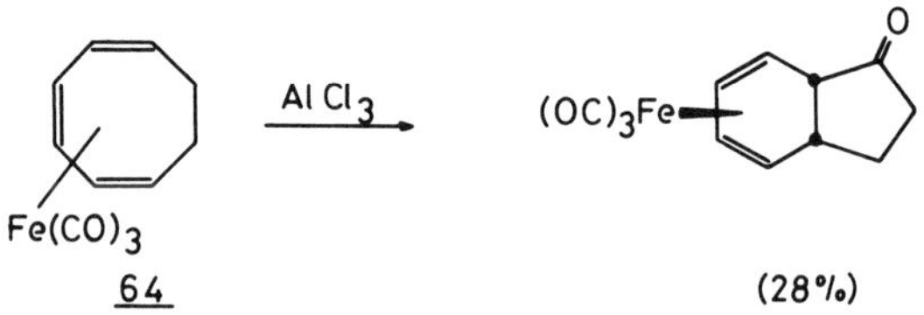

[93]

[93]); 63 does not react under these conditions *(256)*. As the authors note, "the mechanism of formation of this complex is not obvious."

A variety of cyclooctatrienes with additional rings fused to the saturated carbons has been subjected to reaction with carbonyliron reagents, and indeed results from such studies have helped greatly to clarify the picture presented above for cyclooctatriene itself. The fused rings alter the position of equilibrium between the cyclooctatriene and the bicyclooctadiene analogs and thus the results of reaction with carbonyliron reagents.

The reaction of bicyclo[6.1.0]nonatriene with carbonyliron reagents has been studied by two groups independently *(163,347)*. With $Fe_2(CO)_9$ in ether, both get as principal product (46 %) *(347)* 68, the $Fe(CO)_3$ complex of bicyclo-[4.3.0]nonatriene, (see Scheme X). The same complex can be obtained in high yield from the latter triene directly *(169,*

282,347), and is the most stable of all the complexes in this series, since all the others convert to it at about 110°C. (Cyclononatetraene)tricarbonyliron, 69, (12 %) *(347)* which will be discussed in Section IV.F, was also obtained, along with a hexacarbonyldiiron complex, 70, (14 %) *(347)*. In addition to these products, an unstable yellow oil, assigned the extraordinary structure 71, with the $Fe(CO)_3$ group coordina-

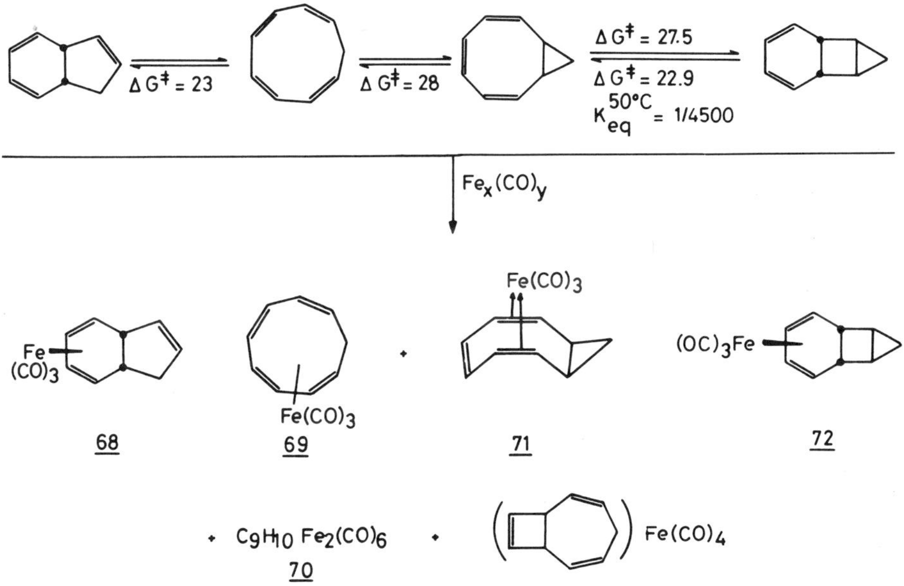

Scheme X: Bicyclo[6.1.0]nonatriene Valence Isomers and Carbonyliron Complexes.

ting two unconjugated double bonds, was obtained *(163,347)*, (15 %). Reaction with $Fe(CO)_5$ under irradiation *(347)* gave qualitatively the same products but in very different ratios, the yields of 68, 69, 70, and 71 being 6 %, 35 %, 1 %, and 15 %, respectively.

In addition to these products reported by both groups, Deganello's group *(163)* has also reported an $Fe(CO)_4$ complex thought to be a complex of bicyclo[5.2.0]nonatriene.

In contrast, (BDA)$Fe(CO)_3$ reacts with bicyclo[6.1.0]nonatriene at 55°C to produce, in addition to 68 (21 %) and 69 (9 %), the otherwise unobtainable product 72 (48 %) *(358,215)*. As in the case of cyclooctatriene itself, these products from (BDA)$Fe(CO)_3$ are thought to arise from direct trapping of the three corresponding unstable valence tautomers. Failure to detect 71 (or another complex of bicyclo[6.1.0]nonatriene) using (BDA)$Fe(CO)_3$ is a further example of the particular unreactivity of (BDA)$Fe(CO)_3$ toward 8-membered rings. It is most curious that (BDA)$Fe(CO)_3$ does react, however, with the 9-membered ring tetraene, forming 69.

Reactions of 9-substituted bicyclo[6.1.0]nonatrienes with $Fe_2(CO)_9$ have been briefly mentioned *(165)*. Both the 9-chloro and 9,9-dimethyl derivatives give $Fe_2(CO)_6$ complexes and dihydroindene complexes analogous to 68. Cyclononatetraene derivatives are apparently not produced. Reaction of the *syn*-9-methyl compound with (BDA)$Fe(CO)_3$ at 70°C gives only the analog of 72, in 68 % yield *(215)*, consistent with the much slower closure of this derivative to the dihydroindene, compared to the parent bicyclo[6.1.0]nonatriene.

Reaction of 9-oxabicyclo[6.1.0]nonatriene with iron carbonyls does not proceed analogously with the reactions of Scheme X, because of carbonyliron-induced opening of the epoxide ring. Thus, the principal product of reaction of the oxide with $Fe_3(CO)_{12}$ at 80°C is the aldehyde complex 46 (19 %). With $Fe_2(CO)_9$ a very complicated mixture of products, including 46 and complexes of cyclooctatetraene and cyclooctatrienone is obtained *(306)*. On treatment with $Fe(CO)_5$ under ultraviolet irradiation at -60°C, a probable intermediate, 73, is detectable *(25)*. 73 rearranges to the $Fe(CO)_3$ complex of 9-oxabicyclo[4.2.1]nonatriene on warming (eq. [94]).

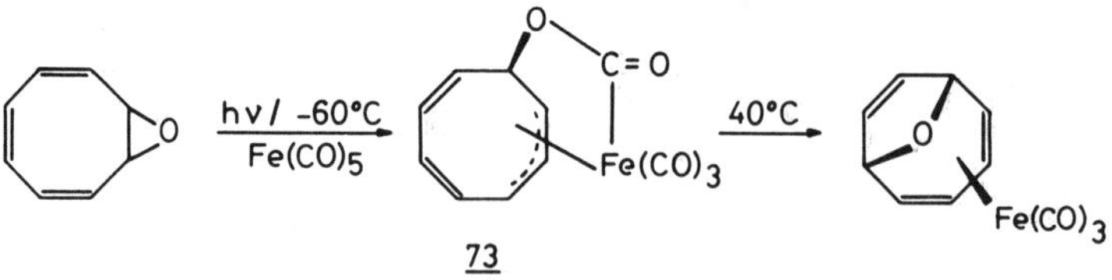

73

[94]

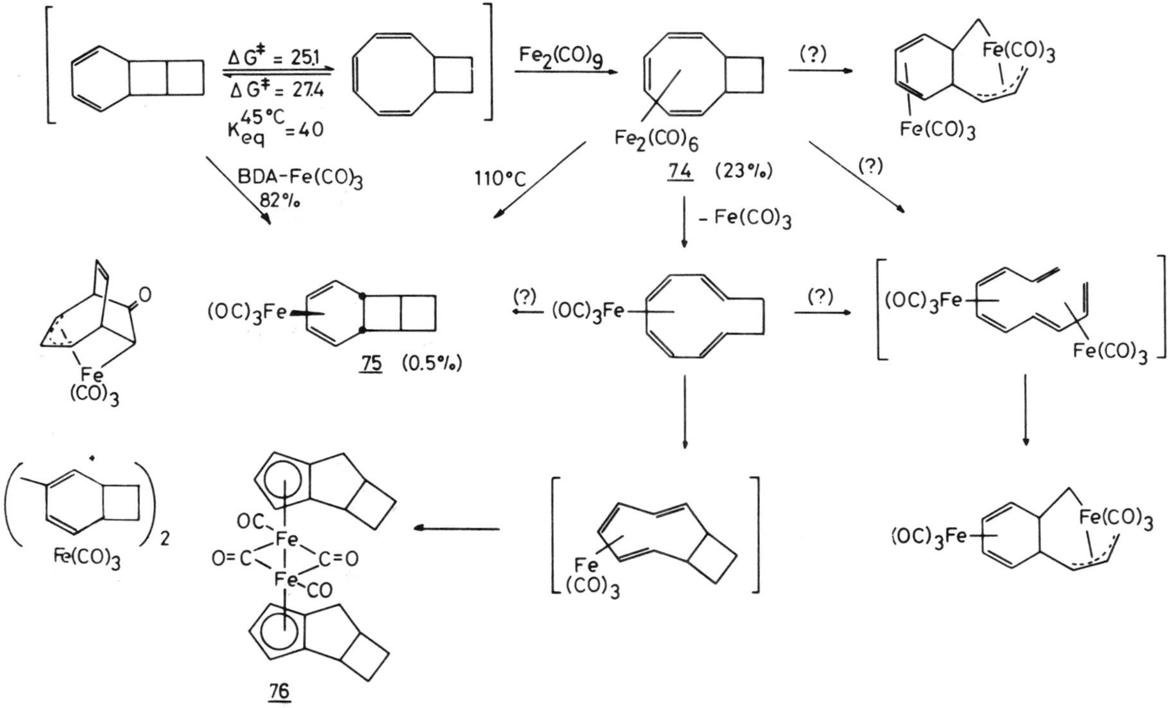

Scheme XI: Products Obtained from Complexation of Bicyclo[6.2.0]decatriene.

Reactions of bicyclo[6.2.0]deca-2,4,6-triene and its derivatives appear to be even more complex than those of bicyclo[6.1.0]nonatrienes, if only because they have been more extensively studied. Reaction with $Fe_2(CO)_9$ in ether gives principally an $Fe_2(CO)_6$ complex, 74, and an $Fe(CO)_3$ complex of the tricyclic valence isomer, although the latter is scarcely detectable in the free hydrocarbon *(126,127)* (Scheme XI). Numerous minor reaction products have been studied by the extraordinary technique of eschewing spectroscopic methods and characterizing all products of which a single crystal can be obtained, by X-ray crystallography *(132,133,143-146)*. Many of the "exotic" products so characterized (Scheme XI) are difficultly related to the putative starting material, and presumably derive from trace impurities in it *(143)*. Some routes of conversion have been proposed *(132,133)* as indicated in Scheme XI. The structure of the minor product 75 has been confirmed by X-ray crystallography *(146)*. Conversion of 74 to 75 and 76 at 110°C has been demonstrated *(126,133)*. Reaction with $(BDA)Fe(CO)_3$ gives overwhelmingly 75 (82 %), in contrast to the complex mixture described in Scheme XI, again by trapping the minor valence isomer *(358)*.

The cyclooctatetraene photodimer, m.p. 53°C, is a derivative of bicyclo[6.2.0]decatriene. Reaction with $Fe(CO)_5$ under irradiation gave two $Fe(CO)_3$ complexes, 77 and 78, both with a coordinated tricyclic system analogous to 75 *(361,362)* (eq. [95]). The formation of 78 was found to result from

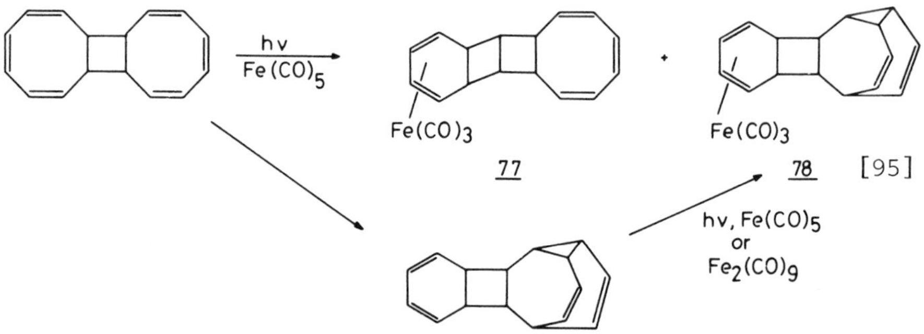

[95]

conversion of the photodimer to the isomeric dimer. The structures of both 77 *(353)* and 78 *(350)* were confirmed by X-ray crystallography. Both were degraded upon reaction with $FeCl_3$ to benzene and naphthalene *(361)*, and both underwent further complexation to give an extraordinary $C_{16}H_{16}Fe_3(CO)_9$ complex, discussed in Section XIII.

Studies of annulated cyclooctatrienes have been extended to those with five- and six-membered rings. The results are summarized in Scheme XII *(124,127,128)*. The structures of the

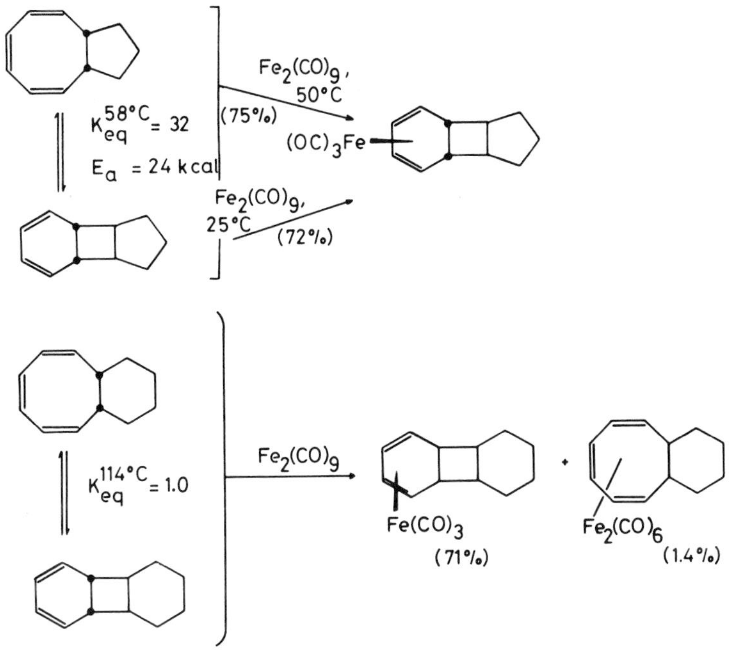

Scheme XII: Complexation and Tautomerism of Annulated Cyclooctatrienes.

tricyclic $Fe(CO)_3$ complexes were confirmed by X-ray crystallography *(124,165,130)*. Reactions with (BDA)$Fe(CO)_3$ have not been reported, but production of other than the tricyclic species would not be expected in view of the previously described results.

The only functionalized cyclooctatriene derivative which has been subjected to complexation is cyclooctatrienone. Reaction of this ketone with $Fe_3(CO)_{12}$ or $Fe_2(CO)_9$ in refluxing benzene gave an $Fe(CO)_3$ complex, 79, in 13 % yield and an $Fe_2(CO)_6$ complex in 2.4 % yield. Photochemical reaction with $Fe(CO)_5$ gave *ca.* 40 % yield *(69)*. 79 has more recently been obtained in better yield by acidic hydrolysis of (methoxycyclooctatetraene)tricarbonyliron *(257a,341)*. NMR studies indicate the 2-5-η-structure shown for 79; the isomeric compound has not been detected. Thermodynamic preference for 79 is consistent with results obtained in complexation of acyclic trienes and with evidence of iron-ketone interaction in (tropone)tricarbonyliron, 52. Reaction of cyclooctatrienone with (BDA)$Fe(CO)_3$ at 55-75°C gave (bicyclo-[4.2.0]octa-2,4-dien-7-one)tricarbonyliron, 80, in 50-59 % yield *(69,71,257a)*. The free ligand could be isolated by

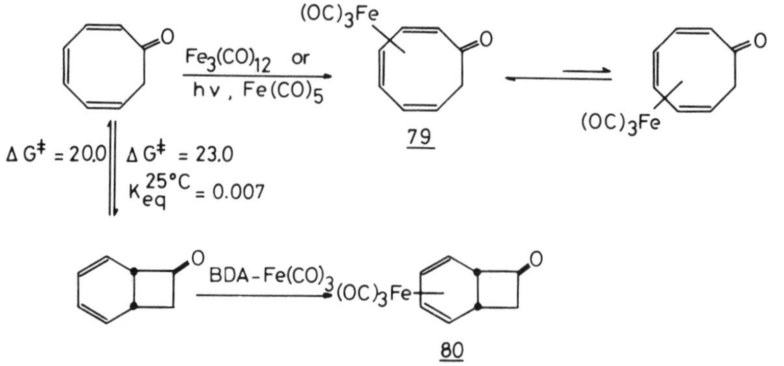

Scheme XIII: Equilibria Involving Cyclooctatrienone and Its $Fe(CO)_3$ Complexes.

reaction of the complex 80 with ceric ammonium nitrate *(69, 71)*, permitting determination of the results in Scheme XIII.

79 shows a rather high barrier (*ca.* 15 kcal/mol) to rotational averaging of the carbonyl groups, as revealed by ^{13}C-NMR *(69)*. It is not, from published data, heavily enolized, but studies of enolization and fluxionality would be of interest. 79 and 80 have not been interconverted directly, and thus their relative stability is unclear. 79 undergoes electrochemical reduction to give a stable radical anion with g = 2.0403, similar to those obtained from phenyltropone complexes *(170a)*. Protonation of 79 has not been reported, but a study would provide a useful comparison with the results on 52, previously discussed.

80 reacts with sodium methoxide in methanol at room temperature with opening of the four-membered ring (eq. [96]),

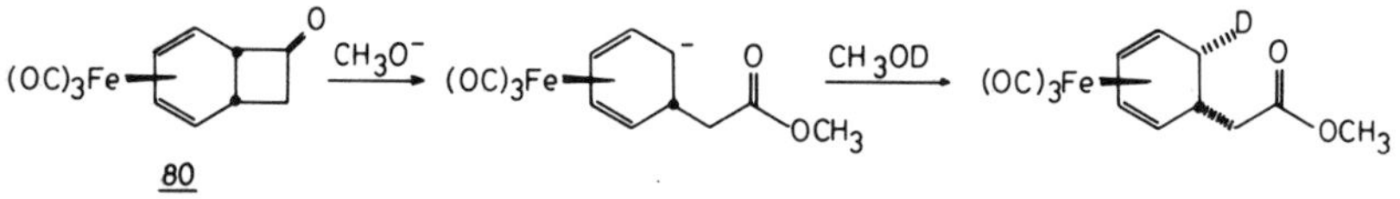

[96]

indicating some stability for the intermediate anion *(69)*. Labelling studies showed specific deuteration of the anion *anti* to the $Fe(CO)_3$ group, and no migration of the $Fe(CO)_3$ group during the lifetime of the anion, although the product is known not to be the most stable isomer. 80 also reacts

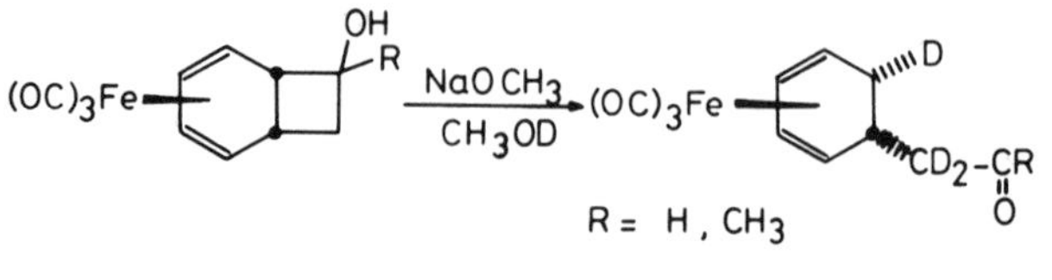

[97]

with methyllithium and sodium borohydride by addition to the carbonyl group. The carbinols also undergo ring opening with sodium methoxide in a consistent manner with the ketone (eq. [97]).

E. $Fe(CO)_3$ COMPLEXES OF CYCLOOCTATETRAENES

Probably no single compound in organotransition metal chemistry has been more extensively investigated and discussed than (cyclooctatetraene)tricarbonyliron, 65. Much of the discussion has resulted from the remarkable history of 65, in which first the static structure then the dynamics and reactions (*e.g.* protonation and cycloadditions) were repetitiously misconstrued at first, then corrected.

One noncontroversial feature of 65 is the ease of its synthesis. Reaction of cyclooctatetraene with any of the iron carbonyls appears to give good yields of 65; yields below 60 % are seldom mentioned in public. The most commonly used synthesis appears to be thermal reaction with $Fe(CO)_5$ in a high-boiling solvent such as ethylcyclohexane; yields of 60-87 % are reported *(221,273,309,310)*. Photochemical reaction with $Fe(CO)_5$ proceeds in 72 % yield *(346,361)*, reaction with $Fe_2(CO)_9$ in benzene proceeds in 86 % yield *(27)*, and with $Fe_3(CO)_{12}$ in 56-88 % yield *(247,302)*. Reaction with $Fe(CO)_5$ and trimethylamine oxide gives 68-80 % of 65, depending on solvent, in only one hour at room temperature *(368)*. A metal atom synthesis of 65 is clearly not required; however, co-condensation of cyclooctatetraene and trifluorophosphine with iron atoms at -196°C gives $(C_8H_8)Fe(PF_3)_3$ in low yield *(407)*. Reaction of cyclooctatetraene with (BDA)$Fe(CO)_3$ has not been reported; the relative unreactivity of that reagent toward eight-membered rings might lead to no reaction or to a complex of the valence tautomer, bicyclo[4.2.0]octatriene.

A variety of substituted cyclooctatetraenes has been subjected to reaction with carbonyliron reagents. Thermal reaction with $Fe(CO)_5$ produced the $Fe(CO)_3$ complexes of benzocyclooctatetraene (no yield given) *(315)* and of a fused-ring lactone *(221)* (eq. [98]). Photochemical reaction with $Fe(CO)_5$

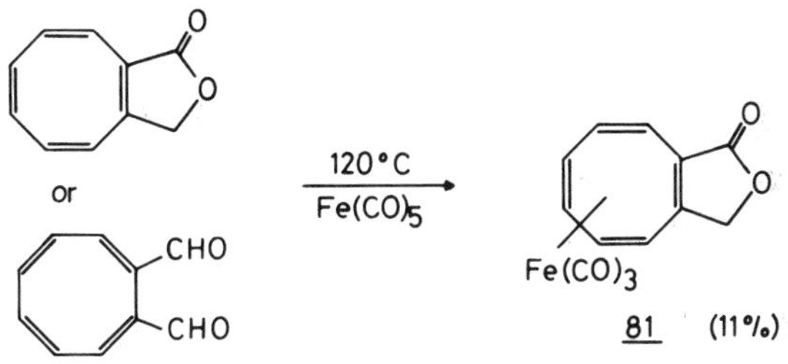

[98]

was used with phenylcyclooctatetraene to give the $Fe(CO)_3$ complex (28 %) *(323,324)*. This was obtained in better yield (50 %) using $Fe_2(CO)_9$ in refluxing hexane *(341)*. In general, $Fe_2(CO)_9$ has been most commonly used with substituted cyclooctatetraenes, C_8H_7R, including R = CH_3 (49 %) *(11,67)*, $Si(CH_3)_3$ (40 %) *(118)*, $Ge(CH_3)_3$ (60 %) *(118)*, $Sn(CH_3)_3$ (10 %) *(118)*, OCH_3 (45 %) *(341)*, CO_2-CH_3 (80 %) *(341)*, and benzocyclooctatetraene (16 %) *(187)*. The latter was obtained in much better yield (62 %) using $Fe_3(CO)_{12}$ *(187,247)*, which was also successful in giving $Fe(CO)_3$ complexes of C_8H_7Cl *(247)*, 1,8-dicarbomethoxy-cyclooctatetraene (*ca*. 30 %) *(221)*, and the bicyclic molecules shown in equation [99] *(364)*. The $Fe(CO)_3$

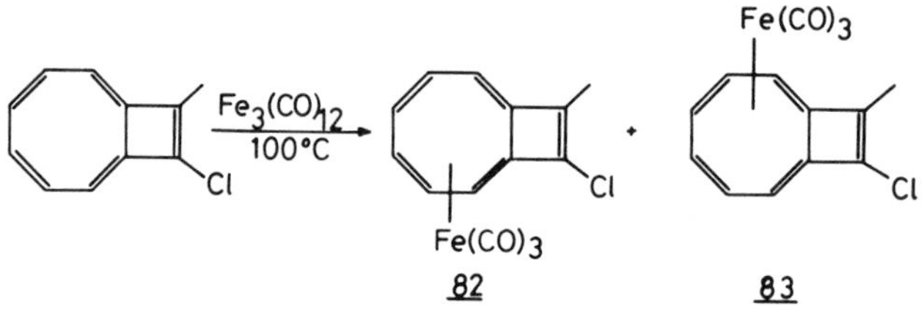

[99]

complexes obtained in these reactions are usually accompanied by smaller amounts of binuclear iron complexes, which are discussed further on in the appropriate sections.

Complexation of bicyclooctatetraenyl with carbonyliron reagents gives, in addition to the $Fe(CO)_3$ complex expected, an unusual $Fe(CO)_2$ complex, and complexes of cyclooctatetraene itself *(175)*. Two other cyclooctatetraenes have given anomalous results (eqs. [100] *(140)* and [101] *(225)*).

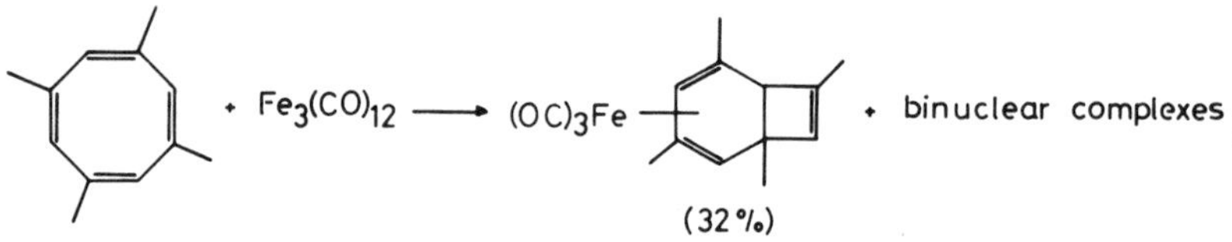

[100]

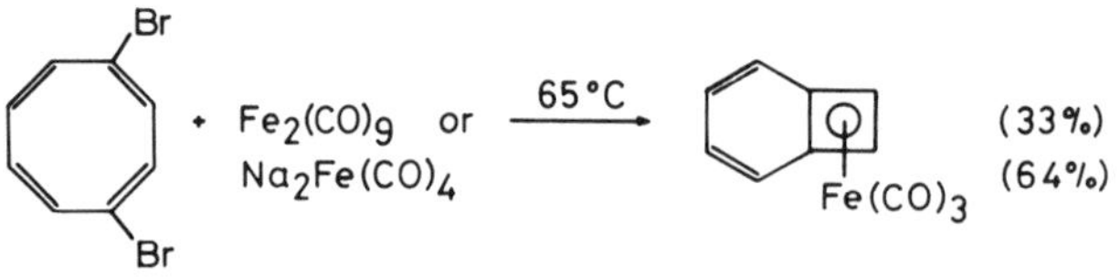

[101]

A few adventitious syntheses of 65 or derivatives have been reported (eqs. [102] *(269)*, [103] *(35)*, and [104] *(224)*). In the first two cases, the cyclooctatetraene products apparently result from initial diene-tricarbonyliron complexes which eject leaving groups to form derivatives of the highly stabilized (bicyclo[5.1.0]heptadienyl)tricarbonyliron cation,

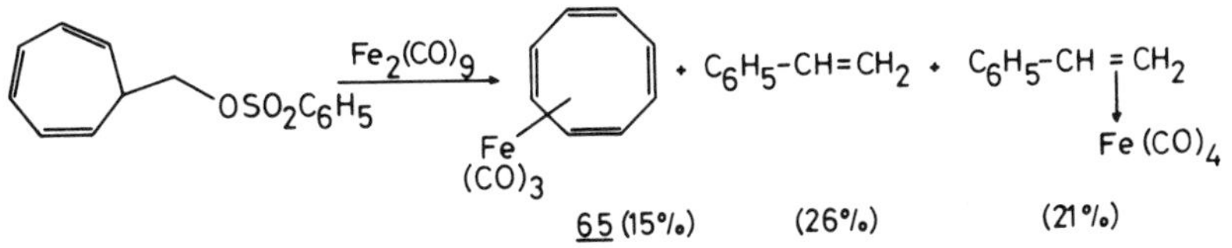

[102]

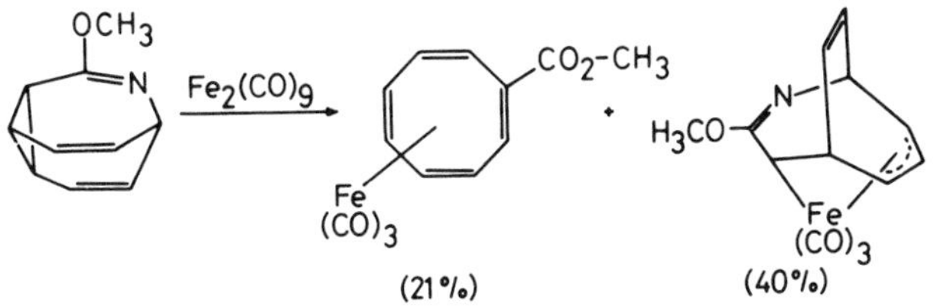

[103]

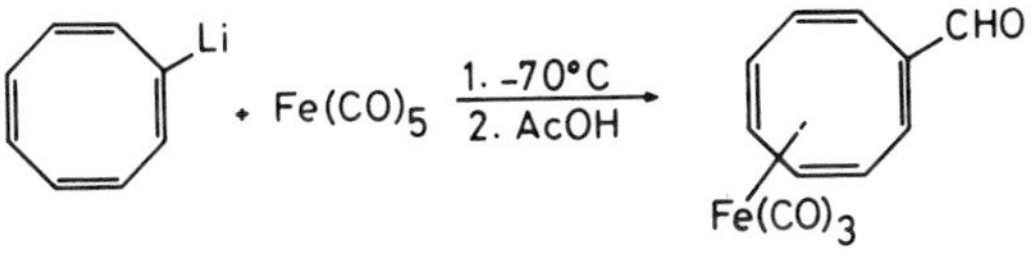

[104]

which give the product on loss of a proton. The latter case evidently results from attack of the carbanionic reagent on a carbonyl group of $Fe(CO)_5$.

The structure of 65 and its derivatives was the subject of much contention in early years, most of which centered about the fact that the proton magnetic resonance spectrum in solution shows only a single sharp line, even down to -100°C *(286)*. This was difficult to reconcile with the crystal structure of Dickens and Lipscomb *(172,173)*, which showed the $Fe(CO)_3$ group bonded to a 1,3-diene unit of a biplanar cyclooctatetraene ring:

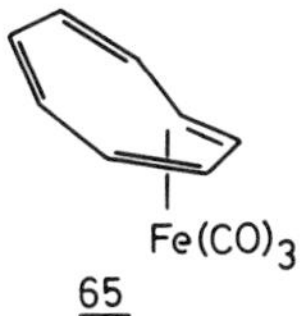

65

Despite Brown's 1959 suggestion *(73)* that "complexes formed from the planar configurations of the higher membered ring systems are likely to be less stable than those of the lower members" and despite Dickens and Lipscomb's 1962 observation that "a dynamical effect amounting to permutation of the carbon atoms of the ring relative to the $Fe(CO)_3$ group does ... satisfy the experimental observations," other solution structures, including a 1,2,5,6-η-coordinated ring *(121,267,322)* and an axially symmetric planar ring *(119,206,309)* complex were propounded. Tools brought to bear on the structural question have included theory *(73,119)*, infrared *(205,206,29)*, Raman *(29)*, ultraviolet *(322-324)*, and Mössbauer *(221,399)* spectroscopy. Mass spectra, both electron-impact *(75,6)* and chemical ionization *(243)*, of 65 have also been reported, as has the dipole moment, 1.79 D *(293)*.

Resolution of this dispute, in favour of a static structure identical to that in the crystal and rapid relative motion of ring and $Fe(CO)_3$ group, followed from low-temperature NMR studies. Sufficient resolution of the proton spectrum occurs at *ca.* -150°C to indicate the 1-4-η-structure *(12, 286,221)*. Initial studies of the mechanism of shift of the $Fe(CO)_3$ group relative to the ring suggested a 1,2-shift *(221, 286)*. This was confirmed first for the ruthenium analog of 65, whose higher activation energy, 9.4 kcal *(123)*, yields a limiting NMR spectrum at -140°C, then later for 65 itself *(135,351)*. The activation energy for shuffling of the $Fe(CO)_3$ group of 65 in solution was found to be only 8.1 kcal *(135)*. This shuffling motion persists even in the solid state, with an activation energy of 8.3 kcal as shown by broad-line NMR studies *(85,86)* or 9.1 kcal from spin-lattice relaxation measurements *(84)*. Thus, lattice interactions in the crystal do not significantly increase the energy barrier in 65.

The energy barrier to CO scrambling in 65 is about the same as the barrier to $Fe(CO)_3$ group 1,2-shifts. Whether this similarity is coincidental or whether the two processes are mechanistically coupled remains an unsolved question.

The ease with which the $Fe(CO)_3$ group makes its way about the cyclooctatetraene ring has allowed measurement of positional equilibria of the $Fe(CO)_3$ group relative to ring substituents. The available data, which are presumably based on NMR measurements, are summarized in Table 3. The energy differences among positional isomers are small, but in general electronwithdrawing substituents show a preference for the 6-position with respect to the 1-4-η-tricarbonyliron group. An interesting contrast thus exists between carbonyl substituents attached to rings, (as in 46 and these cases) which interact best with the $Fe(CO)_3$ group vinylogously, and with keto groups *on* rings (*e.g.*, 52 and 79) which interact best directly (eqs. [105] and [106]). The latter phenomenon apparently results from the relatively poor stabilization provided to the sickle-shaped pentadienyl unit required for direct interaction with carbonyl substituents.

Electron-donating substituents in derivatives of 65 show preference for the 2-position (except for the methoxy group, which appears anomalous).

Disubstituted cyclooctatetraenes in general behave con-

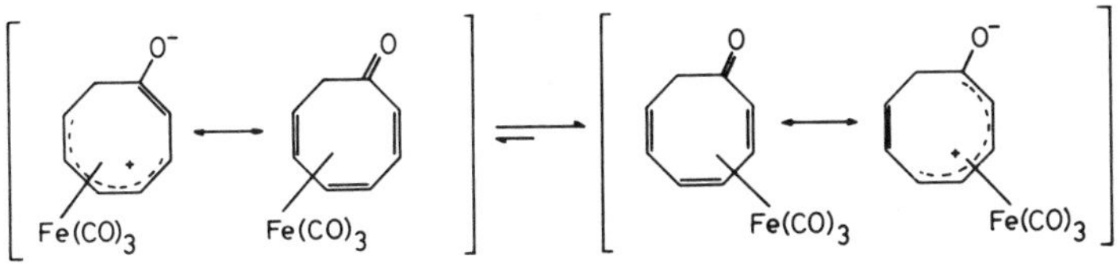

[105]

79

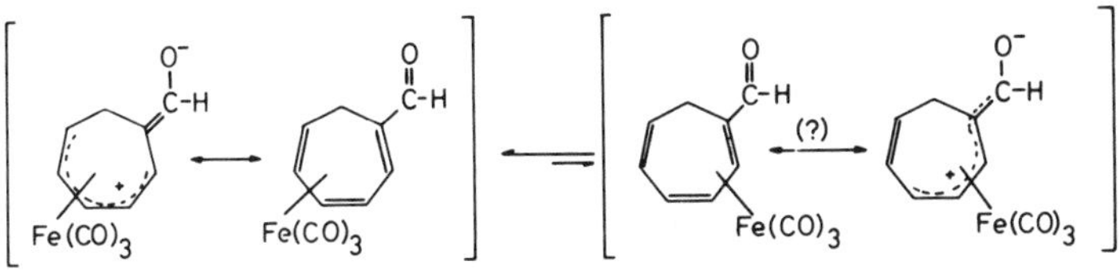

[106]

46

sistently with the pattern set by the individual substituents. 81, for example, exists entirely as shown above, and the tricarbonyliron complex of 1,2-dicarbomethoxy-cyclooctatetraene exists with the ester groups in the center positions of the uncoordinated diene unit *(221)*. The latter two molecules are nonfluxional on the NMR time scale from -60°C to +120°C, indicating a strong positional preference resulting from the combined effects of the two substituents. The barrier to interconversion of the two enantiomers of methylcyclooctatetraene at -12°C, in contrast, is only 7.5 kcal *(11)*. A number of 1,3-disubstituted cyclooctatetraenes with one substituent the triphenylmethyl group and the other a metal have also been prepared *(118)*. All adopt the structure:

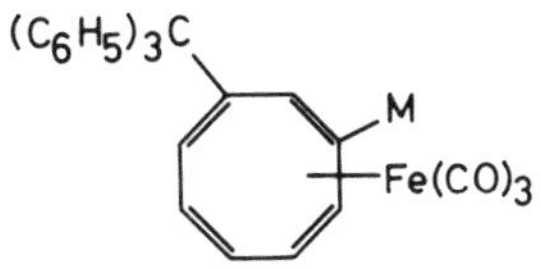

M = $Si(CH_3)_3$, $Ge(CH_3)_3$, $Sn(CH_3)_3$, $Fe(CO)_2Cp$

None of these materials appears to be fluxional on the NMR time scale.

Benzocyclooctatetraene, as may be expected from the 1,2-shift mechanism, has a high barrier, 18.6 kcal, for inter-

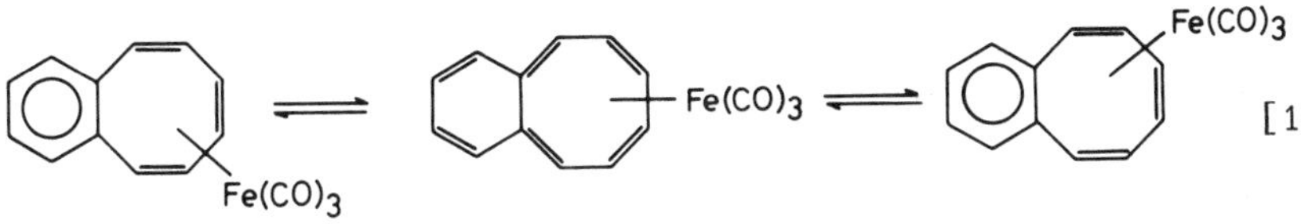

[107]

Table 3: Positional Equilibria in Monosubstituted Cycloocta-tetraene-tricarbonyliron Complexes.

X	T	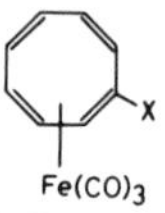	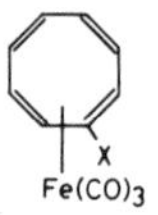	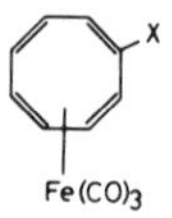	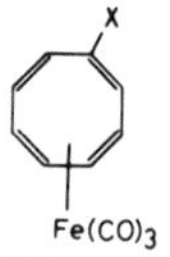	*(Ref.)*
CH_3	-145°C		≥95 %			*267,11,12*
CH_3	25°C		80 %		20 %	*341*
CH_2OH	25°C		50 %	25 %	25 %	*341*
C_6H_5	25°C		50 %		50 %	*341*
OCH_3	25°C		20 %	*ca.* 40 %	*ca.* 40 %	*341*
CHO	25°C			50 %	50 %	*341*
CHO	29°C				predominant	*254*
$COCH_3$	25°C			50 %	50 %	*341*
$COCH_3$	24°C				predominant	*254*
CN	25°C			10 %	90 %	*341*
CO_2R	25°C			10 %	90 %	*341*
$M(CH_3)_3$*			predominant			*118*
$C(C_6H_5)_3$				predominant		*252*
Br				predominant		*217*

*M = Si, Ge, Sn

conversion of its two enantiomers, since an *o*-quinonoid intermediate must be traversed *(404)* (eq. [107]). The β-naphtho analog has a still higher barrier of about 31 kcal.

A comparable barrier is indicated in the fused-ring compounds 82 and 83, which are separable on thin-layer chromatography *(364)*. The barrier to the 1,2-shift in this case is apparently a consequence of the instability of the necessary cyclobutadienoid intermediates to interconversion (eq. [108]).

In marked contrast to 64 and its derivatives, isomerization of 65 to its bicyclo[4.2.0] valence isomer, 84, has not been observed directly, although it has been invoked in a mechanism study *(361)*. 84 has, in fact, only been made indirectly *(370)*, from (*syn*-tricyclo[4.2.0.0^{2,5}]-octa-3,7-diene)-tetracarbonyliron (eq. [109]). The *anti* isomer likewise yields the *endo* isomer of 84. 84 is quite stable, giving no sign of rearrangement to 65 in 10 h at 65°C. No interconversion of 84 and its *endo* isomer are reported.

A number of substituted derivatives of 65 have been found to undergo thermal conversion to derivatives of 84 at 150°C *(117,252)* (eq. [110]). The nature of the bicyclo[4.2.0]octa-

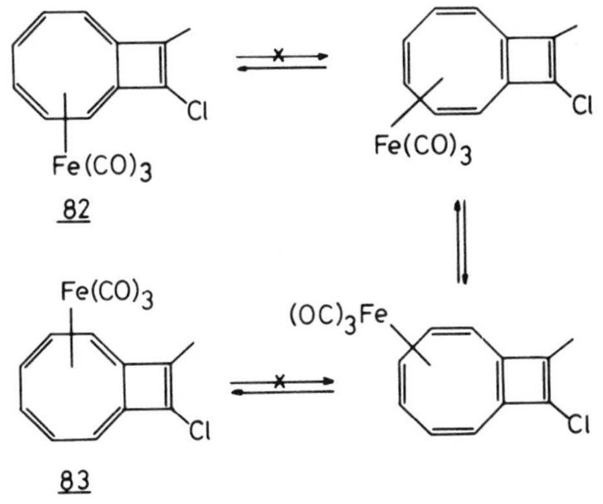

[108]

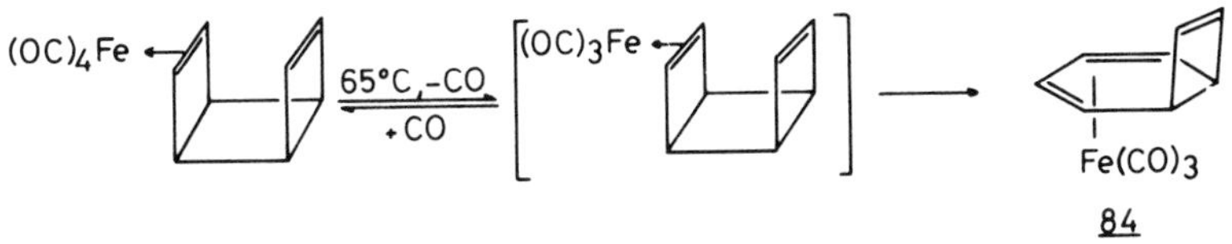

[109]

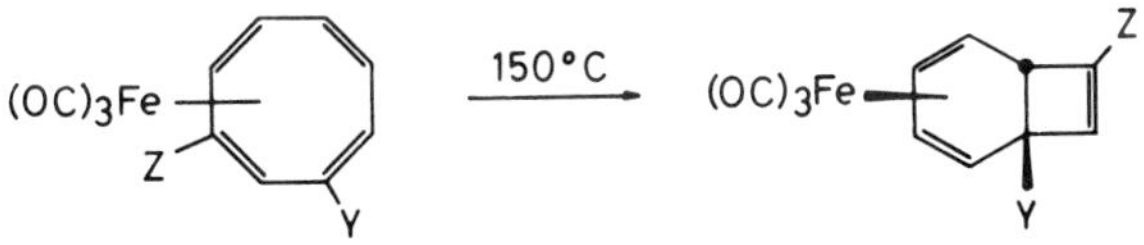

[110]

Z = $Si(CH_3)_3$, $Ge(CH_3)_3$, $(C_6H_5)_3C$, C_6H_5; Y = H
Z = $Si(CH_3)_3$; Y = $(C_6H_5)_3C$

triene complex formed bears no relationship to the more prevalent isomer of the cyclooctatetraene complex as shown in Table 3. Apparently the product formed simply has the largest group Z in the position most remote from the $Fe(CO)_3$ group.

The only other derivative of 84 known results directly from reaction of 1,3,5,7-tetramethylcyclooctatetraene with $Fe_3(CO)_{12}$ at 125°C *(140)*.

From these results it seems probable that 84 may indeed be more stable than 65 (substituted derivatives certainly are more stable), but the rate of conversion of 65 to 84 must be significantly slower than 64 → 63 in the cyclooctatriene case. Quantitative determination of kinetic and thermodynamic parameters relating 65 to 84 would be quite interesting.

Chemical properties of 65 and its derivatives have been very extensively studied. The most characteristic reaction, not surprisingly, is attack of electrophiles on 65. The ease of protonation of 65 was in fact recognized almost as soon as it was first synthesized *(322,155,359)*. The cyclooctatrienium structure 85 was initially proposed for the stable cation *(359)*, but NMR studies *(156,405)* soon showed it to have the bicyclo[5.1.0]octadienium structure, 86. Brookhart was later able to detect 85 by NMR at -80°C and to study its rearrangement at -60°C to 86 *(65-67)* (eq. [111]). Studies of the

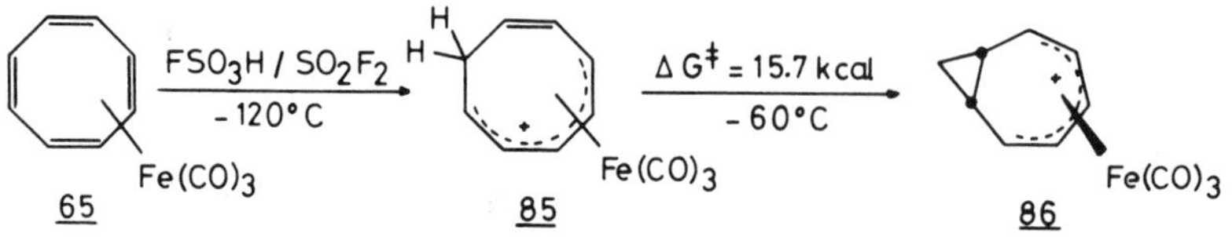

[111]

$Fe(CO)_3$ complex of methylcyclooctatetraene indicated that the protonation occurs on an inner carbon of the uncoordinated diene unit, and confirmed the close relationship between the ring closure and the 1,2-shift in systems such as 85 and 64 (Scheme XIV) *(66,67)*.

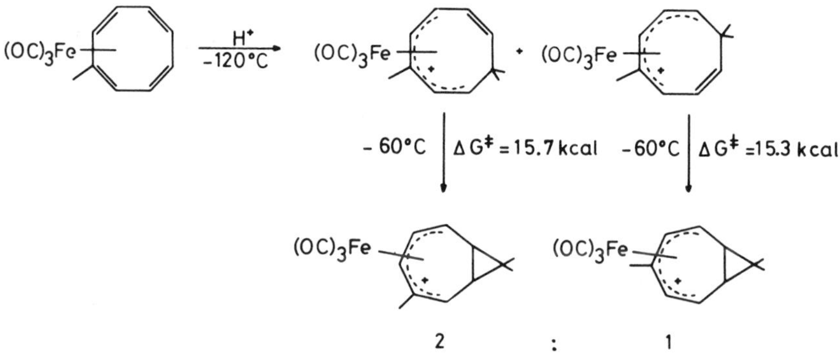

Scheme XIV: Results of Protonation of (Methylcycloocta-tetraene)tricarbonyliron.

In contrast, protonation of (tritylcyclooctatetraene)tri-carbonyliron at 10°C gave the analog of 85, which did not undergo ring closure *(252)*, apparently being inhibited by the very large substituent (eq. [112]). The cation formed on pro-

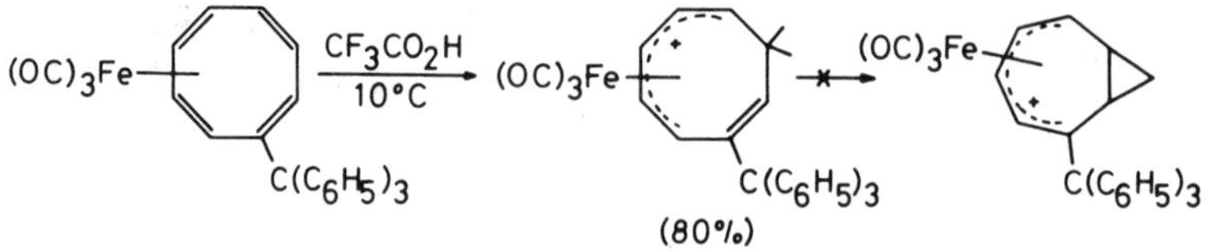

[112]

tonating (benzocyclooctatetraene)tricarbonyliron likewise does not undergo ring closure *(315)*.

Reaction of 65 with HCl at 0°C gives a monocyclic product (eq. [113]) which might be thought to result from trapping of

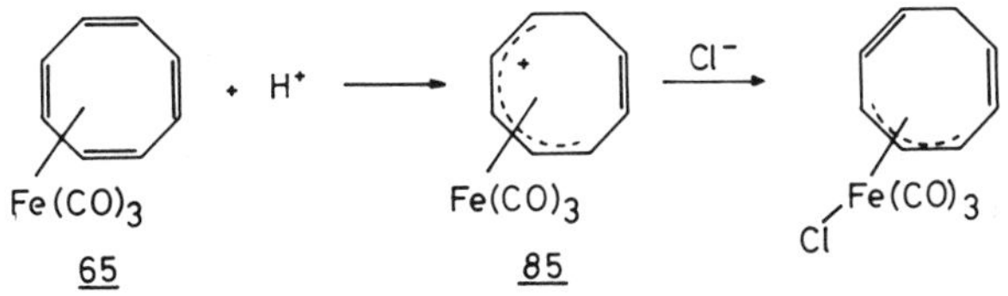

[113]

85 by Cl^- before the rearrangement to 86 occurs *(100)*. However, reactions of 86 itself with nucleophiles give both monocyclic and bicyclic products, depending on nucleophile and conditions. Presumably, kinetic control leads to bicyclic products and thermodynamic control to cyclooctatriene derivatives (Scheme XV).

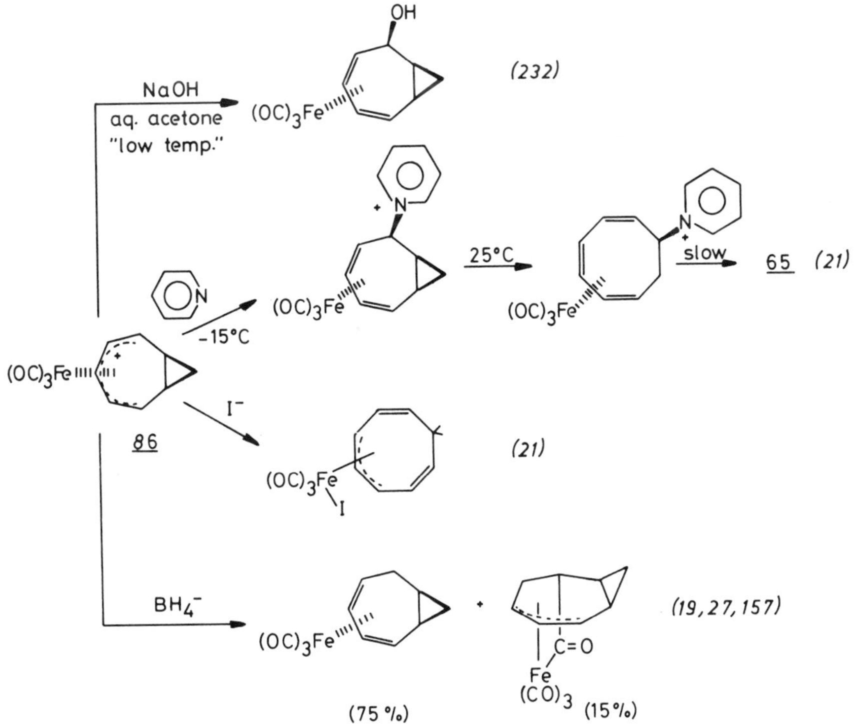

Scheme XV: Reactions of 86 with Nucleophiles.

Reactions of carbenium electrophiles with 65 follow the pattern of the protonation reaction, attack usually being at an inner position of the uncoordinated diene unit. For example, reaction of triphenylcarbenium fluoroborate with 65 gave the tritylated analog of 85, which again did not close to the bicyclic form. Reaction with base gave (tritylcyclooctatetraene)tricarbonyliron *(252)* (eq. [114]). Tritylation of the

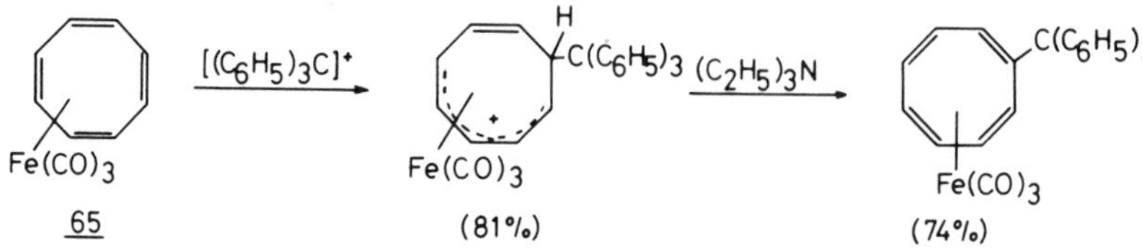

[114]

trimethylsilyl and trimethylgermyl derivatives of 65 does not yield the product expected from analogous attack on the principal fluxomer present in the starting material. The product obtained after deprotonation, 87 (eq. [115]), has the two sub-

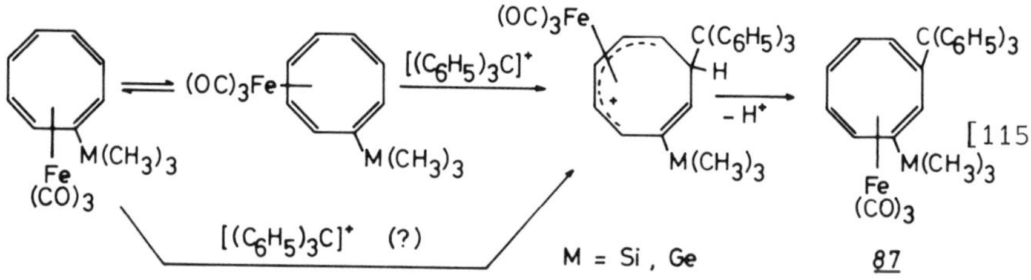

[115]

stituents 1,3 with respect to each other *(118)*, based on NMR. This was interpreted by the authors in terms of attack of the electrophile at an *outer* carbon of the free diene unit, in contrast to all other cases investigated. An alternative interpretation would involve normal attack on a more reactive fluxomer of the starting material (eq. [115]).

65 also undergoes formylation and acetylation reactions with relative ease, giving rise to a variety of substituted derivatives (Scheme XVI) *(6,251,254)*. Alcohols derived from these carbonyl derivatives lost water readily on treatment with acid, giving stable carbenium salts whose NMR spectra could not be obtained and whose structures are therefore unclear. A relationship with the cations formed from (2-hydroxymethyl-butadiene)tricarbonyliron complexes *(57)* may well exist.

Attempts at bromination and nitration *(254)* of 65 led only to decomposition. Reaction of 65 with $F_3B{\cdot}OEt_2$ gave an unstable product thought to be (4-7-η-cycloocta-2,4,6-trienol)-tricarbonyliron *(256)*.

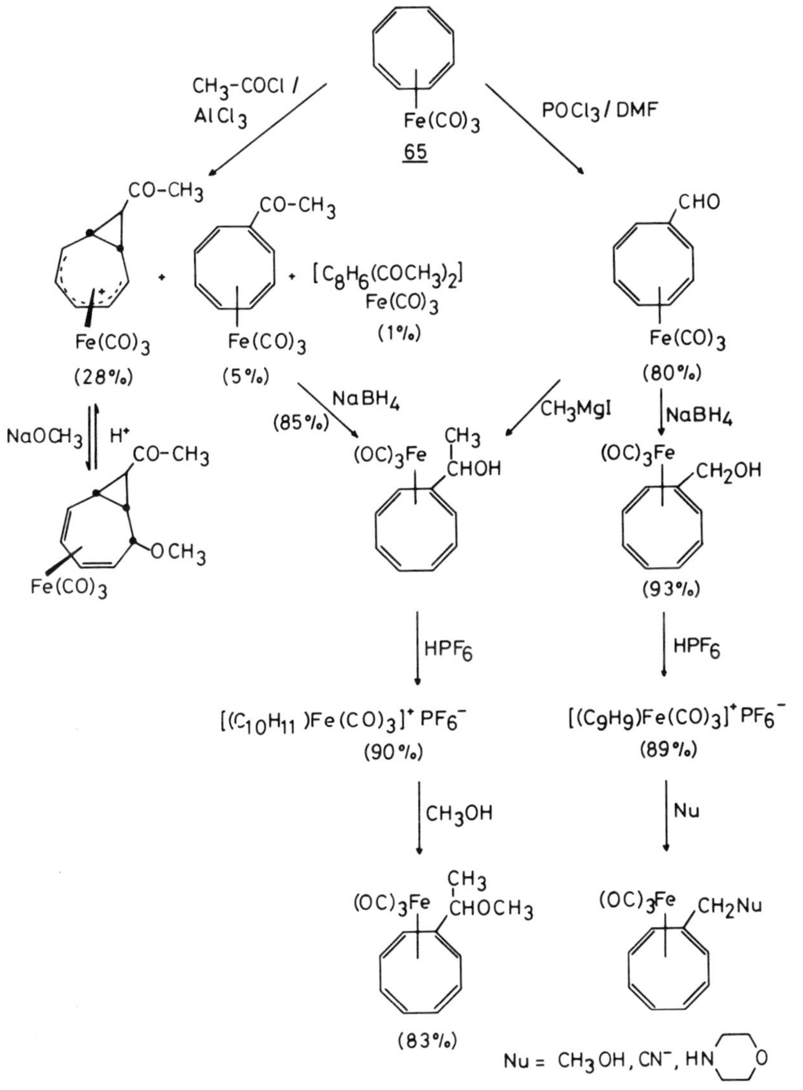

Scheme XVI: Acylation Reactions of (Cyclooctatetraene)tricarbonyliron 65 *(6,254)*.

A further reaction now known to involve electrophilic attack on 65 is the reaction with tetracyanoethylene (TCNE). The rapid formation of a 1:1 adduct was reported at the outset, but the adduct was assigned the structure of a Diels-Alder adduct *(157,360,220,217)*. An analogous adduct of 1,1-dicyano-2,2-bis(trifluoromethyl)ethylene was subsequently assigned the structure of a [2+2] cycloadduct *(219,220)*, based on NMR studies. However, degradation of the TCNE adduct with ceric ammonium nitrate was found to give, not a Diels-Alder adduct of TCNE with cyclooctatetraene, but rather the dihydrotriquinacene derivative, 88 (X = CN) (Scheme XVII), from which the structure shown for the 1:1 adduct was correctly inferred *(178)*. This structure was subsequently confirmed by X-ray crystallography *(339)*. Other highly electrophilic dienophiles, including chlorosulfonyl isocyanate *(339)* and 4-phenyl-1,2,4-triazoline-3,5-dione *(220)* add by a similar dipolar mechanism but give products with a bicyclo[4.2.2]decane skeleton (Scheme XVII); the factors controlling the different modes of zwitterion collapse remain unclear. The reaction of 65 with TCNE has proved useful in syntheses of substituted triquinacenes *(340)*.

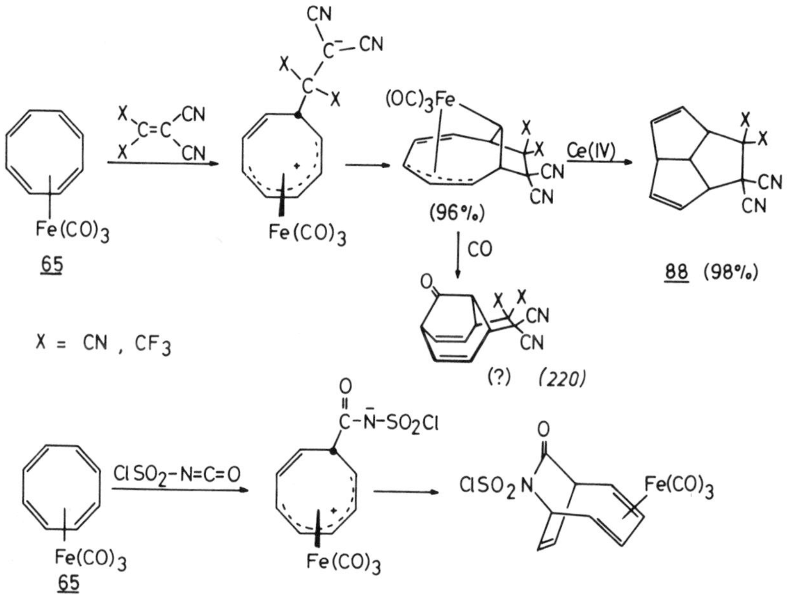

Scheme XVII: Cycloadditions of 65 with Electrophilic Dienophiles.

Reactions of substituted cyclooctatetraene-tricarbonyliron complexes with TCNE have been investigated by two groups of workers, whose results are in general agreement *(217,341)*. The products obtained are those expected from attack on the predominant fluxomer shown in Table 3 (Scheme XVIII). With the methyl compound, the results are in almost quantitative agreement with the protonation results of Scheme XIV. An analogous electrophilic reaction of TCNE with (benzocyclooctatetraene)tricarbonyliron also occurs, as indicated by product analysis *(217)*.

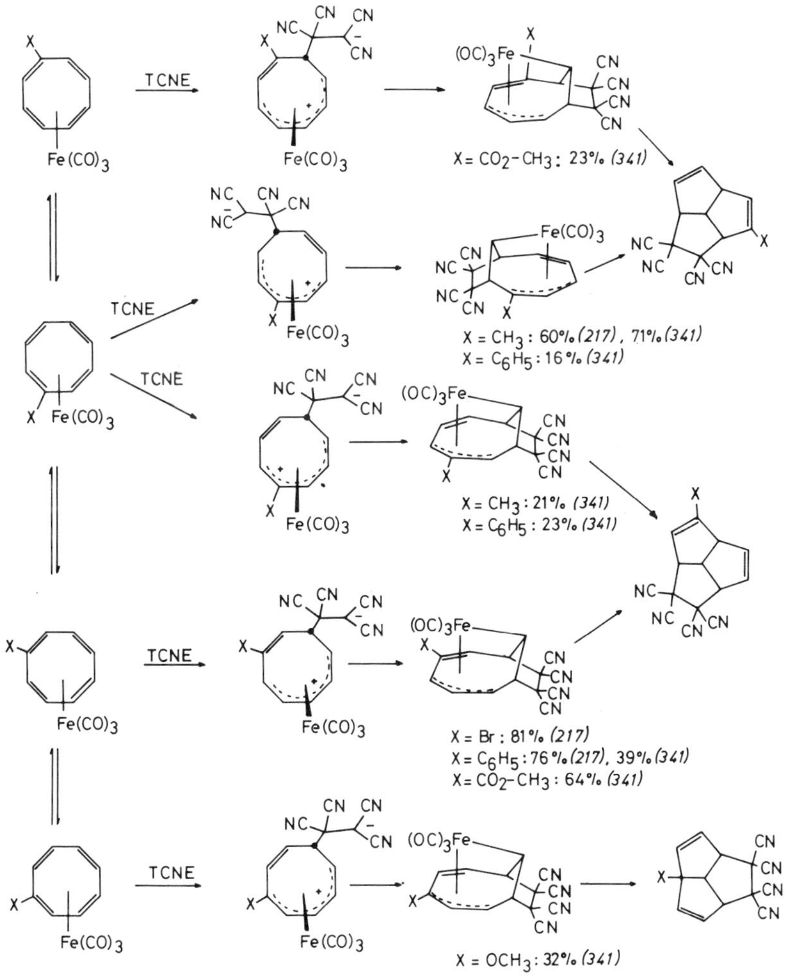

Scheme XVIII: Reactions of TCNE with Substituted Cycloocta-tetraene-tricarbonyliron Complexes.

In contrast to the highly electrophilic dienophiles discussed above, maleic anhydride reacts with 65 only under severe conditions *(346)*. Similarly, at temperatures at which 65 decomposes more or less rapidly, it reacts with diphenylacetylene to give a variety of products, including a cycloadduct, obtainable in 35 % yield in the presence of CO *(289, 326)* (eq. [116]).

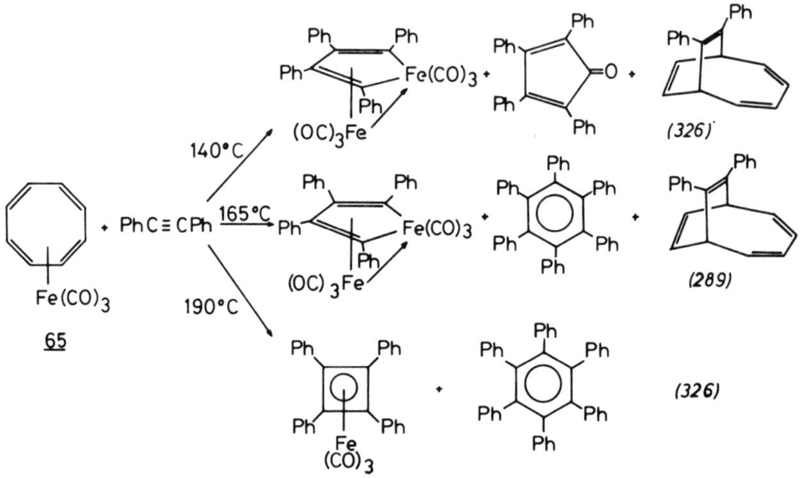

[116]

In addition to the metal-assisted reactions with electrophiles, the free double bonds of 65 undergo a number of apparently normal reactions. For example, reaction of 65 with excess $Fe(CO)_5$ *(310,346)* or $Fe_2(CO)_9$ *(266)* gives hexacarbonyldiiron complexes, discussed in Section IX. 65 undergoes Simmons-Smith methylenation (eq. [117]), but the product is

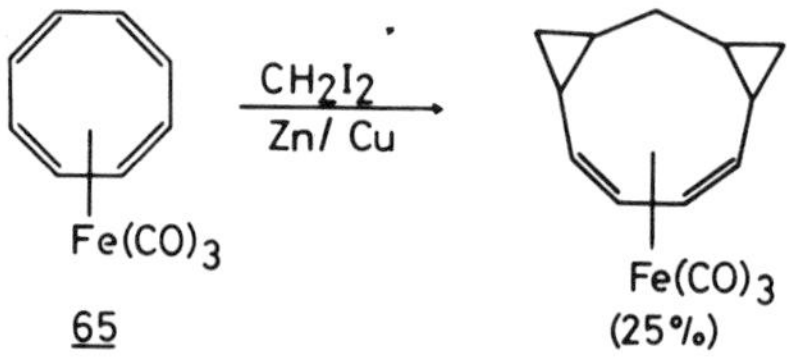

[117]

not the one expected *(349)*. 65 resists catalytic hydrogenation *(157,346)*, although hydrogenation over Raney nickel was reported in one patent *(247)*. 65 undergoes electrochemical reduction *(170,188)*. The radical ion obtained on one-electron reduction is reported to have the low g-value of an organic radical (2.0073) *(170)*, in curious contrast to the radical-anions from substituted tropone-tricarbonyliron complexes and from 79. Addition of water to a solution of the radical anion gives equal amounts of 65 and 64. 64 can be obtained quantitatively by irreversible reduction in the presence of trimethylammonium bromide as proton source *(188)*. Reaction of 65 with excess trimethylamine oxide gives free cyclooctatetraene in 95 % yield *(367)*.

Reaction of 65 with aluminium chloride in benzene gives the same complex, shown below, as is obtained on reaction of barbaralone with $Fe_2(CO)_9$ *(144)*. This complex gives barbaralone in high yield on reaction with CO *(227,256)* (eq. [118]).

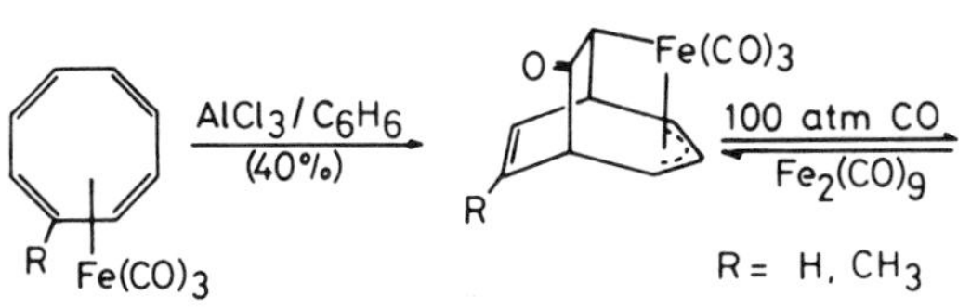

[118]

A mechanism involving electrophilic attack on 65 by another molecule of 65 having an oxygen coordinated to $AlCl_3$ has been proposed *(256)* (eq. [119]).

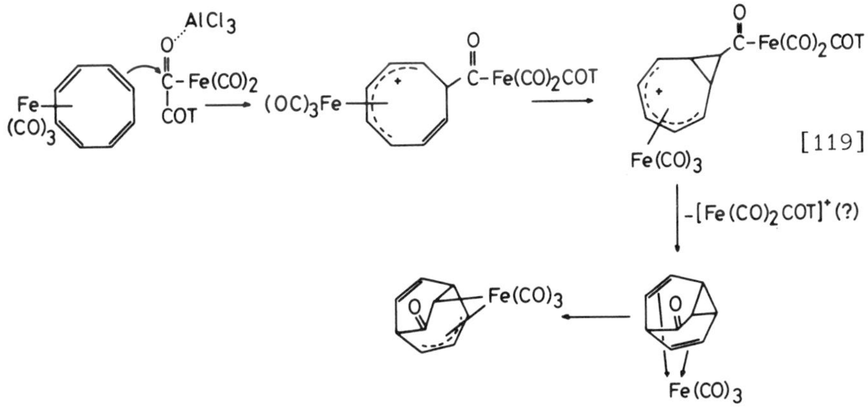

[119]

Another curious reaction ensues when 65 is treated with an olefin metathesis catalyst *(56)* (eq. [120]).

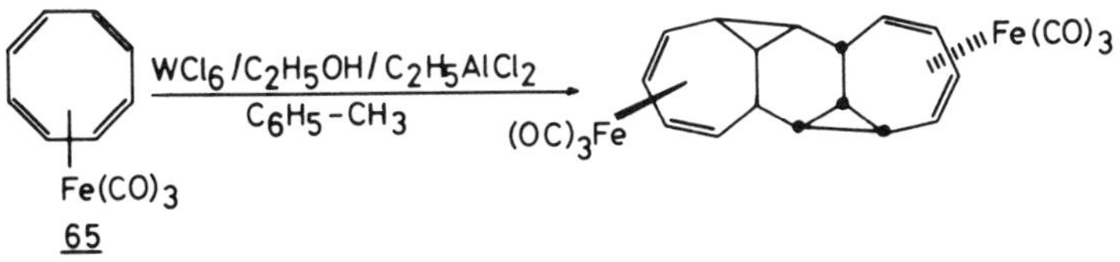

[120]

65 reacts with soft nucleophiles at moderate temperatures predominantly by displacement of the organic ligand rather than a carbonyl group. This behaviour is in contrast to that of 44, which gives predominantly carbonyl displacement. With

CO at 100°C, $Fe(CO)_5$ and free cyclooctatetraene are produced from 65 *(323)*. Reaction of 65 with a triphosphine (TP) at 100°C gave $Fe(CO)_3(TP)$ (eq. [121]), which at 150°C lost a mole

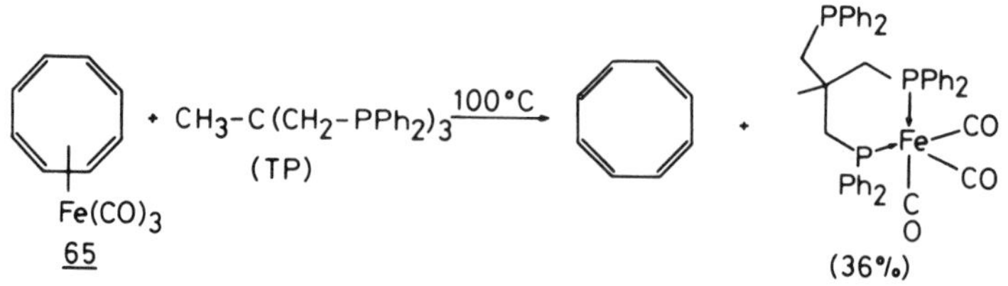

[121]

of CO to form $Fe(CO)_2(TP)$ with all three phosphorus atoms coordinated to iron *(37)*. The tripyridyl ligand gave the analogous product at 150°C *(37)* (eq. [122]).

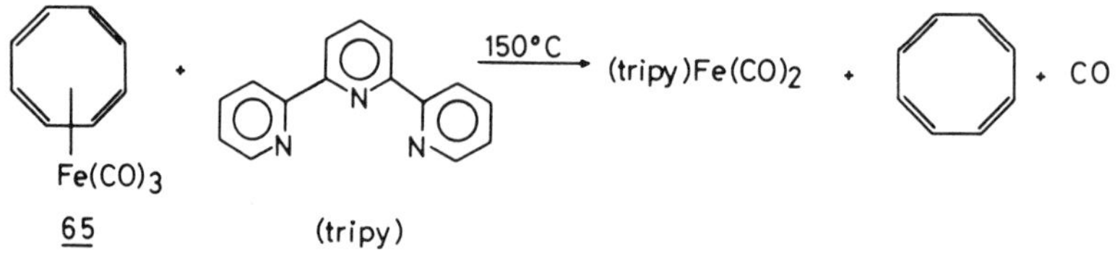

[122]

65 reacts with triphenylphosphine at 80°C to give bis(triphenylphosphine)tricarbonyliron as the only reported iron-containing product *(193,323)*. At 130°C, this is still the main product (48 %) but some (cyclooctatetraene)(triphenylphosphine)dicarbonyliron may also be produced *(310)*. Triphenylarsine and -stibine yield only the latter type of product, by carbonyl displacement *(310)*.

The kinetics of reaction of phosphines with 65 and its ruthenium analog have been investigated *(192,193)*. Diphosphine-tricarbonyliron complexes were the only reported products, and the reactions were chlearly bimolecular. With tributylphosphine and 65 at 90-110°C, the $\Delta H^{\ddagger}$ was 11.4 kcal and the $\Delta S^{\ddagger}$ was -38 cal/K, indicating an associative mechanism *(193)* (eq. [193]). The ruthenium compound was found to react

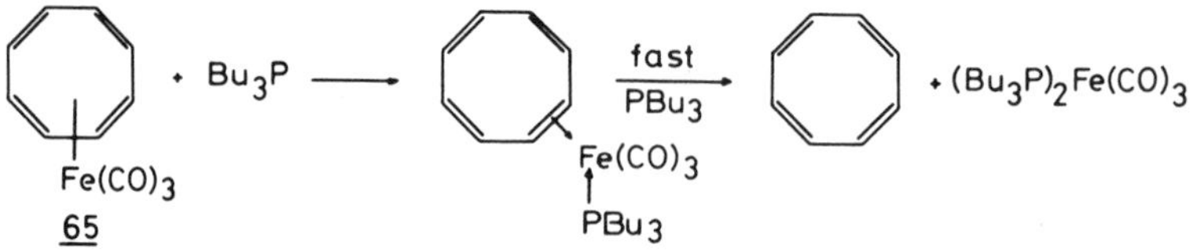

[123]

more rapidly, due to a much less negative $\Delta S^{\ddagger}$ of -24 cal/K *(192)*, a difficult result to interpret.

Evidently, however, the iron in the cyclooctatetraene complex 65 is much more susceptible to direct nucleophilic attack than the iron in the cycloheptatriene complex 44.

Whether this is due to steric or electronic reasons remains unclear.

The photochemistry of 65 has been studied in some detail. In the absence of added nucleophiles, photolysis gives (cyclooctatetraene)pentacarbonyldiiron in at least 65 % yield *(365)*, a result which suggests that 65 itself must be a good nucleophile toward coordinatively unsaturated iron (eq. [124]).

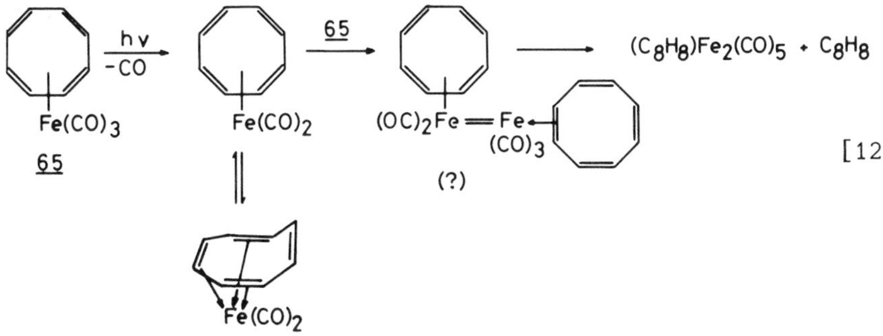

[124]

Irradiation of 65 in the presence of free cyclooctatetraene leads to cycloadducts of composition $(C_{16}H_{16})Fe(CO)_3$ *(360,361)*. The same adducts, 89 and 90, can be obtained by direct complexation of cyclooctatetraene dimers and have been characterized by X-ray crystallography *(350,353)*. Some of the reactions involving these materials are shown in Scheme XIX.

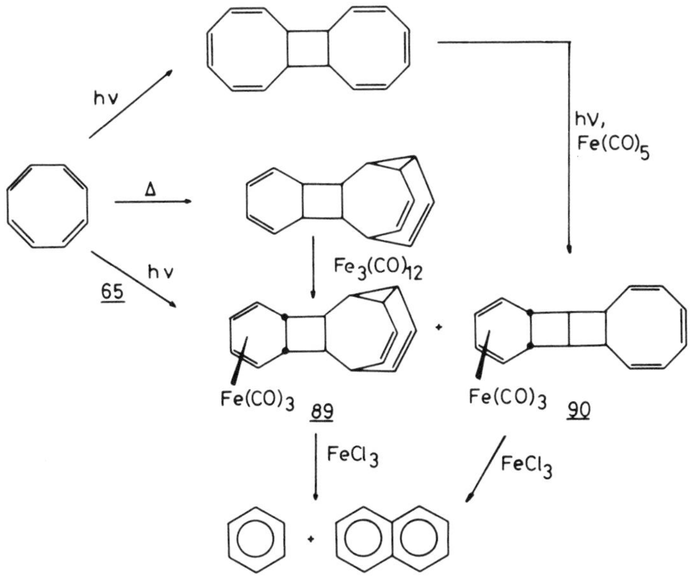

Scheme XIX: $Fe(CO)_3$ Complexes of Cyclooctatetraene Dimers *(360-362)*.

The photochemical reactions of 65 in the presence of phosphines, which should provide evidence of the $(C_8H_8)Fe(CO)_2$ intermediate presumed to be involved in these reactions, do not appear to have been studied.

F. $Fe(CO)_3$ COMPLEXES OF CYCLONONATETRAENE AND OTHER MACROCYCLES

As previously indicated (Section IV.D and Scheme X), (1-4-η-cyclononatetraene)tricarbonyliron, 69, is obtained on reaction of carbonyliron reagents with bicyclo[6.1.0]nona-2,4,6-triene *(163,347)*. The yields are 12 % with $Fe_2(CO)_9$, 35 % with $Fe(CO)_5$ and irradiation *(347)*, and 3 % with (BDA)-$Fe(CO)_3$ *(358)*. The symmetrical 3-6-η-isomer was also reported to be formed in small amounts in one case *(163)*.

Both isomers are thermodynamically unstable, undergoing quantitative conversion to (bicyclo[4.3.0]nona-2,4,7-triene)-tricarbonyliron, 68, at about 100°C. The $\Delta G^{\ddagger}$ for this process 69 → 68 (eq. [125]) is 28.4 kcal, as compared to 23 kcal

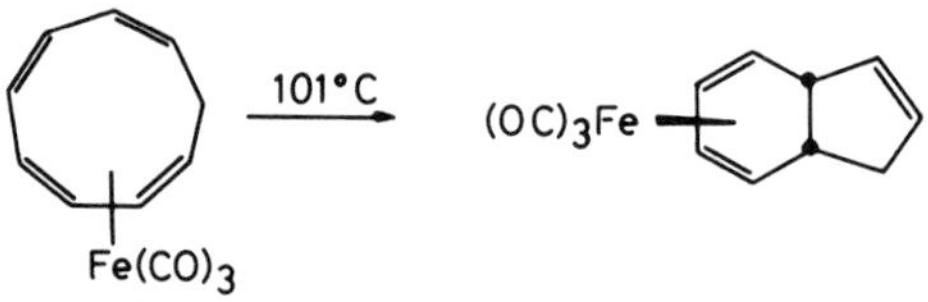

69 68 [125]

for the analogous ring closure of free all-*cis*-cyclononatetraene *(347)*. 68 and derivatives are obtained directly on thermal decomposition of $Fe_2(CO)_6$ complexes of bicyclo[6.1.0]-nonatrienes *(160,162,164,347)*; this process very probably involves ring opening to 69 (or an $Fe_2(CO)_6$ complex of cyclononatetraene) followed by closure to 68.

69 undergoes protonation in the same manner as previously described $Fe(CO)_3$ complexes *(347)* (eq. [126]).

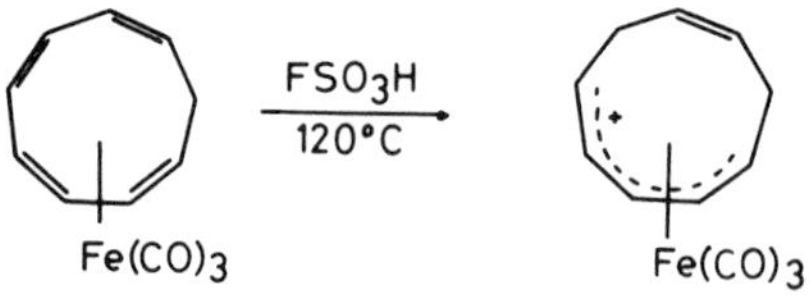

69 [126]

Complexation of cyclooctatetraene oxide, described in Section IV.D, gives no evidence of any oxonin complex as product or intermediate *(25,306)*. A complexed valence isomer of a substituted oxonin was obtained by an indirect route *(46)* (eq. [127]).

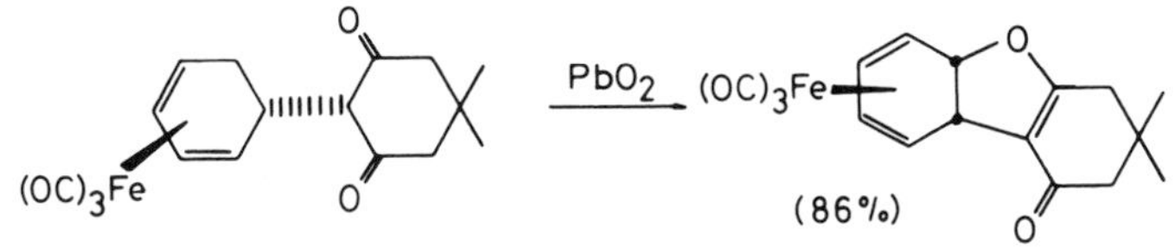

[127]

An analogous $Fe(CO)_3$ complex of a valence isomer of a cyclononatetraenone has also been reported *(77)* (eq. [128]).

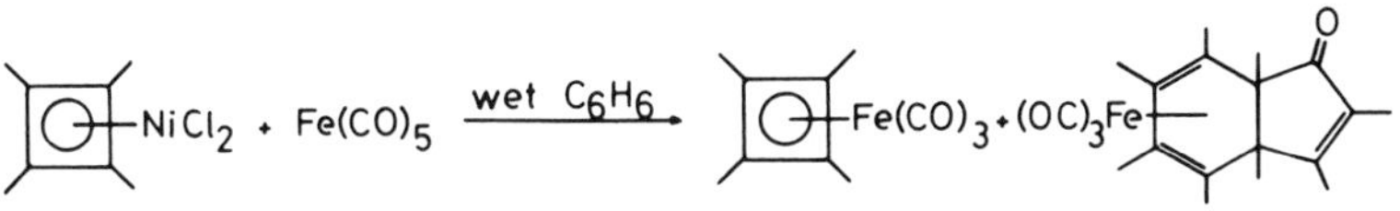

[128]

(Cyclodecatetraene)tricarbonyliron, in this case the symmetrical 3-6-η-isomer, has been briefly mentioned as one product of reaction of $Fe_2(CO)_9$ with bicyclo[6.2.0]decatriene (Scheme XI), along with its valence isomer 75 and a variety of other products *(143)*. No data were reported, although the compound was characterized by X-ray crystallography. It has been suggested as an intermediate in formation of some of the products shown in Scheme XI, but the conversions have not been independently verified.

Complexes of the bicyclo[4.4.0]decatetraene valence isomer of a cyclodecapentaene have been reported *(8,234,289,398)* (eq. [129]). No $Fe(CO)_3$ complexes of cyclodecapentaene

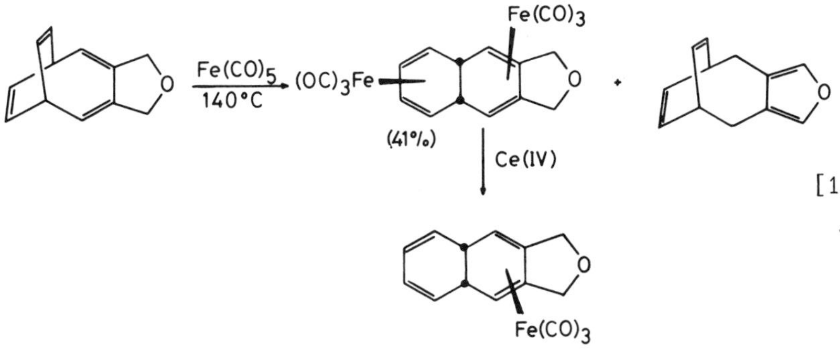

[129]

([10]annulene) or any of the larger annulenes have been reported, although the complex 90 *(353,361)* is a valence tautomer of ([16]annulene)tricarbonyliron.

Reaction of cyclododeca-1,5,9-triene with $Fe(CO)_5$ gives an incompletely described $Fe(CO)_3$ complex, perhaps resulting from hydrogen migration to form a complexed 1,3-diene unit *(165)*.

V. COMPLEXES OF $Fe(CO)_2$

In contrast to the hundreds of $Fe(CO)_3$ complexes known, $Fe(CO)_2$ complexes of trienes or tetraenes are extremely rare. The $Fe(CO)_2$ group would be expected to be able to coordinate a triene to form an organometallic product with an inert gas configuration, as the $Cr(CO)_3$ group does. The rarity of triene-dicarbonyliron complexes may also be contrasted with the extremely numerous class of dicarbonyliron complexes having a cyclopentadienyl and an alkyl ligand.

Photolysis of triene- or tetraene-tricarbonyliron compounds may be expected to proceed with loss of CO *(284)*, giving either a 16-electron η^4-complex or an 18-electron η^6-complex. The latter seem not to be isolable products, but may play roles as intermediates in photochemical reactions of $Fe(CO)_3$ complexes, as described in Sections IV.C and E.

One clear case of intramolecular trapping of a photogenerated $Fe(CO)_2$ intermediate has been reported *(222)* (eq. [130]).

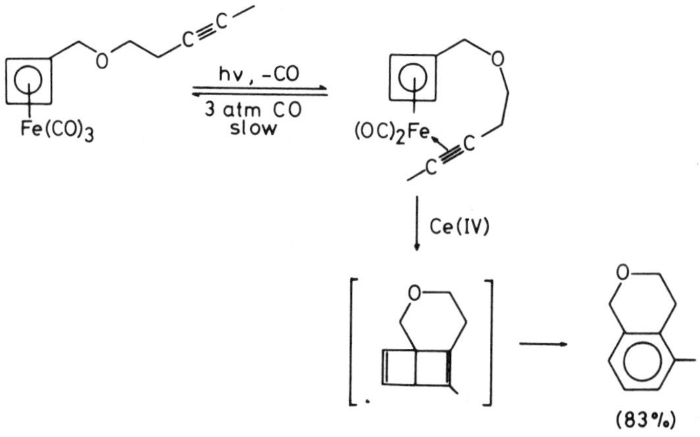

[130]

An intermolecular analog of this process occurs upon photolysis of (cyclobutadiene)tricarbonyliron in the presence of cycloheptatriene *(392)* (eq. [131]). The success of these

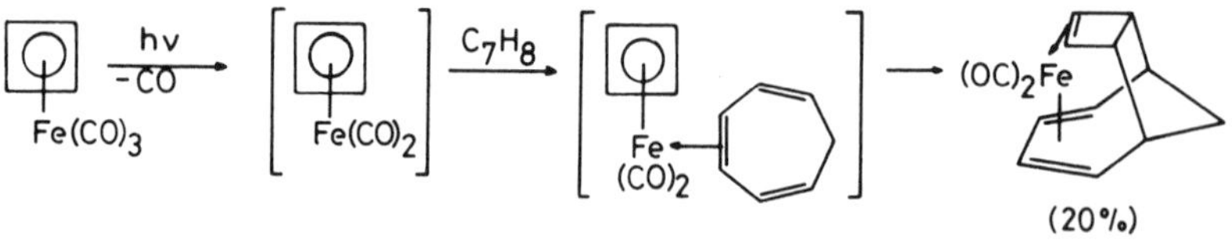

[131]

reactions may result from the very high quantum yield for CO loss on photolysis of (cyclobutadiene)tricarbonyliron as compared to (butadiene)tricarbonyliron *(269a)*.

Some further examples of $Fe(CO)_2$ complexes, which result

from thermal reactions, are known. One is the previously discussed product, 13 (Section III). The other, 91, is obtained on reaction of bicyclooctatetraenyl with iron carbonyls *(175)* (eq. [132]). The structure of 91 was confirmed by X-ray

[132]

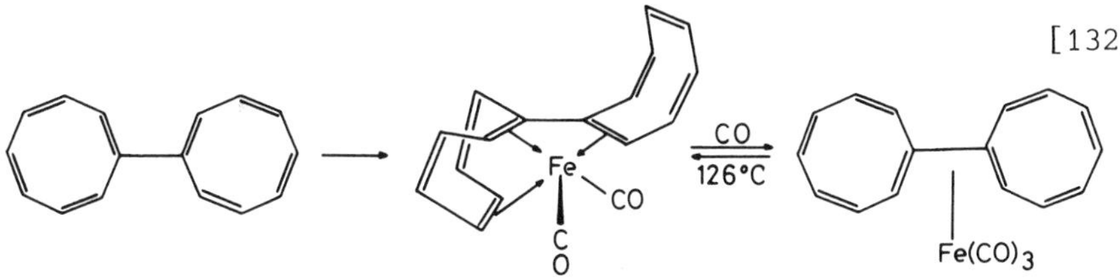

91 (16%)

crystallography, which revealed that the cyclooctatetraene rings, in contrast to 65, maintain the "tub" conformation of uncomplexed cyclooctatetraenes and that the coordination geometry of the iron, in contrast to diene-tricarbonyliron complexes, is approximately trigonal bipyramidal. The third double bond is only weakly bonded to the iron as shown by its easy replacement by CO at room temperature.

Reaction of 6,6-dialkylpentafulvenes with $Fe_2(CO)_9$ gives, among other products, bis(pentafulvene)dicarbonyliron products *(345,399)*. A structure first proposed on mechanistic grounds *(268)* has received support from NMR studies *(39)* and from crystallographic study of the derived $Fe_2(CO)_5$ complex from 6,6-dimethylpentafulvene *(39,355)*. These products, though derived from trienes, have structures related to the well-known stable alkyl(η^5-cyclopentadienyl)dicarbonyliron compounds.

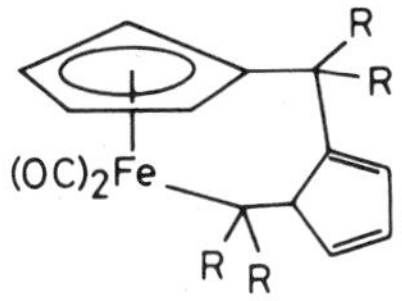

None of these $Fe(CO)_2$ complexes involves coordination of a 1,3,5-triene by the $Fe(CO)_2$ group, perhaps because the stereochemical requirements of such trienes and pentacoordinate iron (unlike hexacoordinate chromium) are not easily reconciled.

VI. COMPLEXES OF Fe(CO)

Fe(CO) complexes of appropriate tetraenes would be consistent with the 18-electron rule, but do not appear to exist. Presumably the stereochemical requirements for a tetraene which would fit four sites in the coordination sphere of an iron atom would be severe.

A few mixed Fe(CO) complexes of cyclooctatetraene and butadienes, derivatives of bis(diene)monocarbonyliron complexes, are, however, known. They may be synthesized either from bis(cyclooctatetraene)iron, 92, by exchange with butadiene at 0°C followed by addition of 2 atm CO (*ca.* 25 % yield) *(89)* or by treatment of $FeCl_3$ with isopropylmagnesium bromide in the presence of both cyclooctatetraene and butadiene followed by photolysis then carbonylation (*ca.* 25 %) *(88)* (eq. [133]).

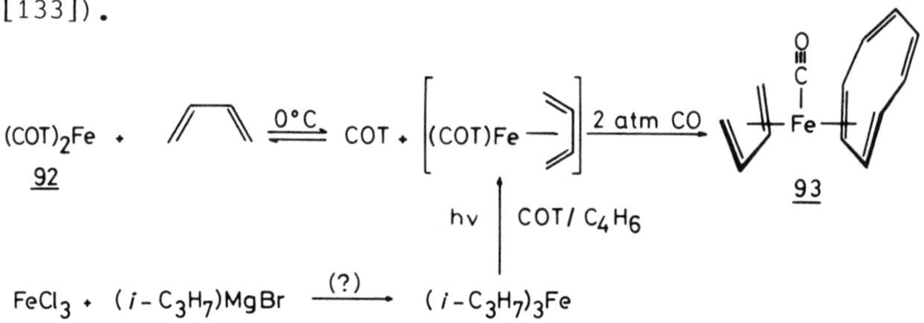

[133]

The structure of 93 has been studied by X-ray crystallography *(33)*, which reveals 1-4-η-coordinated cyclooctatetraene and butadiene coordinated at the basal position of square pyramidal iron, with the CO axial. The "open" ends of the two coodinated diene units are toward the axial CO. The structure is very similar to that of 65, with the two equatorial CO's replaced by the butadiene *(33)*.

Like 65, 93 is fluxional at room temperature. However, decoalescence occurs at a higher temperature for 93, and a limiting spectrum is obtained at -85°C *(13)*, implying a higher barrier to positional shifts in 93 than in 65. However, quantitative activation parameters for 65 have not been reported.

Continuing the analogy, the cyclooctatetraene ring behaves toward electrophiles in essentially the same way in 93 as in 65 *(13)* (eq. [134]). Clearly, the replacement of two

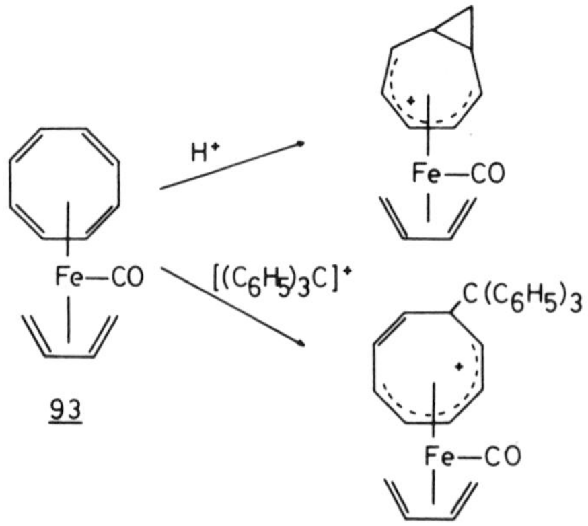

[134]

carbonyl groups in 65 by a butadiene moiety makes no qualita-
tive change in the structural or chemical properties.
The related complex, 3, $(\eta^3\text{-}C_7H_7)(\eta^5\text{-}C_5H_5)FeCO$ *(116)*
(Section II) may also be again noted here.

VII. COMPLEXES OF Fe WITHOUT ANCILLARY LIGANDS

If complexes of this sort are to obey the 18-electron
rule, the iron atom must be enshrouded by organic ligands of
appropriate shape so as to coordinate it in a reasonable
geometry and capable of providing ten electrons, for example
in the form of five double bonds. This has not been accomplished with a single polyene ligand, and has been accomplished in only a small number of cases involving two polyenes, one coordinated as a diene and the other as a triene.
This is in contrast to the ease of formation and large number
of complexes (*e.g.* ferrocenes) in which an iron atom is coordinated to two pentadienyl or cyclopentadienyl ligands.
Presumably the instability of the ligands as free molecules in
the latter cases leads to much greater stability of the complexes with respect to dissociation.
The most extensively studied bis(polyene)iron compound is
bis(cyclooctatetraene)iron, 92. 92 can be prepared by a
variety of routes, all involving generation of iron(O) in the
presence of cyclooctatetraene. Methods of producing iron(O)
have included the Grignard method *(93)*, and reduction of tris-
(acetylacetonyl)iron electrochemically *(290)* and using
triethylaluminium *(207)* (eq. [135]). The latter process gives

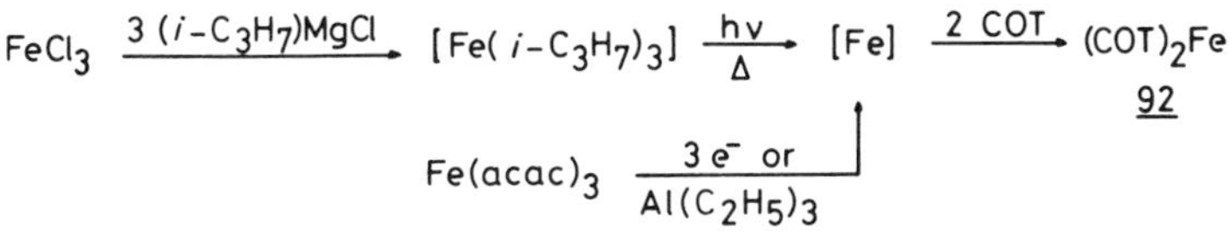

$$FeCl_3 \xrightarrow{3\,(i\text{-}C_3H_7)MgCl} [Fe(i\text{-}C_3H_7)_3] \xrightarrow[\Delta]{h\nu} [Fe] \xrightarrow{2\ COT} (COT)_2Fe$$

$$Fe(acac)_3 \xrightarrow[Al(C_2H_5)_3]{3\ e^-\ or}$$

$$\underline{92} \qquad [135]$$

the best yield (74 %) of 92. 92 has also been synthesized by
an indirect metal atom synthesis *(303,378)* (eq. [136]).

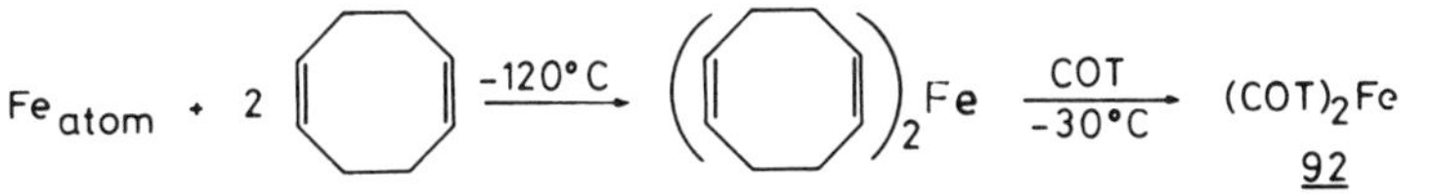

$$Fe_{atom} + 2 \;\bigcirc\; \xrightarrow{-120^\circ C} \;\left(\bigcirc\right)_2 Fe \xrightarrow[-30^\circ C]{COT} (COT)_2Fe \qquad [136]$$

$$\underline{92}$$

The structure of 92 has been studied by X-ray crystallography *(2)*, which shows one ring bonded to the iron as a diene
in the manner of 65 and the other as a triene, with coordination similar to $(\eta^6\text{-cyclooctatetraene})$tricarbonylmolybdenum.
92 is thus the first example we have encountered of a 1,3,5-

triene fully coordinated (η^6) to iron.

In solution at room temperature, 92 shows only a single peak in the ^{1}H-NMR spectrum *(94)*. At -84°C fluxional motion of the η^6-ring is frozen out, but the η^4-ring remains fluxional. Broad-line NMR studies of the crystal indicate rigid molecules with no motion below -185°C. The disorder observed in the X-ray crystallographic results at room temperature *(2)* was indicated as dynamic, resulting from interconversion of η^6- and η^4-rings, for which process an activation energy of 2.6 kcal/mol was indicated *(3,102)*.

92 is a highly reactive compound in solution. In the absence of added nucleophiles it decomposes slowly to iron metal and free cyclooctatetraene *(90)*. In the presence of CO, 65 is produced. Tris(bipyridyl)iron is produced in the presence of 2,2'-bipyridyl *(93)*. Reaction with triethylphosphine under a nitrogen atmosphere gave a nitrogen-containing product which may have been a dinitrogen complex of tetrakis(triethylphosphine)iron but which could not be obtained pure *(87)*.

In all of these reactions, 92 behaves as though it contained a coordinatively unsaturated iron, suggesting that one double bond, presumably on the η^6-ring, is very weakly bound to the iron. The resulting near-vacant coordination site is also responsible for the catalytic properties of 92 with respect to acetylenes and dienes. Thus, 92 catalyses cyclotrimerization of alkynes *(91)* (eq. [137]). A product of compo-

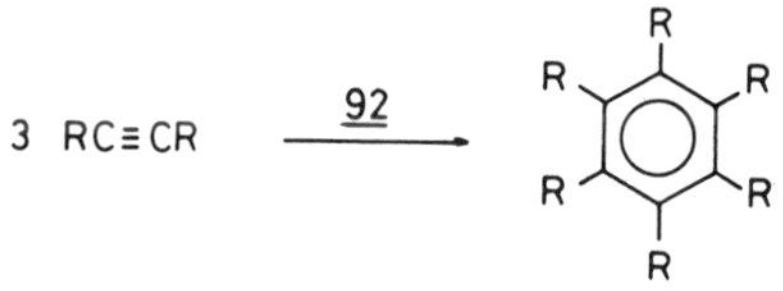

[137]

R = C_6H_5, CH_3

sition $(C_6H_5C_2C_6H_5)_5Fe$ was isolated from such a cyclotrimerization mixture *(93)*. The presence of one hexaphenylbenzene ligand was inferred; the remaining two alkynes may be present as either a tetraphenylcyclobutadiene or as two independent diphenylacetylene ligands.

In the presence of butadiene (eq. [139] or norbonene (eq.

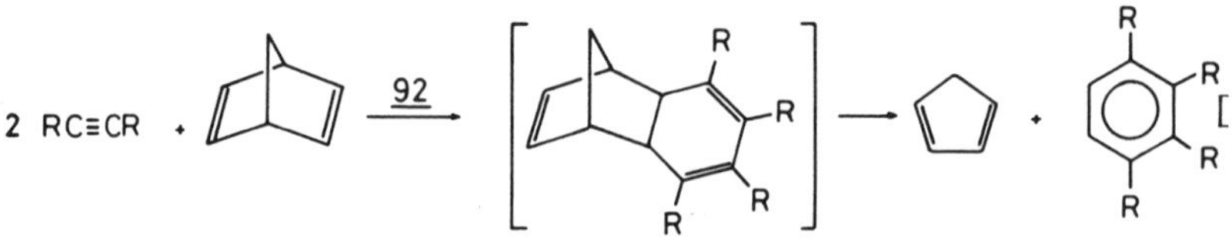

[138]

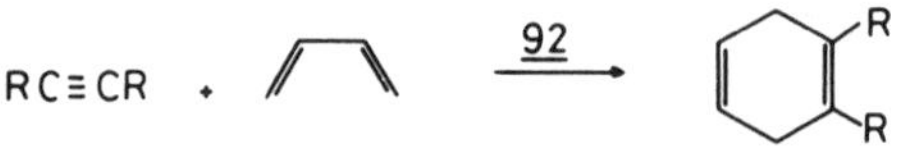

[139]

[138]), mixed sixmembered ring products are obtained *(91,92)*. Butadiene alone is oligomerized by 92, primarily to 1,3,6,10-dodecatetraene *(90)*. 93 is obtained (25 %) when a mixture of butadiene and 92 is carbonylated *(88,89)*.

Reaction of 92 with acetylene in the presence of cyclopentadiene rather than butadiene gives, not the expected Diels-Alder adducts, but instead a mixed bis(dienyl)-iron product *(216)* (eq. [140]). An analogous hydrogen transfer

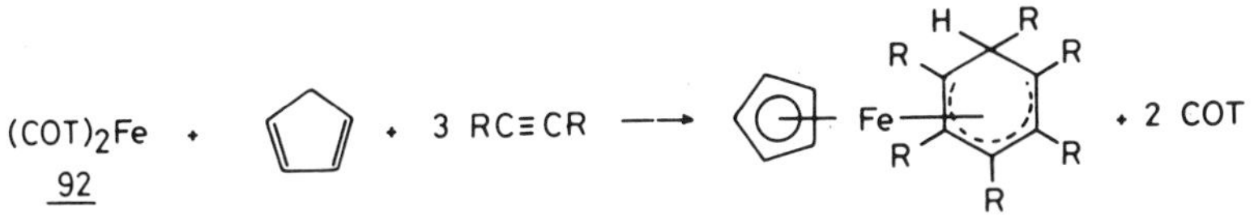
(COT)2Fe
92
3 RC≡CR
Fe
H
R
+ 2 COT

[140]

$R = C_6H_5$

occurs on reaction of 92 with cyclopentadiene alone (eq. [141]). Cyclopentadiene differs from butadiene in these

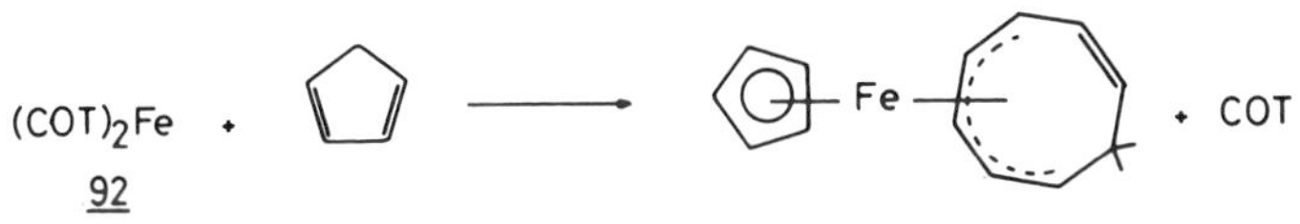
(COT)2Fe
92
Fe
+ COT

[141]

reactions by virtue of its active hydrogen, transfer of which converts a presumed (diene)(triene)iron compound into a more stable bis(dienyl)-iron. No (diene)(triene)-iron complexes which bear such a transferable hydrogen appear to exist.

A possible reaction sequence summarizing these results is given in Scheme XX. The conversions appear to be facilitated by the variable capacity of the cyclooctatetraene ligand to occupy one, two or three coordination sites about the iron, and to associate and dissociate rapidly to displace or make room for an entering ligand.

A related compound unusual for its method of synthesis was reported by Pettit *(392)* (eq. [142]). Under analogous con-

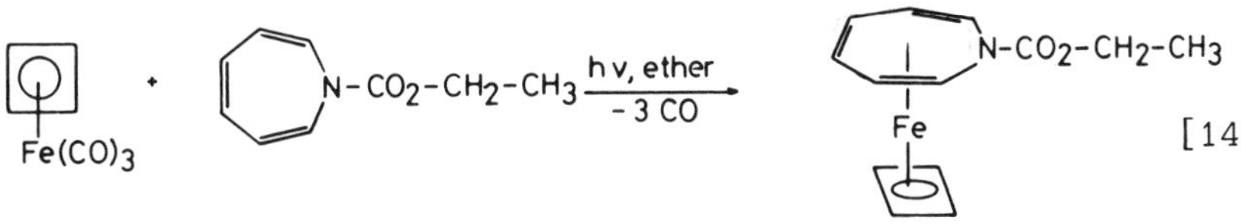
Fe(CO)3
N-CO2-CH2-CH3
hν, ether
- 3 CO
N-CO2-CH2-CH3
Fe

[142]

ditions cycloheptatrienes gave $Fe(CO)_2$ cycloadduct complexes (Section V); the reason for the unusual behaviour of the azepine ligand is unclear.

Reaction of "triisopropyliron" with cyclooctatriene gave, in 29 % yield, a complex of composition $(C_8H_{10})_2Fe$ *(318)*. This was shown by X-ray crystallography to be composed of cyclooctatriene bound through all three double bonds and bicyclo[4.2.0]octa-2,4-diene bound through its diene unit

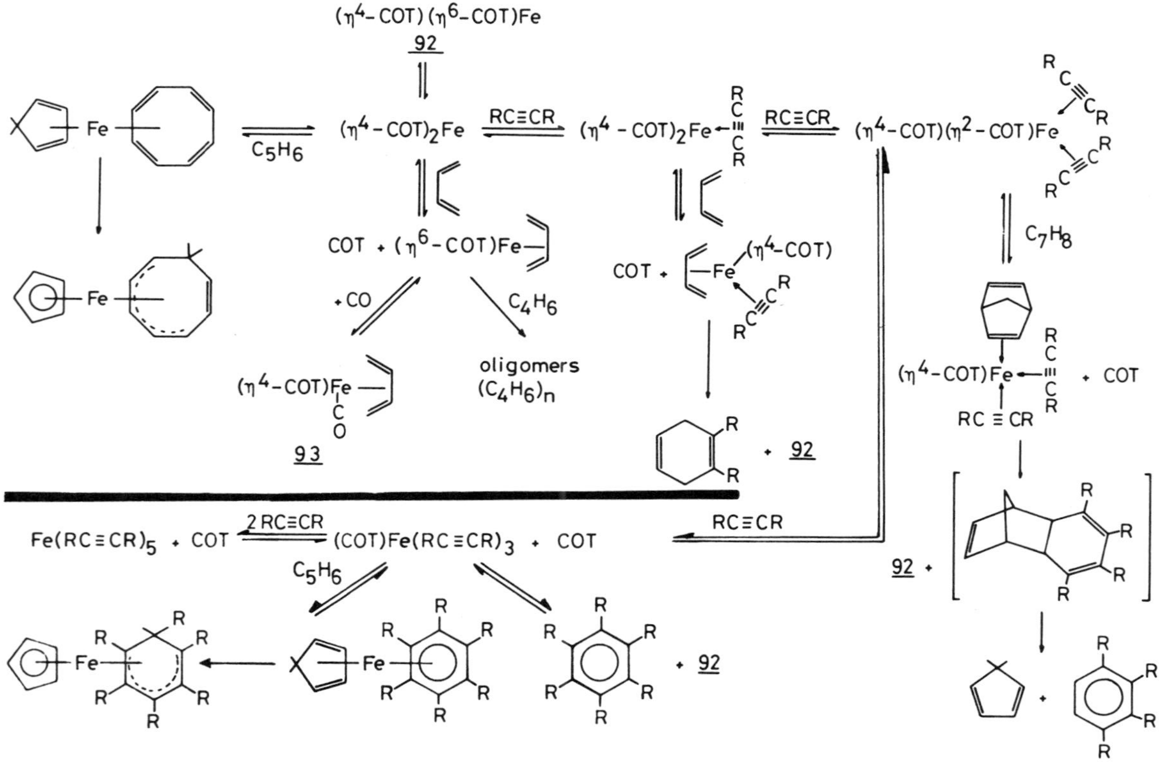

Scheme XX: Possible Reaction Sequences Involving Bis(cyclooctatetraene)iron, 92.

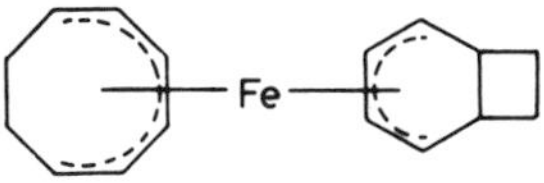

(244). Note that a hypothetical complex containing two monocyclic cyclooctatrienes would probably disproportionate by hydrogen transfer (eq. [143]), as discussed above. In any

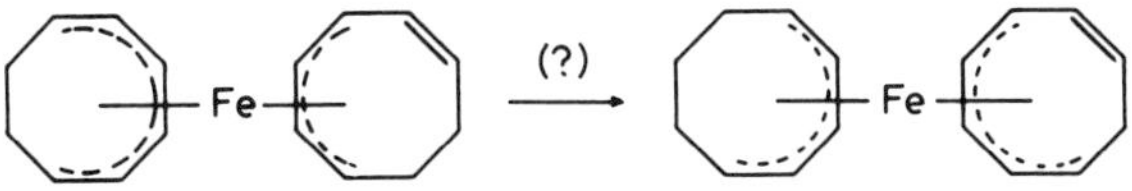

[143]

case, no such product has been reported to date.

Reaction of "triisopropyliron" with 1,5-cyclooctadiene or (better) a 1:1 mixture of cyclooctadiene and -triene gave, in 13 % yield in the latter case, an air-sensitive product of composition $(C_8H_{10})(C_8H_{12})Fe$ *(198,318)*. This was taken to be

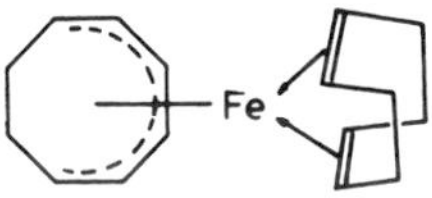

An analogous product from cycloheptatriene (31 %) originally taken to be (η^6-cycloheptatriene)(η^4-cycloheptadiene)-iron *(196)*, was later assigned the structure bis(η^5-cycloheptadienyl)iron, again a product of hydrogen transfer *(319)* (eq. [144]). Based on these precedents, the nature of the

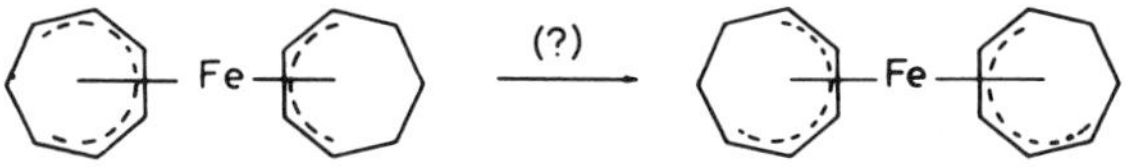

[144]

product from cyclooctadiene may bear reinvestigation.

Cocondensation of iron atoms with cycloheptatriene likewise failed to give bis(cycloheptatriene)iron, (η^5-cycloheptadienyl)(η^5-cycloheptatrienyl)iron being obtained instead *(55)* (eq. [145]).

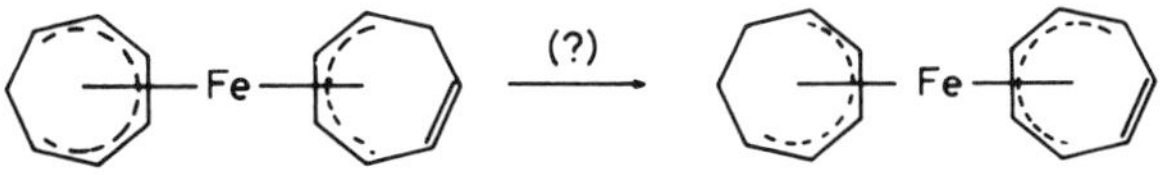

[145]

A bis(polyene)-iron compound can also be converted to a bis(dienyl)-iron by covalency formation between uncoordinated parts of the ligands, if sufficient proximity can be attained. This is illustrated in the reaction of 6,6-dimethylpentafulvene with iron atoms on cocondensation *(377)* (eq. [146]).

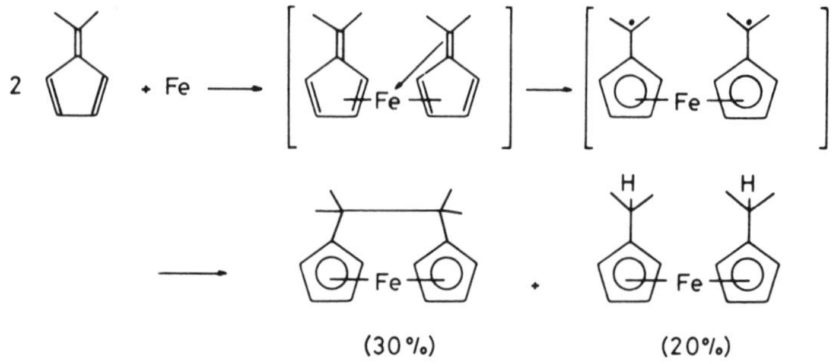

[146]

In a related example, reaction of the lithium salt of pentalene dianion with $FeCl_2$ in THF at -78°C gives an analogous complex, 94, in 18 % yield (12 % using $FeCl_3$) *(261)*. The structure was subsequently shown by X-ray crystallography to have the staggered C_2 structure shown *(110)* (eq. [147]).

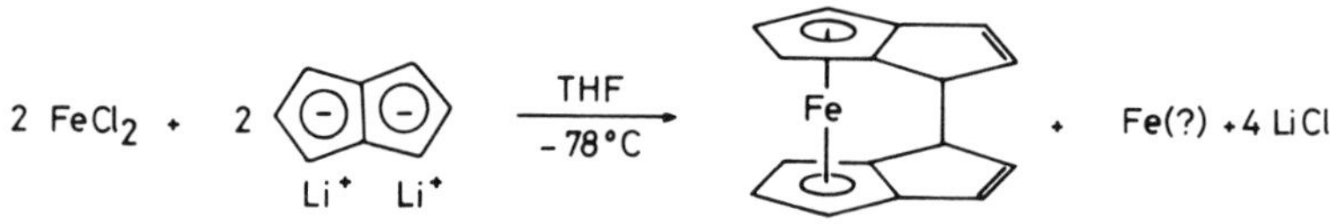

94

[147]

The newly-formed C-C bond has the rather long length of 1.57 Å. Reaction of the monolithium or thallium salt of dihydropentalene in an analogous way with $FeCl_2$ gave bis(hydropentalenyl)iron *(263,264)* (eq. [148]).

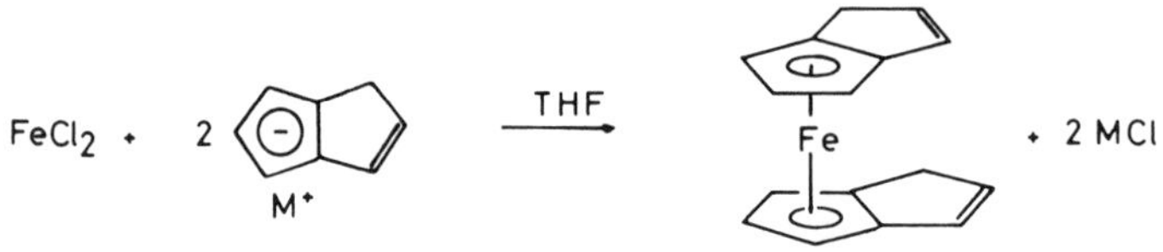

[148]

M = Li (54%),
M = Tl (49%)

Reaction of "triisopropyliron" with azulene under photolysis gave a compound, bis(azulene)iron, originally assigned a (diene)(triene)iron structure *(197)*. This was later found by X-ray crystallography, however, to be a ferrocene derivative analogous to 94 *(111,112)*. Again a long bond

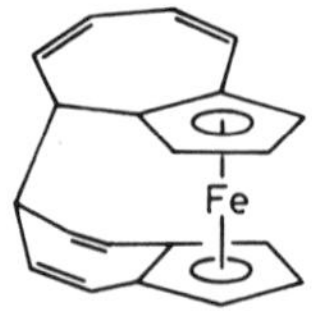

(1.58 Å) between azulene moieties was found; this bond in fact appears to undergo hydrogenation *(197)*.
A product resulting from reaction of (cyclopentadienyl)dicarbonyliron anion with tetrakis(trifluoromethyl)allene at 0°C may also be mentioned here. The structure followed from X-ray crystallography *(332)*.

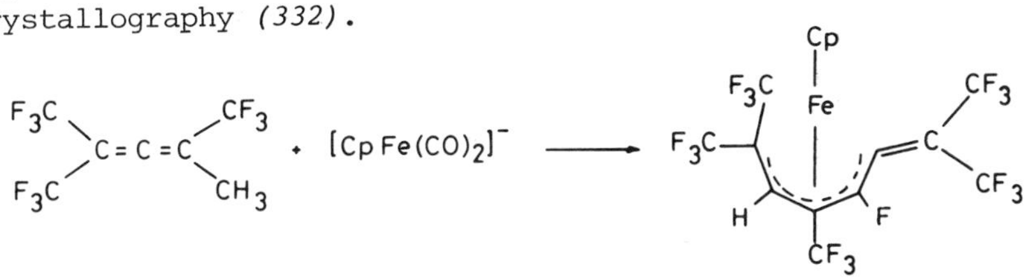

In summary, (diene)(triene)iron complexes are unusual compounds which exist only when routes to bis(dienyl)-iron compounds, such as hydrogen transfer and covalency formation, are excluded. The third double bond of the triene appears to be weakly coordinated in such compounds, the only η^6-1,3,5-triene complexes of iron which seem to exist at all.

The heavily studied and discussed ferrocenylcarbenium ions (eq. [149]), which are (at least formally) complexes of

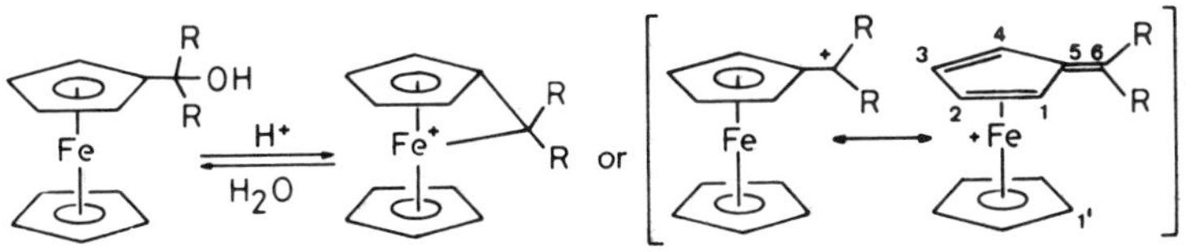

[149]

95 96

pentafulvenes, also fall within the scope of this Section. The manner of stabilization of the positive charge by the ferrocene moiety has been under discussion since 1961, when these species were first reported. Extensive reviews and discussions of earlier literature have appeared in 1966 *(81)*, 1973 *(149)*, and 1974 *(380)*.

The structure of ferrocenylcarbinyl cations has been interpreted by some as involving 95, an 18-electron species with a bond between the Fe and C(6) of the pentafulvene moiety; others have favoured 96, a 16-electron species with no Fe-C(6) bond and reduced Fe-C(5) bond order resulting from hyperconjugative stabilization. Extended Hückel calculations in support of 95 have been published *(212)*.

A compilation of pK_{R+} data for ferrocenylcarbinyl cations is given in Table 4. The ability of 6-substituents to provide stabilization decreases in the order ferrocenyl > phenyl > methyl > hydrogen. The stabilizing effect of a methyl substituent decreases with its position in the order

Table 4: pK_{R+} Values for Ferrocenylcarbinyl Cations in aqueous H_2SO_4.

Cation[a]	*(381)*	*(228,229)*	*(379)*	*(333)*[b]
$CpFeC_5H_4\overset{+}{C}H_2$	-	-1.28	-0.55	-1.75
$CpFeC_5H_4\overset{+}{C}HCH_3$	-0.65	-0.66	-0.10	-0.92
$CpFeC_5H_4\overset{+}{C}(CH_3)_2$	0.00	-0.01	-	-
$CpFeC_5H_4\overset{+}{C}HC_6H_5$	-	+0.4	-	-
$(CpFeC_5H_4)_2\overset{+}{C}H$	-	-	+4.33	-
$CpFe(2\text{-}CH_3\text{-}C_5H_3)\overset{+}{C}(CH_3)_2$	-0.50	-	-	-
$CpFe(3\text{-}CH_3\text{-}C_5H_3)\overset{+}{C}(CH_3)_2$	+1.10	-	-	-
$(CH_3\text{-}C_5H_4)FeC_5H_4\overset{+}{C}(CH_3)_2$	+0.72	-	-	-

[a]Cp = η^5-cyclopentadienyl; [b]In aqueous $HClO_4$ solutions

6 > 2 > 1'. The stability of the bridged cation 97 (pK_{R+} = + 2.50) *(381)* relative to 2,6,6,1'-tetramethylferrocenylcarbinyl cation (pK_{R+} = + 1.40) is inconsistent with a major shift of the incipient pentafulvene ring in forming the cation, since such a shift would be rendered difficult in 97.

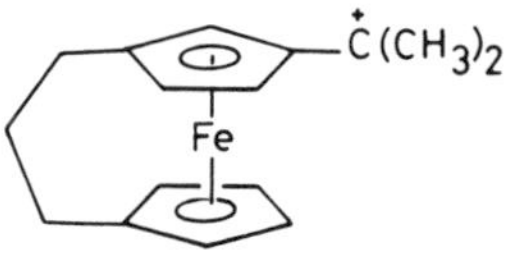

97

Proton NMR studies of secondary ferrocenylcarbinyl cations have indicated hindered rotation about the C(5)-C(6) bond *(149)*, with barriers of about 20 kcal *(380)*. Conversion of an alcohol to a cation causes greater deshielding of 2-hydrogens than of 1-hydrogens. Both of these results are reasonably consistent with structure 95 or 96. ^{13}C-NMR studies of these cations have shown that the downfield shifts attending cation formation from alcohols fall in the order 6 > 2 > 5 > 1 > 1' *(60)*. The $^1J(C,H)$ values for C(6) were 163 Hz ± 6, suggesting sp^2 hybridization of that carbon, in better agreement with structure 96 than 95.

ESCA and Mössbauer studies of diferrocenylcarbinyl cation have also been done. The ESCA results *(213)* show an ionization potential of the Fe $2p_{3/2}$ level in the cation of 709.6 eV, compared to 709.0 for ferrocene itself and 710.8 for ferricenium ion. The Mössbauer isomer shift for the carbenium ion was 0.692 mm/sec, compared to 0.714 for ferrocene and 0.658 for ferricenium ion. The authors concluded "that the two iron atoms in diferrocenylcarbinyl cation have approximately the same charge and that this does not differ appreciably from that in ferrocene". But, one could as well infer from the data that each iron atom bears about one-third of a positive charge.

X-ray crystallographic studies of two highly stabilized ferrocenylcarbinyl cations have been reported. In diferrocenylcarbinyl fluoroborate *(294)*, the two ferrocenyl rings were found to be non-equivalent, the Mössbauer result not withstanding. The carbinyl carbon, with a C-C-C bond angle of 131°, is not coplanar with either of the five-membered rings to which it is attached, being bent toward one iron by 20° and away from the other by 18°. The distances between the two iron atoms and the carbinyl carbon are 2.71 and 2.85 Å. In 1,2-diphenyl-3-ferrocenylcyclopropenium fluoroborate *(369)*, the plane of the 3-membered ring is bent from coplanarity with the 5-membered ring to which it is attached by 15° toward the iron, and the inter-ring bond is 1.445 Å long. The Fe-C(3) distance is again rather long, 2.83 Å.

These results do not allow for the existence of significant direct Fe-C(6) bonding, at least in these stabilized cations, nor do they agree very well with structure 96.

It appears that no simple pictorial representation of these cations currently available suffices to describe their properties. In this, they are very analogous to some related puzzling cations previously described:

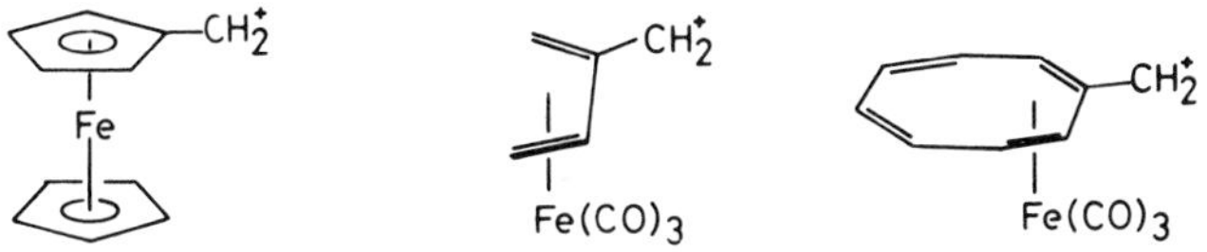

This Section Section IV.A Section IV.E

VIII. COMPLEXES OF $Fe_2(CO)_8$ AND $Fe_2(CO)_7$

Polyene complexes with more than one iron atom offer the additional question (beyond those in mononuclear complexes) of interactions between the metal atoms. The existence or absence of metal-metal bonds should affect not only chemical and physical properties, but also the stoichiometry of the compounds formed.

However, no $Fe_2(CO)_8$ or $Fe_2(CO)_7$ complexes having iron-iron bonds are known. The compounds known all have discrete $Fe(CO)_4$ groups coordinated to double bonds and $Fe(CO)_3$ groups bonded to diene units, but not directly interacting.

Both kinds of complexes have been obtained even from 1,3,5-hexatriene. Reaction with equimolar $Fe_2(CO)_9$ in ether at 40°C gave, in addition to $Fe(CO)_3$ and $Fe(CO)_4$ complexes, an $Fe_2(CO)_8$ complex, 98, in 10 % yield *(320)*. 98 was also produced by disproportionation of the $Fe(CO)_4$ complex on attempted crystallization. Use of 2-3 fold excess $Fe_2(CO)_9$ did not increase the yield of 98, but did give the $Fe_2(CO)_7$ complex 99 in *ca*. 30 % yield.

IR and NMR indicated that the end double bonds were coordinated in 98 *(320)*. 98, like other bis[$Fe(CO)_4$] complexes, appears to be more stable than expected from the properties of mono-[$Fe(CO)_4$] complexes. This is shown most clearly by the disproportionation, and suggests a synergetic interaction between $Fe(CO)_4$ groups.

99 had a low dipole moment of 1.56 D, compared with 2.17 D for 14, (hexatriene)tricarbonyliron, and 1.94 D for (butadiene)tetracarbonyliron *(320)*. A structure with the iron carbonyl groups *trans* with respect to the ligand seems clearly required. The only other acyclic bis[$Fe(CO)_4$] com-

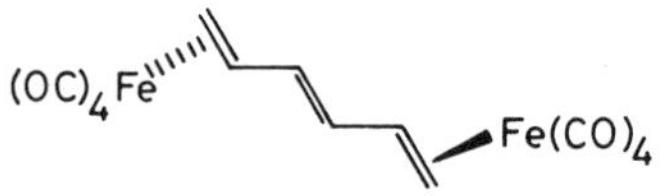

98

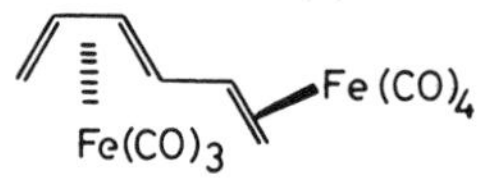

99

plex reported was a very unstable material obtained by reaction of 1,4-bis(trimethylsilyl)-1,3-butadiene and $Fe_3(CO)_{12}$ under photolysis *(337)*. Thermal reaction of diphenylbutadiyne did not give an analogous isolable product *(237)*.

The remaining bis[$Fe(CO)_4$] complexes have all been reported in reactions of cyclic dienes. Thus, reaction of 6,6-diphenylpentafulvene with $Fe_2(CO)_9$ at 40°C gave a bis[$Fe(CO)_4$] complex, 100, in 49-66 % yield *(395,397)*. Spectra and dipole moment (1.7 D) indicated *trans*-coordination of the two endocyclic double bonds, which was subsequently confirmed by X-ray crystallography *(36)*. 6,6-dialkylpentafulvenes gave no isolable $Fe_2(CO)_8$ complexes analogous to 100. 6,6-Bis(*p*-chlorophenyl)pentafulvene gave only an 8 % yield, in addition to a larger amount of $Fe(CO)_3$ complex. 100 decomposed to the $Fe(CO)_3$ complex at *ca*. 65°C. 100 did not give reversible polarographic reduction *(169)*.

A further $Fe_2(CO)_8$ derivative was obtained from spiro-[4.2]hepta-2,4-diene and $Fe_2(CO)_9$ in ether *(167)*. Unlike 100, however, it did not decompose to an $Fe(CO)_3$ complex; instead, it gave (6-methylpentafulvene)hexacarbonyldiiron, 101, on standing in solution for 18 h at room temperature (eq. [150]).

Further examples of interconversions of $Fe_2(CO)_x$ complexes

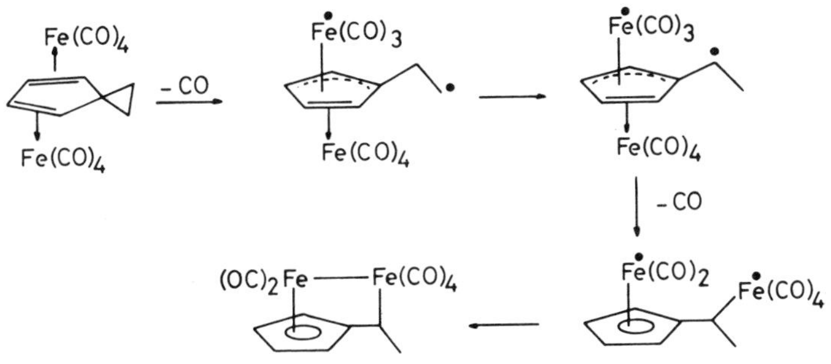

[150]

101

with x = 8, 7, and 6, resulted from study of the reaction of semibullvalene with $Fe_2(CO)_9$ at room temperature *(22,179)* (Scheme XXI).

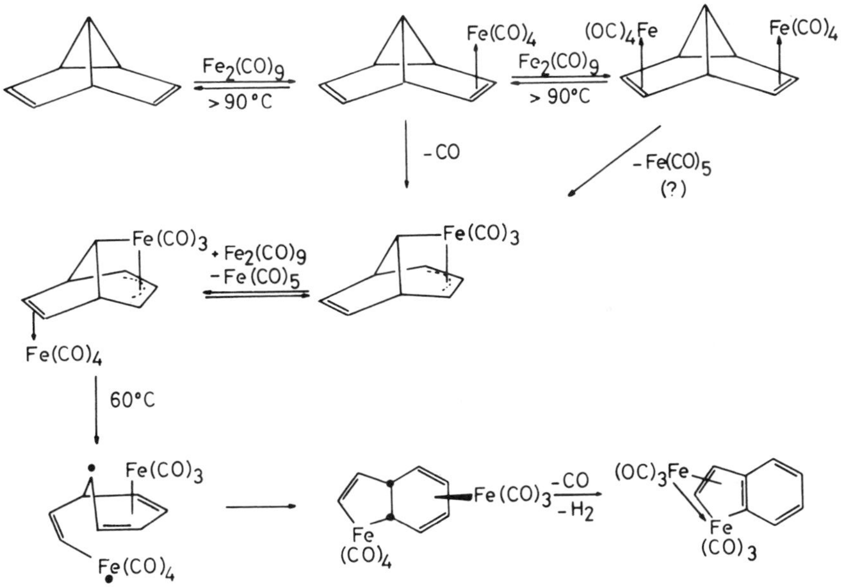

Scheme XXI: Paths in Reaction of Semibullvalene with $Fe_2(CO)_9$ *(179)*.

A related bicyclic $Fe_2(CO)_7$ complex *(25)* in which the $Fe(CO)_3$ group is already bonded to a 1,3-diene (eq. [151])

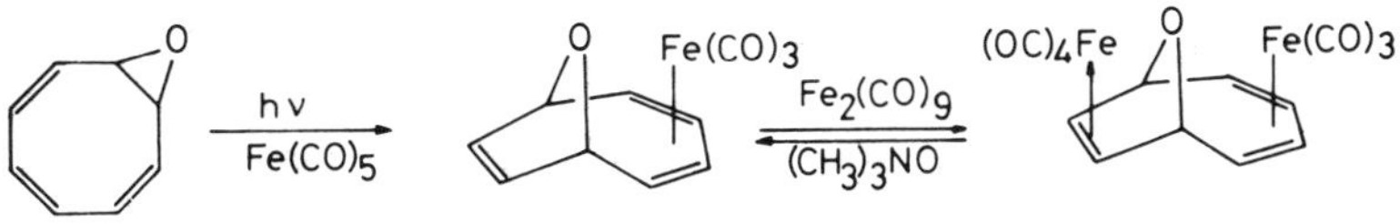

[151]

does not rearrange further as in Scheme XXI; instead, deoxygenation occurs on heating to give cyclooctatetraene complexes *(306)*.

In summary, bis[$Fe(CO)_4$] complexes are occasionally obtained on reaction of $Fe_2(CO)_9$ with trienes at 40°C or below. They may be somewhat more stable than $Fe(CO)_4$ complexes, but decompose in a variety of ways at temperatures above 70°C.

IX. COMPLEXES OF $Fe_2(CO)_6$

In the absence of an Fe-Fe bond, $Fe_2(CO)_6$ complexes are simply bis[$Fe(CO)_3$] complexes, each iron independently coordinated with a diene or other four-electron unit. With an iron-iron bond, only six electrons need be supplied by an organic ligand to provide a closed shell configuration to the irons. Tetraene- or polyene-hexacarbonyldiiron complexes usually assume the bis[$Fe(CO)_3$] structure; triene-hexacarbonyldiiron compounds can only form products with Fe-Fe bonds.

A. *BIS[$Fe(CO)_3$] COMPLEXES WITHOUT Fe-Fe BONDS*

These complexes are commonly available by straightforward complexation of suitable tetraenes. Most tetraenes give only complexes with independent $Fe(CO)_3$ groups. Some, discussed at the end of this section, give products with both independent and Fe-Fe bonded $Fe(CO)_3$ groups. Most bis[$Fe(CO)_3$] complexes are made from the tetraene itself or from an already partially complexed derivative. A particularly interesting example of the latter route was the further complexation of the resolved complex 27, which gave both active and meso products *(312)* (eq. [152]). The loss of optical activity of 103 at 119.4°C

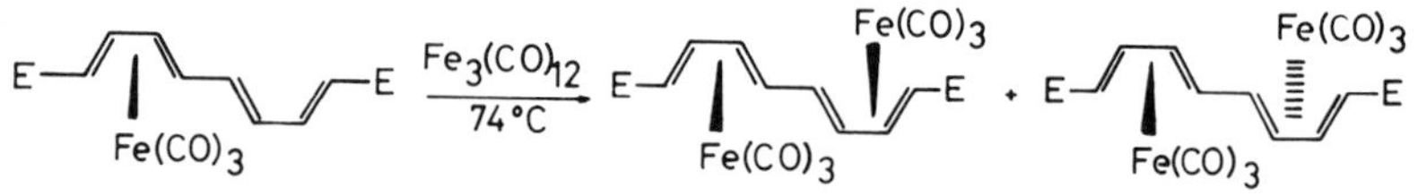

[152]

(-) 27 102 103

E = CO_2-CH_3

proceeded by conversion to 102 rather than by formation of its enantiomer by means of the two $Fe(CO)_3$ groups passing each other on opposite sides of the tetraene chain. The reaction 102 ⇌ 103 had an equilibrium constant of about 2 *(312)*.

Likewise, reaction of (styrene)tricarbonyliron, 104, with $Fe_2(CO)_9$ or with $Fe(CO)_5$ (eq. [153]) under ultraviolet irradiation gave *trans*-bis[$Fe(CO)_3$] complexes *(385,387)*. Irradiation of 104 alone caused disproportionation to styrene and the bis[Fe(CO)$_3$] complex, a reaction whose driving force may be the recovery of benzenoid delocalization in the free styrene. Mass spectra of the bis[Fe(CO)$_3$] complexes have been studied in detail *(388)*. The free diene unit of the heptafulvene complex 58 also underwent complexation to give a (presumably *trans*) bis[Fe(CO)$_3$] complex in 77 % yield *(269)*. Reaction of the *o*-xylylene complex 32 with Fe(CO)$_5$ under irradiation gave both *cis*- and *trans*-C_8H_8[Fe(CO)$_3$]$_2$ complexes *(384)*, in compa-

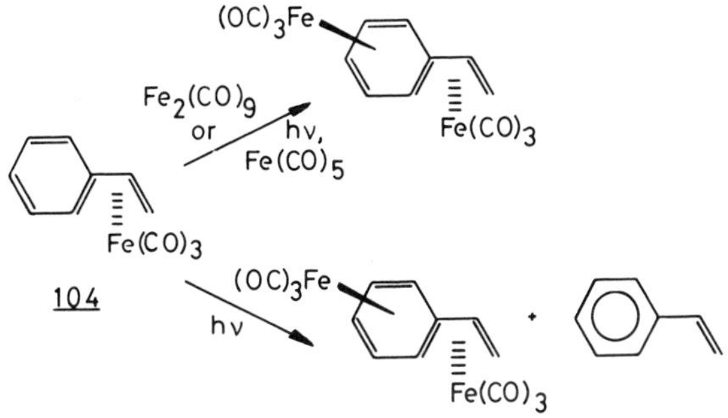

104

[153]

[154]

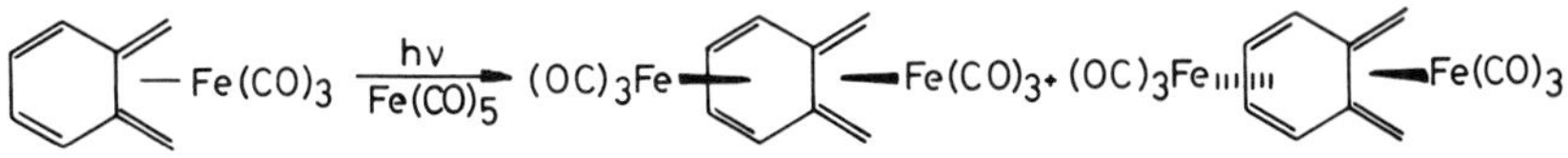

32

rable yields (eq. [154]); reaction of (benzocyclobutadiene)-tricarbonyliron proceeded likewise *(383)*. Reaction of *meta*- or *para*-divinylbenzene with $Fe_3(CO)_{12}$ in refluxing benzene gave only bis[Fe(CO)$_3$] complexes (eq. [155]), the partially

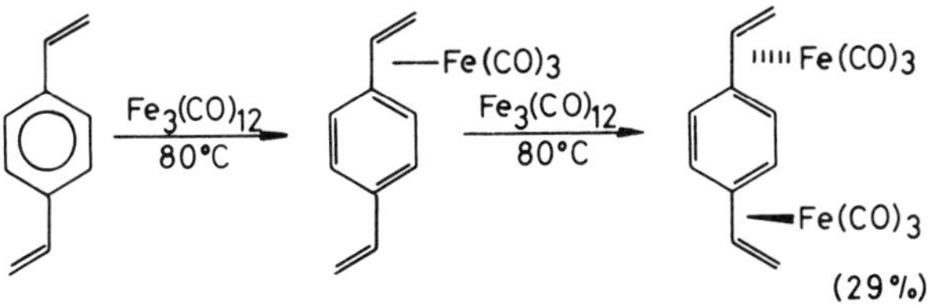

[155]

localized Fe(CO)$_3$ complexes evidently being more reactive than the starting hydrocarbons *(308)*. The *trans* stereochemistry in each case was shown by X-ray crystallography *(153)*.

Complexation reactions of a series of tetraenic propellanes have been extensively studied by Ginsburg and coworkers *(210)*. The propellanes studied are all derivatives of 9,10-dihydronaphthalene having 3-atom bridges across the 9,10-posi-

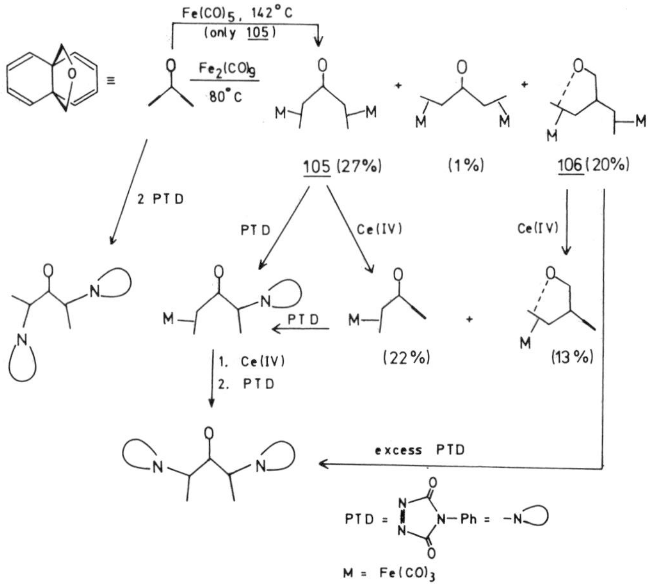

Scheme XXII: Complexation of 12-oxa[3.4.4]propella-2,4,7,9-tetraene *(7,54,210,285)*. %'s shown refer to $Fe_2(CO)_9$ reaction.

tions. All of the these compounds, having $-CH_2-O-CH_2-$, $-\overset{O}{\overset{\|}{C}}-O-\overset{O}{\overset{\|}{C}}-$, $-\overset{O}{\overset{\|}{C}}-NH-\overset{O}{\overset{\|}{C}}-$, $-\overset{O}{\overset{\|}{C}}-N(CH_3)\overset{O}{\overset{\|}{C}}-$ chains as bridges, give primarily bis[$Fe(CO)_3$] derivatives on complexation *(7,54)*. The most heavily studied has been the ether series, in which the results shown in Scheme XXII have been reported.

The obtention of 105 as sole product using $Fe(CO)_5$ at high temperature suggests that it is the most stable of the various complexes. Its identity as the all-*syn* adduct was shown by X-ray crystallography *(53)*. The *syn-anti*-bis-[$Fe(CO)_3$] adduct, 106, was also studied by X-ray crystallography *(52)*, which suggested an interaction between the ether oxygen and the *anti*-complexed diene unit. It is interesting that ceric ammonium nitrate treatment selectively removes the *syn*-$Fe(CO)_3$ group, leaving this interaction undisturbed *(52, 54)*.

In the reactions with 1-phenyl-1,3,4-triazoline-2,5-dione (PTD), Diels-Alder adducts of the diene units are the products *(285)*. The reaction results are rationalizable on the bases that free dienes are more reactive than complexed and that PTD, a strong oxidizing agent, is capable of oxidatively removing an $Fe(CO)_3$ group giving a free diene, which then undergoes Diels-Alder reaction with another PTD.

Complexation and Diels-Alder reactions of the related

propellane with a -C(=O)-NH-C(=O)- bridge have also been investigated, with results generally similar to those in Scheme XXII *(9)*.

Reaction of the complexes $(C_{16}H_{16})Fe(CO)_3$, 89 and 90 (Scheme XIX), with $Fe(CO)_5$ at 180°C (eq. [156]) gave two bis-$[Fe(CO)_3]$ complexes, at least partially *via* an $Fe_3(CO)_9$ complex, 107, which was isolable from photochemical reaction of $Fe(CO)_5$ with 89 and 90, and which gave the two $Fe_2(CO)_6$ products on heating to 180°C *(361,362)*.

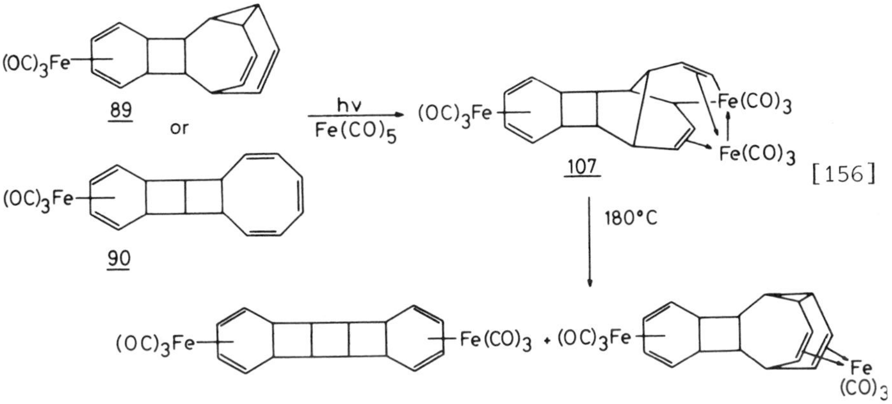

[156]

In addition to direct complexation of tetraenes, synthesis of bis$[Fe(CO)_3]$ complexes has been achieved by combination of diene- or dienyl-tricarbonyliron complexes. Formation of (bitropyl)bis(tricarbonyliron) *(161)* and the dimer of (heptafulvene)tricarbonyliron *(184)* by adventitious dimerization have already been discussed (Sect. IV.C).

Two studies of dimerization of (hexadienyl)tricarbonyliron cation, with somewhat different results, summarized in Scheme XXIII, have been reported *(10,304)*. Zn-induced dimerization of (cycloheptadienyl)tricarbonyliron cation has been reported *(225a)*. These reactions appear to offer some synthetic promise which merits further exploitation.

An ether analogous to that in Scheme XXIII has also been isolated upon hydrolysis of a (cyclohexadienyl)tricarbonyliron cation *(51)* (eq. [157]).

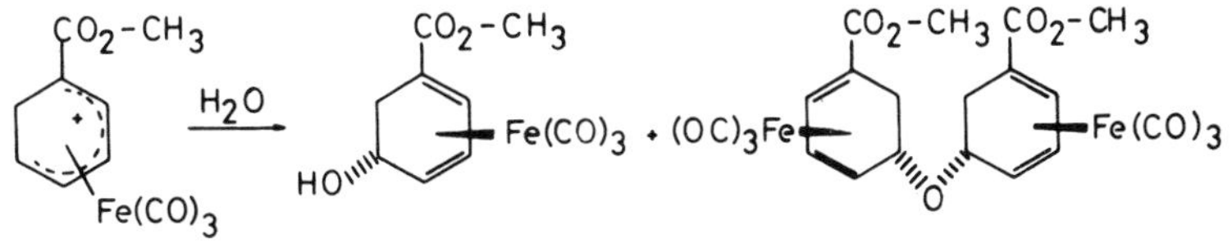

[157]

Some unusual cases of formation of dimeric products upon complexation also merit coverage in this section. One example is the reaction of 1,4-dichlorobut-2-yne with $K_2Fe_2(CO)_8$

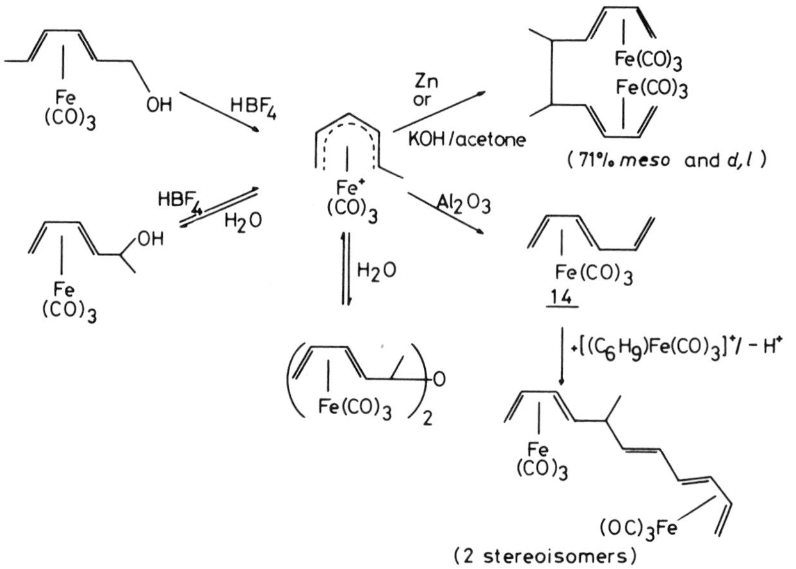

Scheme XXIII: Dimerization Reactions of (Hexadienyl)tricarbonyliron Cation *(10,304)*.

(259) (eq. [158]). The unusual diene-tricarbonyliron products

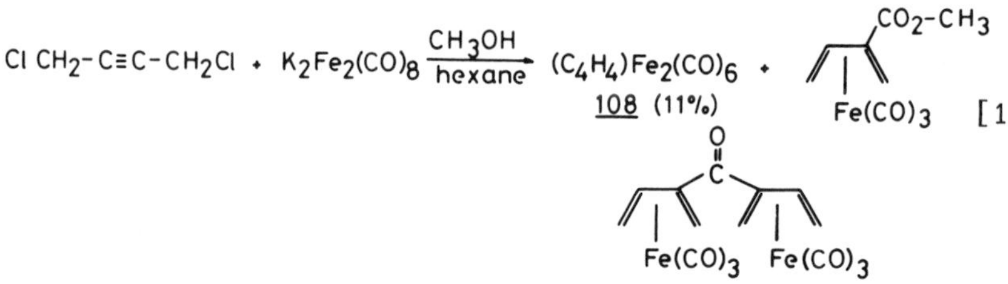

[158]

were not obtained from the dichloride and $Fe_3(CO)_{12}$ in toluene, nor from use of $Na_2Fe(CO)_4$ in THF; only low yields of 108 were reported *(259)*. It is not clear whether the diene products can be formed from 108 under the reaction conditions.

Dimerization also attended photochemical reaction of 1,4-bis(trimethylsilyl)butadiyne with $Fe_3(CO)_{12}$ *(337)* (eq. [159]). This reaction is clearly related to that of diphenylbutadiyne with $Fe_3(CO)_{12}$ *(237)* (eq. [160]). The further closure in the former case probably results from the photolytic conditions.

As previously noted (Sect. IV.A) allylic leaving groups are often lost during complexation reactions. An interesting case involving loss of two such groups has been reported *(357)* (eq. [161]). Reaction of allene with $Fe_3(CO)_{12}$ in saturated

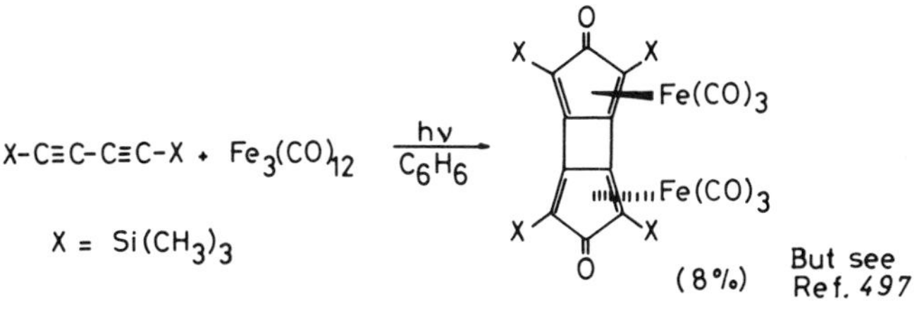

[159]

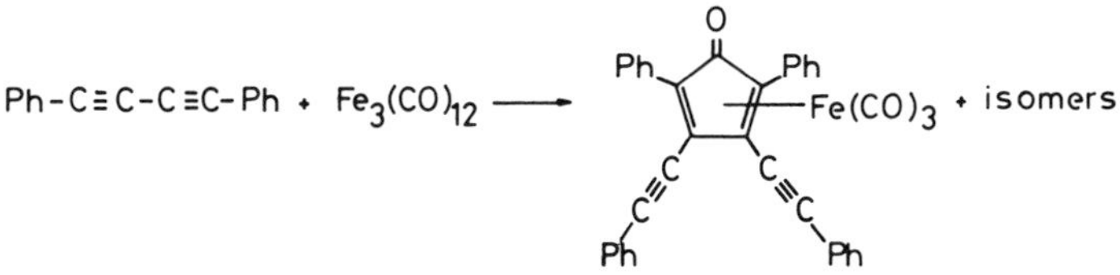

[160]

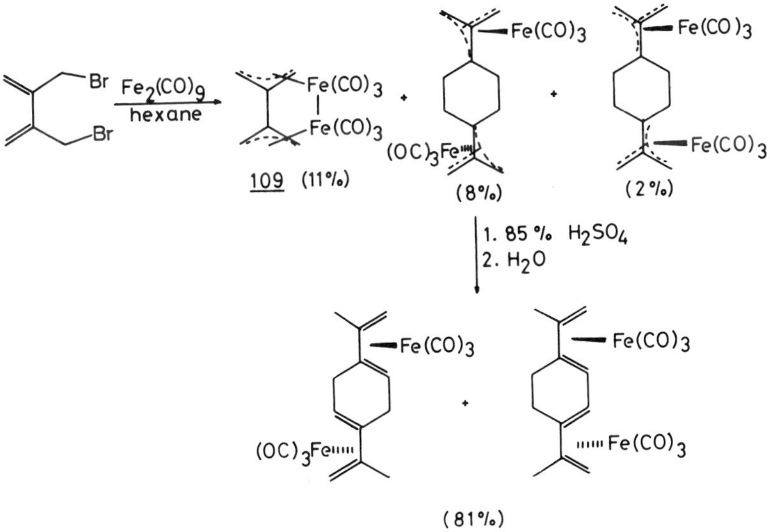

[161]

hydrocarbons also gave 109 *(328,335)*. Reaction of allene with $Fe_3(CO)_{12}$ in aromatic solvents or with 109, a complex of an allene dimer, gave an interesting sequence of $Fe_2(CO)_6$ complexes of allene trimers *(335)* (eq. [162]). The structures of the latter two products were confirmed by X-ray crystallography *(408)*. The mechanisms of formation and rearrangement appear to be homolytic in nature, involving thermal breaking of the Fe-Fe bond in 109 and of the Fe-C σ-bonds in the subsequent intermediate complexes.

Reaction of 7-vinylcycloheptatriene with $Fe(CO)_5$ at 140°C was reported in 1961 to give a not-fully characterized $Fe_2(CO)_6$ derivative, the yellow colour of which suggested a

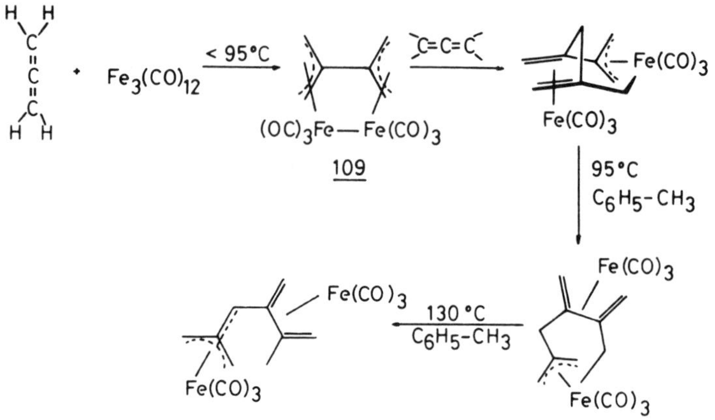

[162]

structure with independent $Fe(CO)_3$ groups *(283)*. A structure such as that shown in equation [163] may have resulted from hydrogen migration.

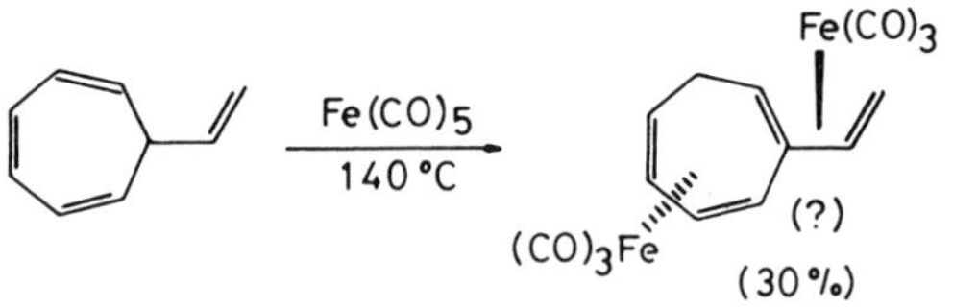

[163]

Some more bizarre $[Fe(CO)_3]_2$ complexes have been obtained in small yields from attemped complexation reactions (eqs. [164] *(194)* and [165] *(132,143,145)*).

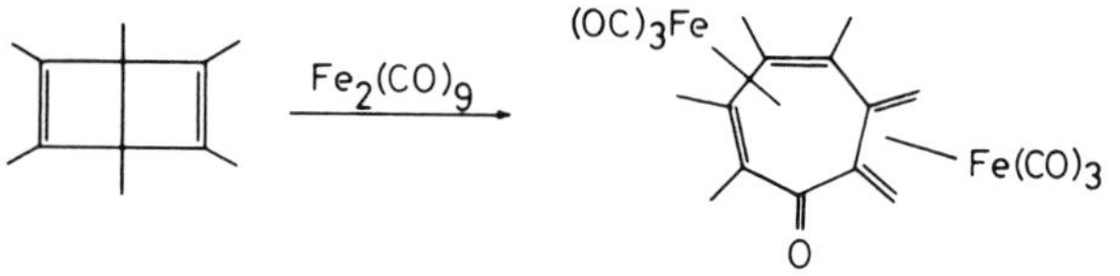

[164]

[165]

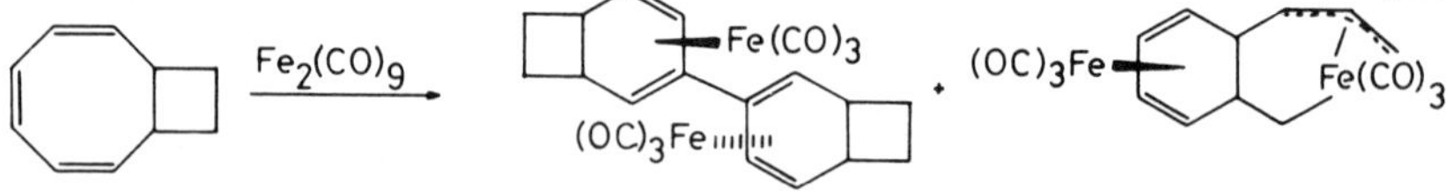

(see also Scheme XI)

Complexation of bullvalene under various conditions leads to about seven isomeric $(C_{10}H_{10})Fe_2(CO)_6$ complexes *(14)*, as reported in a series of publications by Aumann *(14-18,22, 23)*. The relationships among these complexes, and their reactions, are summarized in Scheme XXIV. The key intermediates leading to stable products are thought to be 110 and

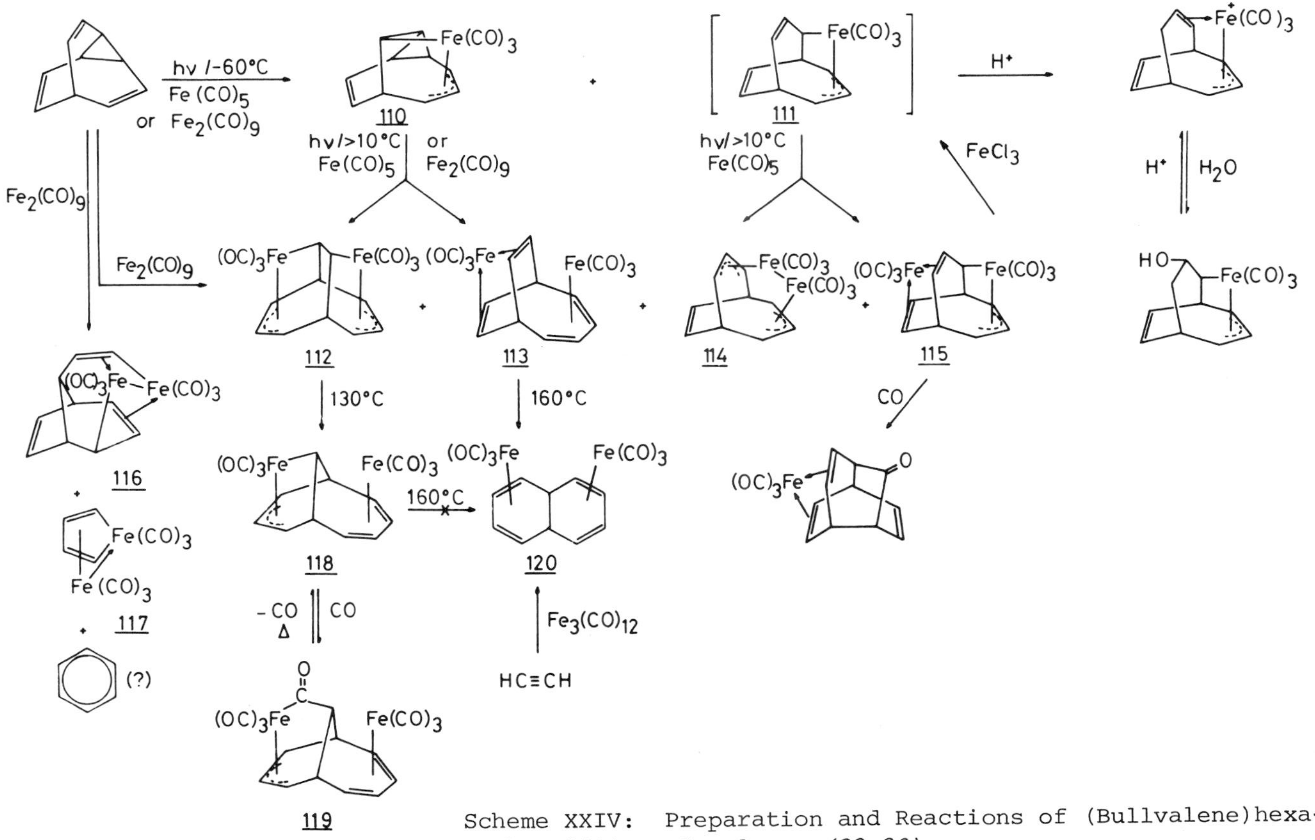

Scheme XXIV: Preparation and Reactions of (Bullvalene)hexacarbonyldiiron Complexes *(22,23)*.

111, the latter not isolable but trappable by acid-catalysed hydration as shown *(22)*. Direct reaction of bullvalene with $Fe_2(CO)_9$ in ether gives primarily 115, 112, and 113 in a 2:1:1 ratio, plus small amounts of 114, 116, and 117, the products with Fe-Fe bonds *(14,22)*. At 80°C, 113 becomes the chief product *(362,22)*. The yield of 114 is highly dependent on conditions, being only 3 % when the reaction is run in ether at 30°C, but much higher in hexane at 60°C, or using (cyclohexene)tetracarbonyliron as the source of carbonyliron groups *(22)*.

The structures of several of the products, including 112 *(390)*, 113 *(363)*, 115 *(245)*, and 119 *(390)*, have been confirmed by X-ray crystallography. Some have fluxional properties. In 112, for example, the two $Fe(CO)_3$ groups exchange attachment to the carbons of the two-carbon bridge. The coalescence temperature for this process is 26°C and the $\Delta G^{\ddagger}$ = 16.0 kcal/mol *(15)*. 115, in contrast, is rigid on the NMR time scale *(14)*. Fluxionality of 116 will be discussed along with that of other complexes with Fe-Fe bonds (Sect. IX.B).

The very stable dihydronaphthalene complex 120 can, as shown, be obtained on heating 113, but not 118, which would otherwise have seemed a plausible intermediate *(16,362)*. The rearrangement does not therefore involve stepwise migrations but rather intramolecular cycloadditions (eq. [166]).

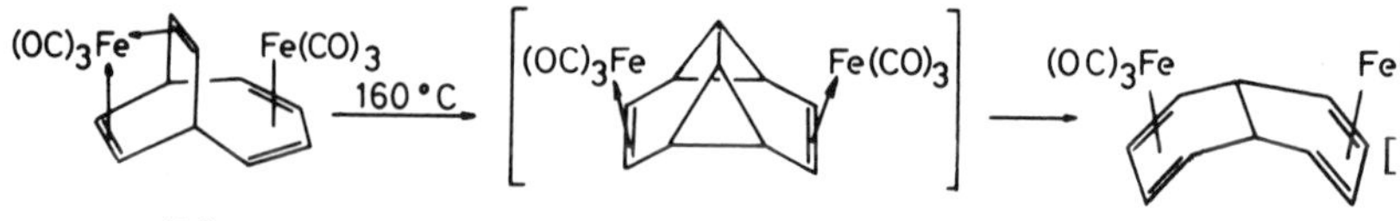

113 120 [166]

A related product has been observed in reaction of 65 with phenylacetylene at 165°C, which gives as main product the 2-phenyl analogue of 120 *(289)* (eq. [167]). 120 has also

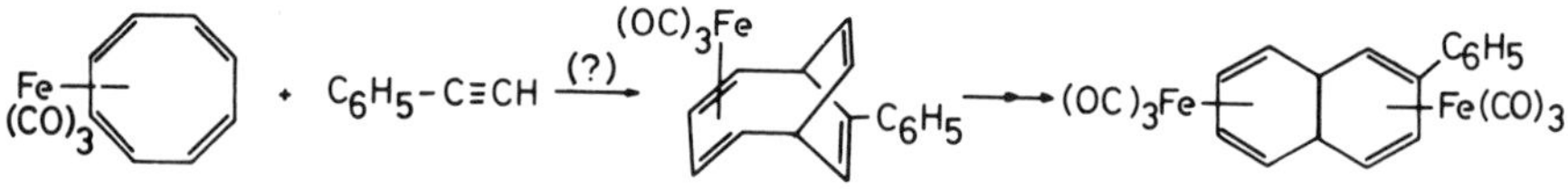

65 [167]

been obtained as one among several products from reaction of acetylene with $Fe_3(CO)_{12}$, in what is evidently a quite complicated process *(234,236,398)*.

Reaction of cyclooctatetraene with iron carbonyls gives, in addition to the main product 65, three different $Fe_2(CO)_6$ complexes. The most stable is a *trans*-bis[$Fe(CO)_3$] complex, 121, 185° dec., which has been extensively characterized by IR, NMR, and UV spectroscopy *(206)*, Mössbauer spectroscopy

(221,399), mass spectroscopy *(75)*, and X-ray crystallography *(171,173)*. It is typically obtained in low yield (3-8 %) in thermal or photochemical reactions with $Fe(CO)_5$ *(309,310,322)*. The yield can be increased to 31 % by use of a large excess of $Fe(CO)_5$ in photochemical reaction *(346)* or by use of $Fe_2(CO)_9$ *(266)*. 121 can also be obtained from 65 under the same conditions (70 %) *(346)*. 121 was also produced as a by-product in reaction of cyclooctatetraene oxide with $Fe_3(CO)_{12}$ *(306)*. Substituted derivatives of 121 have also been obtained from phenylcyclooctatetraene and excess $Fe(CO)_5$ or photochemical reaction (4 %) *(324)* and from 1,3,5,7-tetramethylcyclooctatetraene and $Fe_2(CO)_9$ at 100°C (3 %) *(140)*. 121 and its derivatives are rigid on the NMR time scale.

Reaction of cyclooctatetraene with $Fe_2(CO)_9$ at room temperature also gives two further $Fe_2(CO)_6$ complexes *(266)*, which are not obtained under higher energy conditions. One, 122, is a *cis*-bis[$Fe(CO)_3$] complex, as indicated by the similarity of its NMR spectrum to that of 121. The other has a structure, presumably 123, with an Fe-Fe bond. A bis(η^3-allyl) structure has commonly been written for this compound in the literature, but in view of extensive results on analogous cyclooctatriene-hexacarbonyldiiron complexes to be discussed below, 123 is probably to be preferred *(140)*. The dipole moment of 123 is 3.9 D *(266)*. 122 and 123 form distinct solids, but slowly interconvert on standing in solution. They both still more slowly lose CO, forming (cyclooctatetraene)pentacarbonyldiiron, 124 *(266)*. Tetramethylcy-

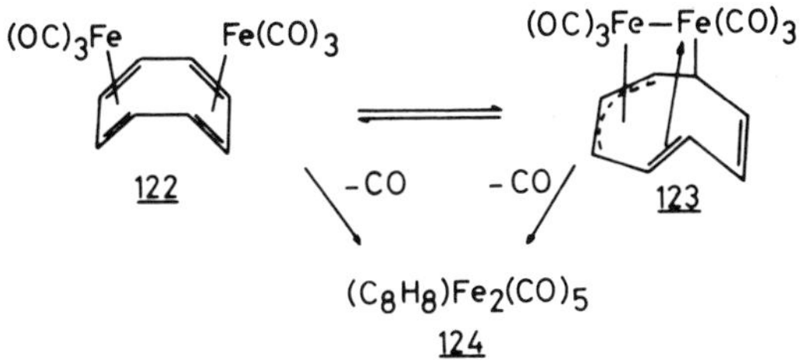

clooctatetraene forms analogues of 123 and 124, but not of 122 *(140)*. Benzocyclooctatetraene also forms an analog of 123, in 4 % yield using $Fe_2(CO)_9$ and 16 % yield using $Fe_3(CO)_{12}$ in refluxing benzene *(187)*. That $Fe_2(CO)_6$ complex decomposes in solution not to an $Fe_2(CO)_5$ complex but instead to the $Fe(CO)_3$ complex *(187)*.

The NMR spectrum of 123 at room temperature indicates two-fold symmetry of the C_8H_8 ligand, indicating limited fluxional character, with rapid enantiomerization but no relative rotation of the $Fe_2(CO)_6$ group and the ring. This process has not been studied quantitatively in 123. It is unclear from the literature whether 123 undergoes protonation on the free double bond or not *(28)*.

B. BIS[Fe(CO)$_3$] COMPLEXES WITH Fe-Fe BONDS

A number of such complexes (107, 108, 109, 114, 116, 117, and 123) have already appeared in Section IX.A, since they have been obtained along with complexes with independent Fe(CO)$_3$ groups.

A compound with a (OC)$_3$Fe-Fe(CO)$_3$ group needs six additional electrons in order for each iron to attain an 18-electron configuration. The most symmetrical way to provide these electrons is *via* coordination of an η^3-allyl group to each iron. The allyl groups may be independent of each other (as in 114), connected 1-1' (as in (cycloheptatriene)hexacarbonyldiiron, 125, discussed below) or 2-2' (as in 109). No examples of 1-2' connection are known, perhaps because the Fe-Fe bond length is inappropriate for such an arrangement.

A few triene-derived analogs of 109, (tetramethyleneethene)-hexacarbonyldiiron, have been reported. The best characterized one results, along with the already discussed bis[Fe(CO)$_3$] complexes, from photochemical reaction of (*o*-xylylene)tricarbonyliron, 32, with Fe(CO)$_5$ (eq. [168]). It

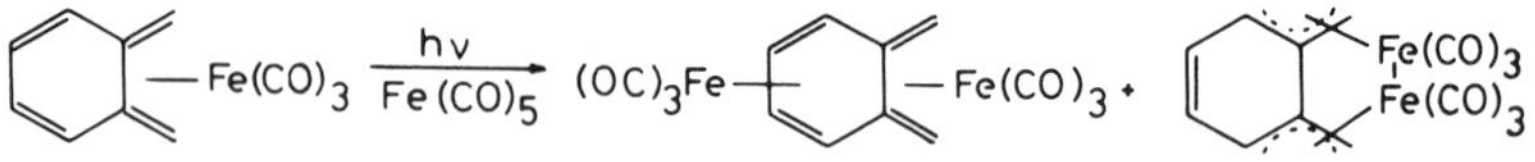

[168]

32

reacts with bromine to give α,α'-dibromo-*o*-xylene *(384)*.

Reaction of bicyclo[3.3.0]octa-1,3,5-triene with Fe$_2$(CO)$_9$ or Fe$_3$(CO)$_{12}$ gave four products, for one of which an analogous structure was suggested *(241,242)*. The product slowly decom-

posed at 45°C to (pentalene)pentacarbonyldiiron, 126.

The best characterized bis(η^3-allyl-tricarbonyliron) complex is 125, which results from reaction of cycloheptatriene with Fe$_2$(CO)$_9$ (8.5 %) *(125,190)*. A similar product is obtained from 7-methoxycycloheptatriene *(190)*. 125 has a dipole

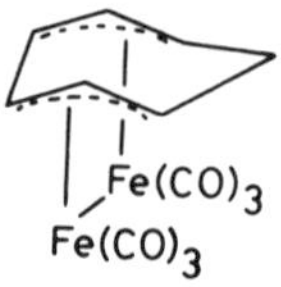

125

moment of 4.8 D. X-ray crystallography *(125)* confirmed the originally proposed *(190)* bis(η^3-allyl) structure, 125, and revealed a long Fe-Fe bond length of 2.866(1) Å. The ^{1}H- and ^{13}C-NMR spectra of 125 are invariant from room temperature to -100°C, as would be expected from the crystal structure *(125, 159)*. The carbonyl ligands on each iron appear to scramble independently without exchanging between the two iron atoms *(136)*.

The *anti* C(7)-H bond in 125 is staggered with respect to the Fe-C(1) and Fe-C(6) bonds, suggesting high reactivity. In fact 125 transfers hydride ion to triphenylcarbenium fluoroborate, giving a cation $[(C_7H_7{}^+)Fe_2(CO)_6]$, also obtainable from reaction of the 7-methoxy compound with acid *(190)*. The cation is very stable, having a pK_{R^+} of 8.0 (*cf*. 4.5 for $[C_7H_7Fe(CO)_3]^+$ and 4.7 for $[C_7H_7]^+$), and shows only a single resonance in the proton magnetic resonance spectrum at room temperature. It is thus a fluxional molecule.

Cyclooctatriene and many of its derivatives also form $Fe_2(CO)_6$ complexes upon treatment with $Fe_2(CO)_9$ or $Fe_3(CO)_{12}$. With cyclooctatriene itself, $Fe_2(CO)_9$ in ether is clearly the reagent of choice, 127 being obtained in 30 % yield *(165)*. Use of $Fe_2(CO)_9$ or $Fe_3(CO)_{12}$ in hydrocarbon solvents consistently gives yields of 1-4 % of 127 *(190,199,272,311,325)*. Benzocyclooctatetraene gives an $Fe_2(CO)_6$ complex in 16 % yield with $Fe_3(CO)_{12}$ in refluxing benzene *(187)*. 5,8-Bis(trimethylsilyl)-1,3,6-cyclooctatriene gives a 0.5 % yield of an $Fe_2(CO)_6$ complex with $Fe_2(CO)_9$ in refluxing methylcyclohexane *(154)* (!). Cyclooctatrienone gives its $Fe_2(CO)_6$ complex, 128, in 2.4 % yield in reaction with $Fe_3(CO)_{12}$ in refluxing benzene *(272)*. Fused-ring cyclooctatrienes form $Fe_2(CO)_6$ complexes most effectively also with $Fe_2(CO)_9$ in ether *(165,162)*; the yield being 14 % of 129 from bicyclo[6.1.0]-nona-2,4,6-triene *(347)*, and 25 % of 130 from bicyclo[6.2.0]deca-2,4,6-triene *(126,134)*. Bicyclo[6.4.0]dodeca-2,4,6-triene gives only 1.4 % $Fe_2(CO)_6$ complex with $Fe_2(CO)_9$ in hexane *(128)*.

Despite the stoichiometric resemblance, 127 and its derivatives do not resemble 125 structurally. Several of the compounds have been studied by X-ray crystallography, including 127 *(129)*, 128 *(341)*, 129 *(376)*, 130 *(130,131)*, and its triethylphosphine adduct *(134)*. The Fe-Fe bond lengths observed were 2.786 Å in 130 *(131)*, and 2.804 Å in the triethylphosphine derivative of 130 *(134)*. These bond lengths are *ca*. 0.09 Å shorter that that in 125 with its bis(η^3-allyl) structure. Presumably such a structure in cyclooctatriene derivatives would require still further stretching and weakening of the Fe-Fe bond, and is therefore not observed.

The structure observed in all cases is the "skew" structure in which one iron is coordinated to an allyl unit comprising C(3) to C(5) of cyclooctatriene, and the other *via* a σ-bond to C(6) and an η^2-bond to C(1) and C(2).

The NMR spectra of the hydrocarbon complexes show, however, planes of symmetry of the organic ligand at room temperature, indicating fluxionality. In view of the stable bis(η^3-allyl)structure for 125, such a structure would appear attractive as an intermediate in the enantiomerization of 127 and its derivatives. However, a number of detailed studies by ^{13}C- and ^{1}H-NMR have shown that the two iron atoms and the carbonyl groups attached to them do not exchange with each other during the fluxional process which exchanges the two sides of the cyclooctatetraene rings. Thus, a μ-(1,6-η:2-5η) intermediate is indicated *(131,134,137,139,376)* (eq. [169]).

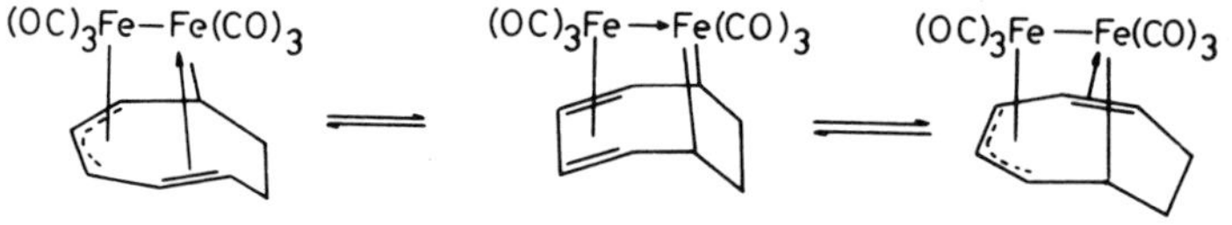

127 127′ [169]

Just such a structure had, in fact, originally been proposed for 127 *(272)*.

The activation energy for this "twitching" process in 127 is 11.6 kcal/mol *(134,137,139)*; in 129, 7.9 *(134)*; in 130, 9.0 *(131,134)*; and in the triethylphosphine derivative of 130, 8.8 *(134)*. In (benzocyclooctatetraene)hexacarbonyldiiron, the E_a is 8.1 kcal *(404)*. Clearly, annulation of the C(7)-C(8) bond of 127 lowers the fluxional barrier. Cotton *(134)* has observed that the lowered barrier correlates with decreased buckling of the eight-membered ring, presumably as a result of strain introduced into the ligand by the fused ring.

In addition to the "twitching" of the $Fe_2(CO)_6$ group in these compounds with respect to the ring, ^{13}C-NMR has also revealed, in order of increasing energy barrier, internal rotation of the CO groups on the η^3-allyl-bound iron (E_a = 8.5-11.4 kcal) and internal rotation of the CO groups on the η^1-η^2-bound iron E_a = 12.9-15.6 kcal) *(134)*.

Two unsymmetrical derivatives of 127 are known. [3,8-bis(trimethylsilyl)cyclooctatriene]hexacarbonyldiiron is said not to be fluxional based on a room-temperature NMR spectrum

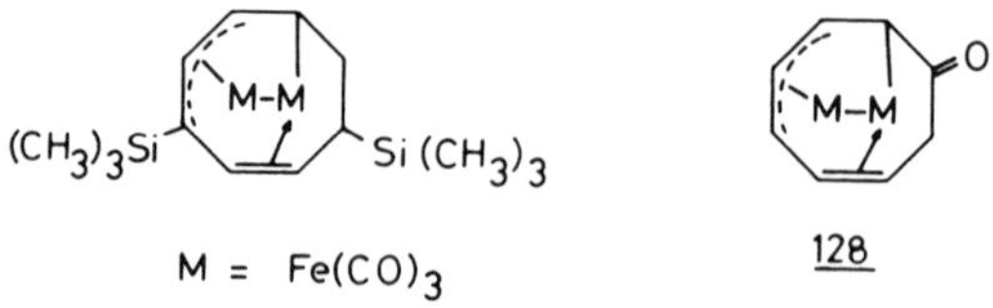

128

(154). The crystal structure of 128 *(341)* shows only one form, that with the σ-bond to C(2); stabilization of this form by hyperconjugative interaction between the Fe-C(2) σ-bond and the carbonyl group (ν(CO) at 1650 cm^{-1}) may be readily envisioned. The NMR spectrum *(341)* is consistent with the same static structure in solution at room temperature. However, variable temperature studies would be required before reaching a firm conclusion that these compounds are non-fluxional at all observable temperatures.

127 has a dipole moment of 3.66 D *(199)*. Reaction with triphenylcarbenium fluoroborate results in hydride abstraction *(272)* to give an interesting cation $[(C_8H_9)Fe_2(CO)_6]^+$ for which little structural information is currently available. Several possible instantaneous structures can be written, such as

The cation adds a variety of nucleophiles, including pyridine, triphenylphosphine, azide, cyanide, and hydride ions, to give substituted derivatives of 127 *(28)*. In all cases the nucleophile appears to add *syn* to the $Fe_2(CO)_6$ group. The cyano and azido products are rigid on the NMR time scale to 50°C. The triphenylphosphine adduct and (more readily) the pyridine adduct show fluxional behaviour qualitatively different from that of 127, involving simultaneous shift of the nucleophile and the proximate $Fe(CO)_3$ group (eq. [170]).

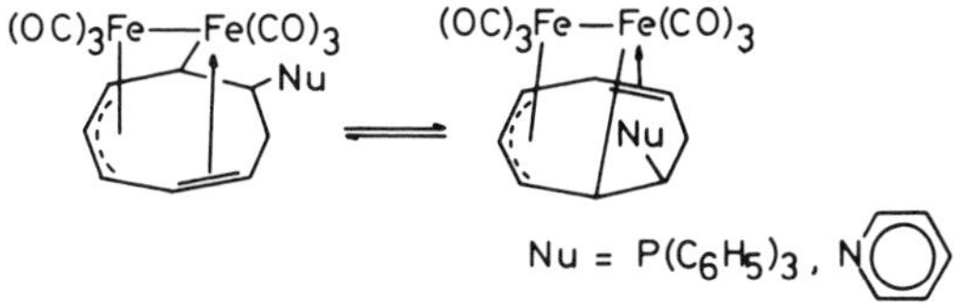

[170]

Reaction of (benzocyclooctatetraene)hexacarbonyldiiron with triphenylphosphine at 100°C *(404)* and of 130 with triethylphosphine at 65°C *(134)* gave in each case a monophosphine product in which the phosphine replaced a carbonyl group on the η^3-allyl-bound iron.

Electrochemical reduction of 127 occurs irreversibly, presumably with cleavage of the Fe-Fe bond *(170a)*. Thermolysis of 127 and its derivatives, usually at about 100°C, results in loss of one $Fe(CO)_3$ group, giving predominantly the most stable $Fe(CO)_3$ complex of the ligand. Thus, 129 forms 68 as the principal product *(160,164,347)*, along with other minor products from ligand rearrangements (eq. [171]). 130

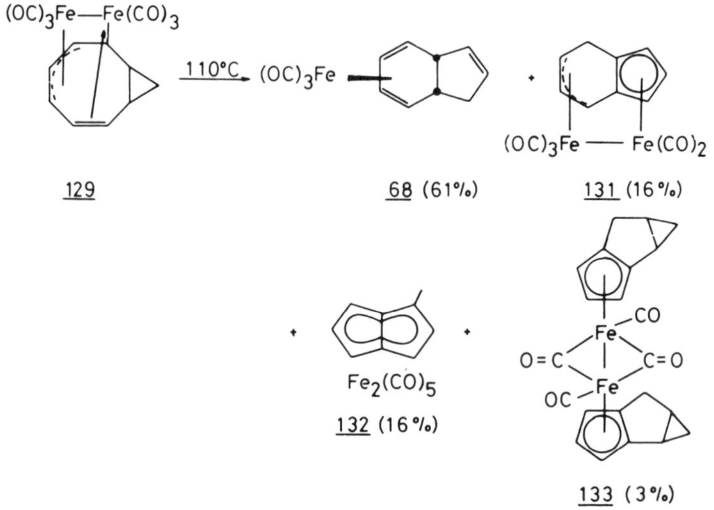

[171]

(and its analogs with fused five- and six-membered rings) forms predominantly the tricyclic $Fe(CO)_3$ complex 75 *(126,133, 160,162)*, along with a dimer analogous to 133 *(126,133)*.

Complexation of 2,6-dimethyloxepin gave, in addition to the already described products, (Sect. IV.C) a nonfluxional $(C_8H_{10}O)Fe_2(CO)_6$ complex in 1 % yield *(145,26)*. This product proved, however, not to be an oxa-analog of 125 but instead a ring-opened complex *(26)*:

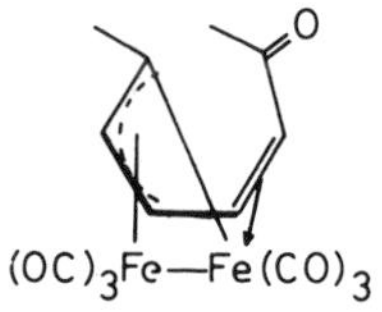

A substantial family of $Fe_2(CO)_6$ complexes, which may be viewed as bis(η^3-allyl) complexes with the allyl groups overlapping, is derived from butatrienes and their higher homologs. In the case of the relatively stable tetraphenylbutatriene the complex, 134, can be obtained directly (26 %) by complexation with $Fe(CO)_5$ at 120°C *(327,328)*. Use of a fivefold excess of $Fe_2(CO)_9$ gives the complex in 94 % yield *(260)*. The existence (see Section III) of an $Fe(CO)_4$ complex 9 with a bent backbone suggests a route for these reactions (eq. [172]). A number of tetrasubstituted butatrienes have undergone complexation using $Fe_3(CO)_{12}$, giving products analogous to 134 *(409)*. The complex 134 can also be obtained in low yield by reaction of 1,1-dibromo-2,2-diphenylethylene with $Fe_3(CO)_{12}$ at 120°C *(259)*. (Scheme XXV). An analogous reaction occurred with (2,2'-biphenyldiyl)dibromoethylene, giving

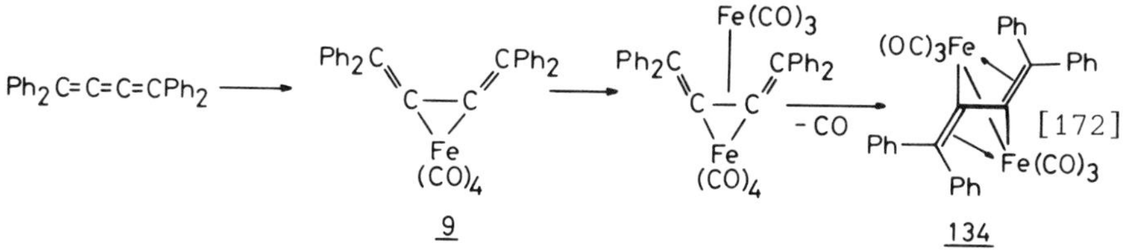

a complex like 134, the structure of which was shown by X-ray crystallography *(63)*.

Many simple butatrienes are insufficiently robust to survive direct complexation, but complexes such as the parent $(C_4H_4)Fe_2(CO)_6$, 108, have been obtained by simultaneous elimination-complexation reactions *(259,328)*. 108 itself can be obtained in 4 % yield from 1,4-dichlorobut-2-yne and $Fe_3(CO)_{12}$ at 100°C *(259)*, in 11 % yield using $K_2Fe_2(CO)_8$ in hexane/methanol *(259)*, or in 18 % yield from the dibromide with zinc and $Fe_3(CO)_{12}$ *(328)*. The tetramethylbutatriene complex is obtained in higher yields (11-36 %) by similar reactions, or directly from 2,5-dimethylhexa-2,3,4-triene (70 %) *(259)*.

The original assignment *(327,328)* of these butatriene complexes as $Fe_2(CO)_5$ complexes was quickly corrected based on mass spectroscopic results *(274,329,336)*. Mössbauer data on 108 have also been reported *(278)*.

A few examples of hexapentaene-hexacarbonyldiiron complexes have also been reported. Tetraphenylhexapentaene gives unstable *(328)* products on reaction with iron carbonyls. However, tetrakis(*tert*-butyl)hexapentaene gives, in addition to the $Fe(CO)_4$ complex, the $Fe_2(CO)_6$ complex 137 in 6 % yield using $Fe_2(CO)_9$ and in 41 % yield using $Fe_3(CO)_{12}$ at 65°C. $Fe(CO)_5$ and $(BDA)Fe(CO)_3$ failed to give isolable products *(279)*. A symmetrical structure is indicated for 137. Reac-

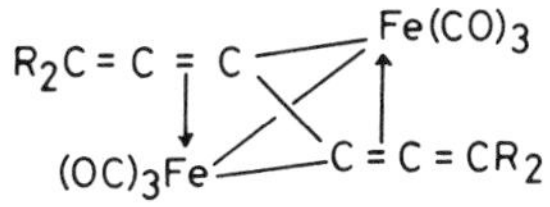

137 R = *tert*-C_4H_9

tion of 2,7-dichloroocta-3,5-diyne *(279)* with $Fe_3(CO)_{12}$ and zinc gave, in very low yield (ca 1 %), what may be a product analogous to 137. 1,6-Dichlorohexa-2,4-diyne, however, gave a $(C_6H_4)Fe_3(CO)_{7-8}$ complex, discussed in Section XIII.

Compounds 134 and 137 can be represented with each iron bound to the ligand by one σ- and one π-bond, as shown above, or with each iron bound to an η^3-allyl moiety, the two central carbons of the ligand being part of each allyl moiety. The average Fe-C distances in the biphenyldiyl analog of 134 are 2.39, 2.04, and 1.94 Å, counting from the end, and the Fe-Fe distance is as short as 2.596 Å *(60)*.

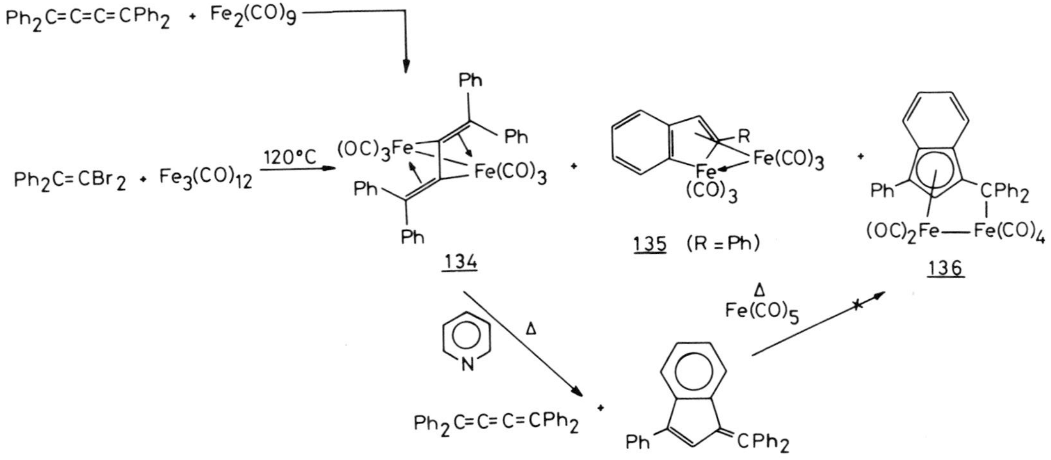

Scheme XXV: Formation and Reactions of (Tetraphenylbutatriene)hexacarbonyldiiron.

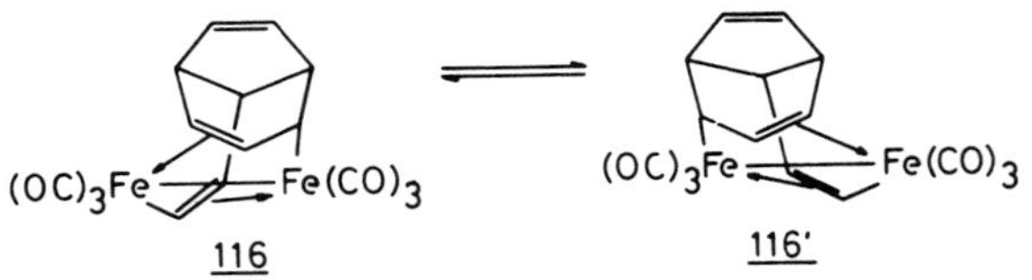

[173]

Many complexes, like 127 and its derivatives, can only be represented with one or more σ-bonds. Among these are two closely related compounds, 116 *(17,22)* formed from bullvalene and iron carbonyls (Scheme XXIV), and 107 *(361,391)* formed from the annulated dihydrobullvalene 89 and iron carbonyls. Variable temperature NMR studies of 116 have related a rapid degenerate valence isomerization which interconverts 116 and its enantiomer, with a $\Delta G^{\ddagger}$ of 16.0 kcal *(22)*. This process (eq. [173]) formally passes the vinylene group from one iron to the other. Curiously, 107 was said to be nonfluxional from its NMR spectrum, which was temperature invariant from -60°C to +100°C *(361)*. This contrast between 107 and 116 is inexplicable at present.

An extensive class of $Fe_2(CO)_6$ complexes with Fe-Fe bonds are the tricarbonylferrole-tricarbonyliron complexes, *e.g.* 117 (Scheme XXIV) and its derivatives. 117 itself was originally obtained as one of a complex mixture of products from reaction of acetylene with $Fe_3(CO)_{12}$ *(236,238)*. Derivatives related to tetraenes include a condensation product *(76)* (eq. [174]) and products from a diacetylene *(277)* (eq. [175]). The analogous reaction of diphenyldiacetylene with iron carbonyls gave related monocyclic products *(237)* (eq. [176]).

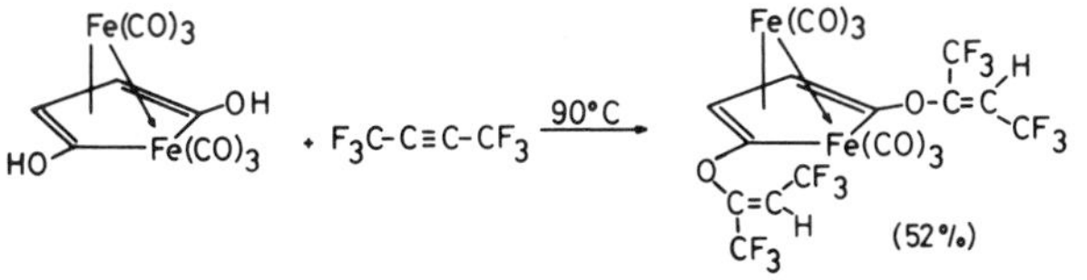

[174]

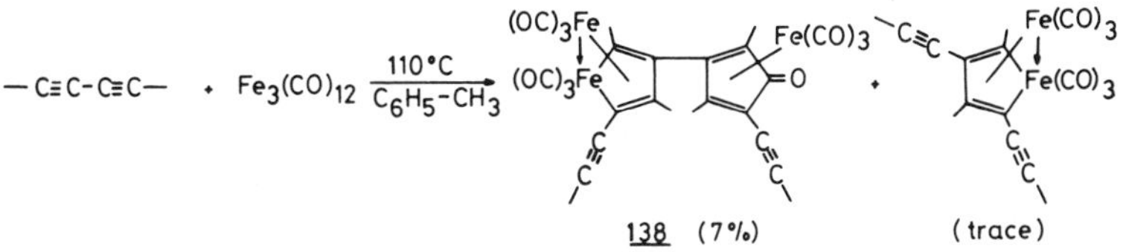

[175]

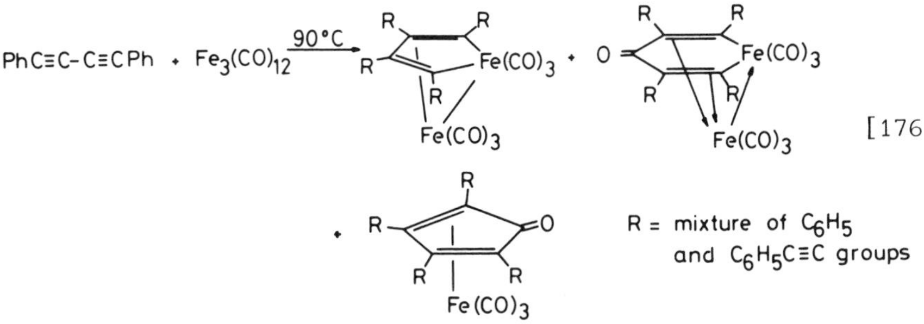

[176]

Similar products (with all R's = C_6H_5) were obtained from diphenylacetylene and various carbonyliron compounds *(234)*, even including 65 *(289,326)*.

The benzo derivatives of 117 constitute a particularly interesting class, whose formation reactions are summarized in Scheme XXVI.

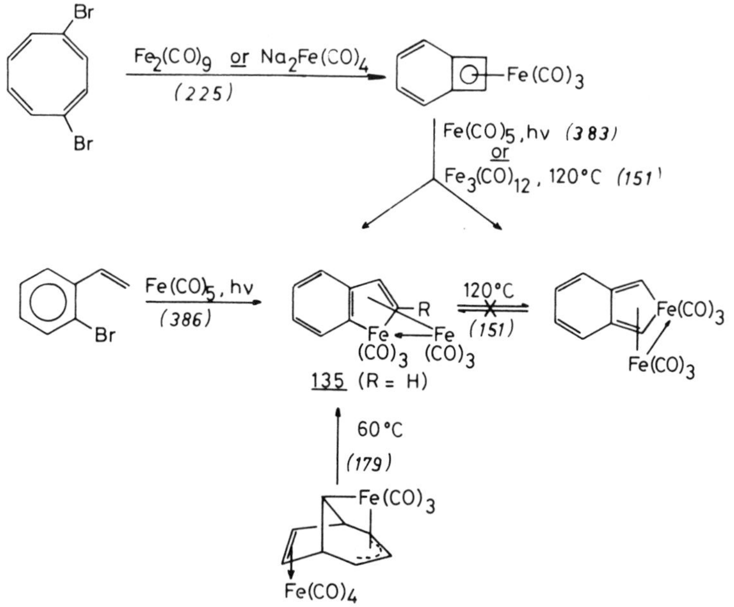

Scheme XXVI: Formation Reactions of Benzoferrole Complexes

135, R = C_6H_5, is one of several products obtained on reaction of diphenylacetylene with iron carbonyls, especially $Fe_3(CO)_{12}$ *(324,235)*. Its structure has been confirmed by X-ray crystallography *(166)*, which shows an Fe-Fe bond length of 2.520 Å and a semibridging CO. Phenylacetylene gives, in addition to the previously described products 53, 54, and 55 (Sect. IV.C), the 2,5- and 3,4-diphenyl derivatives of 117 *(61)*. The former undergoes irreversible electrochemical reduction *(169)*.

Reaction of acetylene and phenylacetylene with iron carbonyls also gives a trienic homolog of the 117 structure, of composition $(RC{\equiv}CH)_3Fe_2(CO)_6$ *(398,234)*. The structure of the product from acetylene was shown to be 139, having each iron bonded to one vinylene unit by a σ-bond and to the other by an η^2-bond *(344,386)*. The possibility of fluxionality in this and most other of these complexes has not been investigated. The Fe-Fe bond length in 139 is 2.53 Å, and the dipole moment of 139 is only 1.70 D *(234)*. On heating under CO pressure, 139 reacts to form a dimeric product *(234)* (eq. [177]).

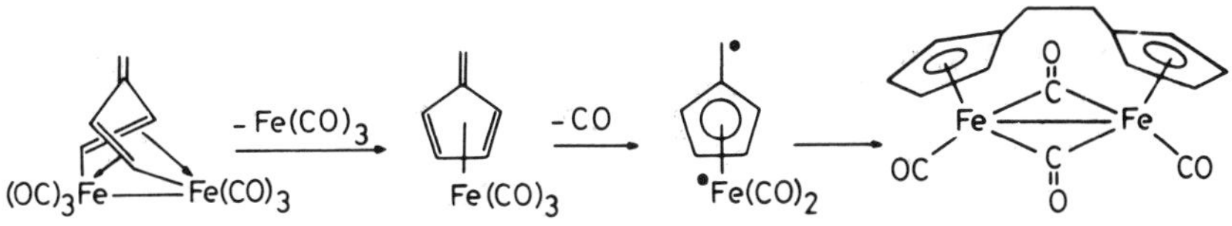

139 [177]

A heteroatom analog of 139 has been reported from reaction of a *(4H)*-diazepine with $Fe_2(CO)_9$ *(98)*. Its formation results in reduction of the weak hydrazine-like N-N bond of the ligand, and the irons are coordinated only to the nitrogens *via* n- and σ- bonds rather than to the π-bonds of the ligand *(98)* (eq. [178]). The formation of this product, in

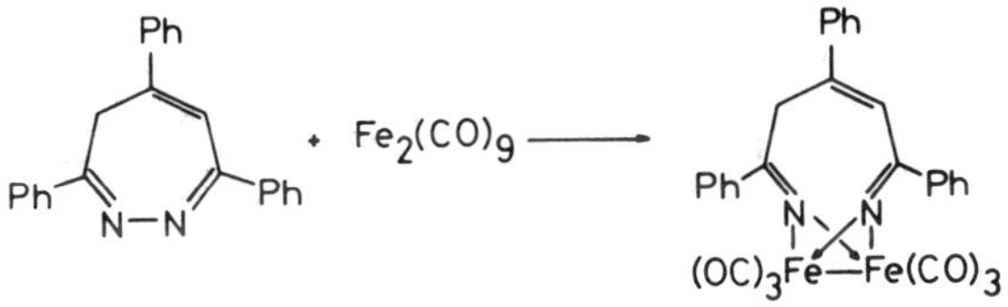

(14%) [178]

which the Fe-Fe bond length is a quite short one, 2.392 Å, contrasts dramatically with the $Fe(CO)_3$ complexes from *(1H)*-diazepines described in Sect. IV.C.

C. *UNSYMMETRICAL* $Fe_2(CO)_6$ *COMPLEXES*

A few unsymmetrical $Fe_2(CO)_6$ complexes, all derivatives of pentafulvenes *(268)*, are known. One iron atom in these complexes is coordinated to the cyclopentadienyl ring and two CO's, and the other to C(6) and four CO's. An X-ray structure *(316,382)* of the parent 140a showed an Fe-Fe bond length of 2.679 Å, the bond being somewhat compressed by the demands of the ligand.

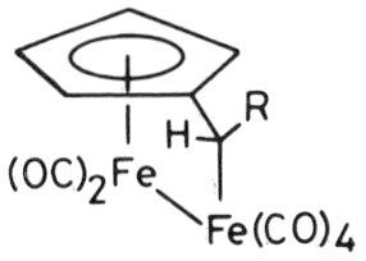

140 a (R=H)
b (R= *n*-C_3H_7)
101 (R=CH_3)

The most straightforward route to these complexes is by direct reaction of the pentafulvene with $Fe_2(CO)_9$ at moderate temperatures. A mixture of products is generally formed, including products of type 140, obtained in 1-5 % yield from 6,6-dialkylpentafulvenes and 6-methoxypentafulvene *(355,395, 397)*. An analogous product was also obtained from 2,3-dihydropentalene *(241,242)* (eq. [179]).

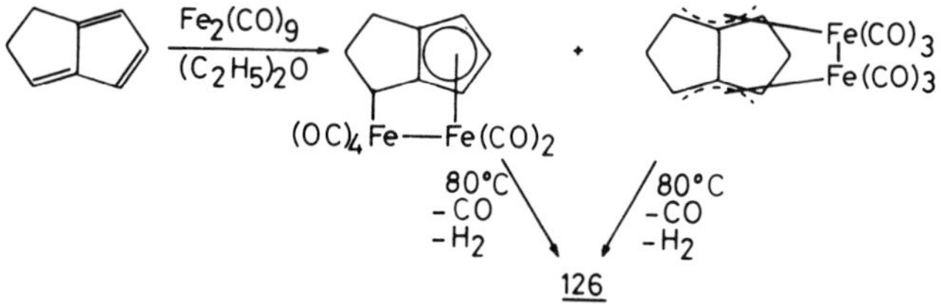

[179]

1,2,3,4-Tetraphenylpentafulvene gave the tetraphenyl derivative of 140a on reaction with $Fe_3(CO)_{12}$ at 105°C (36 %) *(397)*.

6,6-Diarylpentafulvenes do not generally give products of type 140, although a related $Fe_2(CO)_5$ complex 141 was obtained in one case *(39,397)*. 6-Propenylpentafulvene gave a 9 % yield of 140b, which requires a hydrogen source, on reaction with $Fe_2(CO)_9$ in hexane *(175a)*. The main product was a more expected $Fe_2(CO)_5$ complex discussed in Section X along with 141.

A number of derivatives of 140 have been obtained indirectly. 101 has been obtained as a minor product of photoreaction of quadricyclane with $Fe(CO)_5$ *(20)*, and on decomposition of the $Fe_2(CO)_8$ complex from spiro[4.2]heptadiene (section VIII) at room temperature *(167)*. Formation of the complex 136 by indirect means *(259)* (Scheme XXV) has also been mentioned.

The parent 140a has not been obtained from the very labile pentafulvene, but rather from reaction of acetylene at 20 atm pressure with $Fe_3(CO)_{12}$. Phenylacetylene likewise gives a triphenyl derivative of 140 (6 %), along with numerous other products *(61,234)*.

140a has a dipole moment of 3.41 D. Heating under CO pressure converts it to the dimer of (methylcyclopentadienyl)-dicarbonyliron *(234)* (eq. [180]). The triphenyl derivative

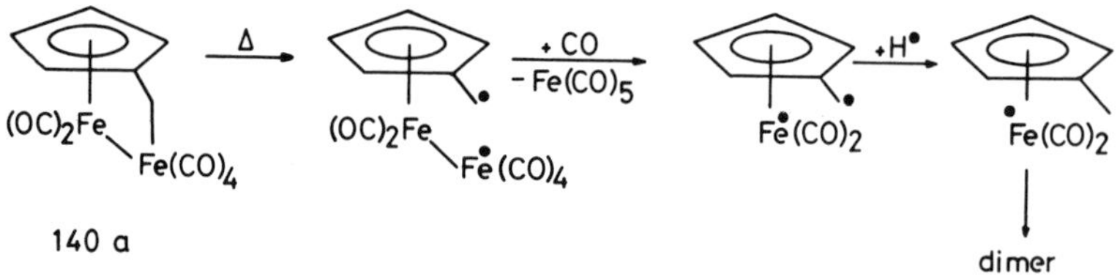

[180]

of 140 undergoes irreversible electrochemical reduction *(169)*.

X. COMPLEXES OF $Fe_2(CO)_5$

In contrast to $Fe_2(CO)_6$ complexes, nearly all $Fe_2(CO)_5$ complexes have Fe-Fe bonds. Many also have a bridging carbonyl group, a feature not seen in $Fe_2(CO)_6$ complexes, which have sufficiently high coordination numbers at each iron without it. The Fe-Fe bonded $Fe_2(CO)_5$ group requires eight additional electrons from an organic ligand to achieve saturation; the ligand must therefore have at least four (formal) double bonds. However, symmetrical coordination of two double bonds to each iron does not occur in any case; instead, unsymmetrical structures with one iron coordinated to an η^5-(cyclo)pentadienyl and the other to an η^3-allyl moiety seem to be preferred.

A. *$Fe_2(CO)_5$ COMPLEXES WITHOUT CARBONYL BRIDGES*

Most of these complexes are related to 140 through replacement of the σ-bonded $Fe(CO)_4$ group by an η^3-allyl-$Fe(CO)_3$ group. This requires a 6-vinylpentafulvene structure in the organic ligand. Indeed, 6-propenylpentafulvene does react with $Fe_2(CO)_9$ at room temperature to give as main product (32 %) the $Fe_2(CO)_5$ complex, 142, along with 9 % 140b *(175a)*.

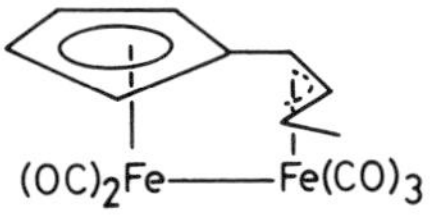

142

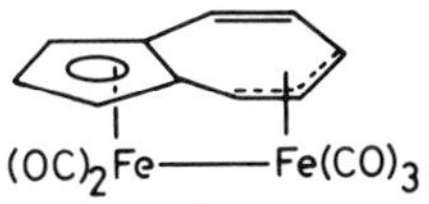

143

An analogous product, 143, had been obtained (16 %) much earlier from azulene on reaction with $Fe(CO)_5$ or $Fe_3(CO)_{12}$ at 120°C *(78,79)*. Several substituted azulenes gave analogous complexes. The high dipole moment of 143 (3.97 D) showed both irons to be on the same side of the azulene ligand *(79)*, despite a calculation supporting a *trans*-structure *(72)*. The structure was eventually shown by X-ray crystallography to have the seven-membered ring bent down toward the iron to accommodate the (slightly stretched) Fe-Fe bond (2.782 Å) *(103,104)*. The mechanism of formation of 143 from azulene parallels that occurring with simpler pentafulvenes *(105)* (eq. [181]). The free double bond of 143 resists hydrogen-

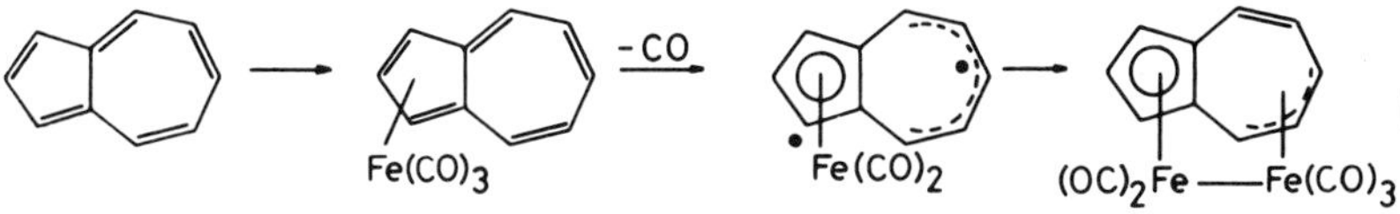

[181]

143

ation *(79)*. Electrochemical reduction has been studied *(170)*.

The compound is not fluxional at ordinary temperatures, the $Fe(CO)_3$ group remaining attached to the same three-carbon unit of the seven-membered ring, giving isomeric complexes from unsymmetrically substituted azulenes *(79)*. Cotton has found, however, that irradiation does cause interconversion of the two isomers of (guiazulene)pentacarbonyldiiron; in the presence of triethylphosphine replacement of a carbonyl group on the η^3-allyl-bound iron occurs. The facts clearly point to the formation of a symmetrical $Fe_2(CO)_4$ complex (eq. [182]).

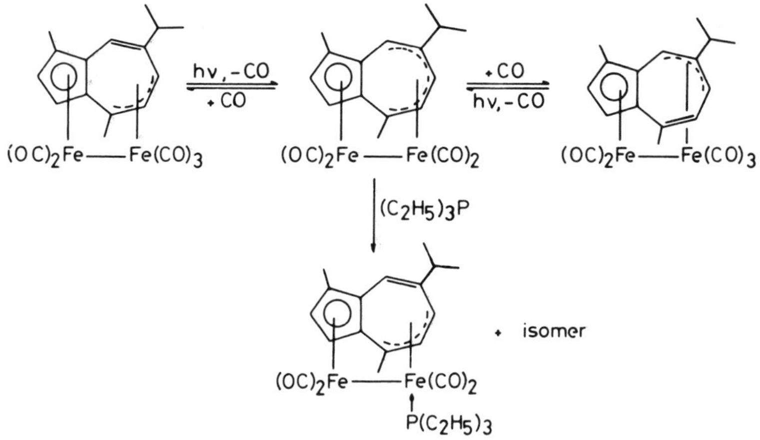

[182]

Cotton has also found that, in contrast to other dinuclear complexes, the CO groups in 143 undergo internuclear scrambling, giving a ^{13}CO singlet at 80°C. At lower temperatures the $Fe(CO)_3$ group pseudorotates independently. The exchange may involve reorganization of the ligand-metal bonding (eq. [183]), not possible in most $Fe_2(CO)_{4-6}$ complexes.

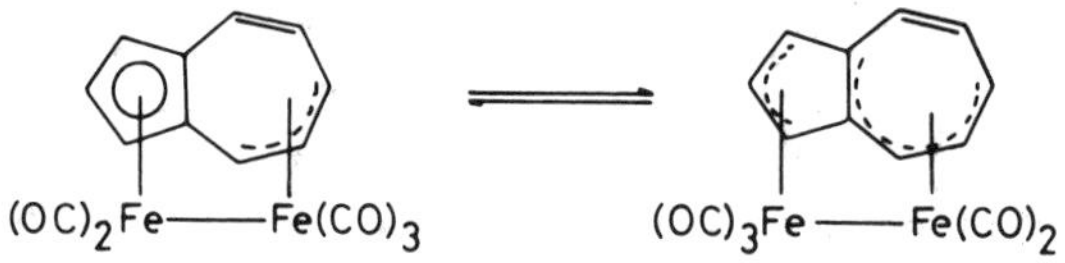

143

[183]

The compound 131 is essentially a lower homolog of 143 with a six-membered ring instead of a seven-membered ring. In addition to its indirect formation from (bicyclo[6.1.0]nonatriene)hexacarbonyldiiron, 129 *(164)*, it can be formed directly by reaction of indene with $Fe_2(CO)_9$ at 65°C *(241)* (eq. [184]).

A related $Fe_2(CO)_5$ complex, 141, is also obtained from 6,6-diphenylpentafulvene, in up to 3.5 % yield using $Fe_2(CO)_9$ in pentane *(395,397,39)*. An X-ray structure *(39)* showed that one of the phenyl groups participates in η^3-allyl bonding to

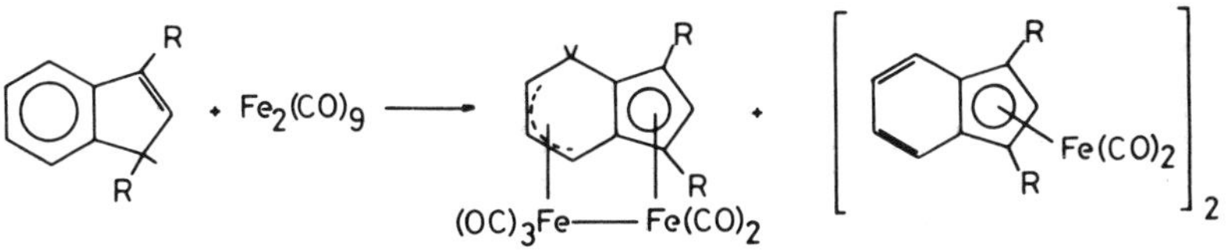

131 R = H, CH3 [184]

an iron atom. The Fe-Fe bond length is 2.765 Å. In the formation of both 131 and 141, therefore, the resonance energy of a benzene ring is substantially sacrificed, an impressive demonstration of the stability of these complexes.

Reaction of 6-(dimethylamino)pentafulvene with $Fe_2(CO)_9$ gives, in addition to a predominant $Fe_2(CO)_4$ complex, 144, about 2 % of an $Fe_2(CO)_5$ complex, 145, which results from combination of two pentafulvene units *(38)*. The Fe-Fe bond in 145 is 2.739 Å in length.

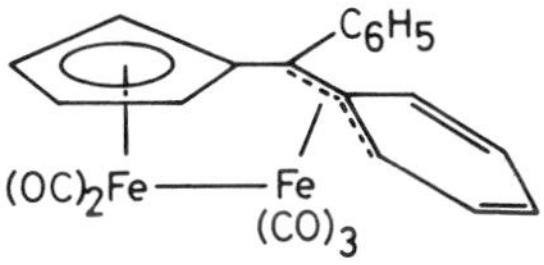

141

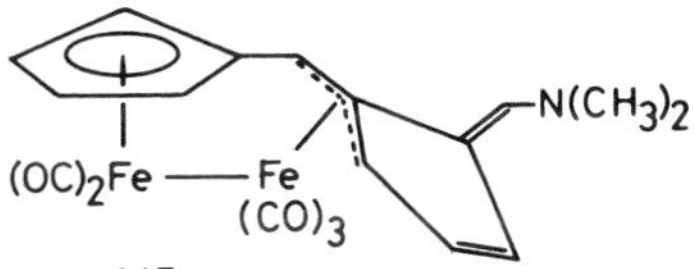

145

One example, 146, also exists in which the η^5-cyclopentadienyl ring is attached to the center carbon of an η^3-allyl unit *(114,282)* (eq. [185]). The yield was 32 % with $Fe_3(CO)_{12}$ in refluxing benzene *(274,282)*.

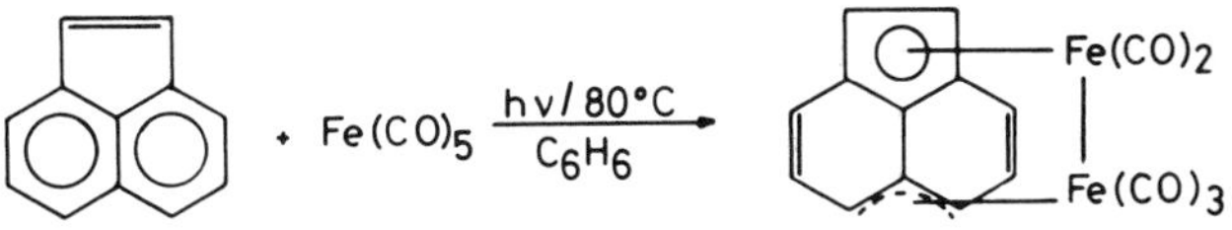

146 (18%) [185]

An alternative structure with the more customary 1'-allyl arrangement, can be envisioned, but the rigidity of the tri-

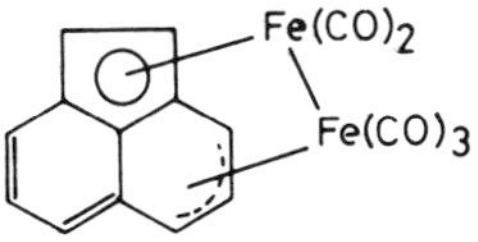

cyclic ligand apparently does not allow a reasonable Fe-Fe bond in this arrangement. The Fe-Fe bond length in 146 is a normal 2.769 Å *(114)*. ^{13}C-NMR studies of 146 have shown that, as in 127 and derivatives *(134)*, the carbonyl groups on the η^3-allyl-bound iron exchange internally. They do not exchange with the carbonyl groups of the $Fe(CO)_2$ group, which are rigid. As with most compounds with Fe-Fe bonds, 146 undergoes irreversible electrochemical reduction *(170a)*.

A compound related to the pentafulvene-derived $Fe_2(CO)_5$ complexes has been obtained in trace amounts on reaction of 1,3,5,7-tetramethylcyclooctatetraene with $Fe_2(CO)_9$ or $Fe_3(CO)_{12}$ at 100°C *(140)* (eq. [186]). X-ray crystallography

[186]

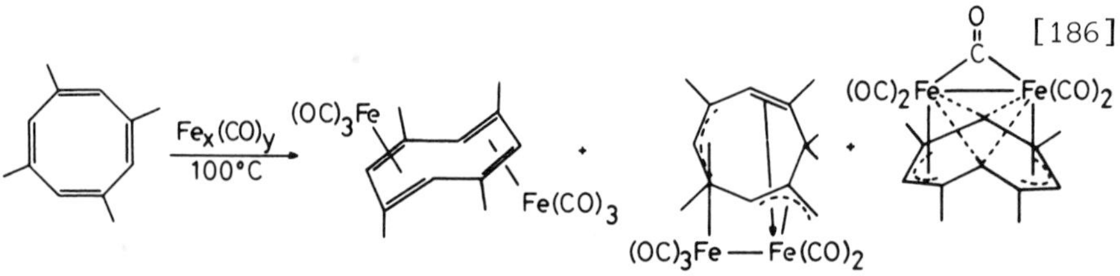

(3%) 147 (trace) 148 (6%)

revealed a structure, 147, in which the cyclopentadienyl ring of 142 to 146 was replaced by a double bond and an allyl unit (generated by a 1,3-hydrogen shift) *(142)*. The Fe-Fe bond length in 147 was a normal 2.766 Å.

In perhaps the strangest $Fe_2(CO)_5$ complex, the major product from reaction of phenylacetylene (and several other acetylenes) with $Fe_3(CO)_{12}$ at 50-70°C *(61,234,235)*, one iron coordinates an allyl unit and a double bond, as in 147, while the other forms three sigma bonds to carbon atoms of the bicyclic ligand assembled from three acetylene molecules. An X-ray study of the product from phenylacetylene was needed to determine the structure 55 *(270)*. The Fe-Fe bond length is a

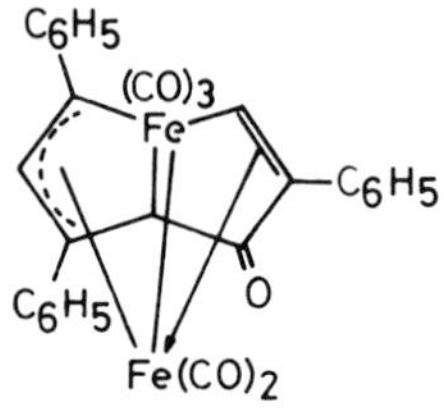

55

short 2.50 Å. As already noted, this compound is converted to triphenyltropone-tricarbonyliron complexes, 53 and 54, on heating in benzene or chromatography on alumina. At its melting point, it decomposes to 1,3,5-triphenylbenzene *(61,234)*. The analogous product from 3-hexyne has a dipole moment of 3.13 D *(234)*.

B. *$Fe_2(CO)_5$ COMPLEXES WITH CARBONYL BRIDGES*

As previously noted, (cyclooctatetraene)pentacarbonyldiiron, 124, forms spontaneously from the *cis*-hexacarbonyldiiron complexes 122 and 123 on standing in solution *(266)*. It is not clear whether this can be reversed at high CO pressures. 124 is obtained as a by-product of complexation of cyclooctatetraene, but is best obtained by photolysis of 65 or 121 in hexane solution, which cleanly leads to its formation *(365)*.

The tetramethyl analog, 148, is produced directly on complexation of tetramethylcyclooctatetraene at 100°C (6 %), though not at 65°C *(140)*. Heating (trimethylsilyl-cyclooctatetraene)tricarbonyliron to 150°C in solution results in formation of some $Fe_2(CO)_5$ complex in addition to bicyclic $Fe(CO)_3$ complex *(117)*. A novel Fe-Ru analog of 124, 149, has also been synthesized by reaction of 65 with $Ru_3(CO)_{12}$ in refluxing xylene *(1)*.

124 was originally believed to have each iron coordinated to a diene unit, with rapid shifts about the ring to account for the single resonance in the ^{1}H-NMR spectrum *(266)*. The crystal structure *(201)* showed, however, a static structure which can be conceptualized as having each iron bound to an η^3-allyl unit, and sharing the remaining two carbons in a delocalized bonding scheme. The Fe-Fe bond distance was

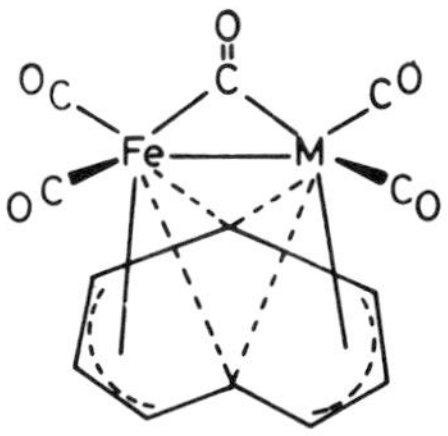

124 (M = Fe)
149 (M = Ru)

2.742 Å, the Fe-η^3-allyl carbon distances 2.11-2.14 Å, and the Fe-bridging carbon distances 2.49 Å. The tetramethyl analog 148 showed a similar structure, with the methyl groups in the terminal positions of the η^3-allyl moieties *(138)*.

The delocalized bonding between the two irons and the bridging CH groups of the ring has been interpreted alternatively as involving two two-electron three-center (Fe-C-Fe) bonds *(201)* or "one fairly stable four-center MO" plus two electrons in "some kind of relatively nonbonding orbital" *(138)*.

124, 148, and 149 are all fluxional, and low-temperature limiting spectra have not in fact been obtained. Broad-line NMR studies of 124 in the crystal in fact indicated rapid motion of the ring even at 77 K *(85,86)*. More recently, T_1 measurements at 77 K indicated a barrier of only 2.0 kcal to rotation of the cyclooctatetraene ring relative to the $Fe_2(CO)_5$ unit *(84)*, making 124 the most fluxional molecule in organometallic chemistry.

A closely related family of CO-bridged $Fe_2(CO)_5$ complexes is the pentalene complexes, such as 126. 126 results from complexation of pentalene dimer *(394)* or 1,2-dihydropentalene *(241,242)* with $Fe_2(CO)_9$ (eq. [187]). Similar reactions gave

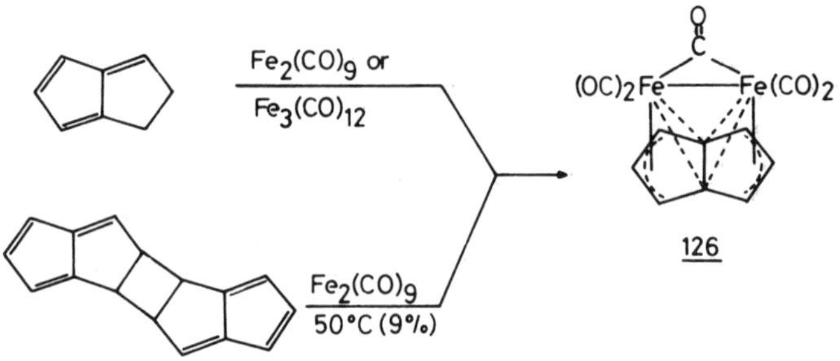

126

[187]

the 1,3-dimethylpentalene complex from its dimer (21 %) *(394)*, and the 1-phenyl *(240)* and 1-dimethylamino *(241)* compounds from the dihydropentalenes and $Fe(CO)_5$ at 105°C (12 %, 11 %). 126 also results from decomposition of the dihydropentalene-hexacarbonyldiiron complexes discussed in Section IX.C, and (1-methylpentalene)pentacarbonyldiiron, 132, is among the products of thermolysis of 129 at 110°C *(160,164,347)*.

The ease with which 126 and derivatives form in these reactions, to the exclusion of other carbonyliron derivatives, suggests substantial thermodynamic stability for this type of complex. They exhibit no fluxionality.

C. *$Fe_2(CO)_5$ COMPLEXES WITHOUT Fe-Fe BONDS*

The only example of this sort is a complex having independent $Fe(CO)_3$ and $Fe(CO)_2$ groups. The relative rarity of polyene-derived compounds with $Fe(CO)_2$ groups (Section V) makes it not surprising that few fall into the current class.

The only one in fact is derived from 6,6-methylpentafulvene on complexation with $Fe_2(CO)_9$ *(395,397)*. It accompanies a bis(pentafulvene)dicarbonyliron complex and presumably is formed from it by complexation of the diene unit *(268)*. A structure proposed on mechanistic grounds *(268)* for bis(6,6-dimethylpentafulvene)pentacarbonyldiiron was found to be correct by X-ray crystallography *(39,355)*:

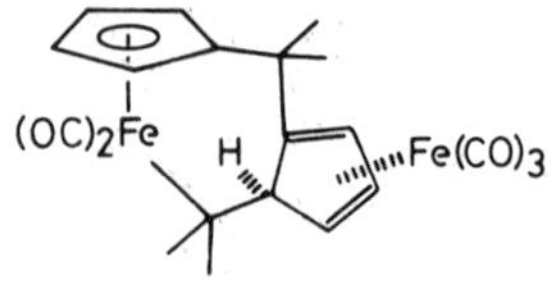

XI. COMPLEXES OF $Fe_2(CO)_4$

In Fe-Fe bonded $Fe_2(CO)_6$ complexes, bridging carbonyl groups were never observed. Many Fe-Fe bonded $Fe_2(CO)_5$ complexes had one carbonyl bridge, but not all. In $Fe_2(CO)_4$ complexes, both Fe-Fe bonds and two bridging carbonyl groups are invariably observed, the bridging evidently serving to complete the coordination spheres of the two iron atoms.

The $Fe_2(CO)_4$ group requires ten additional electrons to achieve inert gas configuration of the iron atoms. This is generally achieved by means of a cyclopentadienyl group coordinated to each iron, giving derivatives of the well-known (cyclopentadienyl)dicarbonyliron dimer.

These products are obtained upon reaction of pentafulvenes with iron carbonyls, especially $Fe(CO)_5$, at high temperatures *(296,395,397)* (eq. [188]). The hydrogens added to the

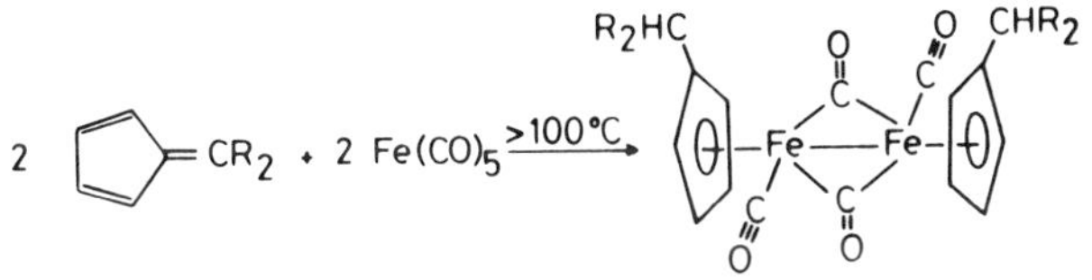

[188]

C(6) of the fulvenes are presumably derived from the solvent in a free-radical process *(268)*. The reaction appears to be quite general for a large variety of fulvenes, with yields maximized in refluxing diglyme (a good hydrogen donor) solution *(296)*. A few pentafulvenes give slightly different results. 6,6-Diethylpentafulvene gives some product resulting from sidechain hydrogen loss *(39)* (eq. [189]). 6-Dimethyl-

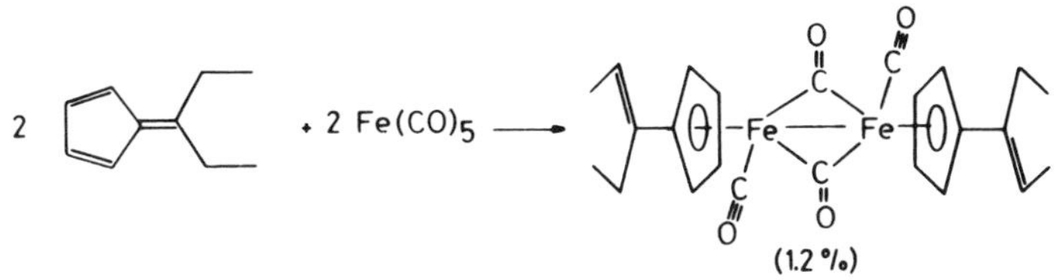

[189]

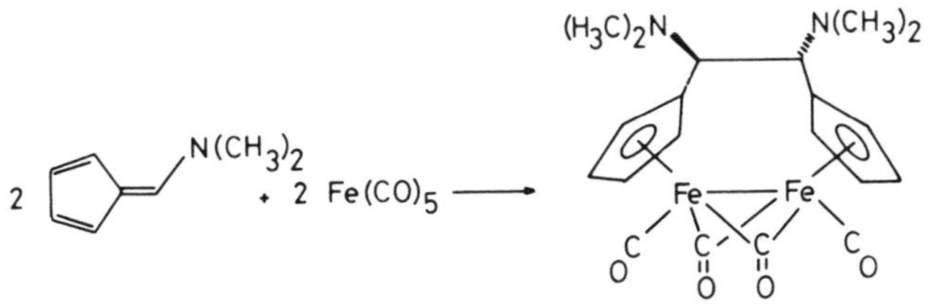

[190]

144 (50%, *cis* and *trans*)

amino-pentafulvene gives only the "linked dimers", 144, having a two-carbon bridge between the two cyclopentadienyl rings *(276,296)* (eq. [190]). The nature of the product was confirmed by Mössbauer spectroscopy *(278)* and X-ray crystallographic study of the principal *(trans)* isomer *(297)*. An Fe-Fe bond length of 2.510 Å was indicated. The reported formation of a small amount of a similar product from 6,6-dimethylpentafulvene *(397)* could not be repeated *(297)*. The "linked dimers" 144 from 6-dimethylamino-pentafulvene presumably result because the intermediate free radical is too well stabilized by the dimethylamino group to abstract hydrogen from the solvent.

An analogous bridged $Fe_2(CO)_4$ complex, 150, has been obtained on complexation of azulene under unclear conditions *(274)*. A bis-$Fe(CO)_3$ complex of 150 can also be obtained using excess metal carbonyl reagent *(79,105-107)* (eq. [191]).

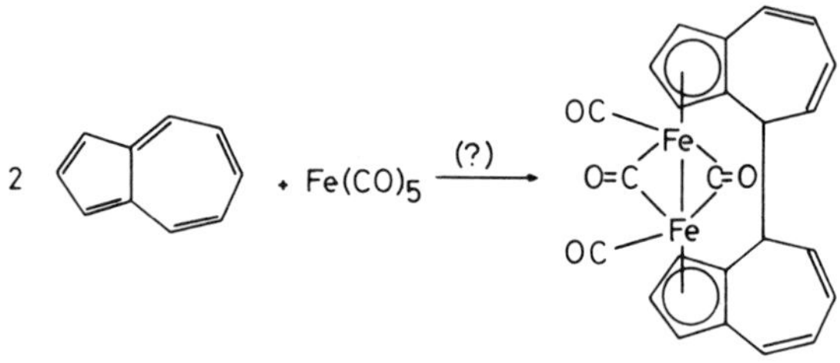

[191]

150

Formation of the unsubstituted "linked dimer" on thermal decomposition of 139 has already been mentioned (Sect. IX.B), as has the formation of (methylcyclopentadienyl)dicarbonyliron dimer from 140a *(234)*.

Complexation of two non-fulvenoid "tetraenes" has also given rise to substituted cyclopentadienyl-dicarbonyliron dimers (eqs. [192] *(317)* and [193] *(291)*). The former reaction may well proceed through the sequence of eq. [194].

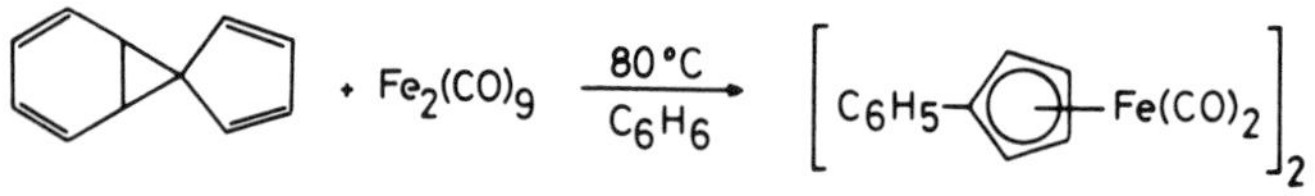

[192]

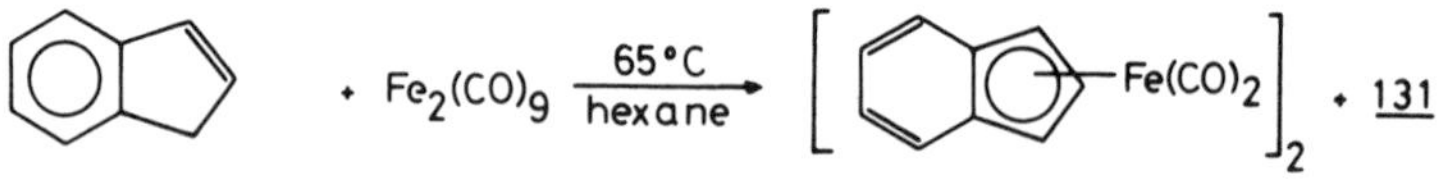
+ Fe2(CO)9
65 °C
hexane
Fe(CO)2
2
+ 131
[193]

Formation of 133 *(162,164)* from thermolysis of 129 at *ca.* 110°C, and similar reaction of 130 *(126,132,143)*, have also been reported. The structure of the cyclobutane dimer from

130 was shown by X-ray crystallography *(133)*.

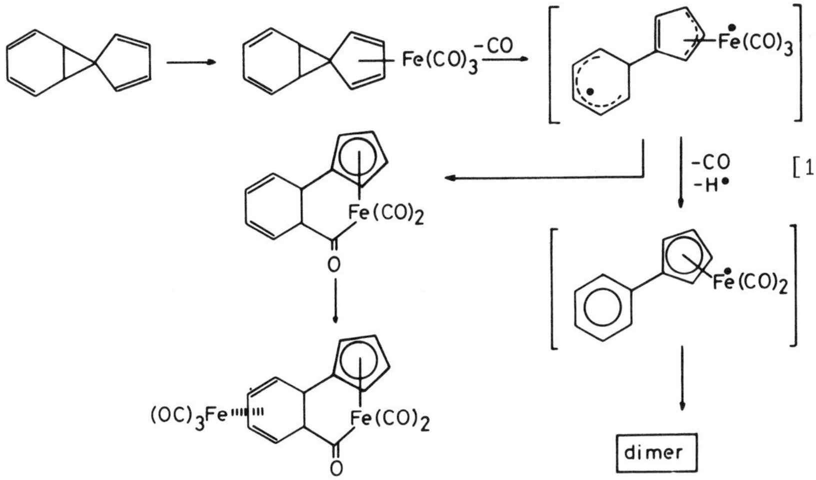

[194]

A dimeric $Fe_2(CO)_4$ complex reportedly formed on thermolysis of an $Fe_2(CO)_7$ complex *(25,306)* is presumably related (eq. [195]).

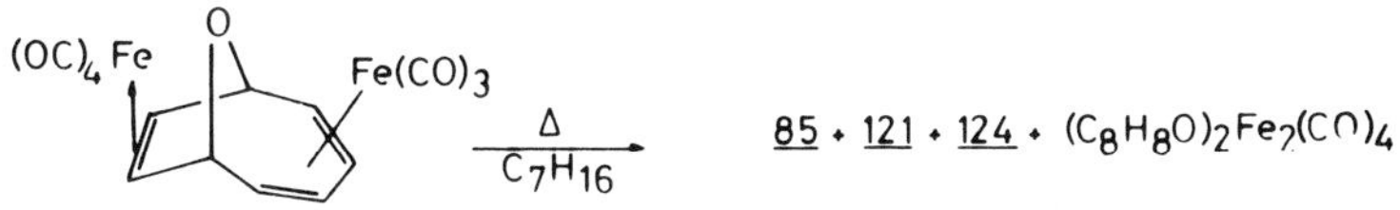

$$\underline{85} + \underline{121} + \underline{124} + (C_8H_8O)_2Fe_2(CO)_4 \qquad [195]$$

The only triene-derived $Fe_2(CO)_4$ complex not having a cyclopentadienyl ring is a reduction product obtained from (cycloheptadienyl)tricarbonyliron cation on reaction with dicyclohexylethylamine in refluxing acetone (eq. [196]),

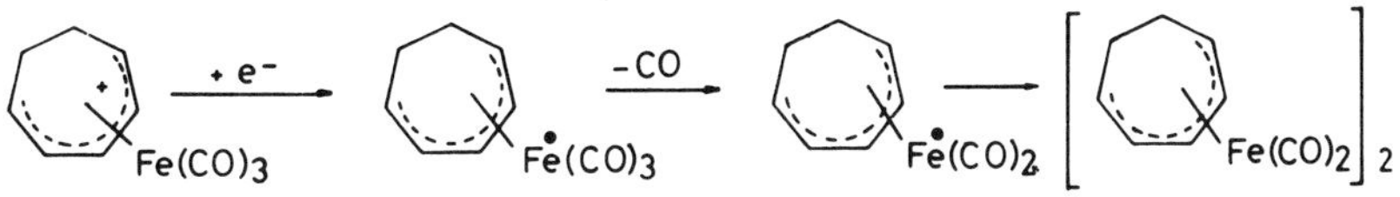

[196]

accompanying the expected main product, $\underline{44}$ *(101)*. The nature of the reducing agent is not perfectly clear; however, these reduction reactions seem to occur only when both acetone and a base are present, implicating the conjugate base of acetone.

XII. COMPLEXES WITH TWO IRON ATOMS WITHOUT ANCILLARY LIGANDS

A modest number of such complexes, all ferrocene derivatives formally without Fe-Fe bonds, have been reported. An early example is the fused-ring pentafulvene complex in equation [197] *(83)*. The nature of the process(es) leading to the product remains somewhat unclear.

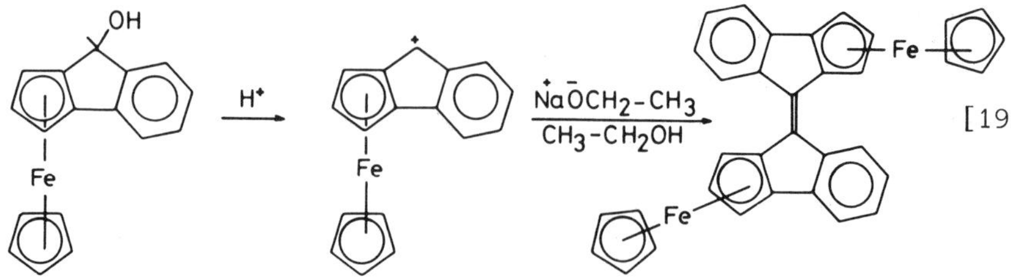

[197]

In bis(fulvalene)diiron, 151, the iron atoms are, of course, constrained to a much more proximate relationship. 151 can be obtained by pyrolysis of poly(1,1'-mercuriferrocenylene) *(345)* or Ullmann-type coupling of 1,1'-diiodoferrocene *(226)* (eq. [198]). The crystal structure of 151 *(113)*

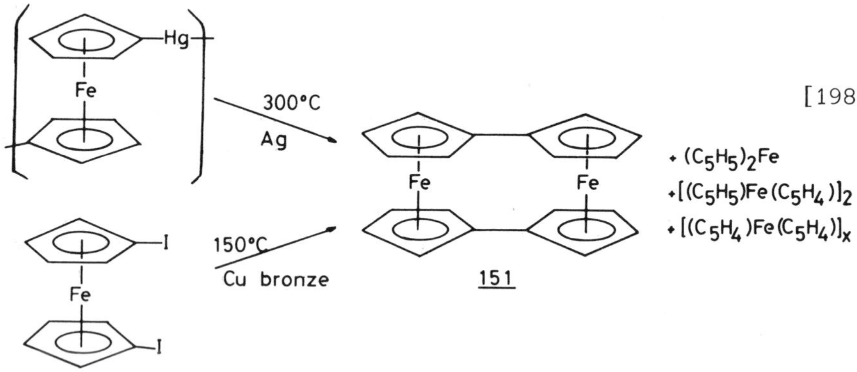

[198]

reveals that the iron atoms are not located between the centers of the rings, but are instead closer to the outer edges, indicating repulsive interaction between the formally nonbonded iron atoms. The Fe-Fe distance is 3.984 Å.

Bis(*as*-indacene)diiron, 152, is a related compound, which can be obtained upon reaction of *as*-indacene dianion with ferrous chloride *(262,265)* (eq. [199]). Again, the

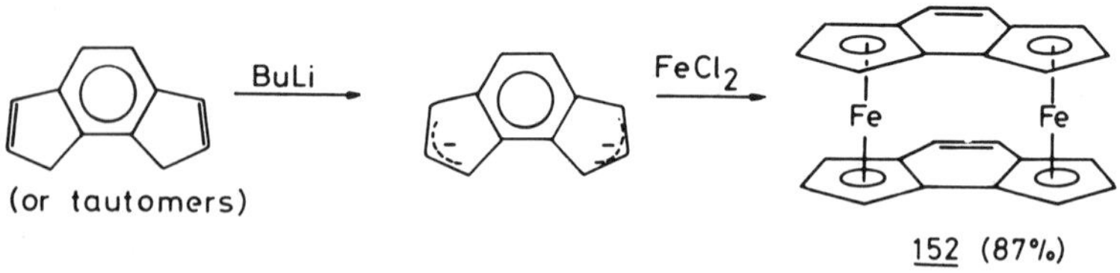

[199]

locations of the iron atoms imply a repulsive interaction, with a resultant Fe-Fe distance of 3.887 Å *(211)*.

XIII. COMPLEXES OF $Fe_3(CO)_x$ AND $Fe_4(CO)_x$

No Fe cluster complexes of trienes or tetraenes are known. All complexes with three or more iron groups have one or all of the irons independent of the rest. Previously described examples include 107 *(361,362,391)* and 138 *(277)*, each of which has one independent $Fe(CO)_3$ group. Another such case is the $Fe(CO)_3$ complexes of benzoferroles *(383,386)* (Scheme XXVI) formed as shown in equation [200].

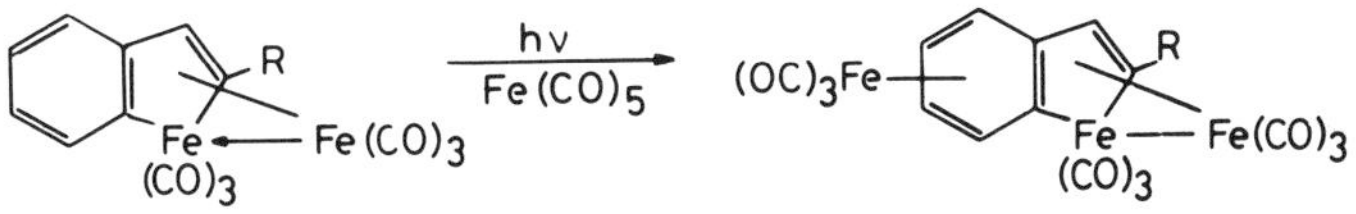

[200]

135 (R = H)

A more interesting case is a compound $(C_7H_8)_2Fe_3(CO)_9$ reported formed on reaction of cycloheptatriene with $Fe(CO)_5$ at 135°C for seven days *(80)*, but not subsequently obtained (under milder conditions) by other workers. The structure shown below was suggested:

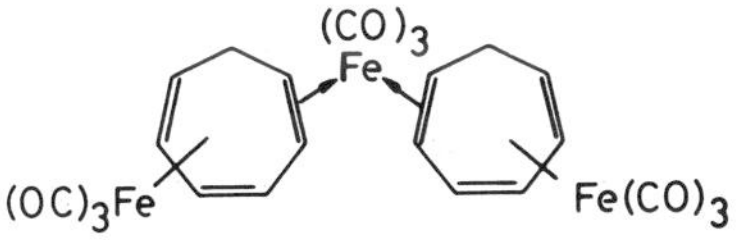

An analogous complex from bi(cycloheptatrienyl) was also mentioned.

Another incompletely characterized Fe_3 complex was obtained on reaction of 1,6-dichlorohexa-2,4-diyne with Zn and $Fe_3(CO)_{12}$ at 80-100°C *(321)*. The product had composition $(C_6H_4)Fe_3(CO)_{7\ \mathrm{or}\ 8}$ and showed two resonances in the ^{1}H-NMR spectrum. The mass spectrum showed no parent ion, but did show an Fe_3^+ peak at m/e 168 *(336)*. Structural investigation of this intriguing material would seem to be justified.

The only well characterized Fe_4 polyene complex is the bis[$Fe(CO)_3$] complex of 150, obtained from azulene and excess metal carbonyl *(79,105)*. An X-ray structure revealed an Fe-Fe bond length of 2.508 Å and an azulene-azulene C-C bond length of 1.58 Å *(106,107)*.

Tetrakis- and pentakis[$Fe(CO)_3$] complexes from β-carotene and lycopene have also been reported *(246)*.

Acknowledgments. I gratefully acknowledge interesting discussions with Profs. Maurice Brookhart and R. Bruce King, and provision of preprints by them and by Prof. Joseph Landesberg.

ADDENDUM, (December 1979)

Over 100 additional papers falling within the scope of this article have appeared in the three years since it was completed. This addendum will attempt to describe some of the highlights of this work.

Quantitative data useful in interpreting chemical properties of organometallics have begun to appear. The heat of formation of (cyclooctatetraene)tricarbonyliron, 65, indicates a dieneiron bond strength of 43 ± 4 kcal, essentially the same as for other dienes *(437)*. Photoelectron spectra of 65 and (cycloheptatriene)tricarbonyliron, 44, have also been measured and interpreted *(462)*.

Studies on fluxionality have continued *(426)*. 44 undergoes racemization by net 1,3-shift of the $Fe(CO)_3$ group with a $\Delta G^{\ddagger}$ of 22.3 kcal/mol, as shown by spin saturation transfer measurements at 60-90° *(482,490)*. A norcaradiene intermediate (path a in Scheme IV) seems best to fit the facts. (Tropone)-tricarbonyliron, 52, was partially resolved by photolysis with circulary polarized light; the rotation of the resulting solution was stable *(490)*, but the barrier to racemization remains unknown. One would expect it to be quite high if a complexed norcaradienone intermediate were obligatory; an oxyallyl intermediate, 153, may be more attractive:

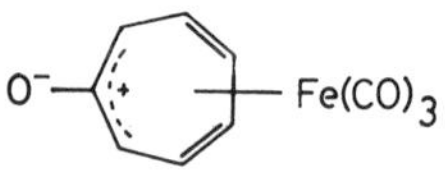

153

A detailed NMR study of fluxionality in bis(cyclooctatetraene)iron, 92, has also appeared *(491)*.

Fluxionality in $Fe_2(CO)_6$ complexes has also received continued attention, with publication of NMR studies of 129 *(445)*, and of 114 and related compounds derived from 89 *(414)*. The complexity of Scheme XXIV has increased with the report of further interconversions *(414)*: [M = $Fe(CO)_3$]

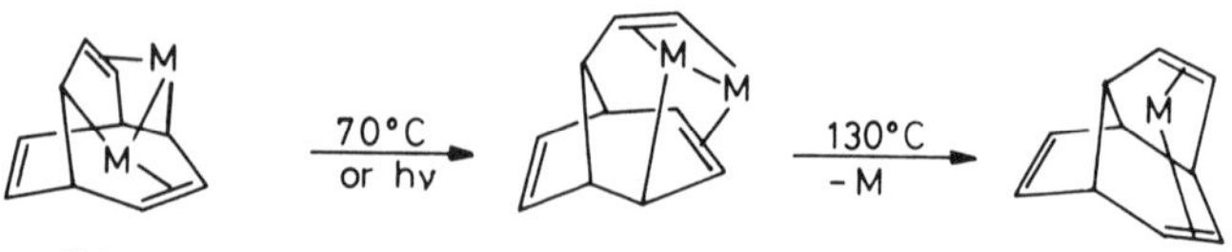

[201]

114 116

Details of X-ray and NMR studies of the azulene-$Fe_2(CO)_5$ complexes, 143 and derivatives *(439,440)*, and of the closely related 131 (R = H) *(438)*, have appeared. Internuclear CO scrambling appears in the latter with coalescence temperature of 120°C, probably *via* a structural rearrangement:

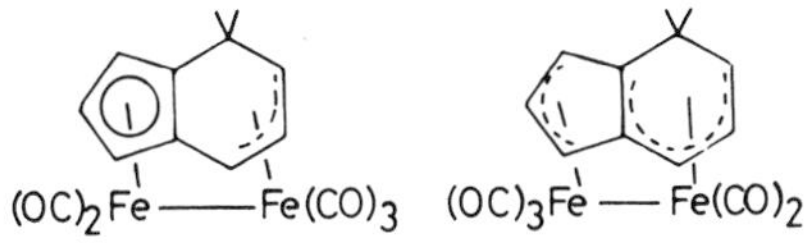

131

It remains unclear why 146 does not behave similarly.

Many *studies of electrophilic attacks* on triene-$Fe(CO)_3$ complexes have appeared. Protonation of 44 occurs exclusively at the 6-*exo* position *(427)*, as previously inferred. O-Protonation of 52 at -78°C has been observed directly, with rearrangement to the more stable C-protonated 56 occurring at 0°C *(428)*, consistent with route b of Scheme VI. Protonation of the anions from (acylcycloheptatriene)tricarbonyliron complexes (Scheme V) gives complexes of 2-acylcycloheptatrienes 155 *(442)* (kinetic control?). Protonation of the latter is said to occur to form the 2-acylcycloheptadienyl cation, 156, based upon NMR. Both this structure and the mechanism of the reaction, which on the surface requires protonation at a coordinated carbon, bear further investigation:

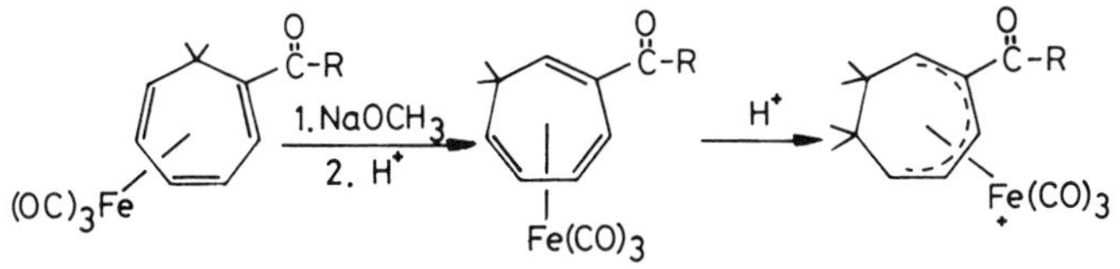

[202]

154 155 156

R = H, CH_3, OCH_2CH_3

Acetylation of 52 and diazoalkane addition to free double bond of the resulting 7-acetyl derivative have initiated a synthesis of β-thujaplicin *(454)*. The cation resulting from acetylation of 65 (Scheme XVI) has been isolated as a PF_6 salt; X-ray crystallography revealed the bicyclo[3.2.1] structure 157 rather than the expected bicyclo[5.1.0] structure *(432)*. This reopens the question of the structures of other species resulting from attack of electrophiles on 65; 86 at least appears secure, however *(432)*.

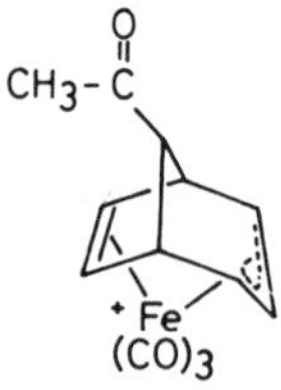

157

Electrophilic cycloadditions to $Fe(CO)_3$ complexes continue to receive attention. In addition to a full description of the reaction of 44 with hexafluorobut-2-yne *(456)*, reaction with diphenylketene has been found to occur by the route *(457)*:

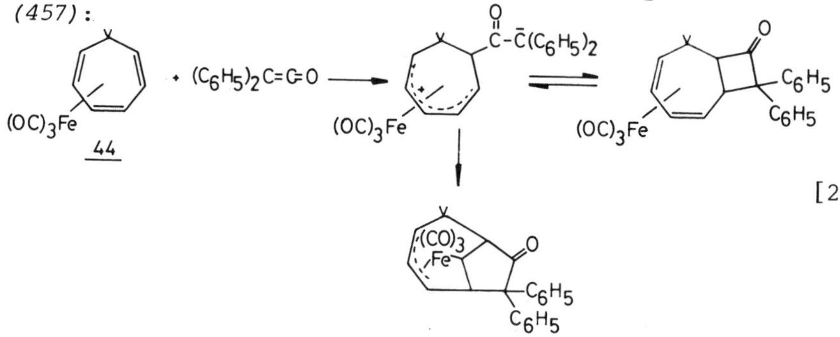

[203]

Sulphur dioxide gives a 1:1 adduct with 44 by analogous attack *(442)*. Both acetyl derivatives of 44, 154 and 155 (R = CH_3), react with TCNE as expected for electrophilic attack *(442)*. 154 (R = H) reacts by attack at a coordinated carbon or by prior isomerization, in contrast *(459)*.

Electrophilic cycloadditions of hexafluoroacetone *(463)*, phenyltriazoline dione *(412,459)* (PTD), and TCNE *(458,463)* to the tropone complex 52 have been studied. In the latter case, deuterium labeling clearly establishes the path:

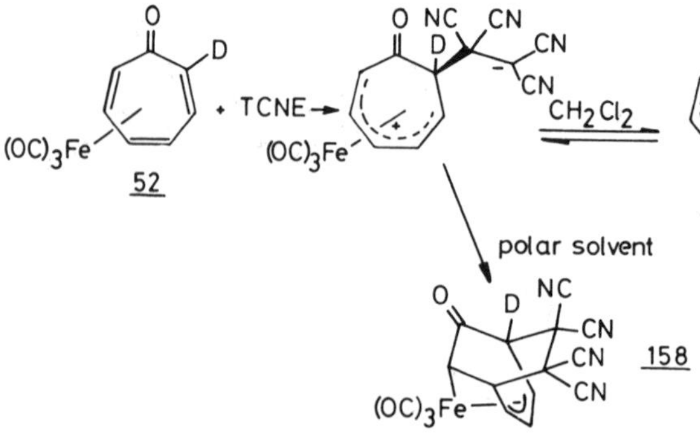

[204]

An ingenious proposal that these reactions may be treated as concerted reactions using orbital symmetry principles *(463)* has been criticized as depending upon seemingly arbitrary choice of initial contributing structure *(488)*, with the cycloadditions of TCNE *(460,461,488)* and PTD *(461)* to the heptafulvene complexes 57 as cases in point:

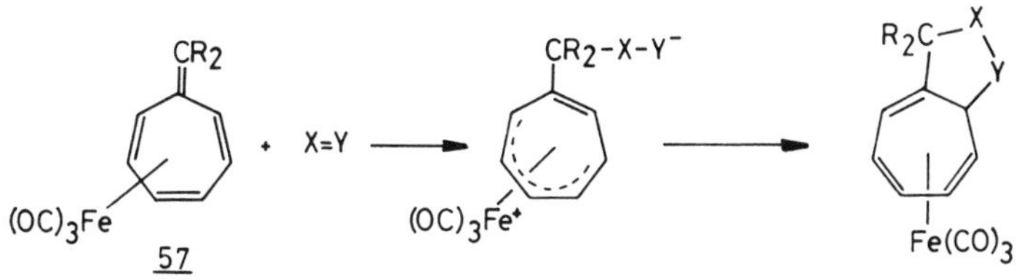

[205]

Cycloadditions of TCNE and hexafluoroacetone to the azepine complex 60 have also been further studied *(463)*, with results still explicable by Scheme VIII. (Cyclooctatrienone)tricarbonyliron, 79, adds TCNE in the expected manner *(459)*. But reaction of PTD with 65 resulted in some new complications (*cf.* Scheme XVII) *(496)*.

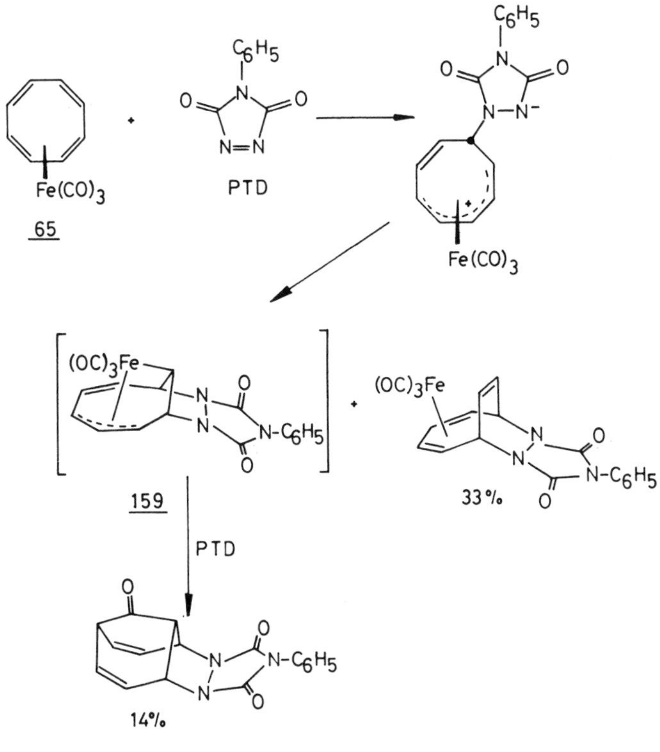

[206]

It was suggested that the adducts of 65 with hexafluoroacetone and $(CF_3)_2C{=}C(CN)_2$ may have structures analogous to 159. In a possibly related reaction, one-electron oxidants convert 44 and 65 to transient radical cations, which abstract H• (44) or dimerize (65) *(435,436)*.

Many reactions resulting from the diene-tricarbonyliron group's ability to stabilize an adjacent carbanion by bond breaking to form an allyl-$Fe(CO)_3$ anion have been reported, particularly in derivatives of 44 *(419,427,442,452,498)*. The η^3- structure of the anion, 48, from deprotonation of 44 has been supported by theory *(469)* and crystallographically confirmed *(503)*. A presumably analogous anion can be formed from a bicyclic analog of 44 *(417)*:

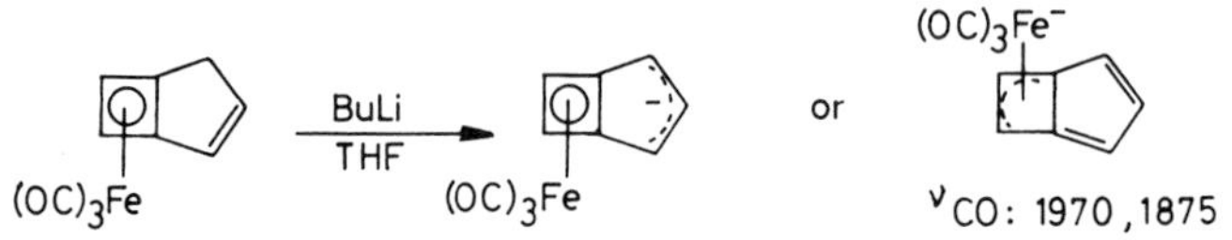

[207]

44 itself shows a molecular radical-anion in its negative-ion mass spectrum, in contrast to diene complexes *(423)*.

Kinetic studies on reactions of $(BDA)Fe(CO)_3$ with polyenes have provided further support for the mechanism shown in Scheme IX *(429)*, and use of phosphite-substituted analogs has been recommended *(477)*.

New complexes of polyenes have appeared with regularity. Some of the more interesting examples include:

Section III:

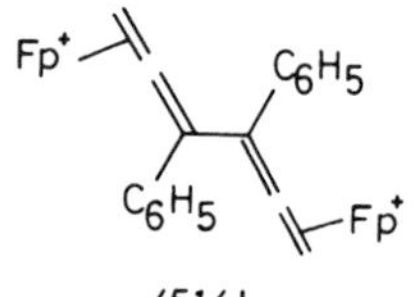

(514)

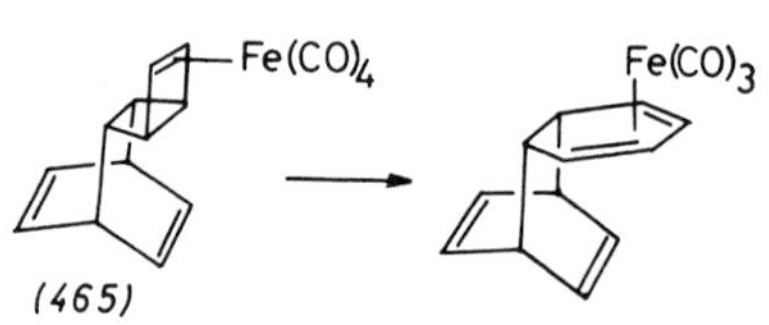

(465)

Section IV.A

Complexes of

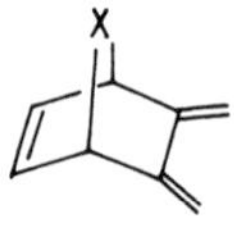

and

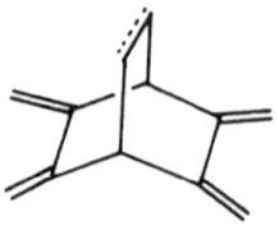

X = CH_2, O
(424,506,507)

(494)

Section IV.B

Derivatives of

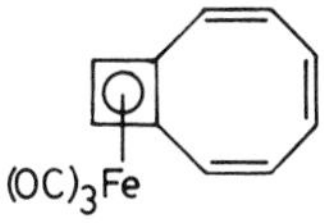

(480, 481, 504)

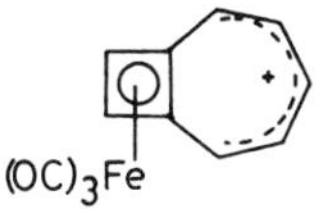

(508)

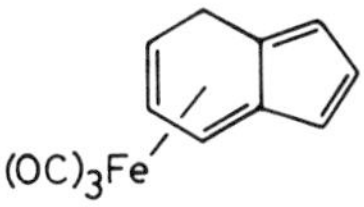

(451)

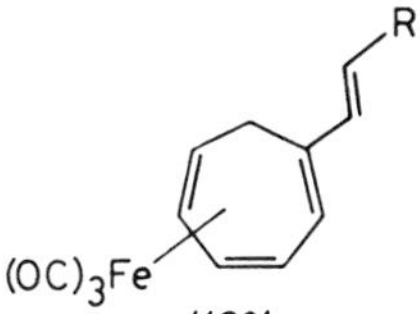

R = C_6H_5, CN, CO_2CH_3

(460)

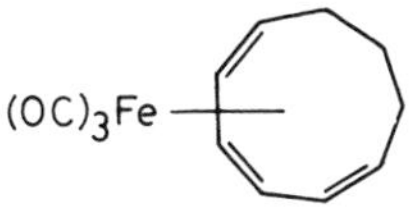

(499)

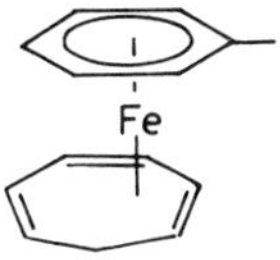

(421, 471)

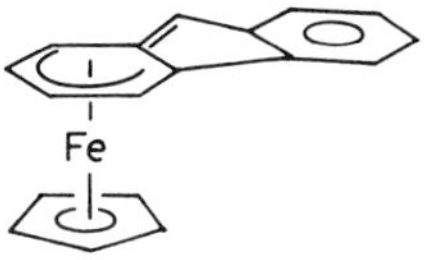

(467, 478)

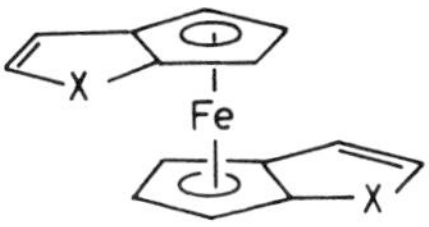

X = S, NCH_3

(511, 512)

Section IX.A

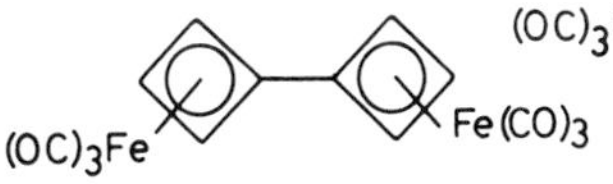

(430)

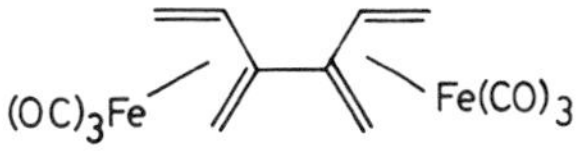

(464)

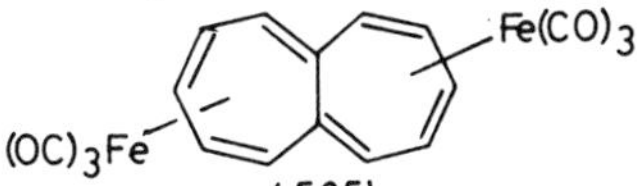

(505)

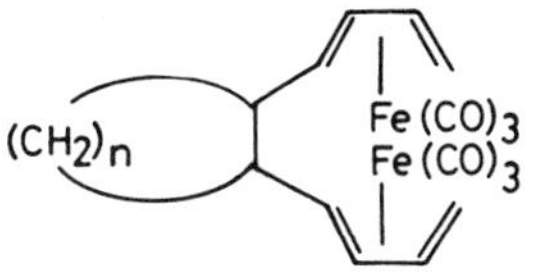

(479, 501)

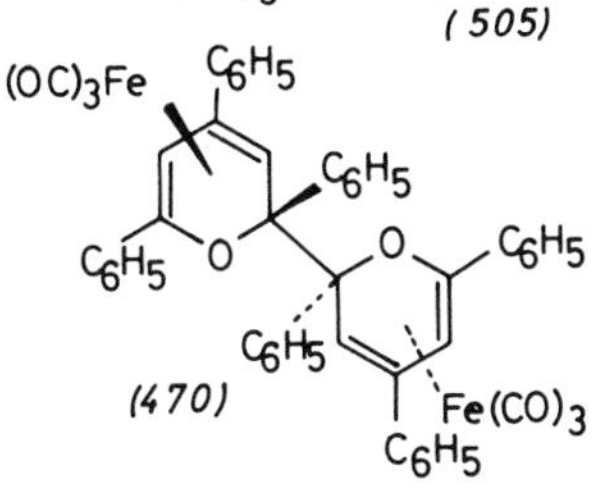

(470)

Section IX.C

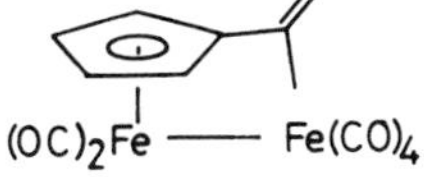

(447)

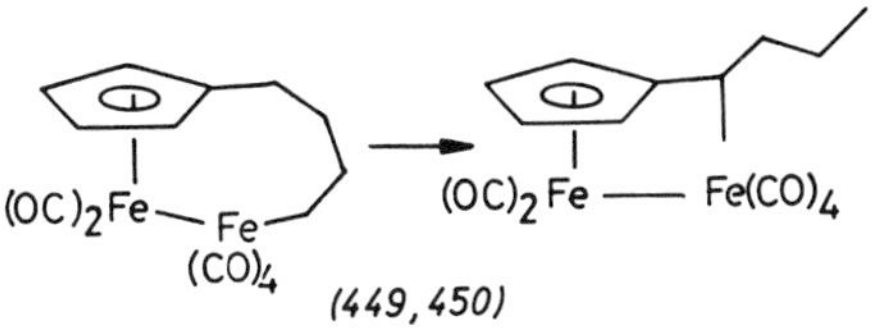

(449, 450)

Section XII

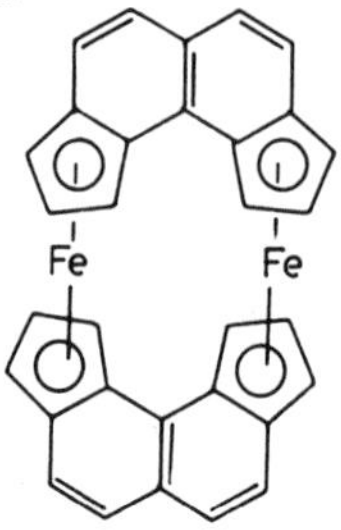

(483)

Section XIII

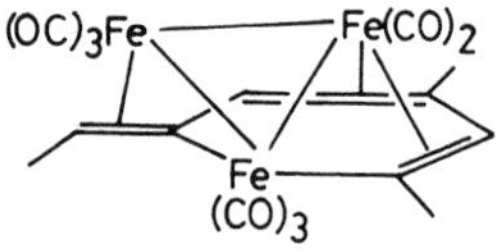

(502)

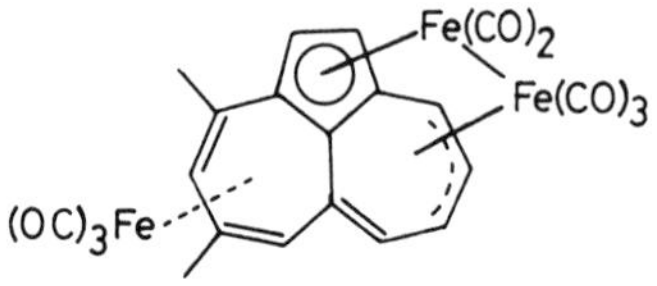

(433,434)

Many additional derivatives of and new reactions of previously described compounds have been described. These may be located from the list of references, where compound numbers from the original article are tabulated with each reference.

References

1. Abel, E.W., and Moorhouse, S., *Inorg. Nucl. Chem. Lett.*, *6*, 621 (1970).
2. Allegra, G., Colombo, A., Immirzi, A., and Bassi, I.W., *J. Amer. Chem. Soc.*, *90*, 4455 (1968).
3. Allegra, G., Colombo, A., and Mognaschi, E.R., *Gazz. Chim. Ital.*, *102*, 1060 (1972); *C.A. 79*, 46634u (1973).
4. Allmann, R., *Angew. Chem.*, *82*, 982 (1970); *Angew. Chem. Int. Ed. Engl.*, *9*, 958 (1970).
5. Alper, H., and Huang, C.-C., *J. Organometal. Chem.*, *50*, 213 (1973).
6. Alsop, J.E., and Davis, R., *J. Chem. Soc. Dalton Trans.*, *1973*, 1686.
7. Altman, J., Cohen, E., Maymon, T., Petersen, J.B., Reshef, N., and Ginsburg, D., *Tetrahedron*, *25*, 5115 (1969).
8. Altman, J., and Ginsburg, D., *Tetrahedron*, *27*, 93 (1971).
9. Amith, C., and Ginsburg, D., *Tetrahedron*, *30*, 3415 (1974).
10. Anderson, M., Clague, A.D.H., Blaauw, L.P., and Couperus, P.A., *J. Organometal. Chem.*, *56*, 307 (1973).
11. Anet, F.A.L., *J. Amer. Chem. Soc.*, *89*, 2491 (1967).
12. Anet, F.A.L., Kaesz, H.D., Maasbol, A., and Winstein, S., *J. Amer. Chem. Soc.*, *89*, 2489 (1967).
13. Ashley-Smith, J., Howe, D.V., Johnson, B.F.G., Lewis, J., and Ryder, I.E., *J. Organometal. Chem.*, *82*, 257 (1974).
14. Aumann, R., *Angew. Chem.*, *83*, 175 (1971); *Angew. Chem. Int. Ed. Engl.*, *10*, 188 (1971).
15. Aumann, R., *Angew. Chem.*, *83*, 176 (1971); *Angew. Chem. Int. Ed. Engl.*, *10*, 189 (1971).
16. Aumann, R., *Angew. Chem.*, *83*, 177 (1971); *Angew. Chem. Int. Ed. Engl.*, *10*, 190 (1971).
17. Aumann, R., *Angew. Chem.*, *83*, 583 (1971); *Angew. Chem. Int. Ed. Engl.*, *10*, 560 (1971).
18. Aumann, R., *Angew. Chem.*, *84*, 583 (1972); *Angew. Chem. Int. Ed. Engl.*, *11*, 522 (1972).
19. Aumann, R., *Angew. Chem.*, *85*, 628 (1973); *Angew. Chem. Int. Ed. Engl.*, *12*, 574 (1973).
20. Aumann, R., *J. Organometal. Chem.*, *76*, C 32 (1974).
21. Aumann, R., *J. Organometal. Chem.*, *78*, C 31 (1974).
22. Aumann, R., *Chem. Ber.*, *108*, 1974 (1975).
23. Aumann, R., *Angew. Chem.*, *88*, 375 (1976); *Angew. Chem. Int. Ed. Engl.*, *15*, 376 (1976).
24. Aumann, R., *Chem. Ber.*, *109*, 168 (1976).
25. Aumann, R., and Averbeck, H., *J. Organometal. Chem.*,

85, C 4 (1975).
26. Aumann, R., Averbeck, H., and Krüger, C., *Chem. Ber.*, *108*, 3336 (1975).
27. Aumann, R., and Knecht, J., *Chem. Ber.*, *109*, 174 (1976).
28. Aumann, R., and Winstein, S., *Angew. Chem.*, *82*, 667 (1970); *Angew. Chem. Int. Ed. Engl.*, *9*, 638 (1970).
29. Bailey, R.T., Lippincott, E.R., and Steele, D., *J. Amer. Chem. Soc.*, *87*, 5346 (1965).
30. Banthorpe, D.V., Fitton, H., and Lewis, J., *J. Chem. Soc. Perkin Trans. I*, *1973*, 2051.
31. Barton, D.H.R., Gunatilaka, A.A.L., Nakanishi, T., Patin, H., Widdowson, D.A., and Worth, B.R., *J. Chem. Soc. Perkin Trans. I*, *1976*, 821.
32. Barton, D.H.R., and Patin, H., *J. Chem. Soc. Perkin Trans. I*, *1976*, 829.
33. Bassi, I.W., and Scordamaglia, R., *J. Organometal. Chem.*, *37*, 353 (1972).
34. Bauch, T.E., Konowitz, H., and Giering, W.P., *J. Organometal. Chem.*, *114*, C 15 (1976).
35. Becker, Y., Eisenstadt, A., and Shvo, Y., *J. Chem. Soc. Chem. Commun.*, *1972*, 1156.
36. Behrens, U., *J. Organometal. Chem.*, *107*, 103 (1976).
37. Behrens, H., Feilner, H.-D., and Lindner, E., *Z. Anorg. Allg. Chem.*, *385*, 321 (1971).
38. Behrens, U., and Weiss, E., *J. Organometal. Chem.*, *59*, 335 (1973).
39. Behrens, U., and Weiss, E., *J. Organometal. Chem.*, *96*, 399 (1975); *J. Organometal. Chem.*, *73*, C 64 (1974).
40. Bennett, M.A., *Advan. Organometal. Chem.*, *4*, 353 (1968).
41. Bennett, M.J., Pratt, J.L., Simpson, K.A., LiShingMan, L.K.K., and Takats, J., *J. Amer. Chem. Soc.*, *98*, 4810 (1976).
42. Berens, G., Kaplan, F., Rimerman, R., Roberts, B.W., and Wissner, A., *J. Amer. Chem. Soc.*, *97*, 7076 (1975).
43. Berezin, R.N., Yablokova, E.P., and Shubin, V.G., *Izv. Akad. Nauk SSSR, Ser. Khim.*, *1973*, 2273; *Bull. Acad. Sci. USSR, Div. Chem. Sci.*, *1973*, 2216.
44. Bertelli, D.J., and Viebrock, J.M., *Inorg. Chem.*, *7*, 1240 (1968).
45. Birch, A.J., Chamberlain, K.B., Haas, M.A., and Thompson, D.J., *J. Chem. Soc. Perkin Trans. I*, *1973*, 1882.
46. Birch, A.J., Chamberlain, K.B., and Thompson, D.J., *J. Chem. Soc. Perkin Trans. I*, *1973*, 1900.
47. Birch, A.J., Cross, P.E., Lewis, J., White, D.A., and Wild, S.B., *J. Chem. Soc. A*, *1968*, 332.
48. Birch, A.J., and Fitton, H., *J. Chem. Soc. C*, *1966*, 2060.

49. Birch, A.J., Fitton, H., Mason, R., Robertson, G.B., and Stangroom, J.E., *Chem. Commun.*, *1966*, 613.
50. Birch, A.J., and Pearson, A.J., *J. Chem. Soc. Chem. Commun.*, *1976*, 601; *Tetrahedron Lett.*, *1975*, 2379.
51. Birch, A.J., and Williamson, D.H., *J. Chem. Soc. Perkin I*, *1973*, 1892.
52. Birnbaum, G.I., *J. Amer. Chem. Soc.*, *94*, 2455 (1972).
53. Birnbaum, K.B., *Acta Crystallogr.*, *28 B*, 161 (1972).
54. Birnbaum, K.B., Altman, J., Maymon, T., and Ginsburg, D., *Tetrahedron Lett.*, *1970*, 2051.
55. Blackborow, J.R., Hildenbrand, K., Koerner von Gustorf, E., Scrivanti, A., Eady, C.R., Ehntholt, D., and Krüger, C., *J. Chem. Soc. Chem. Commun.*, *1976*, 16.
56. Bockmeulen, H.A., Holloway, R.G., Parkins, A.W., and Penfold, B.R., *J. Chem. Soc. Chem. Commun.*, *1976*, 298.
57. Bonazza, B.R., and Lillya, C.P., *J. Amer. Chem. Soc.*, *96*, 2298 (1974).
58. Bottrill, M., Goddard, R., Green, M., Hughes, R.P., Lloyd, M.K., Lewis, B., and Woodward, P., *J. Chem. Soc. Chem. Commun.*, *1975*, 253.
59. Bratton, W.K., Cotton, F.A., Davison, A., Musco, A., and Faller, J.W., *Proc. Nat. Acad. Sci. U.S.*, *58*, 1324 (1967).
60. Braun, S., and Watts, W.E., *J. Organometal. Chem.*, *84*, C 33 (1975).
61. Braye, E.H., and Hübel, W., *J. Organometal. Chem.*, *3*, 25 (1965).
62. Bright, D., and Mills, O.S., *J. Chem. Soc. A*, *1971*, 1979.
63. Bright, D., and Mills, O.S., *J. Chem. Soc. Dalton Trans.*, *1972*, 2465.
64. Brodie, A.M., Johnson, B.F.G., and Lewis, J., *J. Chem. Soc. Dalton Trans.*, *1973*, 1997.
65. Brookhart, M., and Davis, E.R., *J. Amer. Chem. Soc.*, *92*, 7622 (1970).
66. Brookhart, M., and Davis, E.R., *Tetrahedron Lett.*, *1971*, 4349.
67. Brookhart, M., Davis, E.R., and Harris, D.L., *J. Amer. Chem. Soc.*, *94*, 7853 (1972).
68. Brookhart, M., Dedmond, R.E., and Lewis, B.F., *J. Organometal. Chem.*, *72*, 239 (1974).
69. Brookhart, M.S., Koszalka, G.W., Nelson, G.O., Scholes, G., and Watson, R.A., *J. Amer. Chem. Soc.*, *98*, 8155 (1976).
70. Brookhart, M., Lippman, N.M., and Reardon, E.J., Jr.,*J. Organometal. Chem.*, *54*, 247 (1973).
71. Brookhart, M., Nelson, G.O., Scholes, G., and Watson, R.A., *J. Chem. Soc. Chem. Commun.*, *1976*, 195.

71a. Brookhart, M., personal communication.
72. Brown, D.A., *Chem. Ind. (London), 1959,* 126.
73. Brown, D.A., *J. Inorg. Nucl. Chem., 10,* 39 (1959).
74. Brown, D.A., *J. Inorg. Nucl. Chem., 13,* 212 (1960).
75. Bruce, M.I., *Int. J. Mass Spectrom. Ion. Phys., 2,* 349 (1969).
76. Bruce, M.I., and Kuc, T.A., *Aust. J. Chem., 27,* 2487 (1974).
77. Bruce, R., Moseley, K., and Maitlis, P.M., *Can. J. Chem., 45,* 2011 (1967).
78. Burton, R., Green, M.L.H., Abel, E.W., and Wilkinson, G., *Chem. Ind. (London), 1958,* 1592.
79. Burton, R., Pratt, L., and Wilkinson, G., *J. Chem. Soc., 1960,* 4290.
80. Burton, R., Pratt, L., and Wilkinson, G., *J. Chem. Soc., 1961,* 594.
81. Cais, M., *Organometal. Chem. Rev., 1,* 435 (1966); *Record Chem. Progr., 27,* 177 (1966).
82. Cais, M., and Maoz, N., *J. Organometal. Chem., 5,* 370 (1966).
83. Cais, M., Modiano, A., and Raveh, A., *J. Amer. Chem. Soc., 87,* 5607 (1965).
84. Campbell, A.J., Cottrell, C.E., Fyfe, C.A., and Jeffrey, K.R., *Inorg. Chem., 15,* 1321 (1976).
85. Campbell, A.J., Fyfe, C.A., and Maslowsky, E., Jr., *J. Chem. Soc. D, Chem. Commun., 1971,* 1032.
86. Campbell, A.J., Fyfe, C.A., and Maslowsky, E., Jr., *J. Amer. Chem. Soc., 94,* 2690 (1972).
87. Campbell, C.H., Dias, A.R., Green, M.L.H., Saito, T., and Swanwick, M.G., *J. Organometal. Chem., 14,* 349 (1968).
88. Carbonaro, A., and Cambisi, F., *J. Organometal. Chem., 44,* 171 (1972).
89. Carbonaro, A., and Greco, A., *J. Organometal. Chem., 25,* 477 (1970).
90. Carbonaro, A., Greco, A., and Dall'Asta, G., *Tetrahedron Lett., 1967,* 2037.
91. Carbonaro, A., Greco, A., and Dall'Asta, G., *J. Org. Chem., 33,* 3948 (1968).
92. Carbonaro, A., Greco, A., and Dall'Asta, G., *Tetrahedron Lett., 1968,* 5129.
93. Carbonaro, A., Greco, A., and Dall'Asta, G., *J. Organometal. Chem., 20,* 177 (1969).
94. Carbonaro, A., Segre, A.L., Greco, A., Tosi, C., and Dall'Asta, G., *J. Amer. Chem. Soc., 90,* 4453 (1968).
95. Carty, A.J., Hobson, R.F., Patel, H.A., and Snieckus, V., *J. Amer. Chem. Soc., 95,* 6835 (1973).
96. Carty, A.J., Jablonski, C.R., and Snieckus, V., *Inorg.*

Chem., *15*, 601 (1976).

97. Carty, A.J., Kan, G., Madden, D.P., Snieckus, V., Stanton, M., and Birchall, T., *J. Organometal. Chem.*, *32*, 241 (1971).

98. Carty, A.J., Madden, D.P., Mathew, M., Palenik, G.J., and Birchall, T., *J. Chem. Soc. D, Chem. Commun.*, *1970*, 1664.

99. Carty, A.J., Taylor, N.J., and Jablonski, C.R., *Inorg. Chem.*, *15*, 1169 (1976).

100. Charles, A.D., Diversi, P., Johnson, B.F.G., and Lewis, J., *J. Organometal. Chem.*, *116*, C 25 (1976).

101. Chaudhari, F.M., and Pauson, P.L., *J. Organometal. Chem.*, *5*, 73 (1966).

102. Chierico, A., and Mognaschi, E.R., *J. Chem. Soc. Faraday Trans. II*, *1973*, 433.

103. Churchill, M.R., *Chem. Commun.*, *1966*, 450.

104. Churchill, M.R., *Inorg. Chem.*, *6*, 190 (1967).

105. Churchill, M.R., *Progr. Inorg. Chem.*, *11*, 53 (1970).

106. Churchill, M.R., and Bird, P.H., *J. Amer. Chem. Soc.*, *90*, 3241 (1968).

107. Churchill, M.R., and Bird, P.H., *Inorg. Chem.*, *8*, 1941 (1969).

108. Churchill, M.R., and DeBoer, B.G., *Inorg. Chem.*, *12*, 525 (1973).

109. Churchill, M.R., and Fennessey, J.P., *J. Chem. Soc. D, Chem. Commun.*, *1970*, 1056.

110. Churchill, M.R., and Lin, K.-K.G., *Inorg. Chem.*, *12*, 2274 (1973).

111. Churchill, M.R., and Wormald, J., *Chem. Commun.*, *1968*, 1033.

112. Churchill, M.R., and Wormald, J., *Inorg. Chem.*, *8*, 716 (1969).

113. Churchill, M.R., and Wormald, J., *Inorg. Chem.*, *8*, 1970 (1969).

114. Churchill, M.R., and Wormald, J., *Inorg. Chem.*, *9*, 2239 (1970); *Chem. Commun.*, *1968*, 1597.

115. Ciappenelli, D., and Rosenblum, M., *J. Amer. Chem. Soc.*, *91*, 3673 (1969).

116. Ciappenelli, D., and Rosenblum, M., *J. Amer. Chem. Soc.*, *91*, 6876 (1969).

117. Cooke, M., Howard, J.A.K., Russ, C.R., Stone, F.G.A., and Woodward, P., *J. Chem. Soc. Dalton Trans.*, *1976*, 70; *J. Organometal. Chem.*, *78*, C 43 (1974).

118. Cooke, M., Russ, C.R., and Stone, F.G.A., *J. Chem. Soc. Dalton Trans.*, *1975*, 256.

119. Cotton, F.A., *J. Chem. Soc.*, *1960*, 400.

120. Cotton, F.A., *Accounts Chem. Res.*, *1*, 257 (1968).

121. Cotton, F.A., Davison, A., and Faller, J.W., *J. Amer.*

Chem. Soc., *88*, 4507 (1966).

122. Cotton, F.A., *J. Organometal. Chem.*, *100*, 29 (1975).

123. Cotton, F.A., Davison, A., Marks, T.J., and Musco, A., *J. Amer. Chem. Soc.*, *91*, 6598 (1969).
Cotton, F.A., Davison, A., and Musco, A., *J. Amer. Chem. Soc.*, *89*, 6796 (1967).

124. Cotton, F.A., Day, V.W., Frenz, B.A., Hardcastle, K.I., and Troup, J.M., *J. Amer. Chem. Soc.*, *95*, 4522 (1973).

125. Cotton, F.A., DeBoer, B.G., and Marks, T.J., *J. Amer. Chem. Soc.*, *93*, 5069 (1971).

126. Cotton, F.A., and Deganello, G., *J. Organometal. Chem.*, *38*, 147 (1972).

127. Cotton, F.A., and Deganello, G., *J. Amer. Chem. Soc.*, *94*, 2142 (1972).

128. Cotton, F.A., and Deganello, G., *J. Amer. Chem. Soc.*, *95*, 396 (1973).

129. Cotton, F.A., and Edwards, W.T., *J. Amer. Chem. Soc.*, *91*, 843 (1969).

130. Cotton, F.A., and Frenz, B.A., *Tetrahedron*, *30*, 1587 (1974).

131. Cotton, F.A., Frenz, B.A., Deganello, G., and Shaver, A., *J. Organometal. Chem.*, *50*, 227 (1973).

132. Cotton, F.A., Frenz, B.A., and Troup, J.M., *J. Organometal. Chem.*, *61*, 337 (1973).

133. Cotton, F.A., Frenz, B.A., Troup, J.M., and Deganello, G., *J. Organometal. Chem.*, *59*, 317 (1973).

134. Cotton, F.A., and Hunter, D.L., *J. Amer. Chem. Soc.*, *97*, 5739 (1975).

135. Cotton, F.A., and Hunter, D.L., *J. Amer. Chem. Soc.*, *98*, 1413 (1976).

136. Cotton, F.A., Hunter, D.L., and Lahuerta, P., *Inorg. Chem.*, *14*, 511 (1975).

137. Cotton, F.A., Hunter, D.L., and Lahuerta, P., *J. Amer. Chem. Soc.*, *97*, 1046 (1975).

138. Cotton, F.A., and LaPrade, M.D., *J. Amer. Chem. Soc.*, *90*, 2026 (1968).

139. Cotton, F.A., and Marks, T.J., *J. Organometal. Chem.*, *19*, 237 (1969).

140. Cotton, F.A., and Musco, A., *J. Amer. Chem. Soc.*, *90*, 1444 (1968).

141. Cotton, F.A., and Reich, C.R., *J. Amer. Chem. Soc.*, *91*, 847 (1969).

142. Cotton, F.A., and Takats, J., *J. Amer. Chem. Soc.*, *90*, 2031 (1968).

143. Cotton, F.A., and Troup, J.M., *J. Amer. Chem. Soc.*, *95*, 3798 (1973).

144. Cotton, F.A., and Troup, J.M., *J. Organometal. Chem.*, *76*, 81 (1974).

145. Cotton, F.A., and Troup, J.M., *J. Organometal. Chem.*, *77*, 83 (1974).
146. Cotton, F.A., and Troup, J.M., *J. Organometal. Chem.*, *77*, 369 (1974).
147. Cowles, R.J.H., Johnson, B.F.G., Lewis, J., and Parkins, A.W., *J. Chem. Soc. Dalton Trans.*, *1972*, 1768. Lewis, J., and Parkins, A.W., *Chem. Commun.*, *1968*, 1194.
148. Cutler, A., Ehntholt, D., Giering, W.P., Lennon, P., Raghu, S., Rosan, A., Rosenblum, M., Tancrede, J., and Wells, D., *J. Amer. Chem. Soc.*, *98*, 3495 (1976).
149. Dannenberg, J.J., Levenberg, M.K., and Richards, J.H., *Tetrahedron*, *29*, 1575 (1973).
150. Dauben, H.J., and Bertelli, D.J., *J. Amer. Chem. Soc.*, *83*, 497 (1961).
151. Davis, R.E., Barnett, B.L., Amiet, R.G., Merk, W., McKennis, J.S., and Pettit, R., *J. Amer. Chem. Soc.*, *96*, 7108 (1974).
152. Davis, R.E., Dodds, T.A., Hseu, T.-H., Wagnon, J.C., Devon, T., Tancrede, J., McKennis, J.S., and Pettit, R., *J. Amer. Chem. Soc.*, *96*, 7562 (1974).
153. Davis, R.E., and Pettit, R., *J. Amer. Chem. Soc.*, *92*, 716 (1970).
154. Davison, J.B., and Bellama, J.M., *Inorg. Chim. Acta*, *14*, 263 (1975).
155. Davison, A., McFarlane, W., Pratt, L., and Wilkinson, G., *Chem. Ind. (London)*, *1961*, 553.
156. Davison, A., McFarlane, W., Pratt, L., and Wilkinson, G., *J. Chem. Soc.*, *1962*, 4821.
157. Davison, A., McFarlane, W., and Wilkinson, G., *Chem. Ind. (London)*, *1962*, 820.
158. DeCian, A., L'Huillier, P.M., and Weiss, R., *Bull. Soc. Chim. Fr.*, *1973*, 457.
159. Deganello, G., *J. Organometal. Chem.*, *59*, 329 (1973).
160. Deganello, G., *Chim. Ind. (Milan)*, *56*, 306 (1974); *C.A.*, *81*, 17538 m (1974).
161. Deganello, G., Boschi, T., and Toniolo, L., *J. Organometal. Chem.*, *97*, C 46 (1975).
162. Deganello, G., and Croatto, U., *Proceedings VIth Int. Conf. Organometal. Chem., Amherst. (Mass., U.S.A.) 1973*, Comm. 138.
163. Deganello, G., Maltz, H., and Kozarich, J., *J. Organometal. Chem.*, *60*, 323 (1973).
164. Deganello, G., and Toniolo, L., *J. Organometal. Chem.*, *74*, 255 (1974).
165. Deganello, G., Uguagliati, P., Calligaro, L., Sandrini, P.L., and Zingales, F., *Inorg. Chim. Acta*, *13*, 247 (1975).

166. Degrève, Y., Meunier-Piret, J., van Meerssche, M., and Piret, P., *Acta Crystallogr.*, *23*, 119 (1967).

167. DePuy, C.H., Kobal, V.M., and Gibson, D.H., *J. Organometal. Chem.*, *13*, 266 (1968).

168. Dell, D., Maoz, N., and Cais, M., *Israel J. Chem.*, *7*, 783 (1969).

169. Dessy, R.E., and Pohl, R.L., *J. Amer. Chem. Soc.*, *90*, 1995 (1968).

170. Dessy, R.E., Stary, F.E., King, R.B., and Waldrop, M., *J. Amer. Chem. Soc.*, *88*, 471 (1966).

170a. Dessy, R.E., King, R.B., and Waldrop, M., *J. Amer. Chem. Soc.*, *88*, 5112 (1966).

171. Dickens, B., and Lipscomb, W.N., *J. Amer. Chem. Soc.*, *83*, 489 (1961).

172. Dickens, B., and Lipscomb, W.N., *J. Amer. Chem. Soc.*, *83*, 4862 (1961).

173. Dickens, B., and Lipscomb, W.N., *J. Chem. Phys.*, *37*, 2084 (1962).

174. Dodge, R.P., *J. Amer. Chem. Soc.*, *86*, 5429 (1968).

175. Edwards, J.D., Howard, J.A.K., Knox, S.A.R., Riera, V., Stone, F.G.A., and Woodward, P., *J. Chem. Soc. Dalton Trans.*, *1976*, 75.

175a. Edwards, J.D., Knox, S.A.R., and Stone, F.G.A., *J. Chem. Soc. Dalton Trans.*, *1976*, 1813.

176. Ehntholt, D.J., Emerson, G.F., and Kerber, R.C., *J. Amer. Chem. Soc.*, *91*, 7547 (1969).

177. Ehntholt, D.J., and Kerber, R.C., *J. Chem. Soc. D, Chem. Commun.*, *1970*, 1451.

178. Ehntholt, D.J., and Kerber, R.C., *J. Organometal. Chem.*, *38*, 139 (1972).

179. Ehntholt, D., Rosan, A., and Rosenblum, M., *J. Organometal. Chem.*, *56*, 315 (1973).

180. Eilbracht, P., *Chem. Ber.*, *109*, 1429 (1976).

181. Eisenstadt, A., *Tetrahedron Lett.*, *1972*, 2005.

182. Eisenstadt, A., *J. Organometal. Chem.*, *97*, 443 (1975).

183. Eisenstadt, A., *J. Organometal. Chem.*, *113*, 147 (1976).

184. Eisenstadt, A., Guss, J.M., and Mason, R., *J. Organometal. Chem.*, *80*, 245 (1974).

185. Eisenstadt, A., and Winstein, S., *Tetrahedron Lett.*, *1970*, 4603.

186. Eisenstadt, A., and Winstein, S., *Tetrahedron Lett.*, *1971*, 613.

187. Elix, J.A., and Sargent, M.V., *J. Amer. Chem. Soc.*, *91*, 4734 (1969).

188. El Murr, N., Riveccié, M., Laviron, E., and Deganello, G., *Tetrahedron Lett.*, *1976*, 3339.

189. Elzinga, J., and Hogeveen, H., *Tetrahedron Lett.*, *1976*, 2383.

190. Emerson, G.F., Mahler, J.E., Pettit, R., and Collins, R., *J. Amer. Chem. Soc.*, *86*, 3590 (1964).

191. Evans, G., Johnson, B.F.G., and Lewis, J., *J. Organometal. Chem.*, *102*, 507 (1975).

192. Faraone, F., Cusmano, F., and Pietropaolo, R., *J. Organometal. Chem.*, *26*, 147 (1971).

193. Faraone, F., Zingales, F., Uguagliati, P., and Belluco, U., *Inorg. Chem.*, *7*, 2362 (1968).

194. Fischer, E.O., Kreiter, C.G., and Berngruber, W., *J. Organometal. Chem.*, *12*, P 39 (1968).

195. Fischer, E.O., Kreiter, C.G., Rühle, H., and Schwarzhans, K.E., *Chem. Ber.*, *100*, 1905 (1967).

196. Fischer, E.O., and Müller, J., *J. Organometal. Chem.*, *1*, 89 (1963).

197. Fischer, E.O., and Müller, J., *J. Organometal. Chem.*, *1*, 464 (1963).

198. Fischer, E.O., and Müller, J., *Z. Naturforsch.*, *B 18*, 413 (1963).

199. Fischer, E.O., Palm, C., and Fritz, H.P., *Chem. Ber.*, *92*, 2645 (1959).

200. Fischer, E.O., and Rühle, H., *Z. Anorg. Allg. Chem.*, *341*, 137 (1965).

201. Fleischer, E.B., Stone, A.L., Dewar, R.B.K., Wright, J.D., Keller, C.E., and Pettit, R., *J. Amer. Chem. Soc.*, *88*, 3158 (1966).

202. Fischer, E.O., and Werner, H., *Metal-π-Complexes, Vol. I, Complexes with Di- and Oligo-olefinic Ligands*, Elsevier, Amsterdam 1966; *Metall-π-Komplexe mit di- und oligoolefinischen Liganden*, Verlag Chemie, Weinheim 1963.

203. Franck-Neumann, M., and Martina, D., *Tetrahedron Lett.*, *1975*, 1759.

204. Frankel, E.N., Emken, E.A., and Davison, V.L., *J. Org. Chem.*, *30*, 2739 (1965).

205. Fritz, H.P., *Chem. Ber.*, *95*, 820 (1962).

206. Fritz, H.P., and Keller, H., *Chem. Ber.*, *95*, 158 (1962).

207. Gerlach, D.H., and Schunn, R.A., *Inorg. Synth.*, *15*, 2 (1974).

208. Gieren, A., and Hoppe, W., *Acta Crystallogr.*, *28 B*, 2766 (1972).

209. Gill, G.B., Gourlay, N., Johnson, A.W., and Mahendran, M., *J. Chem. Soc. D, Chem. Commun.*, *1969*, 631.

210. Ginsburg, D., *Accounts Chem. Res.*, *7*, 286 (1974).

211. Gitany, R., Paul, I.C., Acton, N., and Katz, T.J., *Tetrahedron Lett.*, *1970*, 2723.

212. Gleiter, R., and Seeger, R., *Helv. Chim. Acta*, *54*, 1217 (1971); *Angew. Chem.* *83*, 903 (1971); *Angew. Chem. Int. Ed. Engl.*, *10*, 830 (1971).

213. Gleiter, R., Seeger, R., Binder, H., Fluck, E., and Cais, M., *Angew. Chem., 84,* 1107 (1972); *Angew. Chem. Int. Ed. Engl., 11,* 1028 (1972).

214. Gompper, R., and Reiser, W., *Tetrahedron Lett., 1976,* 1263.

215. Graham, C.R., Scholes, G., and Brookhart, M., *J. Amer. Chem. Soc., 99,* 1180 (1977).

216. Greco, A., and Carbonaro, A., *Chim. Ind. (Milan), 52,* 877 (1970); *C.A., 74,* 3715a (1971).

217. Green, M., Heathcock, S., and Wood, D.C., *J. Chem. Soc. Dalton Trans., 1973,* 1564.

218. Green, M., Tolson, S., Weaver, J., Wood, D.C., and Woodward, P., *J. Chem. Soc. D, Chem. Commun., 1971,* 222.

219. Green, M., and Wood, D.C., *Chem. Commun., 1967,* 1062.

220. Green, M., and Wood, D.C., *J. Chem. Soc. A, 1969,* 1172.

221. Grubbs, R., Breslow, R., Herber, R., and Lippard, S.J., *J. Amer. Chem. Soc., 89,* 6864 (1967).

222. Grubbs, R.H., Pancoast, T.A., and Grey, R.A., *Tetrahedron Lett., 1974,* 2425.

223. Günther, H., and Wenzl, R., *Tetrahedron Lett., 1967,* 4155.

224. Harmon, C.A., Streitwieser, A., Jr., *J. Org. Chem., 38,* 549 (1973).

225. Harris, P.J., Howard, J.A.K., Knox, S.A.R., Phillips, R.P., Stone, F.G.A., and Woodward, P., *J. Chem. Soc. Dalton Trans., 1976,* 377.

225a. Hashmi, M.A., Munro, J.D., Pauson, P.L., and Williamson, J.M., *J. Chem. Soc. A, 1967,* 240.

226. Hedberg, F.L., and Rosenberg, H., *J. Amer. Chem. Soc., 91,* 1258 (1969).

227. Heil, V., Johnson, B.F.G., Lewis, J., and Thompson, D.J., *J. Chem. Soc. Chem. Commun., 1974,* 270.

228. Hill, E.A., *J. Organometal. Chem., 24,* 457 (1970).

229. Hill, E.A., and Wiesner, R., *J. Amer. Chem. Soc., 91,* 509 (1969).

230. Hine, K.E., Johnson, B.F.G., and Lewis, J., *J. Chem. Soc. Chem. Commun., 1975,* 81.

231. Holland, J.M., and Jones, D.W., *Chem. Commun., 1967,* 946.

232. Holmes, J.D., and Pettit, R., *J. Amer. Chem. Soc., 85,* 2531 (1963).

233. Howell, J.A.S., Johnson, B.F.G., Josty, P.L., and Lewis, J., *J. Organometal. Chem., 39,* 329 (1972).

234. Hübel, W., *Organometallic Derivatives from Metal Carbonyls and Acetylenic Compounds,* in I. Wender and P. Pino (Eds.), *Organic Syntheses via Metal Carbonyls,* Interscience, New York, 1968, pp. 273-342.

235. Hübel, W., and Braye, E.H., *J. Inorg. Nucl. Chem., 10,* 250 (1959).

236. Hübel, W., Braye, E.H., Clauss, A., Weiss, E., Krüerke, U., Brown, D.A., King, G.S.D., and Hoogzand, C., *J. Inorg. Nucl. Chem., 9,* 204 (1959).

237. Hübel, W., and Merényi, R.G., *Chem. Ber., 96,* 930 (1963).

238. Hübel, W., and Weiss, E., *Chem. Ind. (London), 1959,* 703.

239. Hunt, D.F., Farrant, G.C., and Rodeheaver, G.T., *J. Organometal. Chem., 38,* 349 (1972).

240. Hunt, D.F., and Russell, J.W., *J. Organometal. Chem., 46,* C 22 (1972).

241. Hunt, D.F., and Russell, J.W., *J. Amer. Chem. Soc., 94,* 7198 (1972).

242. Hunt, D.F., and Russell, J.W., *Proc. VIth Int. Conf. Organometal. Chem., Amherst (Mass., U.S.A.), 1973,* Comm. 13.

243. Hunt, D.F., Russell, J.W., and Torian, R.L., *J. Organometal. Chem., 43,* 175 (1972).

244. Huttner, G., and Bejenke, V., *Chem. Ber., 107,* 156 (1974).

245. Huttner, G., and Regler, D., *Chem. Ber., 105,* 3936 (1972).

246. Ichikawa, M., Tsutsui, M., and Vohwinkel, F., *Z. Naturforsch., B 22,* 376 (1967).

247. Ihrman, K.G., and Coffield, T.H., U.S. Patent 3 077 489 (1963); *C.A., 59,* 5199b (1963).

248. Johnson, B.F.G., Lewis, J., McArdle, P., and Randall, G.L.P., *J. Chem. Soc. D, Chem. Commun., 1971,* 177.

249. Johnson, B.F.G., Lewis, J., McArdle, P., and Randall, G.L.P., *J. Chem. Soc. Dalton Trans., 1972,* 456.

250. Johnson, B.F.G., Lewis, J., McArdle, P., and Randall, G.L.P., *J. Chem. Soc. Dalton Trans., 1972,* 2076.

251. Johnson, B.F.G., Lewis, J., Parkins, A.W., and Randall, G.L.P., *J. Chem. Soc. D, Chem. Commun., 1969,* 595.

252. Johnson, B.F.G., Lewis, J., and Quail, J.W., *J. Chem. Soc. Dalton Trans., 1975,* 1252.

253. Johnson, B.F.G., Lewis, J., and Randall, G.L.P., *J. Chem. Soc. D, Chem. Commun., 1969,* 1273.

254. Johnson, B.F.G., Lewis, J., and Randall, G.L.P., *J. Chem. Soc. A, 1971,* 422.

255. Johnson, B.F.G., Lewis, J., and Thompson, D.J., *Tetrahedron Lett., 1974,* 3789.

256. Johnson, B.F.G., Lewis, J., Thompson, D.J., and Heil, B., *J. Chem. Soc. Dalton Trans., 1975,* 567.

257. Johnson, B.F.G., Lewis, J., and Twigg, M.V., *J. Chem. Soc. Dalton Trans., 1974,* 2546.

257a. Johnson, B.F.G., Lewis, J., and Wege, D., *J. Chem. Soc. Dalton Trans., 1976*, 1874.

258. Johnson, S.M., and Paul, I.C., *J. Chem. Soc. B, 1970*, 1783.

259. Joshi, K.K., *J. Chem. Soc. A, 1966*, 594.

260. Joshi, K.K., *J. Chem. Soc. A, 1966*, 598.

261. Katz, T.J., Acton, N., and McGinnis, J., *J. Amer. Chem. Soc., 94*, 6205 (1972).

262. Katz, T.J., Balogh, V., and Schulman, J., *J. Amer. Chem. Soc., 90*, 734 (1968).

263. Katz, T.J., and Mrowca, J.J., *J. Amer. Chem. Soc., 89*, 1105 (1967).

264. Katz, T.J., and Rosenberger, M., *J. Amer. Chem. Soc., 85*, 2030 (1963).

265. Katz, T.J., and Schulman, J., *J. Amer. Chem. Soc., 86*, 3169 (1964).

266. Keller, C.E., Emerson, G.F., and Pettit, R., *J. Amer. Chem. Soc., 87*, 1388 (1965).

267. Keller, C.E., Shoulders, B.A., and Pettit, R., *J. Amer. Chem. Soc., 88*, 4760 (1966).

268. Kerber, R.C., and Ehntholt, D.J., *Synthesis, 1970*, 449.

269. Kerber, R.C., and Ehntholt, D.J., *J. Amer. Chem. Soc., 95*, 2927 (1973).

269a. Kerber, R.C., and Koerner von Gustorf, E.A., *J. Organometal. Chem., 110*, 345 (1976).

270. King, G.S.D., *Acta Crystallogr., 15*, 243 (1962).

271. King, R.B., *J. Amer. Chem. Soc., 84*, 4705 (1962).

272. King, R.B., *Inorg. Chem., 2*, 807 (1963).

273. King, R.B., *Organometallic Syntheses, Vol. I, Transition-Metal Compounds*, Academic Press, New York 1965, pp. 126-128.

274. King, R.B., *J. Amer. Chem. Soc., 88*, 2075 (1966).

275. King, R.B., *Appl. Spectrosc., 23*, 536 (1969).

276. King, R.B., and Bisnette, M.B., *Inorg. Chem., 3*, 801 (1964).

277. King, R.B., and Eavenson, C.W., *J. Organometal. Chem., 42*, C 95 (1972).

278. King, R.B., Epstein, L.M., and Gowling, E.W., *J. Inorg. Nucl. Chem., 32*, 441 (1970).

279. King, R.B., and Harmon, C.A., *J. Organometal. Chem., 88*, 93 (1975).

280. King, R.B., and Harmon, C.A., *J. Amer. Chem. Soc., 98*, 2409 (1976).

281. King, R.B., Manuel, T.A., and Stone, F.G.A., *J. Inorg. Nucl. Chem., 16*, 233 (1961).

282. King, R.B., and Stone, F.G.A., *J. Amer. Chem. Soc., 82*, 4557 (1960).

283. King, R.B., and Stone, F.G.A., *J. Amer. Chem. Soc., 83*,

3590 (1961).

284. Koerner von Gustorf, E., and Grevels, F.-W., *Fortschr. Chem. Forsch., Topics in Current Chemistry, 13*, 366 (1970).

285. Korat, M., Tatarsky, D., and Ginsburg, D., *Tetrahedron, 28*, 2315 (1972).

286. Kreiter, C.G., Maasbol, A., Anet, F.A.L., Kaesz, H.D., and Winstein, S., *J. Amer. Chem. Soc., 88*, 3444 (1966).

287. Kruczynski, L., Shing Man, L., and Takats, J., *Proc. VIth Int. Conf. Organometal Chem., Amherst (Mass., U.S.A.) 1973*, Comm. 6.

288. Kruczynski, L., and Takats, J., *J. Amer. Chem. Soc., 96*, 932 (1974).

289. Krüerke, U., *Angew. Chem., 79*, 55 (1967); *Angew. Chem. Int. Ed. Engl., 6*, 79 (1967).

290. Lehmkuhl, H., *Synthesis, 1973*, 377.

291. Leppard, D.G., Hansen, H.-J., Bachmann, K., and v. Philipsborn, W., *J. Organometal. Chem., 110*, 359 (1976).

292. Li Shing Man, L.K.K., and Takats, J., *J. Organometal. Chem., 117*, C 104 (1976).

293. Lumbroso, H., and Bertin, D.M., *J. Organometal. Chem., 108*, 111 (1976).

294. Lupan, S., Kapon, M., Cais, M., and Herbstein, F.H., *Angew. Chem., 84*, 1104 (1972); *Angew. Chem. Int. Ed. Engl., 11*, 1025 (1972).

295. McArdle, P., *J. Chem. Soc. Chem. Commun., 1973*, 482.

296. McArdle, P., and Manning, A.R., *J. Chem. Soc. A, 1970*, 2119.

297. McArdle, P., Manning, A.R., and Stevens, F.S., *J. Chem. Soc. D, Chem. Commun., 1969*, 1310.

298. McArdle, P., and Sherlock, H., *J. Organometal. Chem., 52*, C 29 (1973).

299. McArdle, P., and Sherlock, H., *J. Chem. Soc. Chem. Commun., 1976*, 537.

300. McArdle, P., and Sherlock, H., *J. Organometal. Chem., 116*, C 23 (1976).

301. McFarlane, W., Pratt, L., and Wilkinson, G., *J. Chem. Soc., 1963*, 2162.

302. McFarlane, W., and Wilkinson, G., *Inorg. Synth., 8*, 184 (1966).

303. Mackenzie, R., and Timms, P.L., *J. Chem. Soc. Chem. Commun., 1974*, 650.

304. Mahler, J.E., Gibson, D.H., and Pettit, R., *J. Amer. Chem. Soc., 85*, 3959 (1963).

305. Mahler, J.E., Jones, D.A.K., and Pettit, R., *J. Amer. Chem. Soc., 86*, 3589 (1964).

306. Maltz, H., and Deganello, G., *J. Organometal. Chem.,*

27, 383 (1971).

307. Maltz, H., and Kelly, B.A., *J. Chem. Soc. D, Chem. Commun., 1971*, 1390.

308. Manuel, T.A., Stafford, S.L., and Stone, F.G.A., *J. Amer. Chem. Soc., 83*, 3597 (1961).

309. Manuel, T.A., and Stone, F.G.A., *Proc. Chem. Soc., 1959*, 90.

310. Manuel, T.A., and Stone, F.G.A., *J. Amer. Chem. Soc., 82*, 366 (1960).

311. Manuel, T.A., and Stone, F.G.A., *J. Amer. Chem. Soc., 82*, 6240 (1960).

312. Markezich, R.L., and Whitlock, H.W., Jr., *J. Amer. Chem. Soc., 93*, 5291 (1971).

313. Mason, R., and Robertson, G.B., *J. Chem. Soc. A, 1970*, 1229.

314. Mauldin, C.H., Biehl, E.R., and Reeves, P.C., *Tetrahedron Lett., 1972*, 2955.

315. Merk, W., and Pettit, R., *J. Amer. Chem. Soc., 90*, 814 (1968).

316. Meunier-Piret, J., Piret, P., and van Meerssche, M., *Acta Crystallogr., 19*, 85 (1965).

317. Moriarty, R.M., Chen, K.-N., Churchill, M.R., and Chang, S.W.-Y., *J. Amer. Chem. Soc., 96*, 3661 (1974).

318. Müller, J., and Fischer, E.O., *J. Organometal. Chem., 5*, 275 (1966).

319. Müller, J., and Mertschenk, B., *Chem. Ber., 105*, 3346 (1972).

320. Murdoch, H.D., and Weiss, E., *Helv. Chim. Acta, 46*, 1588 (1963).

321. Nakamura, A., *Bull. Chem. Soc. Japan, 38*, 1868 (1965).

322. Nakamura, A., and Hagihara, N., *Bull. Chem. Soc. Japan, 32*, 880 (1959).

323. Nakamura, A., and Hagihara, N., *Mem. Inst. Sci. and Ind. Res., Osaka Univ., 17*, 187 (1960); *C.A. 55*, 6457f (1961).

324. Nakamura, A., and Hagihara, N., *Nippon Kagaku Zasshi, 82*, 1387 (1961); *C.A. 59*, 2855a (1963).

325. Nakamura, A., and Hagihara, N., *Nippon Kagaku Zasshi, 82*, 1389 (1961); *C.A. 59*, 2855n (1963).

326. Nakamura, A., and Hagihara, N., *Nippon Kagaku Zasshi, 84*, 338 (1963); *C.A. 59*, 14869h (1963).

327. Nakamura, A., Kim, P.-J., and Hagihara, N., *Bull. Chem. Soc. Japan, 37*, 292 (1964).

328. Nakamura, A., Kim, P.-J., and Hagihara, N., *J. Organometal. Chem., 3*, 7 (1965).

329. Nakamura, A., Kim, P.-J., and Hagihara, N., *J. Organometal. Chem., 6*, 420 (1966).

330. Nakamura, A., and Tsutsui, M., *J. Med. Chem., 6*, 796

(1963).

331. Nakamura, A., and Tsutsui, M., *J. Med. Chem., 7,* 335 (1964).

332. Nesmeyanov, A.N., Aleksandrov, G.G., Bokii, N.G., Zlotina, I.B., Struchkov, Yu.T., and Kolobova, N.E., *J. Organometal. Chem. 111,* C 9 (1976).

333. Nesmeyanov, A.N., Kazakova, L.I., Reshetova, M.D., Kazitsina, L.A., and Perevalova, E.G., *Izv. Akad. Nauk SSSR, Ser. Khim., 1970,* 2804; *Bull. Acad. Sci. USSR, Div. Chem. Sci., 1970,* 2639.

334. Nicholson, B.J., *J. Amer. Chem. Soc., 88,* 5156 (1966).

335. Otsuka, S., Nakamura, A., and Tani, K., *J. Chem. Soc. A, 1971,* 154.

336. Otsuka, S., Nakamura, A., and Yoshida, T., *Bull. Chem. Soc. Japan, 40,* 1266 (1967).

337. Pannell, K.H., and Crawford, G.M., *J. Coord. Chem., 2,* 251 (1973).

338. Paquette, L.A., Kuhla, D.E., Barrett, J.H., and Haluska, R.J., *J. Org. Chem., 34,* 2866 (1969).

339. Paquette, L.A., Ley, S.V., Broadhurst, M.J., Truesdell, D., Fayos, J., and Clardy, J., *Tetrahedron Lett., 1973,* 2943.

340. Paquette, L.A., Ley, S.V., and Farnham, W.B., *J. Amer. Chem. Soc., 96,* 312 (1974).

341. Paquette, L.A., Ley, S.V., Maiorana, S., Schneider, D.F., Broadhurst, M.J., and Boggs, R.A., *J. Amer. Chem. Soc., 97,* 4658 (1975).

342. Paul, I.C., Johnson, S.M., Paquette, L.A., Barrett, J.H., and Haluska, R.J., *J. Amer. Chem. Soc., 90,* 5023 (1968).

343. Pelter, A., Gould, K.J., and Kane-Maguire, L.A.P., *J. Chem. Soc. Chem. Commun., 1974,* 1029.

344. Piret, P., Meunier-Piret, J.M., van Meerssche, M., and King, G.S.D., *Acta Crystallogr., 19,* 78 (1965).

345. Rausch, M.D., Kovar, R.F., and Kraihanzel, C.S., *J. Amer. Chem. Soc., 91,* 1259 (1969).

346. Rausch, M.D., and Schrauzer, G.N., *Chem. Ind. (London), 1959,* 957.

347. Reardon, E.J., Jr., and Brookhart, M., *J. Amer. Chem. Soc., 95,* 4311 (1973).

348. Reckziegel, A., and Bigorgne, M., *J. Organometal. Chem., 3,* 341 (1965).

349. Reger, D.L., and Gabrielli, A., *J. Amer. Chem. Soc., 97,* 4421 (1975).

350. Reid, K.I.G., and Paul, I.C., *J. Chem. Soc. D, Chem. Commun., 1970,* 1106.

351. Rigatti, G., Boccalon, G., Ceccon, A., and Giacometti, G., *J. Chem. Soc. Chem. Commun., 1972,* 1165.

352. Roberts, B.W., and Wissner, A., *J. Amer. Chem. Soc.*, *92*, 6382 (1970).
353. Robson, A., and Truter, M.R., *J. Chem. Soc. A*, *1968*, 794; *Tetrahedron Lett.*, *1964*, 3079.
354. Rodeheaver, G.T., Farrant, G.C., and Hunt, D.F., *J. Organometal. Chem.*, *30*, C 22 (1971).
355. Rosenbaum, L., Okaya, Y., and Kerber, R.C., *Proc. VIth Int. Conf. Organometal. Chem., Amherst (Mass., U.S.A.), 1973*, Comm. 14.
356. Roth, W.R., and Meier, J.D., *Tetrahedron Lett.*, *1967*, 2053.
357. Sadeh, S., and Gaoni, Y., *J. Organometal. Chem.*, *93*, C 31 (1975).
357a. Sanders, A., Bauch, T., Magatti, C.V., Lorenc, C., and Giering, W.P., *J. Organometal. Chem.*, *107*, 359 (1976).
Sanders, A., Magatti, C.V., and Giering, W.P., *J. Amer. Chem. Soc.*, *96*, 1610 (1974).
358. Scholes, G., Graham, C.R., and Brookhart, M., *J. Amer. Chem. Soc.*, *96*, 5665 (1974).
359. Schrauzer, G.N., *J. Amer. Chem. Soc.*, *83*, 2966 (1961).
360. Schrauzer, G.N., and Eichler, S., *Angew. Chem.*, *74*, 585 (1962); *Angew. Chem. Int. Ed. Engl.*, *1*, 454 (1962).
361. Schrauzer, G.N., and Glockner, P.W., *J. Amer. Chem. Soc.*, *90*, 2800 (1968).
362. Schrauzer, G.N., Glockner, P., and Merényi, R., *Angew. Chem.*, *76*, 498 (1964); *Angew. Chem. Int. Ed. Engl.*, *3*, 509 (1964).
363. Schrauzer, G.N., Glockner, P., Reid, K.I.G., and Paul, I.C., *J. Amer. Chem. Soc.*, *92*, 4479 (1970).
364. Schröder, G., Ramadas, S.R., and Nikoloff, P., *Chem. Ber.*, *105*, 1072 (1972).
365. Schwartz, J., *J. Chem. Soc. Chem. Commun.*, *1972*, 814.
366. Shubin, V.G., Berezina, R.N., and Piottukh-Peletski, V.N., *J. Organometal. Chem.*, *54*, 239 (1973).
367. Shvo, Y., and Hazum, E., *J. Chem. Soc. Chem. Commun.*, *1974*, 336.
368. Shvo, Y., and Hazum, E., *J. Chem. Soc. Chem. Commun.*, *1975*, 829.
369. Sime, R.L., and Sime, R.J., *J. Amer. Chem. Soc.*, *96*, 892 (1974).
370. Slegeir, W., Case, R., McKennis, J.S., and Pettit, R., *J. Amer. Chem. Soc.*, *96*, 287 (1974).
371. Smith, D.L., and Dahl, L.F., *J. Amer. Chem. Soc.*, *84*, 1743 (1962).
372. Stephenson, T.A., *Metal Compounds Containing Six-electron and Seven-electron Organic Ligands*, in *MTP International Review of Science, Second Series, Volume 6*,

Butterworths (London) - University Park Press (Baltimore), 1975, pp. 287-294.
373. Streith, J., Blind, A., Cassal, J.-M.,and Sigwalt, C., *Bull. Soc. Chim. Fr., 1969,* 948.
374. Streith, J., and Cassal, J.-M., *Angew. Chem., 80,* 117 (1968); *Angew. Chem. Int. Ed. Engl., 7,* 129 (1968).
375. Streith, J., and Cassal, J.-M., *Bull. Soc. Chim. Fr., 1969,* 2175.
376. Takats, J., *J. Organometal. Chem., 90,* 211 (1975).
377. Tan, T.-S., Fletcher, J.L., and McGlinchey, M.J., *J. Chem. Soc. Chem. Commun., 1975,* 771.
378. Timms, P.L., *Angew. Chem., 87,* 295 (1975); *Angew. Chem. Int. Ed. Engl., 14,* 273 (1975).
379. Tirouflet, J., Laviron, E., Moïse, C., and Mugnier, Y., *J. Organometal. Chem., 50,* 241 (1973).
380. Turbitt, T.D., and Watts, W.E., *J. Chem. Soc. Perkin Trans. II, 1974,* 177.
381. Turbitt, T.D., and Watts, W.E., *J. Chem. Soc. Perkin Trans. II, 1974,* 185.
382. van Meerssche, M., Piret, P., Meunier-Piret, J., and Degrève, Y., *Bull. Soc. Chim. Belg., 73,* 824 (1964).
383. Victor, R., and Ben-Shoshan, R., *J. Chem. Soc. Chem. Commun., 1974,* 93.
384. Victor, R., and Ben-Shoshan, R., *J. Organometal. Chem., 80,* C 1 (1974).
385. Victor, R., Ben-Shoshan, R., and Sarel, S., *Tetrahedron Lett., 1970,* 4257.
386. Victor, R., Ben-Shoshan, R., and Sarel, S., *J. Chem. Soc. D, Chem. Commun., 1971,* 1241.
387. Victor, R., Ben-Shoshan, R., and Sarel, S., *J. Org. Chem., 37,* 1930 (1972).
388. Victor, R., Deutsch, J., and Sarel, S., *J. Organometal. Chem., 71,* 65 (1974).
389. Waite, M.G., and Sim, G.A., *J. Chem. Soc. A, 1971,* 1009.
390. Wang, A.H.-J., Paul, I.C., and Aumann, R., *J. Organometal. Chem., 69,* 301 (1974).
391. Wang, A.H.-J., Paul, I.C., and Schrauzer, G.N., *J. Chem. Soc. Chem. Commun., 1972,* 736.
392. Ward, J.S., and Pettit, R., *J. Amer. Chem. Soc., 93,* 262 (1971).
393. Weaver, J., and Woodward, P., *J. Chem. Soc. A, 1971,* 3521.
394. Weidemüller, W., and Hafner, K., *Angew. Chem., 85,* 958 (1973); *Angew. Chem. Int. Ed. Engl., 12,* 925 (1973).
395. Weiss, E., and Hübel, W., *Angew. Chem., 73,* 298 (1961).
396. Weiss, E., and Hübel, W., *Chem. Ber., 95,* 1179 (1962).
397. Weiss, E., and Hübel, W., *Chem. Ber., 95,* 1186 (1962).

398. Weiss, E., Hübel, W., and Merényi, R., *Chem. Ber., 95,* 1155 (1962).
399. Wertheim, G.K., and Herber, R.H., *J. Amer. Chem. Soc., 84,* 2274 (1962).
399a. Whitesides, T.H., and Budnik, R.A., *J. Chem. Soc. D, Chem. Commun., 1971,* 1514.
400. Whitlock, H.W., Jr., and Chuah, Y.N., *Inorg. Chem., 4,* 424 (1965).
401. Whitlock, H.W., Jr., and Chuah, Y.N., *J. Amer. Chem. Soc., 87,* 3605 (1965).
402. Whitlock, H.W., Jr., and Markezich, R.L., *J. Amer. Chem. Soc., 93,* 5290 (1971).
403. Whitlock, H.W., Jr., Reich, C., and Woessner, W.D., *J. Amer. Chem. Soc., 93,* 2483 (1971).
404. Whitlock, H.W., Jr., and Stucki, H., *J. Amer. Chem. Soc., 94,* 8594 (1972).
405. Winstein, S., Kaesz, H.D., Kreiter, C.G., and Friedrich, E.C., *J. Amer. Chem. Soc., 87,* 3267 (1965).
406. Woodhouse, D.I., Sim, G.A., and Sime, J.G., *J. Chem. Soc. Dalton Trans., 1974,* 1331.
407. Xavier, J., Thiel, M., and Lippincott, E.R., *J. Amer. Chem. Soc., 83,* 2403 (1961).
408. Yasuda, N., Kai, Y., Yasuoka, N., Kasai, N., and Kakudo, M., *J. Chem. Soc. Chem. Commun., 1972,* 157.
409. Zimniak, A., and Jasiobedzki, W., *Rocz. Chem., 48,* 365 (1974).

Addendum (compound numbers from the original article are given in parentheses [ ]).

410. Airoldi, M., Deganello, G., and Kozarich, J., *Inorg. Chim. Acta, 20,* L 5 (1976). [68,69]
411. Allison, N.T., Kawada, Y., and Jones W.M., *J. Amer. Chem. Soc., 100,* 5224 (1978). [1,3]
412. Andreetti, G.D., Bocelli, G., and Sgarabotto, P., *J. Organometal. Chem., 150,* 85 (1978). [52]
413. Aumann, R., *Chem. Ber., 110,* 1432 (1977). [112,115]
414. Aumann, R., Averbeck, H., and Krüger, C., *J. Organometal. Chem., 160,* 241 (1978). [89,107,114,116]
415. Aumann, R., Wörmann, H., and Krüger, C., *Chem. Ber., 110,* 1442 (1977). [39]
416. Bachmann, K., von Philipsborn, W., Amith, C., and Ginsburg, D., *Helv. Chim. Acta, 60,* 400 (1977). [105,106]
417. Bamberg, J.T., and Bergmann, R.G., *J. Amer. Chem. Soc., 99,* 3173 (1977).
418. Bauch, T.E., and Giering, W.P., *J. Organometal. Chem., 144,* 335 (1978). [7]
419. Behrens, H., Geibel, K., Kellner, R., Knöchel, H.,

Moll, M., and Sepp, E., *Z. Naturforsch., B 31,* 1021 (1976). [44,48]

420. Behrens, H., Moll, M., and Würstl, P., *Z. Naturforsch., B 31,* 1017 (1976). [65]

421. Bennett, M.A., and Matheson, T.W., *J. Organometal. Chem., 153,* C 25 (1978).

422. Biehl, E.R., and Reeves, P.C., *Synthesis, 1974,* 883. [28]

423. Blake, M.R., Garnett, J.L., Gregor, I.K., and Wild, S.B., *J. Organometal. Chem., 178,* C 37 (1979). [44]

424. Boschi, T., Vogel, P., and Roulet, R., *J. Organometal. Chem., 133,* C 36 (1977). [32]

425. Boudjouk, P., and Lin, S., *J. Organometal. Chem., 155,* C 13 (1978). [104]

426. Brookhart, M., Graham, C.R., Nelson, G.O., and Scholes, G., *Ann. N.Y. Acad. Sci., 295,* 254 (1977).

427. Brookhart, M., Karel, K.J., and Nance, L.E., *J. Organometal. Chem., 140,* 203 (1977). [44,48]

428. Brookhart, M.S., Lewis, C.P., and Eisenstadt, A., *J. Organometal. Chem., 127,* C 14 (1977). [52,56]

429. Brookhart, M., and Nelson, G.O., *J. Organometal. Chem., 164,* 193 (1979). [44,65]

430. Brune, H.A., Horlbeck, G., and Zâhorszky, V.-I., *Z. Naturforsch., B 26,* 222 (1971).

431. Cais, M., Dani, S., Herbstein, F.H., and Kapon, M., *J. Amer. Chem. Soc., 100,* 5554 (1978). [95,96]

432. Charles, A.D., Diversi, P., Johnson, B.F.G., Karlin, K.D., Lewis, J., Rivera, A.V., and Sheldrick, G.M., *J. Organometal. Chem., 128,* C 31 (1977). [65]

433. Churchill, M.R., and Julis, S.A., *Inorg. Chem., 17,* 1453 (1978).

434. Churchill, M.R., Julis, S.A., King, R.B., and Harmon, C.A., *J. Organometal. Chem., 142,* C 52 (1977).

435. Connelly, N.G., and Kelly, R.L., *J. Organometal. Chem., 120,* C 16 (1976). [44,65]

436. Connelly, N.G., Kitchen, M.D., Stansfield, R.F.D., Whiting, S.M., and Woodward, P., *J. Organometal. Chem., 155,* C 34 (1978). [44,65]

437. Connor, J.A., Demain, C.P., Skinner, H.A., and Zafarani-Moattar, M.T., *J. Organometal. Chem., 170,* 117 (1979). [65]

438. Cotton, F.A., and Hanson, B.E., *Inorg. Chem., 16,* 1861 (1977). [131]

439. Cotton, F.A., Hanson, B.E., Kolb, J.R., and Lahuerta, P., *Inorg. Chem., 16,* 89 (1977). [143]

440. Cotton, F.A., Hanson, B.E., Kolb, J.R., Lahuerta, P., Stanley, G.G., Stults, B.R., and White, A.J., *J. Amer. Chem. Soc., 99,* 3673 (1977). [143]

441. Cotton, F.A., Hunter, D.L., Lahuerta, P., and White, A.J., *Inorg. Chem.*, *15*, 557 (1976). [94,144,150]
442. Cunningham, D., McArdle, P., Sherlock, H., Johnson, B.F.G., and Lewis, J., *J. Chem. Soc. Dalton Trans.*, *1977*, 2340. [44,46]
443. Davies, S.G., Green, M.L.H., and Mingos, D.M.P., *Tetrahedron*, *34*, 3047 (1978). [20,47,56,85,86]
444. Davison, A., and Rudie, A.W., *J. Organometal. Chem.*, *169*, 69 (1979). [151,152]
445. Deganello, G., Li Shing Man, L.K.K., and Takats, J., *J. Organometal. Chem.*, *132*, 265 (1977). [129]
446. Dettlaf, G., Behrens, U., Eicher, T., and Weiss, E., *J. Organometal. Chem.*, *152*, 203 (1978). [117,135]
447. Edelmann, F., and Behrens, U., *J. Organometal. Chem.*, *128*, 131 (1977). [101,140,144]
448. Efraty, A., Liebman, D., Sikora, J., and Denney, D.Z., *Inorg. Chem.*, *15*, 886 (1976). [28]
449. Eilbracht, P., and Dahler, P., *J. Organometal. Chem.*, *135*, C 23 (1977). [30,140]
450. Eilbracht, P., and Mayser, U., *J. Organometal. Chem.*, *135*, C 26 (1977). [30,31,101,140]
451. Eisenstadt, A., *Tetrahedron Lett.*, *1976*, 3543. [68,129,131]
452. El Borai, M., Guilard, R., Fournari, P., Dusausoy, Y., and Protas, J., *J. Organometal. Chem.*, *148*, 285 (1978). [45]
453. Felkin, H., Lednor, P.W., Normant, J.-M., and Smith, R.A.J., *J. Organometal. Chem.*, *157*, C 64 (1978). [65,92]
454. Franck- Neumann, M., Brion, F., and Martina, D., *Tetrahedron Lett.*, *1978*, 5033. [52]
455. Genco, N., Marten, D., Raghu, S., and Rosenblum, M., *J. Amer. Chem. Soc.*, *98*, 848 (1976). [47]
456. Goddard, R., and Woodward, P., *J. Chem. Soc. Dalton Trans.*, *1979*, 711. [44]
457. Goldschmidt, Z., and Antebi, S., *Tetrahedron Lett.*, *1978*, 271. [44]
458. Goldschmidt, Z., and Bakal, Y., *Tetrahedron Lett.*, *1976*, 1229. [52]
459. Goldschmidt, Z., and Bakal, Y., *Tetrahedron Lett.*, *1977*, 955. [46,52,79]
460. Goldschmidt, Z., and Bakal, Y., *J. Organometal. Chem.*, *168*, 215 (1979). [44,46,57]
461. Goldschmidt, Z., and Bakal, Y., *J. Organometal. Chem.*, *179*, 197 (1979). [46,57]
462. Green, J.C., Powell, P., and van Tilborg, J., *J. Chem. Soc. Dalton Trans.*, *1976*, 1974. [44,65]
463. Green, M., Heathcock, S.M., Turney, T.W., and Mingos,

D.M.P., *J. Chem. Soc. Dalton Trans.*, *1977*, 204. [52,60]

464. Greene, R.N., DePuy, C.H., and Schroer, T.E., *J. Chem. Soc. C*, *1971*, 3115.

465. Grimme, W., and Schneider, E., *Angew. Chem.*, *89*, 754 (1977); *Angew. Chem. Int. Ed. Engl.*, *16*, 717 (1977).

466. Grubbs, R.H., and Pancoast, T.A., *J. Amer. Chem. Soc.*, *99*, 2382 (1977). [28]

467. Helling, J.F., and Hendrickson, W.A., *J. Organometal. Chem.*, *141*, 99 (1977).

468. Herbstein, F.H., and Reisner, M.G., *Acta Crystallogr.*, *B 33*, 3304 (1977). [104,135]

469. Hofmann, P., *Z. Naturforsch.*, *B 33*, 251 (1978). [48]

470. Hughes, R.P., *J. Organometal. Chem.*, *141*, C 29 (1977).

471. Ittel, S.D., and Tolman, C.A., *J. Organometal. Chem.*, *172*, C 47 (1979). [44,65]

472. Hunt, D.F., and Russell, J.W., *J. Organometal. Chem.*, *104*, 373 (1976). [101,109,126,140]

473. Jablonski, C.R., *J. Organometal. Chem.*, *174*, C 3 (1979). [95,96]

474. Jeffreys, J.A.D., and Metters, C., *J. Chem. Soc. Dalton Trans.*, *1977*, 729. [44]

475. Johnson, B.F.G., Karlin, K.D., and Lewis, J., *J. Organometal. Chem.*, *174*, C 29 (1979). [44,65]

476. Johnson, B.F.G., Lewis, J., Parker, D.G., and Postle, S.R., *J. Chem. Soc. Dalton Trans.*, *1977*, 794. [14,98,99]

477. Johnson, B.F.G., Lewis, J., Stephenson, G.R., and Vichi, E.J.S., *J. Chem. Soc. Dalton Trans.*, *1978*, 369.

478. Johnson, J.W., and Treichel, P.M., *J. Amer. Chem. Soc.*, *99*, 1427 (1977); *J. Chem. Soc. Chem. Commun.*, *1976*, 688.

479. Jotham, R.W., Kettle, S.F.A., Moll, D.B., and Stamper, P.J., *J. Organometal. Chem.*, *118*, 59 (1976).

480. Kaplan, F.A., and Roberts, B.W., *J. Amer. Chem. Soc.*, *99*, 513 (1977). [29]

481. Kaplan, F.A., and Roberts, B.W., *J. Amer. Chem. Soc.*, *99*, 518 (1977). [29]

482. Karel, K.J., and Brookhart, M., *J. Amer. Chem. Soc.*, *100*, 1619 (1978). [44]

483. Katz, T.J., and Slusarek, W., *J. Amer. Chem. Soc.*, *101*, 4259 (1979). [152]

484. Kirchner, R.F., Loew, G.H., and Mueller-Westerhoff, U.T., *Inorg. Chem.*, *15*, 2665 (1976). [151]

485. LeVanda, C., Bechgaard, K., Cowan, D.O., Mueller-Westerhoff, U.T., Eilbracht, P., Candela, G.A., and Collins, R.L., *J. Amer. Chem. Soc.*, *98*, 3181 (1976). [151]

486. Litman, S., Gedanken, A., Goldschmidt, Z., and Bakal,

Y., *J. Chem. Soc. Chem. Commun.*, *1978*, 983. [52]

487. Luh, T.-Y., Lai, C.H., and Tam, S.W., *Tetrahedron Lett.*, *1978*, 5011. [68]

488. McArdle, P., *J. Organometal. Chem.*, *144*, C 31 (1978). [57]

489. McArdle, P., and Sherlock, H., *J. Chem. Soc. Dalton Trans.*, *1978*, 1678. [14]

490. Mann, B.E., *J. Organometal. Chem.*, *141*, C 33, (1977). [44]

491. Mann, B.E., *J. Chem. Soc. Dalton Trans.*, *1978*, 1761. [92]

492. Moll, M., Behrens, H., Kellner, R., Knöchel, H., and Würstl, P., *Z. Naturforsch.*, *B 31*, 1019 (1976). [44]

493. Morrison, Jr., W.H., Krogsrud, S., and Hendrickson, D.N., *Inorg. Chem.*, *12*, 1998 (1973). [151]

494. Narbel, P., Boschi, T., Roulet, R., Vogel, P., Pinkerton, A.A., and Schwarzenbach, D., *Inorg. Chim. Acta*, *36*, 161 (1979); *ibid*, *35*, 197 (1979).

495. Nunn, E.E., *Aust. J. Chem.*, *29*, 2549 (1976). [28]

496. Olsen, H., *Acta Chem. Scand.*, *B 31*, 635 (1977). [65]

497. Pettersen, R.C., and Cash, G.G., *Inorg. Chim. Acta*, *34*, 261 (1979). [134,137]

498. Reuvers, J.G.A., and Takats, J., *J. Organometal. Chem.*, *175*, C 13 (1979). [44,48]

499. Salzer, A., and von Philipsborn, W., *J. Organometal. Chem.*, *161*, 39 (1978). [44,62,63,64,68,69,71,98,99,123,125]

500. Salzer, A., and von Philipsborn, W., *J. Organometal. Chem.*, *170*, 63 (1979). [62]

501. Sapienza, R.S., Riley, P.E., Davis, R.E., and Pettit, R., *J. Organometal. Chem.*, *121*, C 35 (1976).

502. Sappa, E., Milone, L., and Tiripicchio, A., *J. Chem. Soc. Dalton Trans.*, *1976*, 1843.

503. Sepp, E., Pürzer, A., Thiele, G., and Behrens, H., *Z. Naturforsch.*, *B 33*, 261 (1978). [48]

504. Stallings, W., and Donohue, J., *J. Organometal. Chem.*, *139*, 143 (1977). [29]

505. Stegemann, J., and Lindner, H.J., *J. Organometal. Chem.*, *166*, 223 (1979).

506. Steiner, U., and Hansen, H.-J., *Helv. Chim. Acta*, *60*, 191 (1977).

507. Steiner, U., Hansen, H.-J., Bachmann, K., and von Philipsborn, W., *Helv. Chim. Acta*, *60*, 643 (1977).

508. Stringer, M.B., and Wege, D., *Tetrahedron Lett.*, *1977*, 65. [29]

509. Timms, P.L., and Turney, T.W., *J. Chem. Soc. Dalton Trans.*, *1976*, 2021. [44]

510. Victor, R., *J. Organometal. Chem.*, *127*, C 25 (1977).

[108,134]

511. Volz, H., and Draese, R., *Tetrahedron Lett.*, *1975*, 3209. [94]

512. Volz, H., and Kowarsch, H., *J. Organometal. Chem.*, *136*, C 27 (1977). [94]

513. von Büren, M., and Hansen, H.-J., *Helv. Chim. Acta*, *60*, 2717 (1977). [10,11]

514. Waterman, P.S., and Giering, W.P., *J. Organometal. Chem.*, *155*, C 47 (1978). [7]

515. Watts, W.E., *J. Organometal. Chem. Libr.*, *7*, 399 (1979). [95,96]

516. Weber, S.R., and Brintzinger, H.H., *J. Organometal. Chem.*, *127*, 45 (1977). [92]

ARENE COMPLEXES

By R.B. KING

Department of Chemistry, University of Georgia
Athens, Georgia 30602, U.S.A.

TABLE OF CONTENTS

ISBN 0-12-417102-8

I. INTRODUCTION

Several types of compounds containing *hexahapto* benzenoid rings bonded to iron are known. Dications of the type $[(arene)_2Fe]^{2+}$ and the "mixed sandwich" monocations $[(arene)-Fe(C_5H_5)]^+$ have the favoured 18-electron rare gas configuration. Numerous examples of cations of both of these types have been prepared. On the other hand the neutral 20-electron $(arene)_2Fe$ derivatives are very unstable and have only been isolated in the pure state with hexamethylbenzene. Other types of known *hexahapto* benzenoid iron compounds include derivatives of the types (arene)Fe(diene) and $(arene)Fe(CO)_2$, which also have the favoured 18-electron rare gas configuration. However, compounds of these types are comparatively rare.

Compounds are also known in which only a part of a benzenoid system is involved in the bonding to iron. For example, in styrene and other benzenoid systems containing unsaturated side chains, a 1,3-diene system involving one benzenoid double bond and one side-chain double bond can bond to a tricarbonyliron group. Also in condensed polycyclic benzenoid systems such as anthracene or naphthacene, part of the large π-system can bond to a carbonyliron group as either a 1,3-diene or a tetramethyleneethane system. In addition, bonding to iron of a condensed ring system containing both five- and six-membered planar unsaturated carbocyclic rings such as indene or acenaphthylene involves the π-electrons of the benzenoid system as well as the five-membered ring system.

Some iron π-complexes of heterocyclic systems with aromatic properties similar to benzene are known. For example, analogues to bis(arene)iron complexes have been prepared using tetramethylthiophene as well as borabenzene and phosphabenzene derivatives.

This chapter discusses first the iron complexes of benzenoid systems in which all six carbon atoms of the benzenoid system are η^6-coordinated to the iron atom. Next, benzenoid complexes are dicussed in which only two to four of the six benzenoid carbons bond to the iron atom. Finally, iron complexes are mentioned containing heterocyclic systems with π-electron systems similar to those of benzenoid derivatives.

II. IRON COMPLEXES CONTAINING *HEXAHAPTO* MONOCYCLIC BENZENOID LIGANDS

Compounds in which an iron atom is bonded to two *hexahapto* benzenoid rings are known. In addition, numerous examples of very stable cations of the type $[(arene)Fe(C_5H_5)]^+$

have been studied in considerable detail. A limited number of arene-dicarbonyliron complexes and (arene)Fe(diene) derivatives containing a *hexahapto* arene-iron bond have also been prepared.

A. *BIS(ARENE)IRON COMPLEXES AND THEIR DERIVATIVES*

Compounds of the type $[(\text{arene})_2\text{Fe}]^{2+}$ in which the iron has the favoured 18-electron rare gas configuration are relatively stable. In addition paramagnetic salts of the type $[(\text{arene})_2\text{Fe}]^{+}$ with a 19-electron configuration are known in particularly favourable cases such as hexamethylbenzene. Neutral 20-electron complexes of the type $(\text{arene})_2\text{Fe}$ have also been obtained, but these are very unstable and reactive.

1. The $[(\text{Arene})_2\text{Fe}]^{2+}$ Dications

The cations of the type $[(\text{arene})_2\text{Fe}]^{2+}$ are most frequently prepared by reactions of iron halides such as $FeBr_2$ *(34)*, $FeCl_2$ *(46)*, or $FeCl_3$ *(48)* with aluminium trichloride in the presence of the arene at elevated temperatures. The reaction can be run in cyclohexane in cases where the arene cannot be used as a solvent *(46,48)*. The cations $[(\text{arene})_2\text{Fe}]^{2+}$ can be isolated after hydrolysis as salts of a large anion such as tetraphenylborate *(34)*, diamminetetrathiocyanochromate(III) *(34)*, iodide *(34)*, or preferably hexafluorophosphate *(46,48)*. Using this method $[(\text{arene})_2\text{Fe}]^{2+}$ cations can be prepared from benzene *(48)*, toluene *(48)*, *m*-xylene *(46)*, mesitylene *(34,46)*, durene *(46)*, and hexamethylbenzene *(46)*.

A difficulty with these preparations is the tendency for the $[(\text{arene})_2\text{Fe}]^{2+}$ cation to decompose upon hydrolysis to ferrous ion and the free arene. Qualitative tests with solutions of potassium thiocyanate, ferrocyanide, and ferricyanide *(106)* indicate that the hydrolytic stabilities of the $[(\text{arene})_2\text{Fe}]^{2+}$ cations increase gradually as hydrogens are replaced by methyl groups in the benzenoid system. Thus, in the preparation of the bis(benzene)iron(II) dication, the conditions of the hydrolysis of the reaction mixture *(34,46, 48)* are extremely critical in order to avoid complete decomposition of the product. On the other hand, the bis(hexamethylbenzene)iron(II) dication can be reduced to the corresponding iron(I) and iron(O) derivatives in aqueous solution without rupture of the iron-arene bond *(36)*.

Some alternatives have been discovered to the use of iron halides as the starting materials in the formation of $[(\text{arene})_2\text{Fe}]^{2+}$ derivatives upon reaction with the corresponding benzenoid derivatives and aluminium halides. Thus, the preparation of various bis(mesitylene)iron(II) salts by the reaction of pentacarbonyliron with mesitylene in the presence of aluminium chloride has been patented *(25)*. Similarly, reaction of 1,1'-diacetylferrocene with the methylbenzenes $(CH_3)_nC_6H_{6-n}$ in the presence of aluminium chloride at 115 - 160°C results in removal of both acetylcyclopentadienyl rings to give the corresponding $[(arene)_2Fe]^{2+}$ derivatives *(12)*.

Some physical properties of $[(arene)_2Fe]^{2+}$ derivatives have been studied. The diamagnetism of various salts of the bis(mesitylene)iron(II) cation has been measured *(35)*. The wide line NMR spectrum of the bis(hexamethylbenzene)iron(II) cation has been determined and the experimental and theoretical second moments have been compared *(10)*. The bis(arene)-iron(II) hexafluorophosphate salts (arene = durene and hexamethylbenzene) have been shown to form coloured stable 1:1 charge-transfer complexes with the donor molecules benzene, naphthalene, phenanthrene, anthracene, benzo[b]thiophene, indole, aniline, *N,N*-dimethylaniline, *m*-chloroaniline, *p*-bromoaniline, *p*-phenylenediamine, hydroquinone, pyridine, furan, and ferrocene *(16)*. These molecular complexes are stable only in the solid state. When dissolved in polar solvents, an equilibrium mixture is obtained which consists almost entirely of starting materials. The colours of the individual charge-transfer complexes are highly dependent on the electron density in the donor molecule as well as on the complexed arene of the $[(arene)_2Fe]^{2+}$ cation. Thus, the colours range from pink for the benzene complexes to purple for the *N,N*-dimethylaniline complexes, and blue for the *p*-phenylenediamine complexes as the donor ability of the arene is increased *(16)*.

The reaction of bis(arene)iron(II) cations with various nucleophiles have been investigated *(46,47,49)*. Reactions of bis(mesitylene)iron(II) hexafluorophosphate with the alkyllithium compounds RLi (R = C_6H_5, $C(CH_3)_3$, and $CH{=}CH_2$) form successively the ionic monoadducts <u>1</u> and the neutral diadducts <u>2</u> (R = C_6H_5, $C(CH_3)_3$, and $CH{=}CH_2$) *(46)*. Oxidation of the diadducts with ammonium hexanitratocerate(IV) gives the corresponding substituted mesitylene <u>3</u>. This synthetic procedure is of potential value for the synthesis of substituted arenes *(46)*. Reaction of bis(mesitylene)iron(II) hexafluorophosphate with the nucleophiles KCN (in acetone), $LiCH_2{-}NO_2$ (in nitromethane), $LiCH(CH_3){-}NO_2$ (in nitroethane), and $LiCH_2{-}CO_2{-}C(CH_3)_3$ (in diethyl ether) gives only the corresponding monoadducts <u>1</u> (R = CN, $CH_2{-}NO_2$, $CH(CH_3){-}NO_2$, and $CH_2{-}CO_2{-}C(CH_3)_3$, respectively) *(49)*. Reaction of bis(mesitylene)iron(II) hexafluorophosphate with the nucleophiles $LiNH_2$ (in tetrahydrofuran), $LiN(CH_3)_2$ (in benzene), $NaN[Si(CH_3)_3]_2$ (in benzene), $LiOCH_3$ (in methanol), and $KOC(CH_3)_3$ (in tetrahydrofuran) all give the same product <u>4</u> *(49)*.

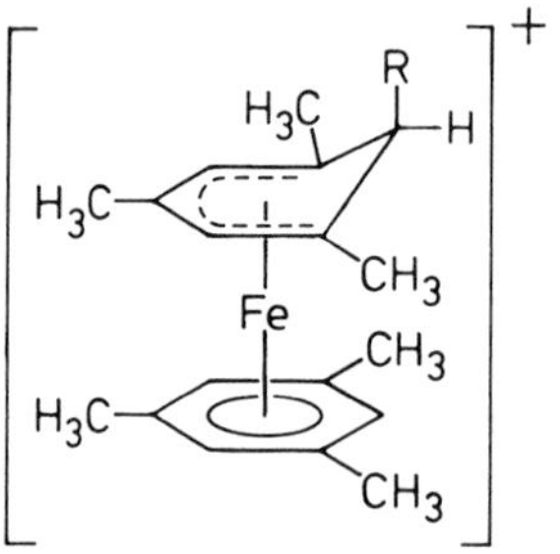

1

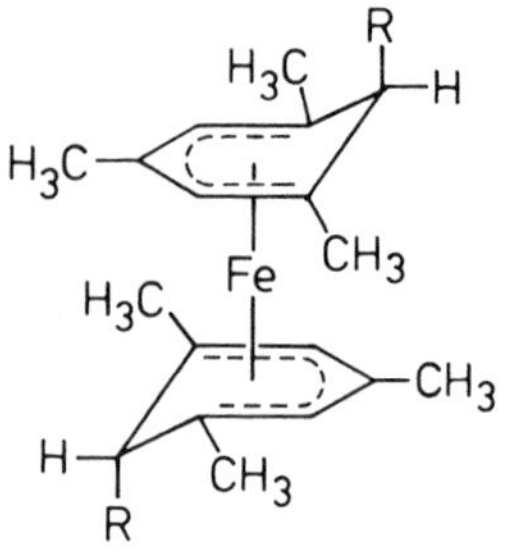

2

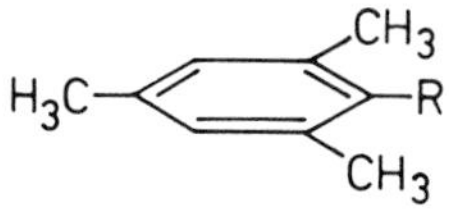

3

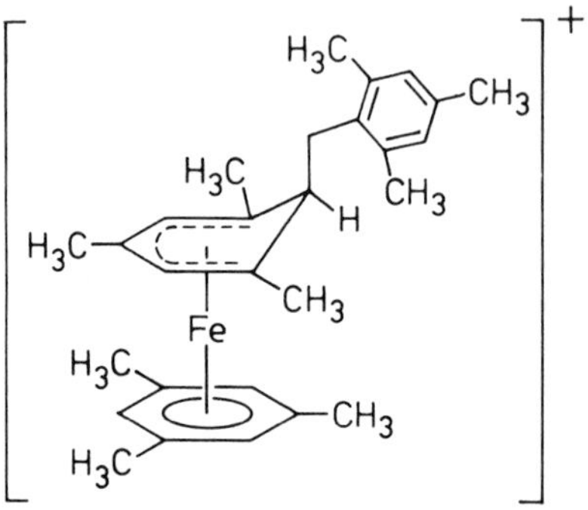

4

The 1:2 adducts from bis(mesitylene)iron(II) hexafluorophosphate and alkyllithium compounds are formulated as the corresponding substituted bis(cyclohexadienyl)iron derivatives 2 on the basis of an X-ray crystallographic study *(69)*. A ^{1}H-NMR study of 2 (R = C_6H_5, $(CH_3)_3C$) indicates fluxional structures with equivalent cyclohexadienyl rings at all temperatures studied. In the low temperature limiting spectrum the substituted cyclohexadienyl rings have no symmetry whereas in the high temperature limiting spectrum the substituted cyclohexadienyl rings have an effective plane of symmetry *(47)*. Free rotation with respect to the iron-ring axis is suggested to occur in the high temperature limiting spectrum in order to generate this effective plane of symmetry. The ^{1}H-NMR spectra of the corresponding monoadducts 1 (R = C_6H_5, $C(CH_3)_3$) are temperature independent.

2. Reduction Products of the $[(Arene)_2Fe]^{2+}$ Dications

If the $[(arene)_2Fe]^{2+}$ cations are sufficiently stable towards hydrolytic cleavage of the arene, they can be reduced to the corresponding bis(arene)iron(I) and -iron(O) derivatives. Thus, reduction of the $[(arene)_2Fe]^{2+}$ hexafluorophosphates (arene = mesitylene *(9)*, hexamethylbenzene *(36,9)*) with sodium dithionite in an aqueous solution buffered with sodium acetate gives the corresponding deep blue to violet iron(I) derivatives $[(arene)_2Fe]^+$ which may be isolated as their hexafluorophosphate salts. The magnetic moments of these iron(I) complexes correspond to the one unpaired electron expected for a 19-electron complex *(36,9)*. The NMR contact shift *(9)* and electron spin resonance *(19,116)* data on the 19-electron iron(I) complex $\{[(CH_3)_6C_6]_2Fe\}[PF_6]$ have been interpreted to indicate that the bis(arene)iron(I) cation has lost the axial symmetry of its 18-electron homologues owing to Jahn-Teller distortion upon addition of the 19th electron.

Further reduction of the hexamethylbenzene-iron(I) complex $\{[(CH_3)_6C_6]_2Fe\}[PF_6]$ by the addition of an extreme excess of potassium hydroxide gives the black crystalline extremely air-, light-, and temperature-sensitive iron(O) complex $[(CH_3)_6C_6]_2Fe$ *(36)*. Magnetic susceptibility measurements on this 20-electron complex indicate two unpaired electrons similar to the 20-electron complex bis(cyclopentadienyl)nickel *(36)*.

An alternative method of possible value for the preparation of neutral $(arene)_2Fe$ derivatives is the co-condensation of iron atoms with various aromatic hydrocarbons *(71,100-102)*. Co-condensation of iron atoms with benzene gives a black solid, which has not been characterized directly since it explodes above -50°C *(100)*. The presence of benzene-iron bonds in this unstable product is suggested by the observation that addition of phosphorus trifluoride below its explosion temperature gives some $(C_6H_6)Fe(PF_3)_2$ *(100)*. A similar co-condensation of iron atoms with hexafluorobenzene also gives a complex which decomposes explosively at relatively low temperatures, this time above -40°C *(62)*. These observations all indicate the relative instability of 20-electron $(arene)_2Fe$ complexes. Co-condensation of iron atoms with mesitylene results in the abstraction of two extra hydrogen atoms to give $(C_9H_{12})Fe(C_9H_{14})$, formulated as an 18-electron arene-cyclohexadiene-iron(O) complex (see below) *(100,102)*.

B. η^6-*ARENE-IRON COMPLEXES CONTAINING ONE BENZENOID RING*

Several types of iron complexes are known containing one *hexahapto* benzenoid ring. In general the iron atom exhibits the usual favoured 18-electron rare gas configuration in these complexes. The most stable iron derivatives of this type are the mixed sandwich cations of the type $[(arene)Fe(cyclopentadienyl)]^+$. In addition, a few examples of neutral 18-electron iron derivatives are known of the types (arene)Fe(diene) and $(arene)FeL_2$ (L = two-electron donor such as CO or PF_3).

1. The $[(Arene)Fe(cyclopentadienyl)]^+$ Cations

Cations of the type $[(arene)Fe(cyclopentadienyl)]^+$ were first prepared by reactions of cyclopentadienyl-dicarbonyl-iron halides with the corresponding aromatic hydrocarbons in the presence of anhydrous aluminium chloride according to equation [1] *(23,24,43,70)*.

$$(C_5H_5)Fe(CO)_2X + AlX_3 + \text{arene} \longrightarrow [(C_5H_5)Fe(\text{arene})]^+[AlX_4]^- + 2\ CO \qquad [1]$$

After hydrolysis the complex cation was precipitated by the addition of appropriate anions such as iodide *(23,24,70)* or tribromide *(43)*. Cations prepared by this method include the benzene derivative $[(C_5H_5)Fe(C_6H_6)]^+$ *(43)* and the mesitylene derivative $\{(C_5H_5)Fe[C_6H_3(CH_3)_3]\}^+$ *(23,23,43)*. In addition, the pentaphenylcyclopentadienyl derivative $\{[(C_6H_5)_5C_5]\text{-}Fe(C_6H_6)\}^+$ has also been prepared by an analogous method *(70)*.

This method for the preparation of $[(C_5H_5)Fe(arene)]^+$ cations was made obsolete through the subsequent discovery by Nesmeyanov, Vol'kenau, and Bolesova of a one step method for converting ferrocene into a wide range of $[(C_5H_5)Fe(arene)]^+$ derivatives *(75)*. Thus, treatment of ferrocene with the aromatic hydrocarbon at 80-165°C in the presence of a two- to four-fold excess of aluminium chloride and a stoichiometric quantity of aluminium powder results in substitution of one of the five-membered rings with a six-membered ring to give the corresponding $[(C_5H_5)Fe(arene)]^+$ cation. Such cations may be isolated after hydrolysis as their tetraphenylborate or preferably hexafluorophosphate salts. Simple benzenoid derivatives forming $[(C_5H_5)Fe(arene)]^+$ derivatives by this method include benzene *(74,75)*, toluene *(11,78)*, ethylbenzene *(11, 96)*, *n*-propylbenzene *(11)*, *i*-propylbenzene *(11)*, *t*-butylbenzene *(11)*, *o*-xylene *(11)*, *m*-xylene *(11)*, *p*-xylene *(11,78)*, mesitylene *(74,75)*, tetrahydronaphthalene *(74,75)*, fluorobenzene *(96)*, chlorobenzene *(78)*, acetanilide *(78)*, *p*-dichlorobenzene *(96)*, biphenyl *(78)*, and several *p*-substituted toluenes of the type $X\text{-}C_6H_4\text{-}CH_3$ ($X = SCH_3$, C_2H_5, $CH_3\text{-}CO\text{-}NH$, F) *(96)*. A study of the relative reactivities of various methylbenzenes in this reaction indicates that reactivity is enhanced by the donor effects of methyl substituents but decreased by their steric hindrance *(11)*. These conflicting effects of methyl substitution are optimized with *p*-xylene. Polycyclic benzenoid hydrocarbons such as naphthalene *(74,28)* and fluorene *(78)* also form the corresponding $[(arene)Fe(C_5H_5)]^+$ derivatives upon treatment with ferrocene, aluminium, and aluminium chloride. Reactions of the polycyclic benzenoid hydrocarbons biphenyl *(65,72)*, diphenylmethane *(65,72)*, fluorene *(65,72)*, 9,10-dihydroanthracene *(72)*, dibenzo[1,2:4,5]cyclohepta-1,4-diene, 3,3',4,4'-tetramethylbiphenyl, *p*-terphenyl, anthracene, phenanthrene, pyrene, chrysene, β-phenylnaphthalene, and dibenzo[1,2:5,6]anthracene with excess ferrocene in the presence of aluminium chloride and aluminium using cyclohexane or decalin as a solvent give $\{(arene)[Fe(C_5H_5)]_2\}^{2+}$ cations in which two $(C_5H_5)Fe$ units are attached to different benzenoid rings of the polycyclic hydrocarbon. However, no compounds have been prepared containing three or more $(C_5H_5)Fe$ units bonded to a system containing the necessary number of benzenoid rings. In the cases of the binuclear dications of *p*-terphenyl, anthracene, phenanthrene, pyrene, chrysene, and β-phenylnaphthalene, the NMR spectra indicate that the cyclopentadienyliron units are bonded to benzenoid rings at the extreme ends of the polycyclic system in all cases *(72)*.

Substituted ferrocenes undergo similar substitutions of cyclopentadienyl rings with benzenoid rings upon treatment with the benzenoid hydrocarbon in the presence of aluminium powder and aluminium chloride. The relative rates of reactivity are $(C_2H_5\text{-}C_5H_4)_2Fe > (C_5H_5)_2Fe > (CH_3\text{-}CO\text{-}C_5H_4)_2Fe$ *(74-76)*. This indicates that electron-donor substituents on the ferrocene ring facilitate ligand exchange whereas electron-acceptor substituents impede ligand exchange. Use of such substituted ferrocenes provides methods for preparing $[(arene)Fe(cyclopentadienyl)]^+$ derivatives containing substituents on the cyclopentadienyl ring. The reaction between 1,1'-diacetylferrocene, various methylbenzenes, aluminium chloride, and aluminium powder can be controlled to replace either one or both acetylcyclopentadienyl rings to give either $[(arene)Fe(C_5H_4\text{-}CO\text{-}CH_3)]^+$ or $[(arene)_2Fe]^{2+}$ derivatives, respectively *(12)*.

Chlorine substituents on either the six- or five-membered rings in $[(arene)Fe(cyclopentadienyl)]^+$ derivatives are very reactive towards nucleophilic substitution in contrast to the chlorine atoms in either chlorobenzene or chloroferrocenes. Thus, reaction of the chlorobenzene complex $[(C_5H_5)Fe(C_6H_5Cl)][BF_4]$ with the nucleophiles sodium ethoxide in ethanol, sodium carbonate in ethanol, sodium phenoxide in acetone, sodium thiophenoxide in ethanol, and sodium *n*-butanethiolate in ethanol results in the nucleophilic substitution of the chlorine to give the corresponding $[(C_5H_5)Fe(C_6H_5X)]^+$ derivatives generally isolated as their tetraphenylborate salts ($X = OC_2H_5$, OC_2H_5, OC_6H_5, SC_6H_5, and SC_4H_9) *(79)*. A similar reaction of $[(C_5H_5)Fe(C_6H_5Cl)][BF_4]$ with potassium phthalimide in acetone followed by decomposition of the resulting *N*-phthalimido derivative with hydrazine hydrate in ethanol gives the aniline derivative $[(C_5H_5)Fe(C_6H_5\text{-}NH_2)][PF_6]$ *(79)*. Similar nucleophilic substitution reactions have been done on the chlorine of the *p*-chlorotoluene complex $[(C_5H_5)Fe(CH_3-C_6H_4Cl)][BF_4]$ *(96)*.

Reaction of 1,1'-dichloroferrocene with benzene in the presence of aluminium chloride followed by addition of tetrafluoroborate gives the salt $[(C_6H_6)Fe(C_5H_4Cl)][BF_4]$ containing a mobile chlorine on the cyclopentadienyl ring *(80)*. Nucleophilic substitutions of this chlorine with ethanol in the presence of sodium carbonate, sodium thio-phenoxide in ethanol, potassium phthalimide in dimethylformamide, piperidine in ethanol, *n*-amylamine in ethanol, and liquid ammonia in an autoclave at 80°C give the corresponding $[(C_6H_6)Fe(C_5H_4X)]^+$ derivatives ($X = OC_2H_5$, SC_6H_5, $N(CO)_2C_6H_4$, $N(CH_2)_5$, $NH-C_5H_{11}$, and NH_2) isolated as their tetrafluoroborate or hexafluorophosphate salts *(80)*. A kinetic study on the relative rates of the reactions of sodium methoxide with the chlorobenzene complex $[(C_5H_5)Fe(C_6H_5Cl)][BF_4]$ and the chlorocyclopentadienyl complex $[(C_6H_6)Fe(C_5H_4Cl)][BF_4]$ indicates the following *(84)*: (a) The chlorine in the chlorobenzene derivative is almost three orders of magnitude more mobile than the chlorine in the chlorocyclopentadienyl derivative; (b) the chlorine in $[(C_5H_5)Fe(C_6H_5Cl)][BF_4]$ is several orders of magnitude more mobile than the chlorine in *p*-chloronitrobenzene and of similar mobility to the chlorine in 2,4-dinitrochlorobenzene; (c) the chlorine in $[(C_5H_5)Fe(C_6H_5Cl)][BF_4]$ is more mobile than the fluorine in (fluorobenzene)tricarbonylchromium and much more mobile than the chlorine in (chlorobenzene)tricarbonylchromium.

Some properties of the aniline derivative $[(C_5H_5)Fe(C_6H_5-NH_2)][BF_4]$ have been investigated *(83,85)*. Alkylations of $[(C_5H_5)Fe(C_6H_5-NH_2)][BF_4]$ with triethyloxonium tetrafluoroborate in acetonitrile at 0°C and with methyl iodide in dimethylformamide at room temperature for 75 days give the corresponding *N*-alkylaniline derivatives $[(C_5H_5)Fe(C_6H_5-NHR)]^+$ ($R = C_2H_5$ and CH_3, respectively) *(83)*. Diazotization of $[(C_5H_5)Fe(C_6H_5-NH_2)][BF_4]$ with sodium nitrite in hydrochloric acid at 0°C gives the corresponding diazonium dication $[(C_5H_5)Fe(C_6H_5-N_2)]^{2+}$. This cation can be isolated as its hexafluorophosphate salt which is stable to 90°C. This diazonium salt couples with β-naphthol in the usual manner. It also undergoes a Sandmeyer reaction with copper(I) chloride to give the chlorobenzene complex $[(C_5H_5)Fe(C_6H_5Cl)]^+$ *(83)*. The amines $[(C_5H_5)Fe(C_6H_5-NH_2)][BF_4]$ and $[(C_6H_6)Fe(C_5H_4-NH_2)][BF_4]$ are both very weak bases with the former amine being a significantly weaker base than the latter amine *(85)*.

Cations of the type $[(\text{arene})Fe(\text{cyclopentadienyl})]^+$ are extremely stable towards oxidizing agents *(82)*. Thus, the cation $[(C_5H_5)Fe(C_6H_6)]^+$ is stable towards treatment with alkaline hydrogen peroxide at room temperature or even towards boiling with chromium trioxide in dilute sulfuric acid for 7 h. For this reason it is possible to oxidize alkyl side chains to carboxyl groups in $[(arene)Fe(cyclopentadienyl)]^+$ derivatives without destroying the sandwich complex. For example, the toluene derivative $[(C_5H_5)Fe(C_6H_5\text{-}CH_3)][BF_4]$ can be oxidized to the benzoic acid derivative $[(C_5H_5)Fe(C_6H_5\text{-}CO_2H)]^+$ in nearly quantitative yield by boiling for 18 to 20 h with excess aqueous potassium permanganate. This benzoic acid derivative, conveniently isolated as its hexafluorophosphate salt, can be converted into the corresponding acid chloride, carboxamide, and nitrile by normal methods. A similar oxidation of the *p*-xylene derivative $\{(C_5H_5)Fe[C_6H_4(CH_3)_2]\}^+$ leads to a mixture of the monocarboxylic and dicarboxylic acids, $[(C_5H_5)Fe(CH_3\text{-}C_6H_4\text{-}CO_2H)]^+$ and $\{(C_5H_5)Fe[C_6H_4(CO_2H)_2]\}^+$, respectively *(82)*. Similar permanganate oxidations of $[(C_6H_6)Fe(C_5H_4\text{-}C_2H_5)]^+$ and $[(C_5H_5)Fe(p\text{-}CH_3\text{-}C_6H_4Cl)]^+$ lead to the corresponding carboxylic acids $[(C_6H_6)Fe(C_5H_4\text{-}CO_2H)]^+$ and $[(C_5H_5)Fe(p\text{-}CO_2H\text{-}C_6H_4Cl)]^+$ *(81)*. The mobile chlorine atom in the latter carboxylic acid can be substituted with thiophenoxide and sodium ethoxide in the normal manner *(81)*.

The acidities of carboxyl substituents in various $[(arene)Fe(cyclopentadienyl)]^+$ derivatives have been determined *(97)*. The benzoic acid derivative $[(C_5H_5)Fe(C_6H_5\text{-}CO_2H)]^+$ is a stronger acid by *ca.* 2.5 orders of magnitude than free benzoic acid. The carboxylic acid $[(C_6H_6)Fe(C_5H_4\text{-}CO_2H)]^+$ is a slightly weaker acid than its isomer but still at least three orders of magnitude stronger than either benzoic acid or ferrocene carboxylic acid.

Some reactions of $[(arene)Fe(cyclopentadienyl)]^+$ derivatives with nucleophiles have been investigated. Hydride additions generally using sodium borohydride in tetrahydrofuran or a similar solvent have been studied in the greatest detail *(43,55,56,65)*. The products were first believed to be (arene)Fe(cyclopentadiene) derivatives *(43)* but were subsequently *(54)* shown to be (cyclohexadienyl)Fe(cyclopentadienyl) derivatives as indicated by structure 5. Reactions of the substituted methylbenzene derivatives $\{(C_5H_5)Fe[C_6(CH_3)_xH_{6-x}]\}^+$ ($0 \leq x \leq 6$; all possible isomers studied) indicate exclusive attack by the hydride on the benzenoid ring in all cases to give cyclohexadienyl derivatives of the general structure 5. In these reactions hydride addition to an unsubstituted arene carbon is preferred. However, if no such carbon is available, as in the case of the hexamethylbenzene complex, the hydride addition, this time to a substituted carbon, can still take place. Similarly, the addition of hydride to a variety of chloroarene-cyclopentadienyl-iron cations, $[(C_5H_5)Fe(R\text{-}C_6H_4Cl)][BF_4]$ (R = H, CH_3, and Cl; all

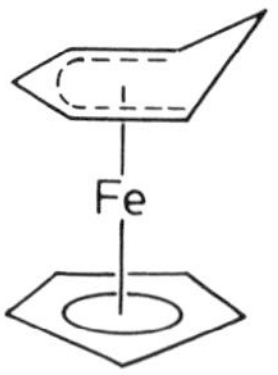

5

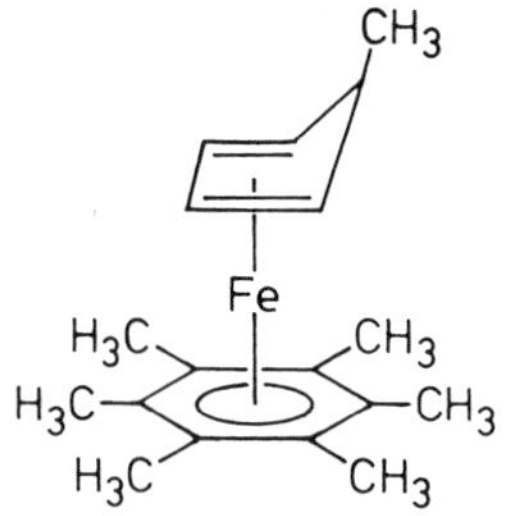

6

possible isomers), also gives exclusively cyclohexadienyl derivatives of the type 5 *(56)*. In this case hydride addition at a position *ortho* to the halogen substituent is favoured *(56)*. Sodium borohydride reductions of the binuclear $\{(\text{arene})[Fe(C_5H_5)]_2\}^{2+}$ dications from biphenyl, diphenylmethane, and fluorene *(65)* give rather unstable volatile binuclear cyclohexadienyl-iron derivatives. The binuclear nature of these hydride addition products as determined by mass spectrometry was used as evidence for the binuclear formulations of their dicationic precursors.

Some similar reactions of $[(\text{arene})Fe(\text{cyclopentadienyl})]^+$ cations with methyllithium to give substituted cyclohexadienyl derivatives have also been investigated *(57)*. In most cases these reactions parallel the corresponding reactions of sodium borohydride. An interesting exception is the reaction of methyllithium with the hexamethylbenzene complex $\{(C_5H_5)Fe[C_6(CH_3)_6]\}[PF_6]$ to give the methylcyclopentadiene complex 6 rather than a substituted cyclohexadienyl derivative of the general type 5. In addition some evidence for production of minor amounts of analogues of 6 was obtained from reactions of methyllithium with the durene and pentamethylbenzene $[(C_5H_5)Fe(\text{arene})]^+$ complexes. Apparently, methyl groups on the benzenoid ring can inhibit its attack by the methyllithium so that the lithium reagent attacks the normally unfavourable cyclopentadienyl ring.

The cyclohexadienyl-cyclopentadienyl-iron derivatives of the type 5 normally react with triphenylmethyl tetrafluoroborate in dichloromethane or with *N*-bromosuccinimide in methanol *(55,56)* with hydride abstraction from the cyclohexadienyl ligand to regenerate the $[(\text{arene})Fe(\text{cyclopentadienyl})]^+$ cation. However, similar reactions of cyclohexadienyl derivatives of the type $(exo\text{-R-}C_6H_6)Fe(C_5H_5)$ (5; R = C_2H_5, $C_6H_5\text{-}CH_2$, or C_5H_5) with triphenylmethyl tetrafluoroborate or *N*-bromosuccinimide can proceed either by *exo*-R abstraction to give $[(C_5H_5)Fe(C_6H_6)]^+$ or by *endo*-H abstraction to give $[(C_5H_5)Fe(C_6H_5\text{-R})]^+$. The tendency for R-abstraction increases along the series R = $C_5H_5 < C_2H_5 < C_6H_5\text{-}CH_2$ *(77)*.

Not all reactions of $[(C_5H_5)Fe(C_6H_6)][BF_4]$ with nucleophiles form stable cyclohexadienyl derivatives of the type 5.

For example, reaction of $[(C_5H_5)Fe(C_6H_6)][BF_4]$ with sodium ethoxide, sodium hydroxide, sodium amide, *etc.*, may simply lead to disproportionation to give inorganic iron(II) and ferrocene *(87)*. Furthermore, heating compounds of the type 5 results in decomposition to ferrocenes at temperatures below 110°C *(77)*.

Earlier in this chapter the reduction of the 18-electron iron(II) complexes $[(arene)_2Fe]^{2+}$ to the less stable 19-electron iron(I) complexes $[(arene)_2Fe]^+$ was mentioned. There is similar evidence for the reduction of the 18-electron complexes $[(C_5H_5)Fe(arene)]^+$ to the unstable neutral 19-electron complexes $(C_5H_5)Fe(arene)$. Polarography indicates that the salt $[(C_5H_5)Fe(C_6H_6)][PF_6]$ undergoes an electrochemically reversible one-electron reduction to give a solution exhibiting an ESR signal which is ill-defined apparently because of metallic iron decomposition products *(31)*. Binuclear complexes of the type $\{(arene)[Fe(C_5H_5)]_2\}^{2+}$ from biphenyl, 1,1'-binaphthyl, biphenylmethane, fluorene, 9,10-dihydroanthracene, anthracene, chrysene, *trans*-stilbene, *p*-terphenyl, *etc.*, undergo reversible polarographic one-electron reduction to a mixed iron(I)-iron(II) monocation *(73)*.

An unstable green solution of neutral (benzene)(cyclopentadienyl)iron, $(C_5H_5)Fe(C_6H_6)$, can be obtained by reduction of $[(C_5H_5)Fe(C_6H_6)][BF_4]$ with excess 1 % sodium amalgam below room temperature at carefully controlled temperatures *(89)*. This dark green solution of $(C_5H_5)Fe(C_6H_6)$ is stable for several days at -80°C but decomposes rapidly at room temperature. Decomposition of $(C_5H_5)Fe(C_6H_6)$ in tetrahydrofuran or acetonitrile at room temperature gives ferrocene. The quantitative oxidation of $(C_5H_5)Fe(C_6H_6)$ to $[(C_5H_5)Fe(C_6H_6)]^+$ by titration with triphenylchloromethane can be used for the determination of its concentration in solution. Reaction of $(C_5H_5)Fe(C_6H_6)$ with naphthalene results in rapid exchange of the benzenoid hydrocarbons to give the neutral naphthalene complex $(C_5H_5)Fe(C_{10}H_8)$ characterized by its air oxidation to the corresponding cation $[(C_5H_5)Fe(C_{10}H_8)]^+$. Reaction of $(C_5H_5)Fe(C_6H_6)$ with carbon monoxide in tetrahydrofuran at room temperature and atmospheric pressure gives $[(C_5H_5)Fe(CO)_2]_2$ in up to 70 % yields. A similar reaction of $(C_5H_5)Fe(C_6H_6)$ with triphenyl phosphite proceeds analogously to give $\{(C_5H_5)Fe[P(OC_6H_5)_3]_2\}_2$. Thus, the unstable and reactive 19-electron system $(C_5H_5)Fe(C_6H_6)$ is a good source of reactive $(C_5H_5)Fe$ units under mild conditions *(89)*.

The cation $[(C_5H_5)Fe(C_6H_6)]^+$ can also be used to prepare some cyclopentadienyliron derivatives of β-diketones *(93,94)*. Treatment of aqueous solutions of this cation with the β-diketones $R\text{-}CO\text{-}CH_2\text{-}CO\text{-}R'$ ($R = R' = CH_3$, C_6H_5; $R = CF_3$, $R' =$ thienyl) results in displacement of the benzene ring to give the corresponding $(C_5H_5)Fe(R\text{-}CO\text{-}CH\text{-}CO\text{-}R')_2$ derivatives.

The infrared spectra of $[(C_5H_5)Fe(C_6H_6)]^+$ *(92)* and some of its substitution products *(86)* have been investigated. In the unsubstituted $[(C_5H_5)Fe(C_6H_6)]^+$ as well as its deuterated derivative $[(C_5D_5)Fe(C_6H_6)]^+$ the vibrations from the $(C_5H_5)Fe$ unit can be treated by C_{5v} local symmetry and the vibrations from the $(C_6H_6)Fe$ unit by C_{6v} local symmetry. The $(C_5H_5)Fe$ bond in $[(C_5H_5)Fe(C_6H_6)]^+$ is weaker than that in ferrocene *(92)*. The vibrations between the two different rings in $[(arene)Fe(C_5H_5)]^+$ derivatives were found to be poorly coupled *(86)*. Thus, the frequencies of the cyclopentadienyl ring are almost unaffected by introduction of substituents into the benzene ring and *vice versa*.

The ^1H-NMR spectra of salts of the types $[(C_5H_5)Fe(C_6H_5X)][PF_6]$, $[(p\text{-}X\text{-}C_6H_4\text{-}CH_3)Fe(C_5H_5)][PF_6]$, and $[(C_6H_6)Fe(C_5H_4X)][PF_6]$ containing various substituents X have been correlated with the sets of Hammett-Taft σ-parameters *(88)*. ^{19}F-NMR data on the *para*- and *meta*-isomers of $[(C_5H_5)Fe(CH_3\text{-}C_6H_4\text{-}X)][PF_6]$ indicate than the benzenoid ring in this complex is a strong electron acceptor *(88)*.

Mössbauer studies on $[(arene)Fe(cyclopentadienyl)]^+$ derivatives have been reported *(98,99)*. In general, compounds of this type have isomer shifts in the range of 0.51 to 0.62 mm/sec and quadrupole splittings in the range of 1.45 to 1.76 mm/sec *(98)*. The temperature dependence of the Mössbauer spectrum of the fluorobenzene complex $[(C_5H_5)Fe(C_6H_5F)][PF_6]$ is normal for sandwich compounds only in the range 80-220 K *(99)*. Above 220 K unusual changes in its Mössbauer spectrum occur. This has been related to an unusual phase transition at these higher temperatures.

Since the $[(arene)Fe(cyclopentadienyl)]^+$ derivatives are ionic, their mass spectra cannot be obtained by conventional techniques because of insufficient volatility. However, the mass spectra of $[(C_5H_5)Fe(arene)][PF_6]$ (arene = benzene, toluene, and mesitylene) have been obtained by field desorption techniques *(42)*. In all cases ions corresponding to the $[(C_5H_5)Fe(arene)]^+$ cation are observed. In the mass spectrum of $[(C_5H_5)Fe(C_6H_6)][PF_6]$ the cluster ion $\{[(C_5H_5)Fe(C_6H_6)]_2\text{-}[PF_6]\}^+$ is also observed.

2. (Arene)$Fe(CO)_2$ and (Arene)$Fe(PF_3)_2$ Derivatives

Complexes of the type (arene)$Fe(CO)_2$ are relatively rare. (*p*-Xylene)dicarbonyliron has been claimed as a low yield photolysis product of (trimethylenemethane)tricarbonyliron under certain conditions *(29)*. Carbonylation of bis(hexamethylbenzene)iron(0) gives red sublimable crystalline $[(CH_3)_6C_6]Fe(CO)_2$ *(20)*. A dimeric (hexamethylbenzene)dicarbonyliron has been isolated in extremely low yield (0.3 %) from the ultraviolet irradiation of pentacarbonyliron with excess hexamethylbicyclo[2.2.0]hexadiene *(38)*.

Ion-molecule reactions leading to benzene-carbonyliron ions have been studied by ion-cyclotron resonance spectroscopy *(40,41)*. The final products from the sequence of ion-molecule reactions in mixtures of benzene and pentacarbonyliron are $[(C_6H_6)Fe(CO)_2]^+$ and $[(C_6H_6)_2Fe]^+$.

The lower volatility of phosphorus trifluoride relative to carbon monoxide allows phosphorus trifluoride to be used as a reagent in iron atom co-condensations. Thus, the co-condensation of iron atoms with an arene and phosphorus trifluoride has been used to prepare (arene)$Fe(PF_3)_2$ derivatives of benzene *(71)* and toluene *(117)*. The products are volatile, air-sensitive red substances. At room temperature $(C_6H_6)Fe(PF_3)_2$ (7, R = H) is a solid but $(CH_3-C_6H_5)Fe(PF_3)_2$ (7, R = CH_3), is a liquid.

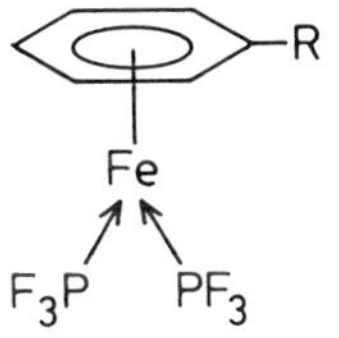

7

3. (Arene)Fe(diene) Derivatives

The first compound of the type (arene)Fe(diene) to be prepared was the complex $(C_6H_6)Fe(C_6H_8)$ (8, R = H). This complex was first obtained *(37)* by the reduction of a mixture of ferric chloride and excess cyclohexa-1,3-diene with excess isopropylmagnesium bromide in the presence of ultraviolet irradiation. This complex 8 (R = H) was later prepared by the co-condensation of iron atoms with cyclohexa-1,3-diene *(117)*. A similar co-condensation of iron atoms with mesitylene gives the related complex *(101,102)* $(C_9H_{12})Fe(C_9H_{14})$ (8, R = CH_3). Apparently in this case the initially formed 20-electron bis(arene)iron intermediate abstracts two hydrogen atoms. Co-condensation of iron vapour with toluene followed

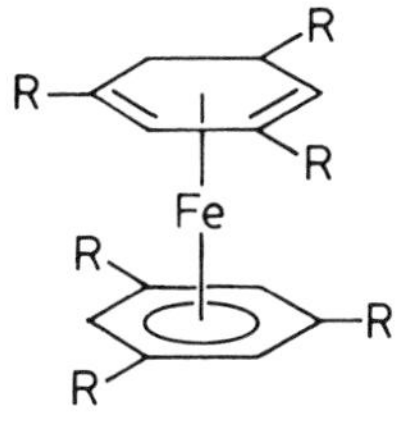

8

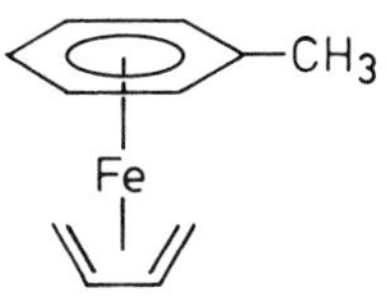

9

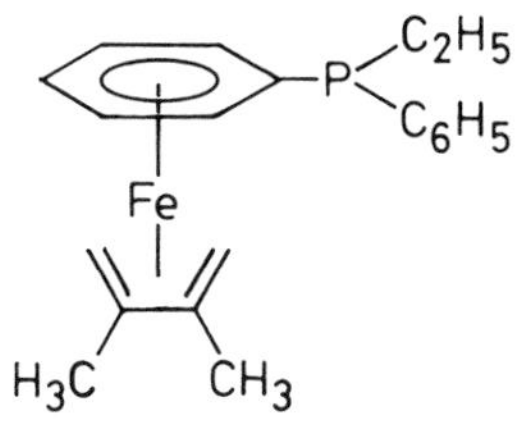

10

by addition of butadiene and warming to room temperature gives the volatile red liquid $(CH_3-C_6H_5)Fe(C_4H_6)$ (9) *(117)*.

Some more unusual (arene)Fe(diene) derivatives are known in which the η^6-arene ligand is a phenyl ring of a phenylphosphine *(39,63)*. Photolysis of $[(C_6H_5)_2PC_2H_5]_3FeH_2N_2$ with 2,3-dimethylbutadiene gives orange crystalline 10. The reactions of 10 are discussed in detail in the chapter on diene-iron complexes.

III. IRON COMPLEXES CONTAINING *DIHAPTO, TRIHAPTO,* AND *TETRAHAPTO* BENZENOID LIGANDS

No complexes are known in which a simple benzenoid ligand bonds to an iron atom using less than all six carbon atoms of the benzene ring. However, if the benzene ring has substituents which can also form metal complexes, then iron complexes can be obtained in which less than six carbon atoms of the benzenoid ring combine with the donor atoms of the substituent to form a *polyhapto* ligand. For example vinylbenzenes can function as dienes using one benzenoid double bond and one double bond of a vinylic substituent. Furthermore, when the five-membered ring of indene forms a *pentahapto* complex similar to η^5-cyclopentadienyl complexes, then two carbon atoms of the benzenoid ring of indene are necessarily involved. More complex organosulfur systems are also known in which a donor sulfur atom and one benzenoid double bond are apparently involved in the bonding to iron.

A. *IRON CARBONYL COMPLEXES OF BENZENOID DERIVATIVES CONTAINING FUSED UNSATURATED GROUPS*

Several types of benzenoid systems containing adjacent planar networks of unsaturated carbon atoms form carbonyliron complexes in which tricarbonyliron groups are bonded only to part of a benzene ring. Certain polycyclic benzenoid systems, notably anthracene, have sufficient dienoid character in one of their benzene rings to form complexes of the $(diene)Fe(CO)_3$ type. Vinylbenzenes also form $(diene)Fe(CO)_3$ derivatives using a vinyl and a benzenoid double bond.

The general theory of tricarbonyliron complexes of benzenoid derivatives with adjacent unsaturated groups has been developed by Nicholson *(91)*. The tendency for such systems to form tricarbonyliron complexes can be satisfactorily explained by a localization energy approximation of the Hückel type which describes the ease with which a 1,3-diene unit can be partially isolated from the π-electron system of the hydrocarbon.

1. Reactions of Iron Carbonyls with Fused Polycyclic Benzenoid Systems

Neither benzene nor naphthalene appear to form stable tricarbonyliron complexes *(68)*. Thus, a compound originally *(45)* believed to be (naphthalene)tricarbonyliron was subsequently *(68)* identified as (thianaphthene)hexacarbonyldiiron involving bonding of a different type. However, reactions of anthracene *(68)* and benz[a]anthracene *(14)* with dodecacarbonyltriiron in boiling benzene or cyclohexane give the corresponding tricarbonyliron derivatives. In both cases the ^{1}H-NMR spectra indicate that the tricarbonyliron group is bonded to an end ring as in 11 and 12, respectively (E = E' = CH).

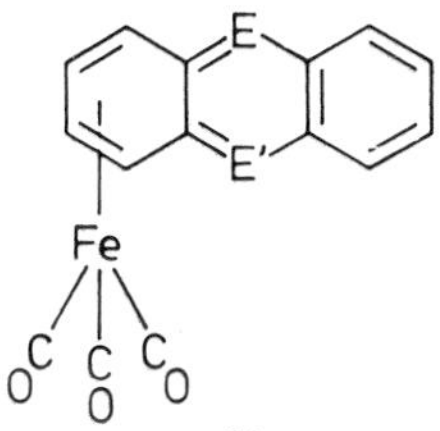

11

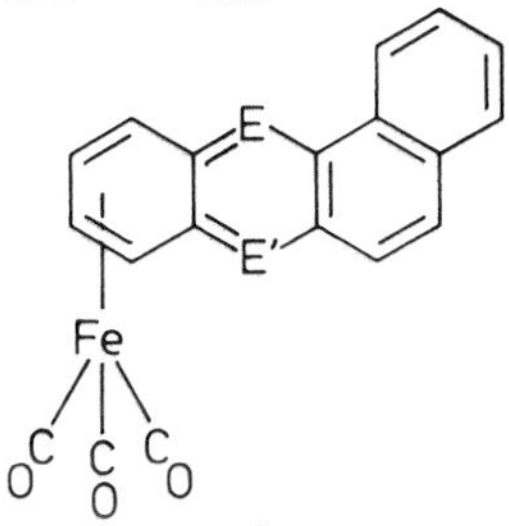

12

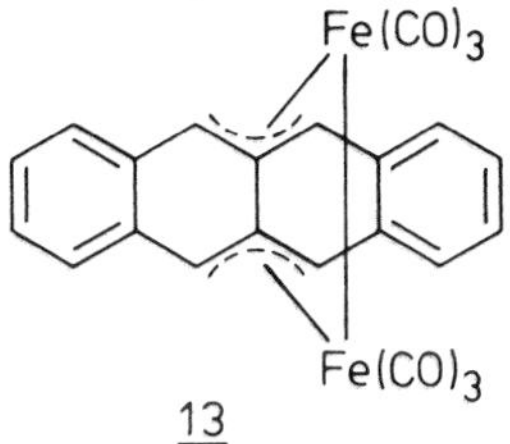

13

Completely analogous compounds have been prepared similarly from the closely related nitrogen heterocyclics (*i.e.* 11 and 12, E = N, E' = CH; E = E' = N) *(14)*. 9-Acetylanthracene also forms a tricarbonyliron complex *(68)*.

Among these systems, (anthracene)tricarbonyliron (11, E = E' = CH) has been investigated in the greatest detail. Its ^{1}H-NMR spectrum has been interpreted to indicate fixation of the 1,3-diene unit bonded to the tricarbonyliron group with enhanced π-delocalization for the residual naphthalene unit *(44)*. Polarographic reduction of (anthracene)tricarbonyliron is both electrochemically and chemically irreversible in contrast to uncomplexed naphthalene and anthracene which readily form the corresponding stable radical anions *(32)*.

The reaction of naphthacene with dodecacarbonyltriiron in boiling benzene gives a hexacarbonyldiiron derivative rather than a tricarbonyliron derivative *(14)*. The ^{1}H-NMR spectrum of this complex exhibits only three resonances indicating maximum symmetry of the naphthacene ligand in this complex. This spectrum has been interpreted on the basis of the tetramethyleneethane-hexacarbonyldiiron structure 13 *(14)*.

2. Carbonyliron Complexes of Vinylbenzenes and Vinylnaphthalenes

The preparation of tricarbonyliron complexes of styrene

(vinylbenzene) derivatives is relatively difficult. Many methods for preparing tricarbonyliron complexes such as thermal reactions of the unsaturated compound with $Fe(CO)_5$, $Fe_3(CO)_{12}$, or $Fe_2(CO)_9$ or ultraviolet irradiation with $Fe(CO)_5$ in benzene apparently fail to convert styrenes and substituted styrenes to their tricarbonyliron derivatives *(110,112)*. However, ultraviolet irradiations of a variety of substituted styrenes with pentacarbonyliron in hexane produce red to violet rather unstable styrene-tricarbonyliron derivatives 14,

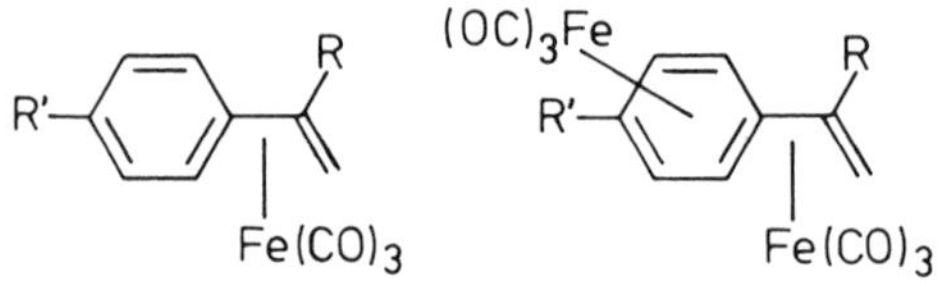

14 15

which generally decompose during their attempted isolation. Only in the cases of the methylstyrene derivatives 14 (R = CH_3, R' = H, Cl, Br, and C_6H_5) could these mononuclear styrene-tricarbonyliron complexes be isolated in the pure state. However, considerably more stable products from the ultraviolet irradiation of styrenes with pentacarbonyliron are the bis(tricarbonyliron) complexes 15 which have been isolated as air-stable orange to red crystalline complexes with a wide variety of styrenes (*e.g.* 15, R = R' = H; R = CH_3, R' = H, CH_3, OCH_3, F, Cl, Br, C_6H_5, $CH(CH_3)_2$, $C(CH_3)_3$; R = C_6H_5, R' = H and OCH_3; R = *p*-CH_3O-C_6H_4, R' = H; R = cyclopropyl, R' = OCH_3). Apparently complexing the first tricarbonyliron unit to styrene to form a derivative of the type 14 destroys the benzenoid aromaticity to leave a free reactive 1,3-diene unit, which easily complexes to a second tricarbonyliron group to form derivatives of the type 15.

Two isomeric hexacarbonyldiiron derivatives 16a and 16b

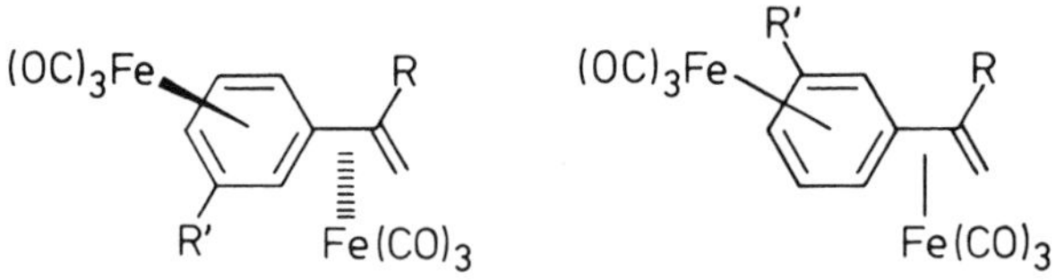

16a 16b

are possible with *m*-substituted styrenes *(108,112)*. The isomers 16a and 16b (R = H, R' = Cl; R = CH_3, R' = Br) can be separated by chromatography from the reaction mixtures obtained by ultraviolet irradiation of pentacarbonyliron with 3-chlorostyrene and 3-bromo-α-methylstyrene, respectively *(112)*. However, separation of the isomers 16a and 16b (R = R' = CH_3) from the reaction mixture obtained by the ultraviolet irradiation of 3,α-dimethylstyrene with pentacarbonyliron required manual sorting of the crystals *(108,112)*. This was feasible since the isomer 16a is orange and the isomer 16b (R = R' = CH_3) is purple. The structure of 16a (R = R' = CH_3) has been confirmed by X-ray crystallography *(51)*. This crystallographic study indicates the *trans*-configuration of the tricarbonyliron groups relative to the styrene. The structure of 16a (R = R' = CH_3) can be considered as two fused isoprenetricarbonyliron units.

Closely related compounds of the types 17 and 18 (R = H,

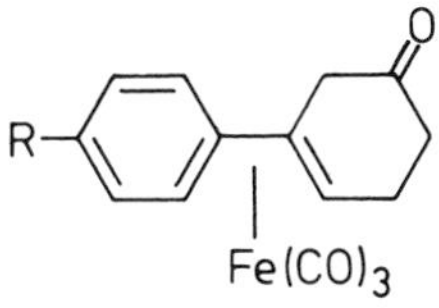

17

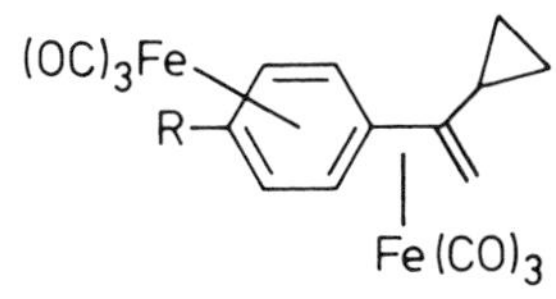

18

Cl, and OCH_3) have been detected in small quantities in the mixtures obtained by the ultraviolet irradiation of 1-aryl-1-cyclopropylethylenes with pentacarbonyliron *(109)*.

The mass spectra of thirty substituted (styrene)bis(tricarbonyliron) derivatives have been reported in detail *(114)*. All of the mass spectra of these organoiron complexes are characterized by the consecutive losses of carbon monoxide followed by loss on one or two iron atoms. Depending upon the nature of the substituent X on the benzenoid ring, the organic moiety in the ion $[(\text{styrene})Fe]^+$ may rupture by four main methods: (a) loss of neutral methane when X = H, CH_3, C_6H_5, CH_3O, and F; (b) loss of neutral acetylene when X = H and OCH_3; (c) loss of HX when X is halogen; (d) loss of iron.

Reactions of *m*- and *p*-divinylbenzenes with iron carbonyls to form the corresponding hexacarbonyldiiron complexes proceed considerably more easily than the corresponding reactions with styrenes discussed above *(67)*. Thus, treatment of either *m*- or *p*-divinylbenzene with dodecacarbonyltriiron in boiling benzene readily gives the corresponding orange crystalline hexacarbonyldiiron complexes *(67)*. X-ray crystallography *(28)* indicates that the complex from *m*-divinylbenzene has the structure 19 whereas the complex from *p*-divinylbenzene has the structure 20. In both cases the tricarbonyliron groups are in *trans*-positions relative to the ring. The carbon-carbon bond lengths for the uncomplexed double bond in 19 and 20 are 1.32 ± 0.01 Å in accord with that expected for a normal carbon-carbon double bond. This indicates that complexation of the divinylbenzene as an η^8-ligand results in localization of the bonds in the benzenoid ring *(28)*.

Vinylnaphthalenes react readily with dodecacarbonyltri-

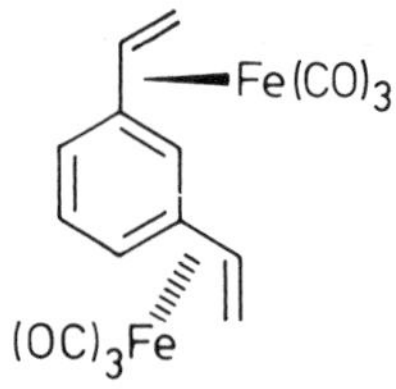

19

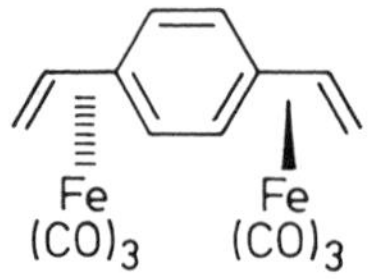

20

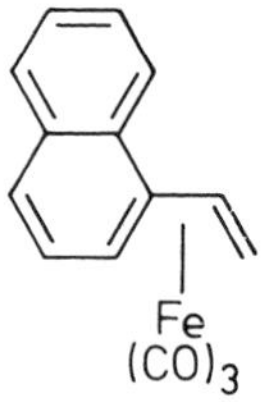

21

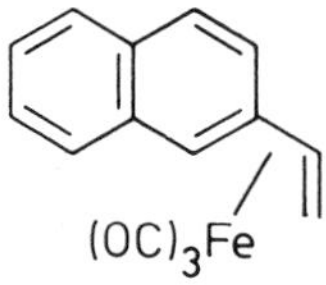

22

iron in boiling cyclohexane to give the corresponding tricarbonyliron complexes *(28,68)*. X-ray crystallography *(28)* indicates structure 21 for the 1-vinylnaphthalene and structure 22 for the 2-vinylnaphthalene complexes. The bond lengths of the uncomplexed portion of the vinylnaphthalene ligand in 21 and 22 resemble those in an uncomplexed styrene ligand.

3. Carbonyliron Complexes of Ferraindene, Benzocyclobutadiene, and *o*-Quinodimethane Derivatives

The carbonyliron chemistry of the diverse ligands ferraindene, benzocyclobutadiene, and *o*-quinodimethane is related. This is best seen from the ultraviolet irradiation of (benzocyclobutadiene)tricarbonyliron (23) with excess pentacarbonyl-

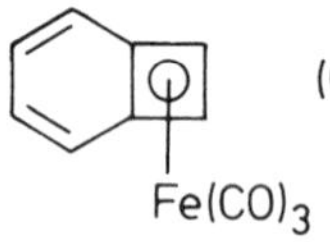

23

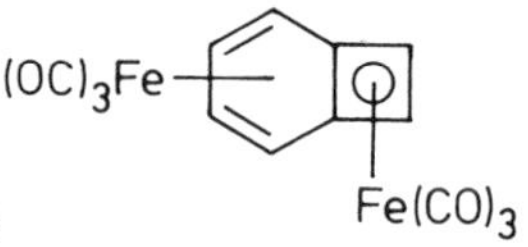

24

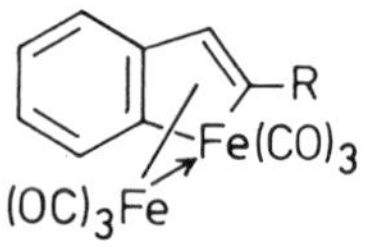

25

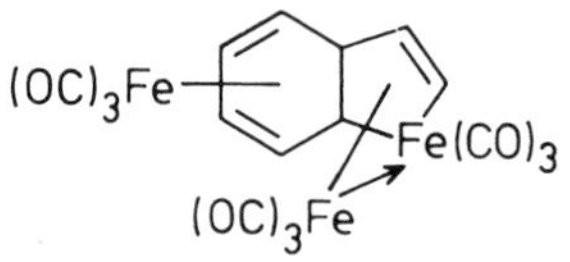

26

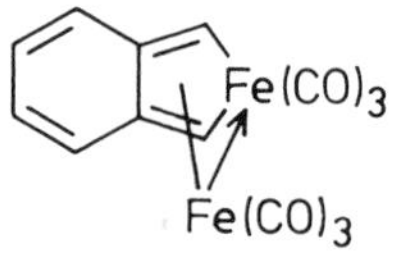

27

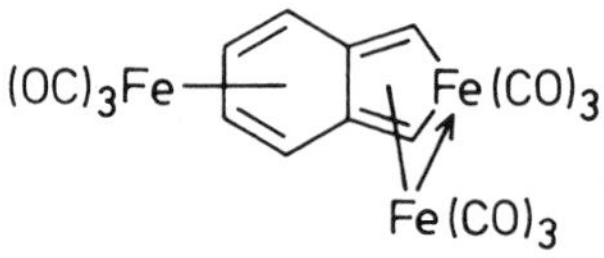

28

iron to give five identifiable carbonyliron complexes *(113)*. One of these products is the expected tricarbonyliron adduct (24) to the two uncomplexed 1,3-double bonds in the benzocyclobutadiene system. Ring opening of the cyclobutadiene ring can also occur in two different ways in this reaction. One mode of ring opening gives the ferraindene complexes 25 (R = H) and 26. The second method of ring opening gives the ferraisoindene complexes 27 and 28 derived from an *o*-quinodimethane type ligand.

Alternate routes to the ferraindene complexes 25 (R = H)

and 26 are also available. Ultraviolet irradiation of *o*-bromostyrene with pentacarbonyliron leads not only to the *o*-bromostyrene-tricarbonyliron and -hexacarbonyldiiron complexes as discussed above but also results in dehydrobromination to give 25 (R = H) and 26 *(111)*. Similar aryl substituted tri-carbonylferraindene-tricarbonyliron derivatives (*e.g.* 25, R = C_6H_5) have also been isolated along with many other products from the complex reactions between diarylacetylenes and iron carbonyls *(18)*. The structures of both the unsubstituted (25, R = H) *(13)* and the phenyl substituted (25, R = C_6H_5) *(30,107)* ferraindene derivatives have been confirmed by X-ray crystallography.

Some other studies have been done on reactions of *o*-quinodimethane-carbonyliron complexes related to 27 and 28. Thus, ultraviolet irradiation of the tricarbonyliron complex 29 with pentacarbonyliron gives three isomeric hexacarbonyl-

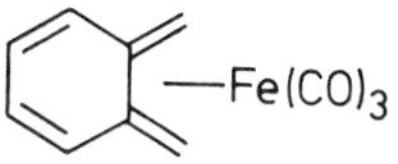

29

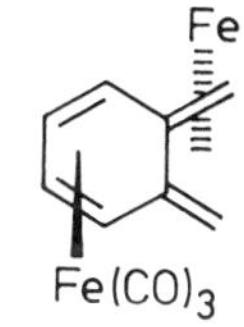

30

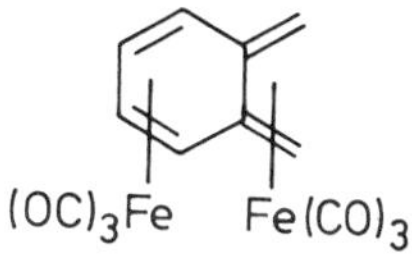

31

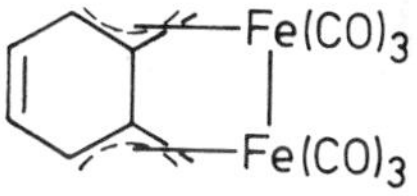

32

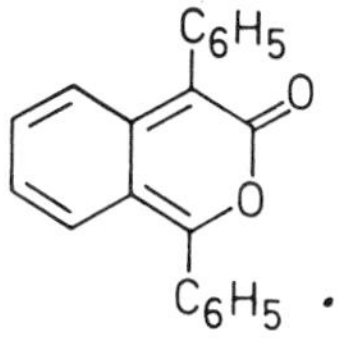

33

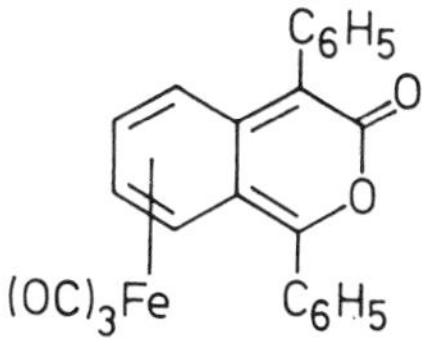

34

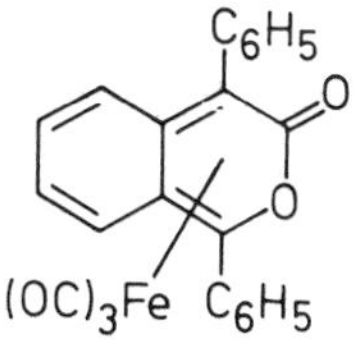

35

diiron complexes: the red *trans*-isomer 30, the purple *cis*-isomer 31, and the tetramethyleneethane complex 32 *(115)*. Reaction of the benzopyrone derivative 33 with dodecacarbonyl-triiron in boiling toluene gives a total 50 % yield of the two isomeric tricarbonyliron complexes 34 and 35 *(52)*.

B. *CARBONYLIRON COMPLEXES OF BENZENOID DERIVATIVES CONTAINING FUSED FIVE-MEMBERED RINGS*

The chemistry of iron complexes of benzenoid derivatives containing fused five-membered rings is somewhat different than the chemistry of benzenoid derivatives fused to other unsaturated groups because of the tendency for the five-membered ring to form a ring-iron bond of the η^5-cyclopentadienyl type. Indene and acenaphthylene are two examples of fused systems of six- and five-membered rings, the iron complexes of which have been studied in some detail.

Shortly after the discovery of ferrocene the closely related bis(indenyl)iron (*sym*-dibenzoferrocene) (36) was pre-

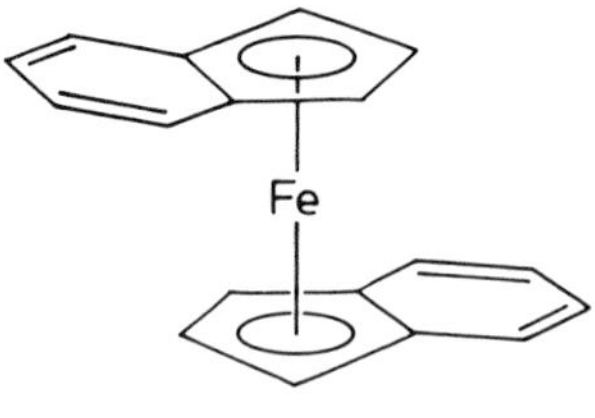

36

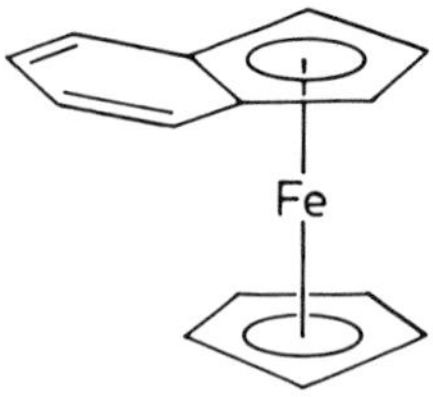

37

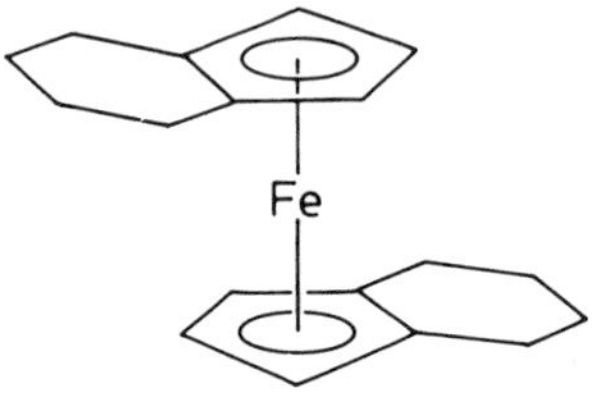

38

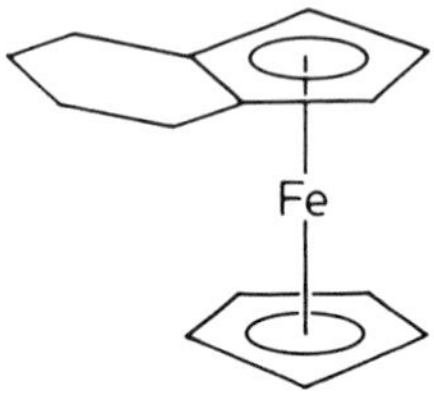

39

pared by reaction of indenylmagnesium bromide with ferric chloride *(33)*. Subsequent preparations of bis(indenyl)iron (36) have also used reactions of iron halides with indenyllithium *(104)* and sodium indenide *(61)*. Bis(indenyl)iron is a purple-black volatile solid considerably more air-sensitive and less thermally stable than ferrocene. The red-violet mixed sandwich compound (cyclopentadienyl)(indenyl)iron (benzoferrocene, 37) has been prepared in 9 % yield by the simultaneous interaction of a mixture of sodium cyclopentadienide and sodium indenide on a solution of ferrous chloride in tetrahydrofuran *(59,60)*. Alternatively 37 may be prepared in 78 % yield from 36 by boiling it with sodium cyclopentadienide in tetrahydrofuran *(103)*.

The bonding of the *pentahapto* five-membered ring of the indene system to iron in 36 and 37 necessarily involves two of the benzenoid carbons and thus destroys the benzenoid six-π-electron system. The olefinic properties of the two uncomplexed carbon-carbon double bonds in each of the six-membered rings of 36 and 37 are indicated by their hydrogenations at atmospheric pressure in the presence of palladium on charcoal to give the corresponding tetrahydroindenyl derivatives 38 and 39, respectively as air-stable orange compounds *(33,59,60)*.

The preparations of $[(C_5H_5)Fe(arene)]^+$ derivatives from ferrocene discussed above involve displacement of one of the cyclopentadienyl rings of ferrocene by a benzenoid ligand in the presence of a Lewis acid catalyst in an intermolecular reaction. An analogous intramolecular reaction occurs in the case of bis(indenyl)iron *(64)*. Thus, treatment of bis-(indenyl)iron (36) with boron trifluoride diethyl etherate in the cold followed by quenching with aqueous ammonium hexafluorophosphate gives the crimson red salt $[(C_9H_7)Fe(C_9H_8)]$ $[PF_6]$. This salt appears to have structure 40 with an indenyl

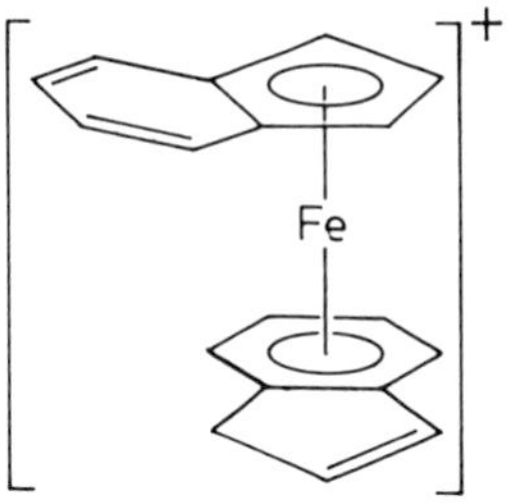

40

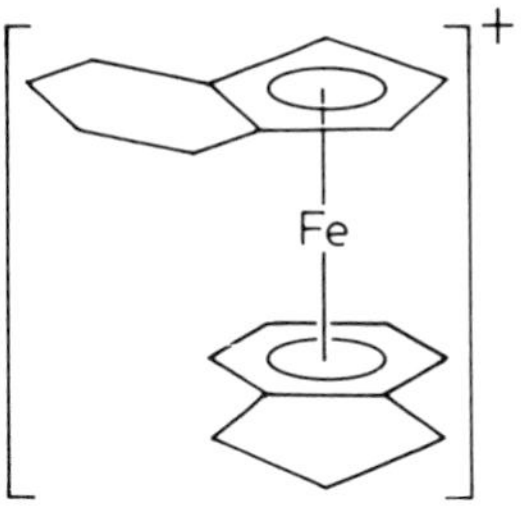

41

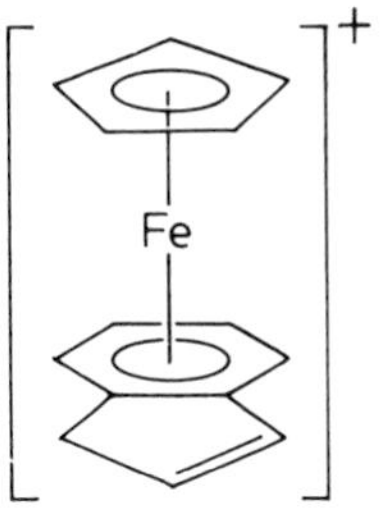

42

ligand bonded to iron through the five-membered ring and an indene unit bonded to iron through the six-membered ring. Hydrogenation of 40 at atmospheric pressure in the presence of palladium on charcoal results in the uptake of three moles of hydrogen to give golden 41. Chemical proof of the structure of 41 is indicated from its preparation by heating bis(tetrahydroindenyl)iron (38) with indane, aluminium chloride, and aluminium powder in boiling heptane followed by addition of aqueous ammonium hexafluorophosphate. An analogous protonation of (cyclopentadienyl)(indenyl)iron (37) with hydrogen chloride in benzene followed by addition of aqueous ammonium hexafluorophosphate also results in migration of the iron from the five-membered ring to the six-membered ring of the indene system to give 42 *(103)*. Studies on the protonation of bis-(indenyl)iron with DCl suggest stereospecific *endo* protonation *(103)*.

Several iron complexes of acenaphthylene have been prepared and studied in some detail. The complex $[(C_5H_5)Fe(CO)_2(C_{12}H_8)]^+$, in which the acenaphthylene is bonded to the iron only through the carbon-carbon double bond in the five-membered ring *(90)*, is discussed in detail in the chapter on olefin-iron complexes. Another complex in which the acenaphthylene is bonded to the iron only through the carbon-carbon double bond of the five-membered ring is $(C_{12}H_8)Fe(CO)_4$. This complex is obtained by reaction of acenaphthylene with enneacarbonyldiiron in tetrahydrofuran at room temperature *(26)*. This complex is also discussed in further detail in the chapter on olefin-iron complexes. Under more vigorous conditions, acenaphthylene reacts with iron carbonyls (*e.g.* $Fe_3(CO)_{12}$ in boiling benzene *(58)*) to give a pentacarbonyldiiron complex $(C_{12}H_8)Fe_2(CO)_5$ shown by X-ray crystallography *(21,22)* to have structure 43. In structure 43 the $Fe(CO)_2$ group is η^5-coordinated to the five-membered ring, and the $Fe(CO)_3$ group is bonded to three benzenoid carbons as an η^3-allylic system. Further details on this chemistry are given in the chapter on π-allyl complexes.

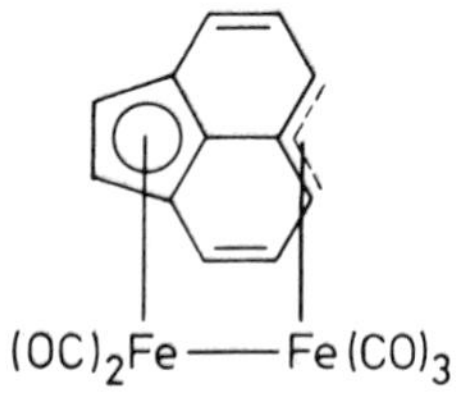

43

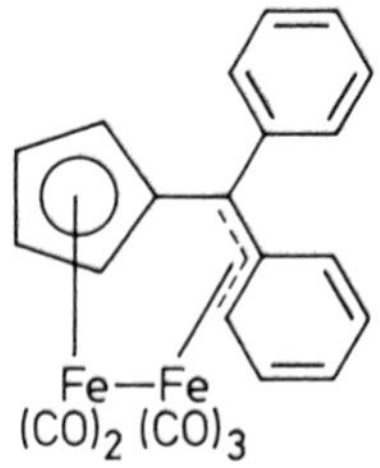

44

Reaction of 1,2,5,6-tetrabromopyracene or pyracylene with enneacarbonyldiiron at room temperature gives the red-black pyracylene complex $(C_{14}H_8)Fe_2(CO)_7$ of unknown structure *(105)*. X-ray crystallography *(15)* of the 6,6-diphenylfulvene complex $[C_5H_4{=}C(C_6H_5)_2]Fe_2(CO)_5$ indicates structure 44 in which one of the phenyl rings is part of an η^3-benzylic ligand.

C. CARBONYLIRON COMPLEXES OF BENZENOID RINGS WITH ADJACENT SULFUR ATOMS

Some benzenoid derivatives with an adjacent carbon-sulfur double bond react with iron carbonyls to form products in which one benzene double bond appears to be bonded to iron. Reactions of the thiobenzophenones *p,p'*-R-C_6H_4CS-C_6H_4-R with enneacarbonyldiiron at room temperature give the *ortho*-metalated complexes 45 (R = R' = H, OCH_3, and $N(CH_3)_2$; R = OCH_3, R' = H; R = H, R' = OCH_3; R = H, R' = CF_3) *(1,2)*. The complex 45 (R = R' = OCH_3) has also been prepared by the deoxygenation of the corresponding sulfine *p,p'*-CH_3O-C_6H_4-C(SO)-C_6H_4-OCH_3 with enneacarbonyldiiron at room temperature *(8)*. Degradation of the complexes 45 with hydrogen peroxide *(3,7)*, cerium(IV) *(2)*, bromine *(2)*, amines *(4,7)*, phosphines *(4,7)*, alcohols *(4,7)*, thiocyanate *(4,7)*, hydroxide *(4)*, methoxide *(4)*, or azide *(4)* under sufficiently vigorous conditions removes the iron to give lactone or thiolactone derivatives. Cleavage of 45 with mercury(II) derivatives results in *ortho*-mercuration *(5,7)*. Reactions of the *O*-methyl aryl thioesters, R-C_6H_4-CS-OR', with enneacarbonyldiiron gives the *ortho*-metalated products of the structure 46 (R = H, R' = CH_3, C_2H_5, $CH(CH_3)_2$, and 1-adamantyl; R' = CH_3, R = *p*-CH_3, *m*-CH_3, *p*-OCH_3, *p*-CF_3, and *p*-Br) *(6)*.

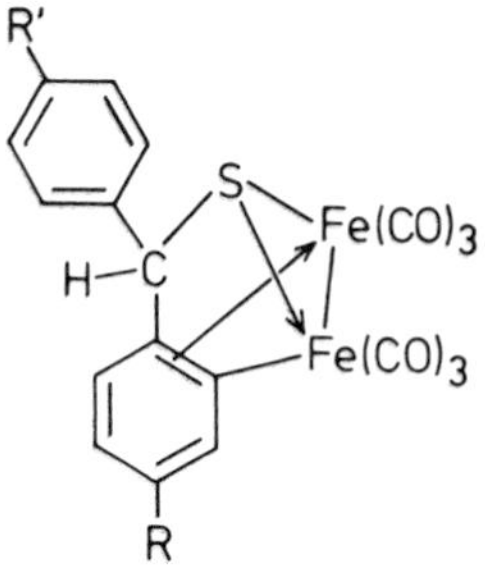

45

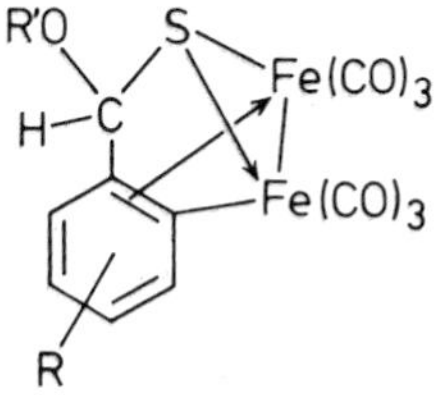

46

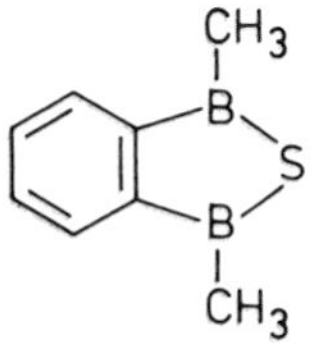

47

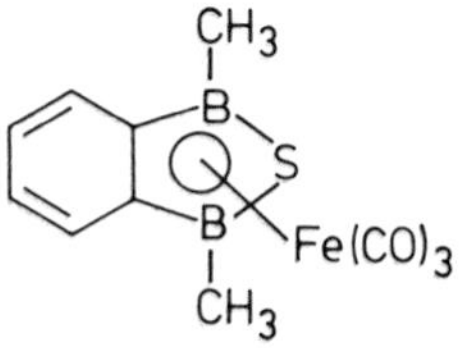

48

Benzothiadiborole also forms a carbonyliron complex shown by X-ray crystallography to involve one carbon-carbon double bond of the benzene ring *(95)*. Reaction of the heterocycle 47 with dodecacarbonyltriiron in boiling toluene in a 1:1 molar ratio gives a volatile deep red tricarbonyliron complex. X-ray crystallography indicates this compound to have structure 48.

IV. IRON COMPLEXES CONTAINING HETEROCYCLIC SYSTEMS RESEMBLING BENZENOID DERIVATIVES

This section discusses iron complexes from unsaturated six-membered heterocycles of boron and phosphorus as well as thiophene and related derivatives.

A. BORABENZENE (BORINATE) DERIVATIVES

Some 1-methyl- and 1-phenylborinato iron complexes have been prepared *(50)*. Reaction of the cobalt derivatives $Co(C_5H_5B\text{-}R)_2$ ($R = CH_3$ and C_6H_5) with excess enneacarbonyldiiron in boiling toluene results in transfer of the borinato ligand from cobalt to iron to give the very stable red-violet $[(R\text{-}BC_5H_5)Fe(CO)_2]_2$ (49, $R = CH_3$ and C_6H_5). The structure of $[(CH_3\text{-}BC_5H_5)Fe(CO)_2]_2$ (49, $R = CH_3$) has been shown by X-ray crystallography *(53)* to be the indicated *cis*-isomer. Pyrolyses of 49 at 230°C/1 atm give the corresponding very stable bis-(borinato)iron complexes $(R\text{-}BC_5H_5)_2Fe$ (50, $R = CH_3$ and C_6H_5). This chemistry resembles very much the corresponding chemistry of cyclopentadienyl-iron compounds *(50)*.

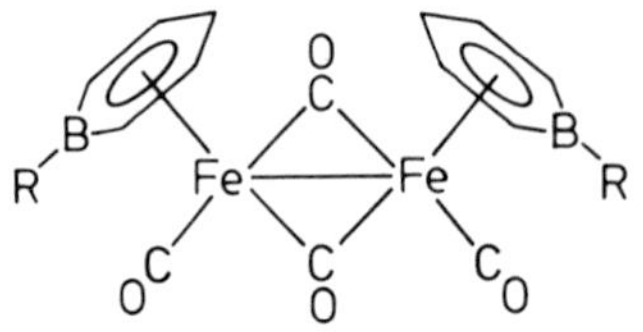

49

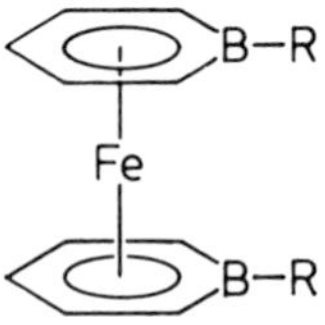

50

B. PHOSPHABENZENE (PHOSPHORINATO) DERIVATIVES

Reaction of 1-alkyl-2,4,6-triphenylphosphorinyllithium with ferrous chloride in tetrahydrofuran gives the neutral brown red bis(phosphorinato)iron complexes $[(C_6H_5)_3C_5H_2P\text{-}R]_2Fe$ (51, R = CH_3 and C_6H_5) *(66)*. On the basis of the limited

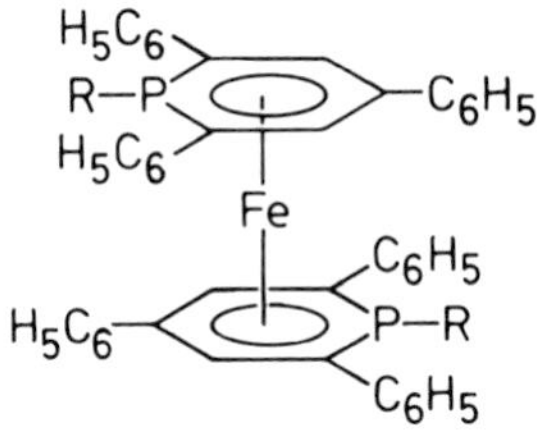

51

information available, these complexes appear to be relatively stable.

C. THIOPHENE DERIVATIVES

Reaction of a mixture of anhydrous iron(II) chloride, aluminium chloride, and tetramethylthiophene in boiling cyclohexane followed by hydrolysis with aqueous ammonium hexafluorophosphate gives the red bis(tetramethylthiophene)iron-(II) salt $\{[(CH_3)_4C_4S]_2Fe\}[PF_6]_2$ *(17)*. Analogous compounds with fewer than four methyl substituents on the thiophene could not be prepared. Electrochemical evidence has been obtained for the reversible one-electron reduction to an iron-(I) derivative as well as for a further irreversible one-electron reduction to an iron(O) derivative.

An iron complex of what formally may be regarded as a *S*-perfluoroarylthiophenium derivative has also been prepared *(27)*. Reaction of $(C_5H_5)Fe(CO)_2SC_6F_5$ with hexafluorobut-2-yne at 75°C gives yellow air-stable $C_6F_5\text{-}S\text{-}C(CF_3)\text{=}C(CF_3)\text{-}Fe(CO)_2(C_5H_5)$. Photochemical decarbonylation of this complex in pentane gives 52, which reacts photochemically with additional hexafluorobut-2-yne in pentane to give yellow-brown crystalline $(C_5H_5)Fe[C_4(CF_3)_4S\text{-}C_6F_5]$ formulated as 53.

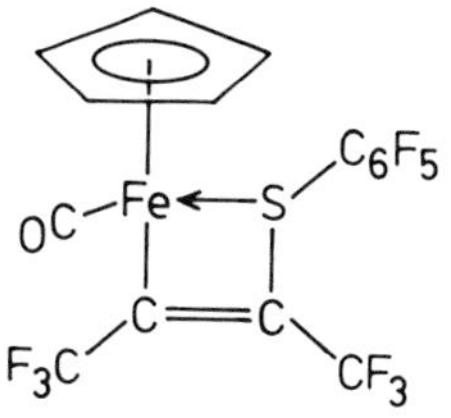

52

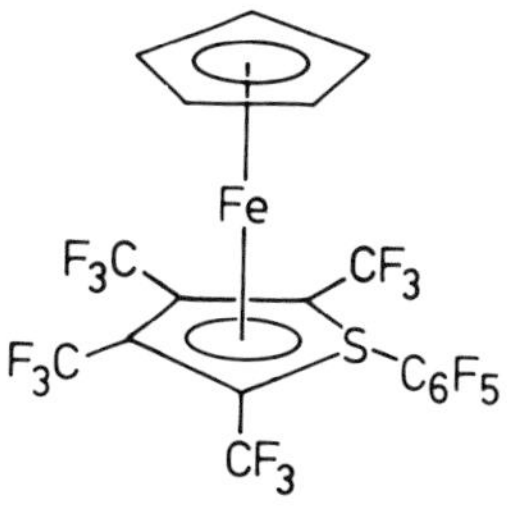

53

Acknowledgement

A fellowship of the Max-Planck-Gesellschaft during the time this article was written at the Institut für Strahlenchemie im Max-Planck-Institut für Kohlenforschung (Mülheim a.d. Ruhr, Germany) is gratefully acknowledged.

REFERENCES

1. Alper, H., and Chan, A.S.K., *J. Chem. Soc. D, Chem. Commun., 1971,* 1203.
2. Alper, H., and Chan, A.S.K., *J. Amer. Chem. Soc., 95,* 4905 (1973).
3. Alper, H., and Root, W.G., *J. Chem. Soc. Chem. Commun., 1974,* 956.
4. Alper, H., Root, W.G., and Chan, A.S.K., *J. Organometal. Chem., 71,* C 14 (1974).
5. Alper, H., and Root, W.G., *Tetrahedron Lett., 1974,* 1611.
6. Alper, H., and Foo, C.K., *Inorg. Chem., 14,* 2928 (1975).
7. Alper, H., and Root, W.G., *J. Amer. Chem. Soc., 97,* 4251 (1975).
8. Alper, H., *J. Organometal. Chem., 84,* 347 (1975).
9. Anderson, S.E., Jr., and Drago, R.S., *J. Amer. Chem. Soc., 92,* 4244 (1970).
10. Anderson, S.E., *J. Organometal. Chem., 71,* 263 (1974).
11. Astruc, D., and Dabard, R., *C.R. Acad. Sci., Ser. C, 272,* 1337 (1971).
12. Astruc, D., *Tetrahedron Lett., 1973,* 3437.
13. Barnett, B.L., and Davis, R.E., *Amer. Cryst. Assoc., Winter Meeting 1970,* New Orleans, Abstracts, p. 45.
14. Bauer, R.A., Fischer, E.O., and Kreiter, C.G., *J. Organometal. Chem., 24,* 737 (1970).
15. Behrens, U., and Weiss, E., *J. Organometal. Chem., 73,* C 64 (1974).
16. Braitsch, D.M., *J. Chem. Soc. Chem. Commun., 1974,* 460.
17. Braitsch, D.M., and Kumarappan, R., *J. Organometal. Chem., 84,* C 37 (1975).
18. Braye, E.H., and Hübel, W., *J. Organometal. Chem., 3,* 38 (1965).
19. Brintzinger, H., Palmer, G., and Sands, R.H., *J. Amer. Chem. Soc., 88,* 623 (1966).
20. Weber, S.R., and Brintzinger, H.H., *J. Organometal. Chem., 127,* 45 (1977).
21. Churchill, M.R., and Wormald, J., *Chem. Commun., 1968,* 1597.
22. Churchill, M.R., and Wormald, J., *Inorg. Chem., 9,* 2239 (1970).
23. Coffield, T.H., Sandel, V., and Closson, R.D., *J. Amer. Chem. Soc., 79,* 5826 (1957).
24. Coffield, T.H., and Closson, R.D., U.S. Patent 3 130 214 (1964).
25. Coffield, T.H., and Burcal, G.J., U.S. Patent 3 190 902 (1965).

26. Cotton, F.A., and Lahuerta, P., *Inorg. Chem., 14,* 116 (1975).
27. Davidson, J.L., and Sharp, D.W.A., *J. Chem. Soc. Dalton Trans., 1975,* 2283.
28. Davis, R.E., and Pettit, R., *J. Amer. Chem. Soc., 92,* 716 (1970).
29. Day, A.C., and Powell, J.T., *Chem. Commun., 1968,* 1241.
30. Degrève, Y., Meunier-Piret, J., Van Meerssche, M., and Piret, P., *Acta Crystallogr., 23,* 119 (1967).
31. Dessy, R.E., Stary, F.E., King, R.B., and Waldrop, M., *J. Amer. Chem. Soc., 88,* 471 (1966).
32. Dessy, R.E., and Pohl, R.L., *J. Amer. Chem. Soc., 90,* 1995 (1968).
33. Fischer, E.O., and Seus, D., *Z. Naturforsch., B 9,* 386 (1954).
34. Fischer, E.O., and Böttcher, R., *Chem. Ber., 89,* 2397 (1956).
35. Fischer, E.O., Joos, G., and Meer, W., *Z. Naturforsch., B 13,* 456 (1958).
36. Fischer, E.O., and Röhrscheid, F., *Z. Naturforsch., B 17,* 483 (1962).
37. Fischer, E.O., and Müller, J., *Z. Naturforsch., B 17,* 776 (1962).
38. Fischer, E.O., Berngruber, W., and Kreiter, C.G., *J. Organometal. Chem., 14,* P 25 (1968).
39. Fischler, I., and Koerner von Gustorf, E.A., *Z. Naturforsch., B 30,* 291 (1975).
40. Foster, M.S., and Beauchamp, J.L., *J. Amer. Chem. Soc., 93,* 4924 (1971).
41. Foster, M.S., and Beauchamp, J.L., *J. Amer. Chem. Soc., 97,* 4808 (1975).
42. Games, D.E., Jackson, A.H., Kane-Maguire, L.A.P., and Taylor, K., *J. Organometal. Chem., 88,* 345 (1975).
43. Green, M.L.H., Pratt, L., and Wilkinson, G., *J. Chem. Soc., 1960,* 989.
44. Günther, H., Wenzl, R., and Klose, H., *J. Chem. Soc. D, Chem. Commun., 1970,* 605.
45. Harper, R.J., U. S. Patent 3 073 855 (1963).
46. Helling, J.F., and Braitsch, D.M., *J. Amer. Chem. Soc., 92,* 7207 (1970).
47. Helling, J.F., and Braitsch, D.M., *J. Amer. Chem. Soc., 92,* 7209 (1970).
48. Helling, J.F., Rice, S.L., Braitsch, D.M., and Mayer, T., *J. Chem. Soc. D, Chem. Commun., 1971,* 930.
49. Helling, J.F., and Cash, G.G., *J. Organometal. Chem., 73,* C 10 (1974).
50. Herberich, G.E., Becker, H.J., and Greiß, G., *Chem. Ber., 107,* 3780 (1974).

51. Herbstein, F.H., and Reisner, M.G., *J. Chem. Soc. Chem. Commun., 1972,* 1077.
52. Holland, J.M., and Jones, D.W., *Chem. Commun., 1967,* 946.
53. Huttner, G., and Gartzke, W., *Chem. Ber., 107,* 3786 (1974).
54. Jones, D., Pratt, L., and Wilkinson, G., *J. Chem. Soc., 1962,* 4458.
55. Khand, I.U., Pauson, P.L., and Watts, W.E., *J. Chem. Soc. C, 1968,* 2257.
56. Khand, I.U., Pauson, P.L., and Watts, W.E., *J. Chem. Soc. C, 1968,* 2261.
57. Khand, I.U., Pauson, P.L., and Watts, W.E., *J. Chem. Soc. C, 1969,* 2024.
58. King, R.B., and Stone, F.G.A., *J. Amer. Chem. Soc., 82,* 4557 (1960).
59. King, R.B., and Bisnette, M.B., *Angew. Chem., 75,* 642 (1963); *Angew. Chem. Int. Ed. Engl., 2,* 400 (1963).
60. King, R.B., and Bisnette, M.B., *Inorg. Chem., 3,* 796 (1964).
61. King, R.B., *Organometallic Syntheses Vol. I, Transition-Metal Compounds,* Academic Press, New York 1965, p. 73.
62. Klabunde, K.J., and Efner, H.F., *J. Fluor. Chem., 4,* 114 (1974).
63. Koerner von Gustorf, E., Fischler, I., Leitich, J., and Dreeskamp, H., *Angew. Chem., 84,* 1143 (1972); *Angew. Chem. Int. Ed. Engl., 11,* 1088 (1972).
64. Lee, C.C., Sutherland, R.G., and Thomson, B.J., *J. Chem. Soc. D, Chem. Commun., 1971,* 1071.
65. Lee, C.C., Sutherland, R.G., and Thomson, B.J., *J. Chem. Soc. Chem. Commun., 1972,* 907.
66. Märkl, G., and Martin, C., *Angew. Chem., 86,* 445 (1974); *Angew. Chem. Int. Ed. Engl., 13,* 408 (1974).
67. Manuel, T.A., Stafford, S.L., and Stone, F.G.A., *J. Amer. Chem. Soc., 83,* 3597 (1961).
68. Manuel, T.A., *Inorg. Chem., 3,* 1794 (1964).
69. Mathew, M., and Palenik, G.J., *Inorg. Chem., 11,* 2809 (1972).
70. McVey, S., and Pauson, P.L., *J. Chem. Soc., 1965,* 4312.
71. Middleton, R., Hull, J.R., Simpson, S.R., Tomlinson, C.H., and Timms, P.L., *J. Chem. Soc. Dalton Trans., 1973,* 120.
72. Morrison, W.H., Jr., Ho, E.Y., and Hendrickson, D.N., *J. Amer. Chem. Soc., 96,* 3603 (1974).
73. Morrison, W.H., Jr., Ho, E.Y., and Hendrickson, D.N., *Inorg. Chem., 14,* 500 (1975).
74. Nesmeyanov, A.N., Vol'kenau, N.A., and Bolesova, I.N.,

Dokl. Akad. Nauk SSSR, 149, 615 (1963); *Dokl. Chem., 149,* 267 (1963).

75. Nesmeyanov, A.N., Vol'kenau, N.A., and Bolesova, I.N., *Tetrahedron Lett., 1963,* 1725.

76. Nesmeyanov, A.N., Vol'kenau, N.A., and Shilovtseva, L.S., *Dokl. Akad. Nauk SSSR, 160,* 1327 (1965); *Dokl. Chem., 160,* 203 (1965).

77. Nesmeyanov, A.N., Vol'kenau, N.A., Shilovtseva, L.S., and Petrakova, V.A., *J. Organometal. Chem., 85,* 365 (1965).

78. Nesmeyanov, A.N., Vol'kenau, N.A., and Bolesova, I.N., *Dokl. Akad. Nauk SSSR, 166,* 607 (1966); *Dokl. Chem., 166,* 116 (1966).

79. Nesmeyanov, A.N., Vol'kenau, N.A., and Bolesova, I.N., *Dokl. Akad. Nauk SSSR, 175,* 606 (1967); *Dokl. Chem., 175,* 661 (1967).

80. Nesmeyanov, A.N., Vol'kenau, N.A., and Isaeva, L.S., *Dokl. Akad. Nauk SSSR, 176,* 106 (1967); *Dokl. Chem., 176,* 772 (1967).

81. Nesmeyanov, A.N., Vol'kenau, N.A., Sirotkina, E.I., and Deryabin, V.V., *Dokl. Akad. Nauk SSSR, 177,* 1110 (1967); *Dokl. Chem., 177,* 1170 (1967).

82. Nesmeyanov, A.N., Vol'kenau, N.A., and Sirotkina, E.I., *Izv. Akad. Nauk SSSR, Ser. Khim., 1967,* 1170; *Bull. Acad. Sci. USSR, Div. Chem. Ser., 1967,* 1142.

83. Nesmeyanov, A.N., Vol'kenau, N.A., and Isaeva, L.S., *Dokl. Akad. Nauk SSSR, 183,* 606 (1968); *Dokl. Chem., 183,* 1026 (1968).

84. Nesmeyanov, A.N., Vol'kenau, N.A., Isaeva, L.S., and Bolesova, I.N., *Dokl. Akad. Nauk SSSR, 183,* 834 (1968); *Dokl. Chem., 183,* 1042 (1968).

85. Nesmeyanov, A.N., Vol'kenau, N.A., Bolesova, I.N., and Isaeva, L.S., *Izv. Akad. Nauk SSSR, Ser. Khim., 1968,* 2416; *Bull. Acad. Sci. USSR, Div. Chem. Ser., 1968,* 2296.

86. Nesmeyanov, A.N., Lokshin, B.V., Vol'kenau, N.A., Bolesova, I.N., and Isaeva, L.S., *Dokl. Akad. Nauk SSSR, 184,* 358 (1969); *Dokl. Chem., 184,* 40 (1969).

87. Nemeyanov, A.N., Vol'kenau, N.A., and Shilovtseva, L.S., *Izv. Akad. Nauk SSSR, Ser. Khim., 1969,* 726; *Bull. Acad. Sci. USSR, Div. Chem. Ser., 1969,* 664.

88. Nesmeyanov, A.N., Leshchova, I.F., Ustynyuk, Yu.A., Sirotkina, Ye.I., Bolesova, I.N., Isayeva, L.S., and Vol'kenau, N.A., *J. Organometal. Chem., 22,* 689 (1970).

89. Nesmeyanov, A.N., Vol'kenau, N.A., Shilovtseva, L.S., and Petrakova, V.A., *J. Organometal. Chem., 61,* 329 (1973).

90. Nicholas, K.M., and Rosan, A.M., *J. Organometal. Chem.,*

84, 351 (1975).

91. Nicholson, B.J., *J. Amer. Chem. Soc.*, *88*, 5156 (1966).
92. Pavlîk, I., and Křiž, P., *Collect. Czech. Chem. Commun.*, *31*, 4412 (1966).
93. Pavlycheva, A.V., Domrachev, G.A., Razuvaev, G.A., and Suvorova, O.N., *Dokl. Akad. Nauk SSSR*, *184*, 105 (1969); *Dokl. Chem.*, *184*, 10 (1969).
94. Razuvaev, G.A., Domrachev, G.A., Suvorova, O.N., and Abakumova, L.G., *J. Organometal. Chem.*, *32*, 113 (1971).
95. Siebert, W., Augustin, G., Full, R., Krüger, C., and Tsay, Y.-H., *Angew. Chem.*, *87*, 286 (1975); *Angew. Chem. Int. Ed. Engl.*, *14*, 262 (1975).
96. Sirotkina, E.I., Nesmeyanov, A.N., and Vol'kenau, N.A., *Izv. Akad. Nauk SSSR, Ser. Khim.*, *1969*, 1524; *Bull. Acad. Sci. USSR, Div. Chem. Ser.*, *1969*, 1413.
97. Sirotkina, E.I., Nesmeyanov, A.N., and Vol'kenau, N.A., *Izv. Akad. Nauk SSSR, Ser Khim.*, *1969*, 1605; *Bull. Acad. Sci. USSR, Chem. Ser.*, *1969*, 1488.
98. Stukan, R.A., Vol'kenau, N.A., Nesmeyanov, A.N., and Gol'danskii, V.I., *Izv. Akad. Nauk SSSR, Ser. Khim.*, *1966*, 1472; *Bull. Acad. Sci. USSR, Div. Chem. Ser.*, *1966*, 1416.
99. Stukan, R.A., Turta, K.I., Gol'danskii, V.I., Kaplan, A.M., Vol'kenau, N.A., and Sirotkina, E.I., *Teor. Eksp. Khim.*, *7*, 74 (1971); *Theor. Exp. Chem.*, *7*, 60 (1971).
100. Timms, P.L., *J. Chem. Soc. D, Chem. Commun.*, *1969*, 1033.
101. Timms, P.L., *Faraday Symp. Chem. Soc.*, *8*, 68 (1974).
102. Timms, P.L., *Angew. Chem.*, *87*, 295 (1975); *Angew. Chem. Int. Ed. Engl.*, *14*, 273 (1975).
103. Treichel, P.M., and Johnson, J.W., *J. Organometal. Chem.*, *88*, 207 (1975).
104. Treichel, P.M., Johnson, J.W., and Calabrese, J.C., *J. Organometal. Chem.*, *88*, 215 (1975).
105. Trost, B.M., and Bright, G.M., *J. Amer. Chem. Soc.*, *91*, 3689 (1969).
106. Tsutsui, M., and Zeiss, H.H., *Naturwiss.*, *44*, 420 (1957).
107. Van Meerssche, M., Piret, P., Meunier-Piret, J., and Degrêve, Y., *Bull. Soc. Chim. Belg.*, *73*, 824 (1964).
108. Victor, R., Ben-Shoshan, R., and Sarel, S., *J. Chem. Soc. D, Chem. Commun.*, *1970*, 1680.
109. Victor, R., Ben-Shoshan, R., and Sarel, S., *Tetrahedron Lett.*, *1970*, 4253.
110. Victor, R., Ben-Shoshan, R., and Sarel, S., *Tetrahedron Lett.*, *1970*, 4257.
111. Victor, R., Ben-Shoshan, R., and Sarel, S., *J. Chem.*

Soc. D, Chem. Commun., 1971, 1241.
112. Victor, R., Ben-Shoshan, R., and Sarel, S., *J. Org. Chem., 37,* 1930 (1972).
113. Victor, R., and Ben-Shoshan, R., *J. Chem. Soc. Chem. Commun., 1974,* 93.
114. Victor, R., Deutsch, J., and Sarel, S., *J. Organometal. Chem., 71,* 65 (1974).
115. Victor, R., and Ben-Shoshan, R., *J. Organometal. Chem., 80,* C 1 (1974).
116. Warren, K.D., *Inorg. Chem., 13,* 1317 (1974).
117. Williams-Smith, D.L., Wolf, L.R., and Skell, P.S., *J. Amer. Chem. Soc., 94,* 4042 (1972).

COMPOUNDS WITH IRON-METAL BONDS AND CLUSTERS

By P. CHINI *

Istituto di Chimica Generale ed Inorganica dell'Università,
Via G. Venezian 21, Milano 20133, Italy

TABLE OF CONTENTS

* Deceased

ISBN 0-12-417102-8

I. INTRODUCTION

The first compound with an iron-iron bond, "$Fe_2(CO)_7$", later recognized to be $Fe_2(CO)_9$, had been obtained in 1891 by Mond and Langer *(441)*, while in 1928 Hock and Stuhlman reported $(ClHg)_2Fe(CO)_4$ *(339,340)*, the first example of a bond between iron and a non transition metal. Much later in 1959 the first example was reported of a bond between iron and another transition metal, $FeCo_3(CO)_{12}H$ *(147)*. As it appears from these historical examples, the organometallic chemistry of iron is frequently complicated by the occurrence of iron-iron or iron-metal bonds, and it seems therefore appropriate to give a discussion of these particular bonds. The principal scope of this discussion is to point out the peculiar chemical properties which are associated with iron-metal bonds, but unfortunately, at least in this field, there are no magic generalizations and most of the useful information must be obtained through patient correlation of analogous cases. Therefore, emphasis is placed on discussion of syntheses and reactivity.

This review is a non comprehensive one, it should be considered only as an introduction to the basic chemistry of the compounds containing iron-metal bonds. In order to help the reader to expand the appropriate information a list of titles of the principal reviews pertinent to the field is given at the end. Because of several delays in publication (from 1972 to 1979) the more relevant literature has been covered up to 1975 and later references have been restricted to a minimum. However, in order to be concise much material has been included only in the tables, and careful information can be extracted only by consideration of the tables themselves.

In order to simplify the identification of groups on each metal atom in bimetallic compounds, often I have indicated the bridging groups by writing them between the two metals:

$$Cp(CO)Fe(PPh_2)(CO)Fe(CO)_2(PPh_3)$$

instead of

$$Fe_2Cp(CO)_3(PPh_3)(\mu_2\text{-}CO)(\mu_2\text{-}PPh_2)$$

In these cases the notation μ_2 results redundant.

The following <u>abbreviations</u> have been used in the text:

acac	Acetylacetonate
bipy	2,2'-bipyridyl
Bu	*n*-butyl
Bz	benzyl
COD	1,5-cyclooctadiene

Cp	(η^5-C_5H_5), *pentahapto*-cyclopentadienyl
Cy	cyclohexyl
DMF	dimethylformamide
dmpe	1,2-bis(dimethylphosphino)ethane
DMSO	dimethylsulphoxide
dppe	1,2-bis(diphenylphosphino)ethane
dppm	bis(diphenylphosphino)methane
E	metallic non-transition element
Et	ethyl
$f_4$ars	1,2-bis(dimethylarsino)tetrafluorocyclobutene
$f_4$asp	1-(dimethylarsino)-2-(diphenylphosphino)-tetrafluorocyclobutene
$f_4$fos	1,2-bis(diphenylphosphino)tetrafluorocyclobutene
$f_6$fosf	1,2-bis(diphenylphosphino)perfluorocyclopentene
Ind	*pentahapto*-indenyl
M	transition metal atom
Me	methyl
Ph	phenyl
Phen	1,10-phenanthroline
Pr	*n*-propyl
p-tol	*para*-tolyl
py	pyridine
triphos	$CH_3C(CH_2PPh_2)_3$
THF	tetrahydrofuran
X	halogen or pseudohalogen

II. COMPOUNDS WITH Fe-E BONDS (E = METALLIC NON-TRANSITION ELEMENT)

A. *GENERAL PROPERTIES AND CLASSIFICATION*

The presence of ligands of high π acidity, such as carbon monoxide, imparts a high electronegativity to the iron atoms. Although this electronegativity changes considerably on changing the ligands (Section II.*C*), as a rough approximation we can assume an average value of the order of the halogens. This high electronegativity results in highly polar metal-iron bonds (E^+)(Fe^-) at the left side of the periodic table, and in bonds which are mainly covalent (E-Fe) with elements on the right side of the periodic table.

A real difficulty arises only at the extreme left side of the periodic table where the elements are highly electropositive; here the polar character of the iron-metal bonds is more pronounced and the compounds of these elements are discussed in a simpler way just by considering the chemistry of the related iron anions, *e.g.* $[Fe(CO)_4]^{2-}$. Although an excellent review of these particular compounds is available *(28)*, a brief discussion will be given later (Section III.*D*).

Then, on moving towards more electronegative elements, one does not observe any abrupt change in the mainly covalent character of the iron-metal bond, and indeed every possible division between most of the remaining metallic elements and the non-metallic ones seems to be more or less artificial. On this basis, and considering the non-conventionality of the Fe-B, Fe-Si, Fe-Ge, and Fe-As bonds, we decided to include also these bonds in the present discussion.

Figure 1 summarizes the non-transition metal elements which will be considered in the following discussion and their formal oxidation numbers assuming metallic character. Additional information on the type of ligands bonded to the iron atoms can be obtained from comparison with Table 1.

In Table 1 we have classified the compounds with Fe-E bonds either according to the parent hydride derivative from which they can be derived by substitution of the hydrogen atoms, or according to the neutral carbonyl complex from which they can be derived by substitution of carbonyl groups. When both possibilities are present, as in the case of $Fe(CO)_4H_2$

Table 1: Structural classification of the compounds with Fe-E bonds

Class	Derivatives of	Types of compounds	
a	$Fe(CO)_3(NO)H$	$-Fe(CO)_3NO$;	$-Fe(CO)_2(L)NO$;
b	$CpFe(CO)_2H$	$-Fe(CO)_2Cp$; $-Fe(L)_2Cp$; $[=Fe(CO)Cp]^-$	$-Fe(CO)(L)Cp$; $=Fe(CO)(H)Cp$;
c	$Fe(CO)_4H_2$	$=Fe(CO)_4$; $-Fe(CO)_4X$; $-Fe(CO)_4H$;	$=Fe(CO)_3L$; $-Fe(CO)_3(L)X$; $[-Fe(CO)_4]^-$
d	$Fe_2(CO)_9$	$=Fe_2(CO)_8$;	$Fe_2(CO)_7$[a] $Fe_2(CO)_6$[b];
	$Cp_2Fe_2(CO)_4$	$=Fe_2(CO)_3Cp_2$	
e	miscellaneous	$\leftarrow Fe(CO)_5$; $=Fe(RNC)_4$; $[-Fe(RNC)_5]^+$; Fe-carboranes	$\leftarrow Fe(CO)_3L_2$; $-Fe(RNC)_4X$;

a) four free positions for coordination; *b)* six free positions for coordination.

		B (III) a b e		
		Al e	Si (IV) b c e	
Cu (I) c	Zn (II) b c	Ga (III) c e	Ge (IV) a b c d	As (III) b c e
Ag (I) c	Cd (II) b c	In (III) b c e	Sn (IV) a b c d e	Sb (III)-(V) b c e
Au (I) a c	Hg (II) a b c d e	Tl a b c e	Pb (IV) a b c e	Bi (III) b e

Fig. 1: Non-transition metal elements considered in the present discussion (the letters refer to classes in Table 1)

and $Fe(CO)_5$, we have preferred the hydride correlation.

Group IIB *(20)*, IIIB *(21,39,46)*, and IVB *(10,19,30,46)* derivatives have been comprehensively reviewed.

1. Structural Description

Class a.

On the basis of IR spectra structures 1 and 2 have been assigned to the derivatives of $Fe(CO)_3(NO)H$ (class a) *(140)*

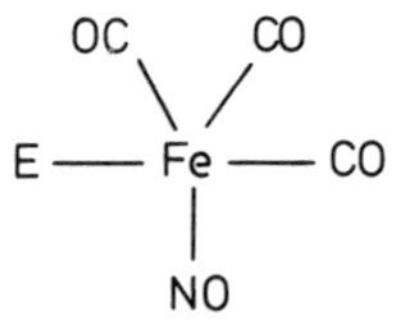

1

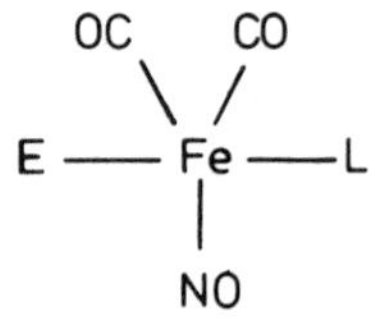

2

where L is a ligand such as PPh_3, $P(OPh)_3$, $AsEt_2Ph$ *(141)*, and $P(SiMe_3)_3$ *(349)*. Structure 2 has now been found *(501)* for $Hg[Fe(CO)_2NO(PEt_3)]_2$. It has been possible to bind up to four such groups to the same metallic center E *(161)*.

Class b.

The more common structures of the derivatives of $CpFe(CO)_2H$, class b, are:

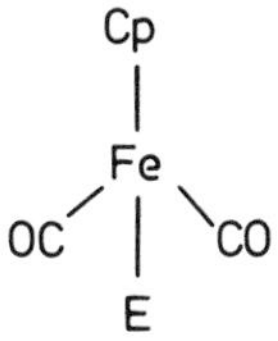

3

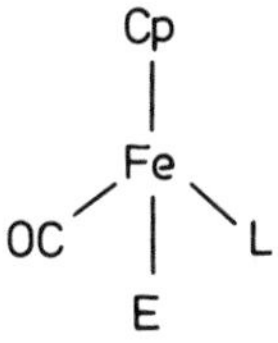

4

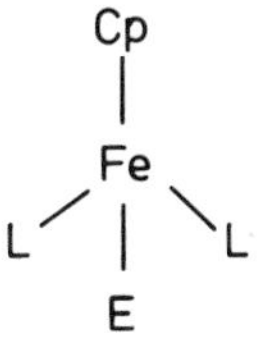

5

Compounds of structure 4 exist in enantiomeric forms *(287)*. Examples of ligands L are PPh_3, $AsPh_3$, $SbPh_3$, PPh_2CF_3, $AsPh_2CF_3$, PPh_2Me, $PPhMe_2$ *(198,268)*, while in 5 bidentate ligands such as dppe and *cis*-1,2-bis(diphenylphosphino)ethylene *(387)* or butadiene *(67,69,168)* can be present. Recent exotic examples in this class are $(Me_3Sn)_3Sn$-$Fe[P(OPh)_3]_2Cp$ *(392)* and Ph_3Sn-$Fe(C_2Ph_2)(CO)Cp$ *(496)*; also the covalent $(THF)_2BrMg$-$Fe(dppe)Cp$, with a magnesium-iron distance of 2.59 Å, has been recently reported *(263)*.

Also in this class it has been possible to bind four such groups to the same metallic centre in the compound $Sn[Fe(CO)_2Cp]_4$ *(449)*.

The related alkoxides $[(ROZn)Fe(CO)_2Cp]_4$ (R = Me, Et) *(129)* have probably cubane structures in which zinc and oxygen atoms are present at the apeces of the cube.

In the same class we can include some compounds derived from substitution of a carbonyl group in $CpFe(CO)_2H$ by two metallic centres E such as $(Cl_3Si)_2Fe(CO)(H)Cp$ *(420)* and the

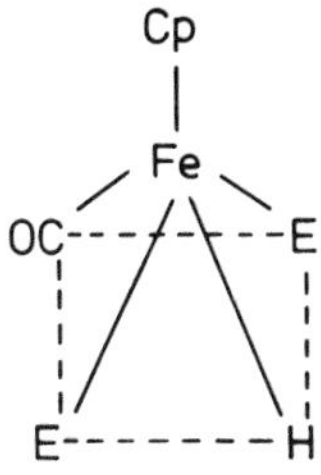

6

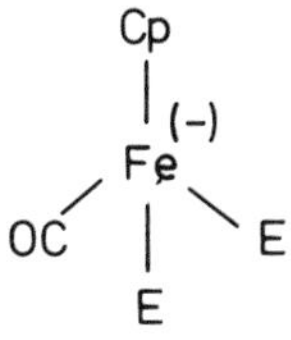

7

related anion $[(Cl_3Si)_2Fe(CO)Cp]^-$ *(361)*.

Class c.

The derivatives of $Fe(CO)_4H_2$ form the most usual class. Bisubstitution can give rise to *cis*- and *trans*-isomers (8 and 9), as in the case of $(Cl_3Ge)_2Fe(CO)_4$ *(406)*. Generally the *cis*-isomer is the more common, and structure 8 has been established by X-ray analysis for the compounds $(BrHg)_2Fe(CO)_4$

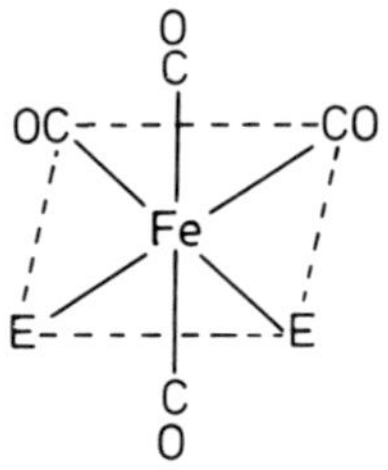

8

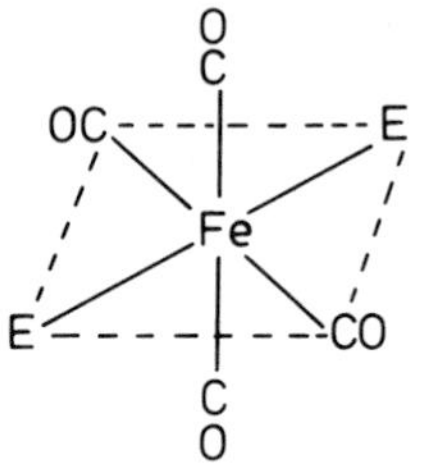

9

(81), $[ClHgPy]_2Fe(CO)_4$ *(82)* and $[AsPh_4][Fe(CO)_4(HgCl)(HgCl_2)]$ *(115)*. It is worth mentioning that the presence of two bonding positions has been used for bonding two different metals, as in $(Cl_3Si)(Cl_3Sn)Fe(CO)_4$ *(362)*. Particularly interesting cases are the compounds $[(C_2F_5)_2As_2]Fe(CO)_4$ (10) *(252)*, and $Cp_2(CO)_2Co_2(GeCl_2)_2Fe(CO)_4$ (11) *(251)*.

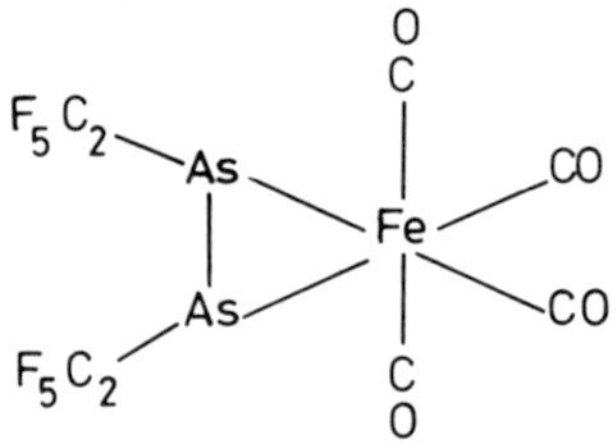

10

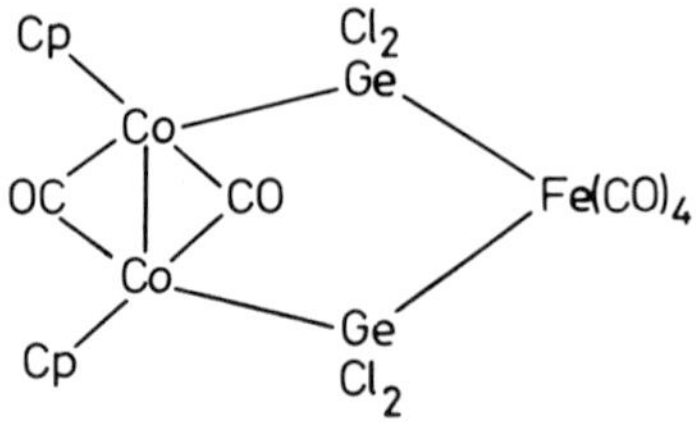

11

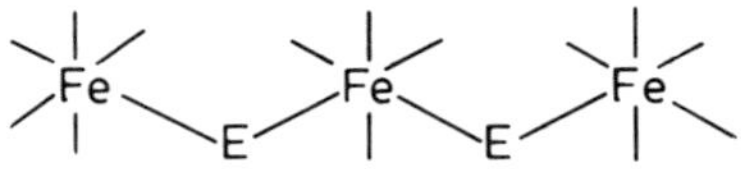

12

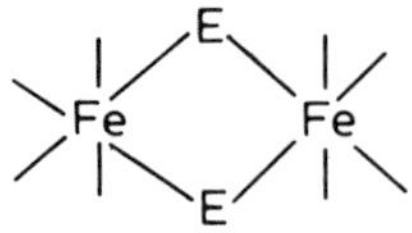

13

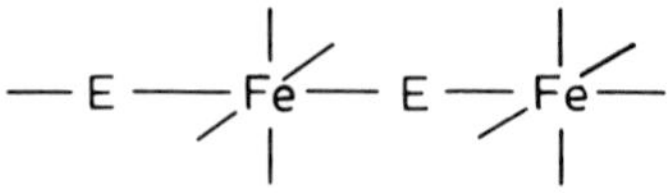

14

When the $Fe(CO)_4$ group is bonded to a divalent element (or to an equivalent group) in a *cis* arrangement, the two different structures 12 and 13 are expected *(344)*, however, a *cis* arrangement resulting either in hexanuclear or octanuclear rings has been recently discovered in $[(bipy)CdFe(CO)_4]_3$ and in $[CdFe(CO)_4]_4 \cdot 2\ CH_3COCH_3$ *(255,256)*. Structure 13 represents a simple example of a metallocycle and it has been found in

$[(Et_2Ge)Fe(CO)_4]_2$ *(539)*, in $[(Me_2Sn)Fe(CO)_4]_2$ *(289)*, and in $[(C_5H_5)_2SnFe(CO)_4]_2$ *(309)*; it is believed to be the usual structure when there are other similar groups R_2E and X_2E (E = Si, Ge, Sn, Pb) *(46,54,360)*, although $[(Cl_2Sn)Fe(CO)_4]_3$ is trimeric and probably contains a six membered ring *(488)*.

Previous speculations *(445)* concerning the structure of $(Me_3Si)_2Fe(CO)_4$ have been found incorrect. These early preparations have been shown to contain a ferracyclopentadiene system *(97)*, but it is now possible to prepare authentic $(Me_3Si)_2Fe(CO)_4$ *(364)*. In structure 13 it is also possible to substitute one or two carbon monoxide molecules with tertiary phosphines, as for instance in $[(Bu_2Sn)Fe(CO)_3(PR_3)]_2$ (R = Et, Ph) *(355,372)*.

A situation related to 13 is present in the compounds $R_4E_3[Fe(CO)_4]_4$ (E = Sn, R = Me, Et, Bu; E = Pb, R = Me) *(190)*. $Me_4Sn_3[Fe(CO)_4]_4$ has been shown by X-ray analysis to have structure 15 *(506)*. Here the central tin atom connects two

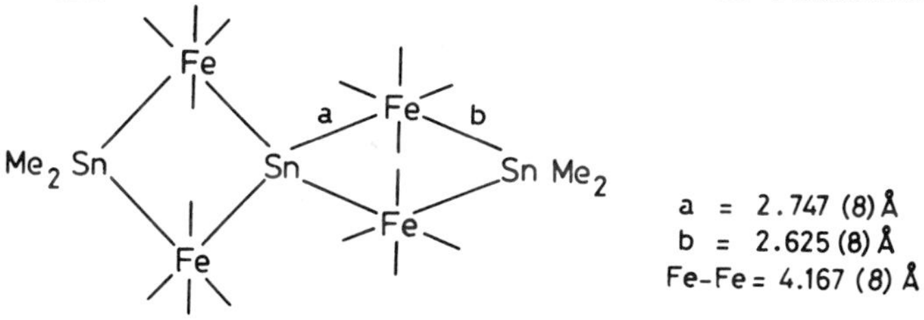

15

rings of type 13, which are situated in perpendicular planes, and because of constraints induced by the valence angles very long Fe-Sn distances are produced.

Structure 14 is probably present in $[HgFe(CO)_4]_n$ since it accounts for the high insolubility of the compound and is in keeping with the preferred *sp* hybridisation of the mercury atoms. The *trans* arrangement 9 has been found in the sterically crowded $Ph_3Sb\text{-}Fe(CO)_3Ph\text{-}SbPh_2\text{-}Fe(CO)_4$ *(134)*.

Finally in this class it is also possible to have mixed compounds (16, 17, 18, and 19) in which one of the bonding positions of the $Fe(CO)_4$ group is occupied by a halogen or a hydrogen atom, or, formally, by a negative charge. Examples

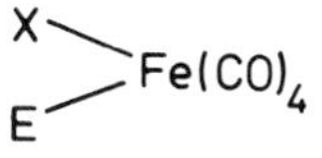

16

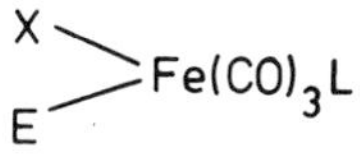

17

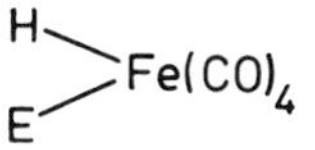

18

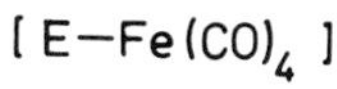

19

of these compounds are $(I_3Ge)Fe(CO)_4I$ *(406)*, $(Cl_3Sn)Fe(CO)_3(PPh_3)Cl$ *(406)*, $(Ph_3Si)Fe(CO)_4H$ *(362)*, $[(Br_2In)Fe(CO)_4]^-$ *(484)*, and $[(Cl_3Si)Fe(CO)_4]^-$ *(358)*; the last anion has a very short iron-silicon distance (2.224(9) Å).

Class d.

The proposed structures 20 *(360)* and 21 *(114)* can be formally related to $Fe_2(CO)_9$, although there is much more resemblance to the anion $[Fe_2(CO)_8]^{2-}$ *(144,499)*. The last type of

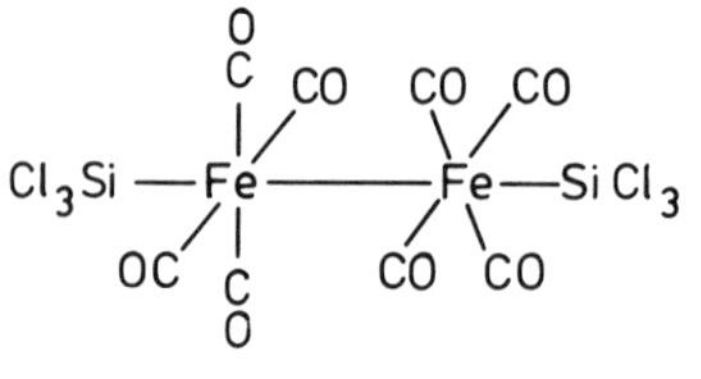

20

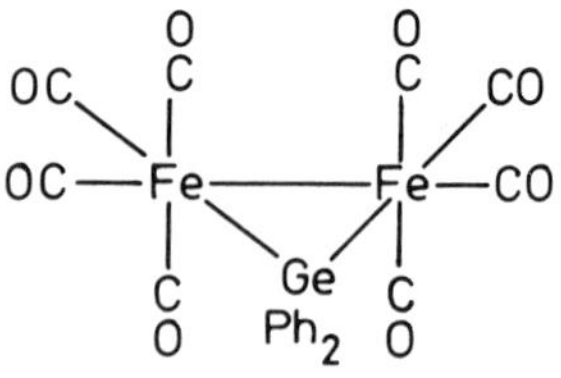

21

structure has been shown to be that of the compound $Sn[Fe_2(CO)_8]_2$ (22) *(412)* and it is common to the other compounds $E[Fe_2(CO)_8]_2$ (E = Ge *(191)*, Sn *(190)*, and Pb *(190)*) of this type. Here also the two $Fe_2(CO)_8$ units are situated in perpendicular planes, and there is considerable angular strain which causes long iron-iron distances. An intermediate struc-

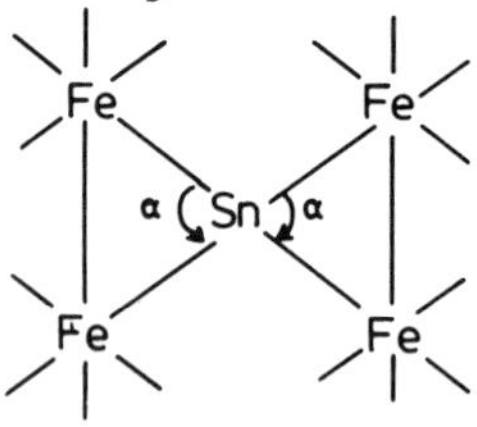

Sn—Fe = 2.54 (1) Å
Fe—Fe = 2.87 (1) Å
α = 69°

22

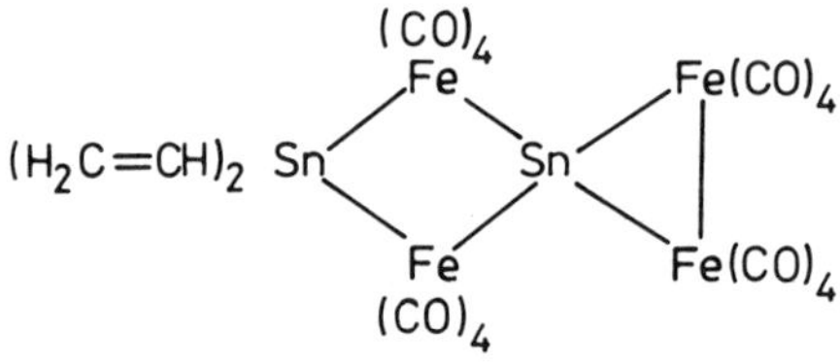

23

ture between 15 and 22 has been assumed in the compound 23 *(283)*.

Strictly related to the structure of $Fe_2(CO)_9$ are the compounds $(R_2E)_2Fe_2(CO)_7$ (R_2E = Ph_2Ge *(250)*, Me_2Si *(405)*, and $(Me_2Ge)_3Fe_2(CO)_6$ *(249)*) where, as shown in 24, there is increasing substitution of the bridging carbonyl groups. A similar substitution of a bridging carbonyl group is probably present in the compounds $(Ph_2E)Fe_2(CO)_3Cp_2$ (E = Ge, Sn) *(162)*, structure 25, and in the related $(Me_2Ge)_2Fe_2(CO)_2Cp_2$ *(366)*. Structure 44 (III.C.4) has been found in $\{[CpFe(CO)_2]_2Sn_2Fe_3(CO)_9\}$ *(435)*.

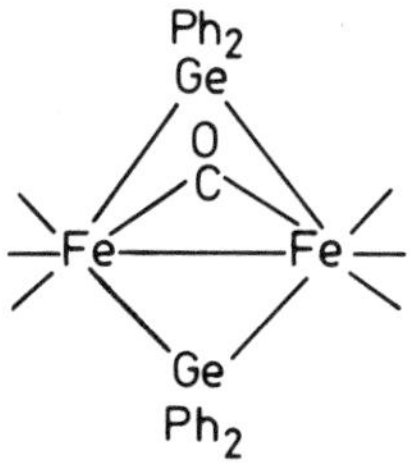

24

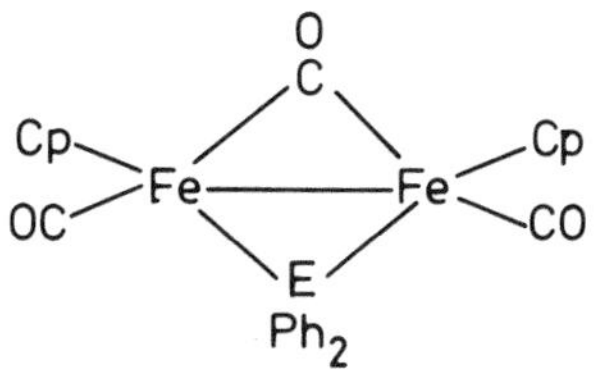

25

Class e.

Finally the principal types of miscellaneous compounds of class e are:

(1) Lewis adducts such as $Fe(CO)_5 \cdot HgCl_2$, $Fe(CO)_5 \cdot SbCl_5$ *(31)*, $\{E[Fe(CO)_5]_2\}(E'F_6)_2$, (E = Cd, SnFe; E' = As, Sb) *(216)*, and $[(XHg)Fe(CO)_3(L)_2]^+$ *(221)*, which probably have octahedral structures related to the similar cobalt adduct $Cp(CO)_2Co \cdot HgCl_2$ *(456)*.

(2) Some isonitrile derivatives, such as $(Cl_3Sn)_2Fe(p\text{-}MeO\text{-}C_6H_4\text{-}NC)_4$ *(430)*.

(3) Carborane π complexes, which have been very recently reviewed *(39,42)* and will not be considered here owing to the unique type of systems involved.

(4) Several compounds of unknown structure, such as $Tl_2Fe_3(CO)_{12}$ and $Bi_2Fe_5(CO)_{20}$ prepared by Hieber, Gruber, and Lux in 1959 *(325)*.

2. Bonding Considerations

One of the more debated problems of compounds containing bonds between transition and non-transition metal atoms concerns the description of this bond. Several authors prefer to represent this bond as a simple σ-bond *(295,519)*. They explain systematic variations of the properties of this bond, such as the contraction in the tin-iron distance in the series of compounds 1 to 4 and 5 to 6 of Table 2, assuming a reasonable variability of the covalent radii of the two metallic elements, which depend on the effective electronic density on the atoms *(295)*. Conversely other authors prefer to represent the same bond assuming the presence of an additional d_π-d_π component from the filled iron *d* orbitals to the vacant orbitals of the other metal *(8,46)*. This d_π - d_π component easily accounts for the same trends, although recent results of Mössbauer spectroscopy seem to disagree with this last picture *(103,198)*.

Table 2: Some values of Sn-Fe and Sb-Fe distances

	Compound	(Sn,Sb)-Fe distance [Å]	*Ref.*
1	$Ph_3Sn\text{-}Fe(CO)_2Cp$	2.536(3)	*(525)*
2	$Ph_2ClSn\text{-}Fe(CO)_2Cp$	2.504(3)	*(295)*
3	$PhCl_2Sn\text{-}Fe(CO)_2Cp$	2.467(2)	*(295)*
4	$Cl_3Sn\text{-}Fe(CO)_2Cp$	2.467(2)	*(294)*
5	$Me_2Sn[Fe(CO)_2Cp]_2$	2.603(5)	*(104)*
6	$Cl_2Sn[Fe(CO)_2Cp]_2$	2.492(8)	*(457)*
7	$ClSn[Fe(CO)_2Cp]_2[Mo(CO)_3Cp]$	2.590(7)	*(458)*
8	$\{Cl_2Sb[Fe(CO)_2Cp]_2\}^+$	2.440	*(248)*
9	$\{ClSb[Fe(CO)_2Cp]_3\}^+$	2.527-2.540(3)	*(516)*

Although there is still much controversy concerning these two ideas, the π bonding hypothesis provides a useful working tool for predicting and accounting for the majority of chemical and physical facts. For instance it easily accounts for fundamental trends such as the increase in "stability" in the series

$$Me_3E < Ph_3E < X_3E \quad (E = Si, Ge, Sn, Pb) \quad \textit{(46)}$$

and

$$Fe(CO)_2Cp < Fe(CO)(L)Cp < Fe(L)_2Cp \quad (L = PR_3, AsR_3, \textit{etc.})$$

In fact d_π-d_π donation is expected to increase in the same way.

Comparison of Sn-Fe distances for compounds 4-6-7 and 8-9 of Table 2 shows the effect of increasing steric hindrance due to the high steric requirements of $Fe(CO)_2Cp$ and of other similar large groups. The continuous increase of the iron-metal distance is in good agreement with the progressive chemical labilization of the bond. For instance it is well known that "stability" decreases in the series *(161)*

$$(Ph_3Sn)Fe(CO)_3NO > Ph_2Sn[Fe(CO)_3NO]_2 > PhSn[Fe(CO)_3NO]_3 > Sn[Fe(CO)_3NO]_4$$

B. SYNTHESES

1. Double Exchange Reactions

This method is based on the process [1] where the more

important by-reactions are the metal-halogen interchange *(194)* (eq. [2]) and the metal-metal redistribution reactions *(224,225)* (eq. [3]). For instance a strong nucleophilic M_1^-

$$M_1^- + M_2X \longrightarrow M_1{-}M_2 + X^- \quad [1]$$

$$M_1^- + M_2X \longrightarrow M_1X + M_2^- \quad [2]$$

$$M_1^- + M_1{-}M_2 \longrightarrow M_1{-}M_1 + M_2^- \quad [3]$$

will favour not only the synthesis (eq. [1]), but also the redistribution (eq. [3]).

Generally this type of synthesis is carried out in ethereal solvents. The use of iron halides is rather uncommon and mainly restricted to Group IV derivatives (eqs. [4] *(224)* and [5] *(62)*). On the contrary, reactions involving carbonyl-

$$[PbPh_3]^- + CpFe(CO)_2I \xrightarrow[\text{glyme}]{25°C} (Ph_3Pb)Fe(CO)_2Cp + I^- \quad [4]$$

$$K[SiH_3] + CpFe(CO)_2Br \xrightarrow[\text{glyme}]{-40°C} (H_3Si)Fe(CO)_2Cp + KBr \quad [5]$$

ferrates have been used with every group of the Periodic System, for instance reactions [6] *(141)* and [7] *(170)*. The

$$(Ph_3P)AuCl + Na[Fe(CO)_3NO] \xrightarrow[\text{THF}]{25°C} [Au(PPh_3)]Fe(CO)_3NO + NaCl \quad [6]$$

$$(C_6F_5)_2AsCl + [Fe(CO)_2Cp]^- \xrightarrow[\text{THF}]{-65°C} [(C_6F_5)_2As]Fe(CO)_2Cp + Cl^- \quad [7]$$

great potential of this method appears also in the synthesis of compounds containing three different metals (*e.g.*, reactions [8] *(362)* and [9] *(231)*). It is interesting to point

$$SnCl_4 + 2\ [(Cl_3Si)Fe(CO)_4]^- \xrightarrow[CH_2Cl_2]{-78°C} Cl_2Sn\ [(Cl_3Si)Fe(CO)_4]_2 + 2\ Cl^- \quad [8]$$

$$Cl_2Sn\ [Fe(CO)_2Cp]_2 + 2\ [Mo(CO)_3Cp]^- \xrightarrow[\text{THF}]{\text{reflux}} Sn[Fe(CO)_2Cp]_2[Mo(CO)_3Cp]_2 + 2\ Cl^- \quad [9]$$

out that sometimes in reactions of this type the equilibrium lies on the reactant side of the equation, as in the case of reaction [10] *(362)* with $K < 1$. Possible elimination of a

$$SiCl_4 + [(Cl_3Si)Fe(CO)_4]^- \xrightarrow[CH_2Cl_2]{} (Cl_3Si)_2Fe(CO)_4 + Cl^- \quad [10]$$

carbonyl group should also be considered when working with *gem*-dihalides (*e.g.*, eq. [11] *(201)*).

$$Me_2GeX_2 + 2\ Na[Fe(CO)_3NO] \longrightarrow Me_2Ge(\mu\text{-}X)Fe(CO)_2NO + CO + NaX \qquad (X = Cl,\ Br) \qquad [11]$$

When using the anions $[Fe(CO)_4]^{2-}$ and $[Fe(CO)_4H]^-$ secondary reactions are common as is shown by the synthesis of $Me_4Sn_3[Fe(CO)_4]_4$ starting from $MeSnCl_3$ *(506)* and of $Sn[Fe_2(CO)_8]_2$ starting from $SnCl_4$ *(190)*.

2. Oxidative Addition

Simple oxidative addition to iron-iron bonds is rather uncommon because usually this process requires contemporary formation of iron-halogen bonds. Examples of this type of synthesis are shown in eqs. [12] *(242)* and [13] *(199,200)*. The possibility of a second oxidative addition is clearly

$$GeX_4 + Cp_2Fe_2(CO)_4 \xrightarrow[C_6H_6]{reflux} (X_3Ge)Fe(CO)_2Cp + CpFe(CO)_2X \qquad [12]$$

$$(X = Cl,\ Br,\ I)$$

$$SbX_3 + 2\ Cp_2Fe_2(CO)_4 \xrightarrow[THF]{} XSb[Fe(CO)_2Cp]_2 + 2\ CpFe(CO)_2X \qquad [13]$$

$$(X = Cl,\ Br,\ I)$$

exemplified by reaction [14] *(512)*. Simple oxidative addition

$$(X_3E)Fe(CO)_2Cp + Cp_2Ni_2(CO)_2$$

$$\xrightarrow[reflux]{C_6H_6} (X_2E)[Fe(CO)_2Cp][Ni(CO)Cp] + CpNi(CO)X \qquad [14]$$

$$(E = Ge,\ Sn;\ X = Cl,\ Br)$$

to the metallic element E is also restricted by the availability of suitable derivatives, and has been so far applied only to germanium(II), tin(II), and indium(I) (*e.g.*, eqs. [15] *(106,431)*, [16] *(191,450)*, and [17] *(346)*). Here it is interesting to note that there is good evidence for oxidative addition of CpFe(dppe)Br and $CpFe(CO)_2Cl$ to magnesium metal in THF *(127,263)*.

$$SnCl_2 + CpFe(CO)_2Cl \xrightarrow[MeOH]{reflux} (Cl_3Sn)Fe(CO)_2Cp \qquad [15]$$

$$Cs[GeCl_3] + MeFe(CO)_2Cp \longrightarrow (MeCl_2Ge)Fe(CO)_2Cp + CsCl \qquad [16]$$

$$InX + CpFe(CO)_2X \xrightarrow[20°C]{THF} (X_2In)Fe(CO)_2Cp \qquad [17]$$

$(X = Cl, Br)$

More often the oxidative addition involves simultaneous breaking of an iron-iron bond and addition to a metallic derivative in a low oxidation state. Examples are the reactions [18] *(103,299)* and [19]. Magnesium metal has been added to

$$SnX_2 + Cp_2Fe_2(CO)_4 \xrightarrow[THF]{65°C} X_2Sn[Fe(CO)_2Cp]_2 \qquad [18]$$

(X = F, Cl, Br, I, NCS, HCO_2, CH_3CO_2; yields 80-90 %)

$$Hg + XFe(CO)_2Cp \xrightarrow[THF]{25°C} XHg[Fe(CO)_2Cp] \qquad [19]$$

(X = $Co(CO)_4$ *(230)*, $Fe(CO)_2Cp$ *(451)*)

$Cp_2Fe_2(CO)_4$ when working in THF, pyridine, and tetramethylethylendiamine *(436,437)*, while zinc, cadmium, and mercury have been added to the dianion $[Fe_2(CO)_8]^{2-}$ to give the species $[(CO)_4Fe\text{-}E\text{-}Fe(CO)_4]^{2-}$ *(95)*.

Finally, some examples of oxidative addition involving both breaking of Fe-Fe and E-E bonds *(47,170)* can be considered as redistributions of metallic bonds (*e.g.*, eq. [20] *(47)*).

$$Me_6Sn_2 + Cp_2Fe_2(CO)_4 \xrightarrow[xylene]{reflux} 2\ (Me_3Sn)Fe(CO)_2Cp \qquad [20]$$

(yield 60 %)

3. Oxidative Addition with Ligand Elimination

Very often the oxidative addition takes place together with elimination of ligands, and often the relative order of these two steps is unknown.

In some cases oxidative addition prior to elimination can be safely assumed. This is the case of the reaction between $Fe(CO)_5$ and mercury(II) halides, which is well known to give Lewis adducts such as $Fe(CO)_5 \cdot HgCl_2$ *(49,339,340)* and $Fe(CO)_3(L)_2 \cdot HgX_2$, where L = $P(OMe)_3$, PMe_3 and X = Cl, Br, I *(221)*. Here the whole process probably corresponds to the stoichiometry given by eq. [21] *(410,464)*. Another similar

$$Fe(CO)_5 + 2\ HgCl_2 \xrightarrow[EtOH]{25°C} (ClHg)_2Fe(CO)_4 + COCl_2 \qquad [21]$$

case is the reaction between $Fe(CO)_5$ and $SnCl_4$ or GeX_4 (X = Cl, Br, I) *(406)* where the first reaction product is shown in eq. [22]. At higher temperature this reaction is followed by redistribution

$$GeI_4 + Fe(CO)_5 \xrightarrow{25°C} cis\text{-}(I_3Ge)Fe(CO)_4I + CO \qquad [22]$$

(eq. [23]), and finally at 80-90°C there is a new cycle of similar processes according to reaction [24].

$$2\ (I_3Ge)Fe(CO)_4I \longrightarrow trans\text{-}(I_3Ge)_2Fe(CO)_4 + FeI_2 + 4\ CO \qquad [23]$$

$$(I_3Ge)_2Fe(CO)_4 + 2\ Fe(CO)_5 \longrightarrow [(I_2Ge)Fe(CO)_4]_2 + FeI_2 + 6\ CO \qquad [24]$$

The opposite case, in which an elimination reaction is followed by an oxidative addition, can be assumed in several cases where there is photochemical excitation of $Fe(CO)_5$. Examples are shown in equations [25] *(360)*, [26] *(483)*, and [27] *(160)*.

$$Fe(CO)_5 + R_3SiH \xrightarrow[hexane]{h\nu,\ -15°C} (R_3Si)Fe(CO)_4H + CO \qquad [25]$$

(R = Me, Ph, Cl)

$$Fe(CO)_5 + [AsPh_4][ECl_3] \xrightarrow[CH_2Cl_2]{h\nu,\ -25°C} [AsPh_4][(Cl_3E)Fe(CO)_4] + CO \qquad [26]$$

(E = Ge, Sn)

$$2\ Ph_{3-n}Cl_nSiH + Cp_2Fe_2(CO)_4 \xrightarrow{h\nu} 2\ (Ph_{3-n}Cl_nSi)Fe(CO)_2Cp + H_2 \qquad [27]$$

(n = 1,2)

A final relevant example, where the mechanism is uncertain, is the reaction between $Fe(CO)_5$ and organo-tin(IV) derivatives such as R_3SnCl *(190)*, R_2SnCl_2 *(190)*, $R_2Sn(CH{=}CH_2)_2$ *(377)*, R_2SnPh_2 *(377)*, R_3SnH *(190)*, and R_2SnH_2 *(190)* or

$Fe_3(CO)_{12}$ and $R_2Sn(C{\equiv}CR)_2$ *(355)* (R = Me, Et, Bu, etc.); see, for instance, eq. [28] *(190)*. Probably reactions such as

$$Me_3SnH + Fe(CO)_5 \xrightarrow[2\ h]{100\text{-}110°C} \underset{(70\ \%)}{(Me_3Sn)_2Fe(CO)_4} + \underset{(10\ \%)}{[(Me_2Sn)Fe(CO)_4]_2} + \underset{(2\ \%)}{Me_4Sn_3[Fe(CO)_4]_4} + \underset{(2\ \%)}{Sn[Fe_2(CO)_8]_2} \quad [28]$$

[28] are complicated by redistribution processes involving both starting materials and reaction products.

4. Nucleophilic Attack

The only example of this type of synthesis is the reaction [29] *(400)*. Excess of triphenylsilyl anion should be

$$[SiPh_3]^- + Fe(CO)_5 \xrightarrow[THF]{25°C} [(Ph_3Si)Fe(CO)_4]^- + CO \quad [29]$$

avoided, owing to competition with reaction [39]. It is interesting that the nucleophilic attack does not take place on a carbonyl group, an observation which can be related to the well known low stability of acyl derivatives of non-transition metals.

C. *REACTIVITY*

At the moment reactivity of Fe-E bonds can be discussed only from the points of view of their strength and polarity.

We have already pointed out (Section II.*A*) that, probably for steric reasons, the strength of this bond diminishes on increasing the number of iron atoms bonded to the same metallic center E. Other general considerations are difficult *(8, 46)*, although some limited relationships, which appear later in the present section, indicate $\bar{D}$(Fe-C) ≃ $\bar{D}$(Fe-Si) and $\bar{D}$(Fe-Sn) > $\bar{D}$(Fe-Pb) > $\bar{D}$(Fe-Hg).

The situation is slightly better from the polarity point of view. Polarity can exercise control of reactivity, and it is therefore significant to realize how easily the polarity of the E-Fe bond can be inverted from

$$E^{\delta+}\text{—}Fe^{\delta-} \text{ to } E^{\delta-}\text{—}Fe^{\delta+}.$$

This change takes place not only on changing the metal E, but also on changing the ligands on iron or the substituents of the metallic element E. Comparison of the ^{1}H-NMR spectra is particularly useful here, because the reference points can be unequivocally assigned to the related homonuclear dimeric compounds *(361)*. Some data of this type have been summarized in Fig. 2. The left part of this figure exemplifies the dependence of the polarity on the metal E and its substituents in the case of cyclopentadienyldicarbonyliron derivatives: for instance the polarity of the Sn-Fe bond firstly disappears in going from Me_3Sn to Ph_3Sn, and then reverses with Ph_2ClSn. In the right part of the same figure the same dependence on the ligands bonded to iron is exemplified in the case of trimethyltin derivatives: here also the polarity reverses in going from $Fe(CO)_2Cp$ to $Fe(CO)(PPh_3)Cp$. Clearly we are faced with an extremely subtle and complicated situation.

1. Ionic Dissociation and Attack of the Fe-E Bond

The transformation of a covalent Fe-E bond into separate ions according to equation [30] is expected to depend not only

$$L_xE\text{-}FeL'_y \rightleftharpoons [L_xE]^+ + [FeL'_y]^- \qquad [30]$$

on E and on the ligands L and L', but also on the polarity of the solvent. This ionization is easily recognized using IR spectroscopy, owing to the change of frequency of the carbonyl groups associated with the negative charge.

It has been recently shown that the covalent Fe-Sn bond ionizes according to the equilibrium [31] *(140)*. In THF this

$$(Ph_3Sn)Fe(CO)_3NO \rightleftharpoons [Ph_3Sn]^+ + [Fe(CO)_3NO]^- \qquad [31]$$

$\tilde{\nu}(CO)$ = 2065 vs, 2050 s, 1980 vs, 1773 s cm^{-1} in heptane

$\tilde{\nu}(CO)$ = 1986 m, 1884 vs, 1686 cm^{-1} in CH_3CN

equilibrium is mainly on the left side, but in acetonitrile dissociation is practically complete *(140)*. On the contrary, $(Ph_3Sn)Fe(CO)_2(PPh_3)NO$ in CH_3CN *(140)* and $(Ph_3Sn)Fe(CO)_2Cp$ in DMF *(126)* are not dissociated.

A similar ionization is also known for the Fe-Si bond (eq. [32] *(75,362)*). We may also note that $Zn[Fe(CO)_2Cp]_2$ is

$$(Cl_3Si)_2Fe(CO)_4 + 2\ NEt_3 \xrightleftharpoons[CH_2Cl_2]{} [Cl_3Si(NEt_3)_2]^+ + [(Cl_3Si)Fe(CO)_4]^- \qquad [32]$$

not dissociated in DMF *(126)* while in the same solvent $[CdFe(CO)_4]$ is dissociated with probable formation of zwitterions *(344)*, and $Hg[Fe(CO)_3NO]_2$ is partially dissociated in acetonitrile and completely dissociated in DMSO *(139)*.

Also the metallocycles $[(R_2E)Fe(CO)_4]_2$, structure <u>13</u>, in basic solvents undergo bridge cleavage reactions to give ad-

Cp, τ[ppm]		$E^{\delta-}-Fe^{\delta+}$	Me, τ[ppm]
4.29	$[CpFe(CO)_2(THF)]^{+f}$ *(481)*		
4.74	$(Cl_3Sn)Fe(CO)_2Cp^a$ *(268)*		
4.78	$(Cl_3Ge)Fe(CO)_2Cp^a$ *(278)*		
4.87	$(Cl_3Si)Fe(CO)_2Cp^b$ *(361)*	$(Me_3Sn)Fe(diphos)Cp^a$ *(387)*	10.55
4.87	$(PhCl_2Sn)Fe(CO)_2Cp^a$ *(268)*		
4.93	$Cl_2Sn[Fe(CO)_2Cp]_2^a$ *(512)*		
4.97	$(ClHg)Fe(CO)_2Cp^a$ *(428)*		
4.99	$ClFe(CO)_2Cp^b$ *(361)*		
5.07	$(Ph_2ClSn)Fe(CO)_2Cp^a$ *(268)*	$(Me_3Sn)Fe(CO)(PPh_3)Cp^a$ *(387)*	10.11
5.08	$(CF_3)_2As\ Fe(CO)_2Cp^c$ *(195)*		
5.13	$(Me_2ClSi)Fe(CO)_2Cp^a$ *(388)*		
5.21	$(Ph_3Sn)Fe(CO)_2Cp^a$ *(268)*		
5.22	$Cp_2Fe_2(CO)_4^a$ *(122,123)*	Me_6Sn_2 *(118)*	9.78
5.23	$Hg[Fe(CO)_2Cp]_2^a$ *(428)*		
5.26	$HFe(CO)_2Cp^d$ *(215)*	$(Me_3Sn)Fe(CO)_2Cp^a$ *(387)*	9.68
5.30	$MeFe(CO)_2Cp^e$ *(215)*		
5.32	$(Me_3Sn)Fe(CO)_2Cp^a$ *(387)*		
5.36	$(Me_3Si)Fe(CO)_2Cp^e$ *(387)*	$(Me_3Sn)_2Fe(CO)_4^f$ *(364)*	9.53
		Me_3SnCl *(117)*	9.37
		$E^{\delta+}-Fe^{\delta-}$	

[a] in $CDCl_3$; [b] in CH_3CN; [c] in CCl_3F; [d] in C_6H_{12}; [e] in CCl_4; [f] in C_6D_6

Fig. 2: Examples of change of polarity of the Fe-E bond as indicated by ^{1}H-NMR data.

ducts which are probably best represented as zwitterions 26 *(298)*. The ease of this transformation decreases in the order Ge > Sn > Pb and py > acetone > THF > diethyl ether *(423)*.

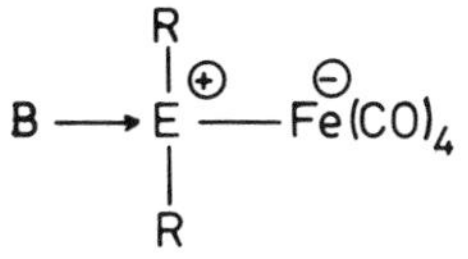

26

A comparison of the available data shows that the ease of ionization diminishes in the series

$$E\text{-}Fe(CO)_4X \simeq E\text{-}Fe(CO)_3NO \gg E\text{-}Fe(CO)_2Cp \simeq E\text{-}Fe(CO)_2(PPh_3)NO$$

a trend which is consistent with a progressive fall of the average π-acidity of the ligands.

The different polar character of the Fe-E bond leads to a related different reactivity. This is well exemplified by the following different behaviour towards acetic acid (eqs. [33]

and [34] *(344)*). Moreover, both these compounds transform to

$$[Cd(NH_3)_2][Fe(CO)_4] + 2\ CH_3COOH \xrightarrow{H_2O} \frac{1}{n}\ [CdFe(CO)_4]_n + 2\ NH_4CH_3CO_2 \qquad [33]$$

$$[Zn(NH_3)_3][Fe(CO)_4] + 5\ CH_3COOH \xrightarrow{H_2O} Fe(CO)_4H_2 + Zn(CH_3CO_2)_2 + 3\ NH_4CH_3CO_2 \qquad [34]$$

$Fe(CO)_4H_2$ when using strong mineral acids, but the less polar $[HgFe(CO)_4]_n$ is stable *(315)*.

Similar differences have been found in the Group IV derivatives (eqs. [35] *(444)* and [36] *(290)*). An intermediate

$$(Me_3Si)Fe(CO)_2Cp + HCl \xrightarrow{0°C} Me_3SiCl + CpFe(CO)_2H \qquad [35]$$

$$(Ph_3Sn)Fe(CO)_2Cp + 3\ HCl \xrightarrow{CH_2Cl_2} (Cl_3Sn)Fe(CO)_2Cp + 3\ C_6H_6 \qquad [36]$$

situation has been found when the Fe-Sn bonds are labile for steric reasons (eq. [37] *(449)*).

$$PhSn[Fe(CO)_2Cp]_3 + 3\ HCl \longrightarrow Cl_2Sn[Fe(CO)_2Cp]_2 + CpFe(CO)_2Cl + C_6H_6 + H_2 \qquad [37]$$

The ease of nucleophilic attack on the Fe-E bond is also emerging at present. Some recent examples are given in equations [38] *(444)*, [39] *(400)*, and [40] *(223)* where the last

$$(Me_3Si)Fe(CO)_2Cp + MeO^- \xrightarrow[MeOH]{25°C} Me_3SiOMe + [Fe(CO)_2Cp]^- \qquad [38]$$

$$[SiPh_3]^- + [(Ph_3Si)Fe(CO)_4]^- \xrightarrow[THF]{25°C} Ph_6Si_2 + [Fe(CO)_4]^{2-} \qquad [39]$$

$$[SnPh_3]^- + (Ph_3Pb)Fe(CO)_2Cp \xrightarrow[glyme]{25°C} [PbPh_3]^- + (Ph_3Sn)Fe(CO)_2Cp \qquad [40]$$

reaction agrees with the higher strength of the Sn-Fe bond *(224,225)*. Similar conclusions have been obtained by controlled-potential electrolysis (eqs. [41] - [43]) *(222)*.

$$Hg[Fe(CO)_2Cp]_2 \xrightarrow[\text{glyme}]{-2.0 \text{ V}} 2 \; [Fe(CO)_2Cp]^- + Hg \qquad [41]$$

$$(Ph_3Pb)Fe(CO)_2Cp \xrightarrow[\text{glyme}]{-2.1 \text{ V}} [Fe(CO)_2Cp]^- + [PbPh_3]^- \qquad [42]$$

$$2 \; (Ph_3Sn)Fe(CO)_2Cp \xrightarrow[\text{glyme}]{-2.6 \text{ V}} 2 \; [Fe(CO)_2Cp]^- + Ph_6Sn_2 \qquad [43]$$

Here, as a first approximation, a more negative potential corresponds to a stronger Fe-E bond.

Finally we can add that generally the compounds with Fe-E bonds are stable or moderately stable to air, while they easily react with halogens *(478)*.

2. Insertion in the Fe-E Bond

Insertion of unsaturated species in the Fe-E bond has been little studied. It has been found that perfluorobutyne adds to the compounds $(Me_3E)Fe(CO)_2Cp$ (E = Si, Ge, Sn) under UV irradiation, and that in this reaction the silicon derivative is the least reactive *(102)*. Using 3,3,3-trifluoropropyne it has been possible to show that this reaction results in *cis*-addition to the alkyne (eq. [44]) *(102)*. The same silicon derivative failed to react with liquid SO_2, but

$$(Me_3Si)Fe(CO)_2Cp + F_3C-C\equiv CH \xrightarrow{h\nu} \begin{array}{c} F_3C \\ Me_3Si \end{array} \!\!\! C=C \!\!\! \begin{array}{c} H \\ Fe(CO)_2Cp \end{array} \qquad [44]$$

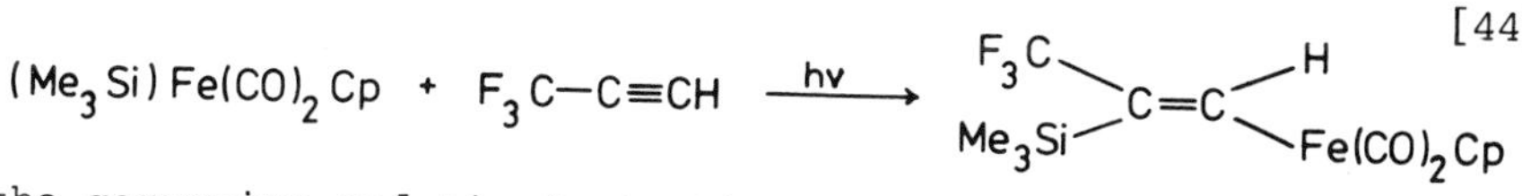

the germanium and tin derivatives gave insertion according to equation [45] *(101)*. It is worth noting that the phenyl

$$(Me_3E)Fe(CO)_2Cp + SO_2 \xrightarrow[\text{SO}_2 \text{ liq.}]{25^\circ C} \underset{O}{\overset{O}{Me_3E-\!\!\overset{\|}{\underset{\|}{S}}\!\!-Fe(CO)_2Cp}} \quad (E = Ge, Sn) \qquad [45]$$

derivative, $Ph_2Sn[Fe(CO)_2Cp]_2$, reacts in a completely different way (eq. [46]) *(241,269)* where insertion in the phenyl

$$Ph_2Sn[Fe(CO)_2Cp]_2 + 2 \; SO_2 \xrightarrow[\text{benzene}]{25^\circ C} (Ph-\overset{O}{\overset{\|}{S}}-O)_2Sn[Fe(CO)_2Cp]_2 \qquad [46]$$

group has been confirmed by X-ray diffraction *(476)*.

3. Substitution on E

The substitution of halogen substituents by organic ra-
dicals or hydrogen atoms is of important synthetic significance (*e.g.*, equations [47] *(278)* and [48] *(160)*). Nucleo-

$$X_2E[Fe(CO)_2Cp]_2 \begin{cases} \xrightarrow[\text{THF, }-80°C]{\text{MeLi}} & Me_2E[Fe(CO)_2Cp]_2 \\ & (X_2E = I_2Ge : 74\%) \\ & (X_2E = Cl_2Sn : 86\%) \\ \\ \xrightarrow[\text{THF, }0°C]{\text{NaBH}_4} & H_2E[Fe(CO)_2Cp]_2 \\ & (X_2E = Cl_2Ge : 55\%) \end{cases}$$

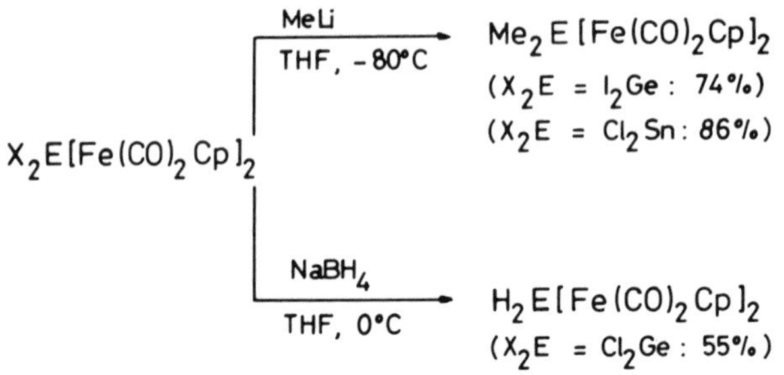

[47]

$$(Cl_3E)Fe(CO)_2Cp + 3\ C_6F_5Li \xrightarrow[-78°C]{Et_2O}$$

$$[(C_6F_5)_3E]Fe(CO)_2Cp + 3\ LiCl$$

$$(E = Si,\ Ge,\ Sn)$$

[48]

philic attack on the carbonyl groups or on the Fe-E bonds has
not been observed in these cases.

Other interesting reactions are [49] *(422)* and [50]
(342). More conventional, but of frequent synthetic importan-

$$(Cl_3Si)Fe(CO)_2Cp + 3\ AgBF_4 \xrightarrow[\text{acetone}]{25°C}$$

$$(F_3Si)Fe(CO)_2Cp + 3\ AgCl + 3\ BF_3\cdot acetone$$

[49]

$$(Cl_3Si)Fe(CO)_2Cp + 3\ NaOR \longrightarrow$$

$$[(RO)_3Si]Fe(CO)_2Cp + 3\ NaCl$$

$$(R = Me,\ Et,\ Pr,\ i\text{-}Pr)$$

[50]

$$(Cl_3Sn)Fe(CO)_2Cp \begin{cases} \xrightarrow{3\ X^-} & (X_3Sn)Fe(CO)_2Cp + 3Cl^- \\ & (X = NCS,\ HCO_2,\ CH_3CO_2)\ (103) \\ \\ \xrightarrow{ox} & (ox_2ClSn)Fe(CO)_2Cp \\ & (ox = 8\text{-hydroxyquinoline})\ (108) \\ \\ \xrightarrow{bipy} & [Cl_3Sn(bipy)]Fe(CO)_2Cp\ (108) \end{cases}$$

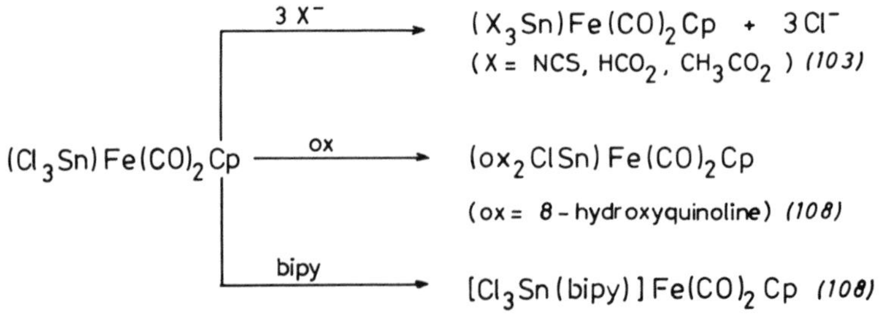

[51]

ce, are processes such as those exemplified in scheme [51].

A particular class of ionic reactions, which involves formation of a new Fe-E bond by addition to E, has recently been found by Burtlich (eq. [52]) *(164)*. This behaviour is

$$[Fe(CO)_2Cp]^- + Hg[Fe(CO)_2Cp]_2 \xrightarrow[THF]{25°C} \{Hg[Fe(CO)_2Cp]_3\}^- \quad [52]$$

quite unexpected on steric grounds, but can be rationalized assuming a better redistribution of the negative charge over the π-acidic ligands (compare with Section III.*E*.4.).

4. Substitution on Fe

Most of these reactions are conventional examples of the well known substitution on metal carbonyls *(32)* and often involve ligands such as tertiary phosphines. It is only sufficient to note here that the ease of substitution seems to follow a trend similar to that found in the ease of ionic dissociation, a not unreasonable coincidence owing to the expected effect of the average π-acidity of the ligands on the S_N2 mechanism. For instance reactions such as [53] *(141)* and [54] *(372)* take place very easily, whereas substitution is

$$[Au(PPh_3)]Fe(CO)_3NO + L \xrightarrow[benzene]{25°C}$$

$$[Au(PPh_3)]Fe(CO)_2(L)NO + CO \quad [53]$$

$$(L = PPh_3,\ P(OPh)_3,\ AsEt_2Ph)$$

$$[(Bu_2Sn)Fe(CO)_4]_2 + PEt_3 \xrightarrow{25°C}$$

$$(Bu_2Sn)_2Fe_2(CO)_7(PEt_3) + CO \quad [54]$$

much more difficult in the E-Fe(CO)$_2$Cp derivatives. This latter reaction does not take place by simple heating in boiling toluene but requires UV irradiation (eq. [55]) *(198, 387)*. Further substitution to give derivatives of type

$$(Me_3Sn)Fe(CO)_2Cp + L \xrightarrow[benzene]{h\nu,\ 25°C} (Me_3Sn)Fe(CO)(L)Cp + CO \quad [55]$$

$$(L = PPh_3,\ AsPh_3,\ PMePh_2,\ PMe_2Ph,\ \textit{etc.})$$

E-Fe(L)$_2$Cp is uncommon although such compounds have been prepared when (L)$_2$ = 2 SbPh$_3$ *(198)*, dppe *(387)*, and *cis*-1,2-bis-(diphenylphosphino)ethylene *(387)*.

5. Redistribution Reactions

We will consider as a redistribution reaction every process in which the number of Fe-E bonds remains constant, while they change their relative distribution. Processes of this type are very common *(190,299,312)*, some representative examples being shown in equations [56] *(313)*, [57] *(331)*, [58] *(106)*, and [59] *(428)*. It should be noted that the equilibrium

$$2\ (R_3Pb)_2Fe(CO)_4 \xrightarrow{25°C} [(R_2Pb)Fe(CO)_4]_2 + 2\ R_4Pb \qquad [56]$$

(R = Me, Et, Pr, Bu, Ph)

$$Hg[Fe(CO)_3NO]_2 \xrightarrow{25°C} \frac{1}{n}\ [HgFe(CO)_4]_n + Fe(CO)_2(NO)_2 \qquad [57]$$

$$Cl_2Sn[Fe(CO)_2Cp]_2 + SnCl_4 \xrightarrow[toluene]{110°C} 2\ (Cl_3Sn)Fe(CO)_2Cp \qquad [58]$$

$$Hg[Fe(CO)_2Cp]_2 + HgX_2 \underset{acetone}{\overset{25°C}{\rightleftharpoons}} 2\ (XHg)Fe(CO)_2Cp \qquad [59]$$

(X = Cl, Br, I, NCS, $Co(CO)_4$)

[59] is on the right side when using halide derivatives, but on the left when using the alkyls HgR_2 (R = Bu, C_6F_5) *(164)*.

Sometimes this type of process involves two different metals as in the cases [60] *(140)*, [61] *(107)*, and [62] *(128)*.

$$Hg[Fe(CO)_2(L)NO]_2 + 2\ SnX_4 \xrightarrow[benzene]{25°C} 2\ (X_3Sn)Fe(CO)_2(L)NO + HgX_2 \qquad [60]$$

(L = CO, PPh_3, $P(OPh)_3$; X_4 = Cl_4, Br_4, $PhCl_3$, Ph_2Cl_2)

$$Hg[Fe(CO)_2Cp]_2 + SnCl_2 \xrightarrow[acetone]{reflux} Cl_2Sn[Fe(CO)_2Cp]_2 + Hg \qquad [61]$$

$$Hg[Fe(CO)_2Cp]_2 + E' \underset{toluene}{\overset{20°C}{\rightleftharpoons}} E'[Fe(CO)_2Cp]_2 + Hg \qquad [62]$$

(E' = Cd, Zn)

Here the constant use of mercury derivatives seems to reflect a lower energy of the Fe-Hg bond.

More recently it has also been pointed out that rearrangements such as [63] strongly indicate similar values for the iron-silicon and iron-carbon bond energies *(531)*.

$$[(ClCH_2)(CH_3)_2Si]Fe(CO)_2Cp \xrightarrow{100°C} ClSi(CH_3)_2\text{-}CH_2\text{-}Fe(CO)_2Cp \qquad [63]$$

6. Elimination Reactions

This is a broad interesting class of reactions which often, as in the previous section, is strictly related to the thermal behaviour of these compounds. A first case concerns elimination of ligands from the iron atoms, with formation of new iron-iron bonds, for instance reactions [64] *(162)* and [65] *(250)*. In other cases this formation of iron-iron bonds

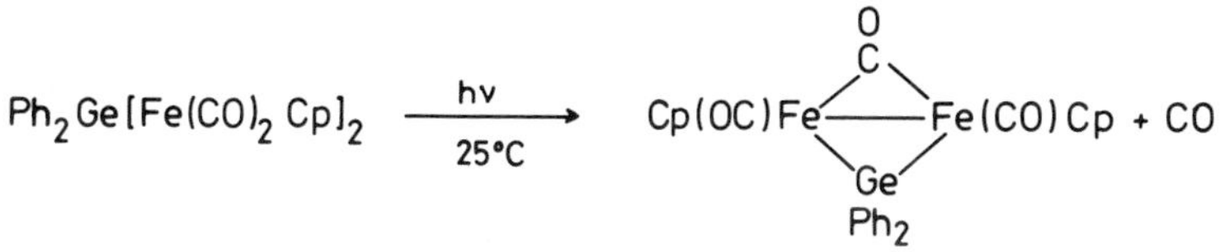

[64]

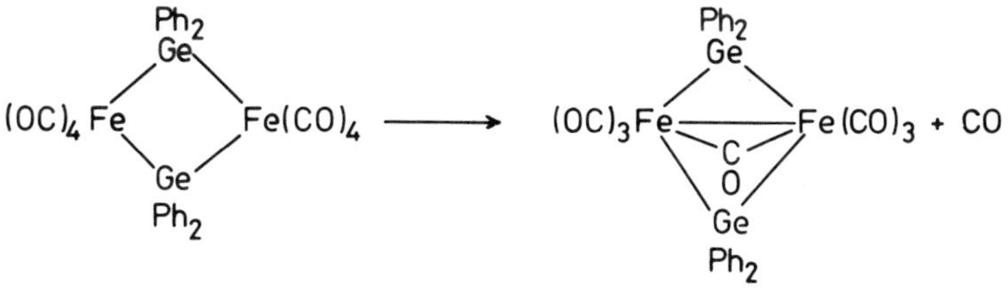

[65]

takes place together with elimination of Fe-E bonds, *e.g.*, reactions [66] *(270)* and [67] *(360)*. In this last case the

$$Hg[Fe(CO)_2Cp]_2 \xrightarrow{80\text{-}90°C} Hg + Cp_2Fe_2(CO)_4 \qquad [66]$$

$$3(Ph_3Si)Fe(CO)_4H \xrightarrow[\text{hexane}]{60°C} 3\ Ph_3SiH + Fe_3(CO)_{12}\ (+\ Fe(CO)_5) \qquad [67]$$

analogous compound $(Cl_3Si)Fe(CO)_4H$ is more stable *(360)*, as indeed would be expected from d_π-d_π bonding considerations.

III. COMPOUNDS WITH Fe-M BONDS (M = TRANSITION METAL).

A. *GENERAL REMARKS*

Over 70 binuclear compounds of iron have been characterized by X-ray analyses, and they show extreme variations in the Fe-Fe distance (from 2.177 *(443)* to 3.05 Å *(208)*). It seems probable that the expanded orbitals of low valent iron can suffer considerable dimensional variations on changing the effective electron density, and therefore give different bond lengths. However, when an extreme difference in the iron-iron distance is observed it indicates different bond orders. This is substantiated by the Raman data reported in Table 3.

The iron-iron single bond mean energy can be approximated

Table 3: Comparison of metal-metal stretching frequencies and distances in some cluster compounds of iron.

Compound	$\tilde{\nu}$(Fe-Fe) [cm^{-1}] *(Ref.)*	k(Fe-Fe) [mdyn $Å^{-1}$] *(Ref.)*	$\bar{d}$(Fe-Fe) [Å] *(Ref.)*	"bond order"
$(t\text{-}Bu_2C_2)_2Fe_2(CO)_4$	289 *(494)*	3.0±0.4 *(121)*	2.215 *(455)*	∿ 2
$Cp_2Fe_2(CO)_4$	226 *(486)*	-	2.531 *(cis)* 2.534 *(trans)* *(122,123)*	∿ 1
$Fe_2(CO)_9$	225 *(486)*	-	2.523 *(186)*	∿ 1
$Cp_4Fe_4(CO)_4$	221 *(486)*	∿1.3 *(404)*	2.520 *(453)*	∿ 1
$Fe_3(CO)_{12}$	219 *(486)*	-	2.636 *(183)*	∿ 1
$Fe_2(CO)_6S_2$	191 *(494)*	1.3±0.2 *(494)*	2.552 *(526)*	∿ 1

on the base of the standard heat of formation of the gaseous iron atom (99.5 kcal mol^{-1}) *(193)* it amounts to about only 17.9 kcal. mol^{-1} *(310)*. Other available thermodynamic data are the following:

Compound	ΔH_f^o	$\Delta H_{sub.}$	$\Delta H_f^o(g)$	$\bar{D}_{Fe-CO}$	$\bar{D}_{Fe-Fe}$
		(kcal.mol^{-1}; 298 K)			
$Fe(CO)_5$ (liq.)	-183±2 *(193)*	9.6±0.2 *(193)*	-173±2 *(193)*	27.9 *(17)*	--
$Fe_2(CO)_9$ (cryst.)	-337±3 *(174)*	18±5 *(174)*	-319±6 *(174)*	29.4 *(310)*	16.7 *(310)*
$Fe_3(CO)_{12}$ (cryst.)	-442±4 *(174)*	23±5 *(174)*	-419±7 *(174)*	30.1 *(310)*	15.5 *(310)*

Using these data it can be shown that both $Fe_2(CO)_9$ and $Fe_3(CO)_{12}$ are expected to transform into $Fe(CO)_5$ under carbon monoxide (298 K, 1 atm), although we are limited to enthalpic approximations and we cannot yet calculate the required free energy balances.

This coarse prediction, about low thermodynamic stability of iron-iron bonds in polynuclear iron carbonyls, seems to be fulfilled in the available experimental data: iron-iron bonds are either in metastable condition (see Section III.*E*.1.), or stabilized through additional bonding contributions due to the presence of bridging ligands.

It is therefore of major significance that, with rare exceptions, the binuclear compounds of iron are characterized by the presence of bridging ligands along the metal-metal bond. Moreover on increasing the number of electrons donated by such a bridging ligand to the metal atoms, there is an "artificial" strengthening of the metal-metal bond, until the extreme case in which the reactivity of the metal-metal bond could become extremely low. On the contrary, although the presence of particular bridging ligands has a great influence on the geometry of the polynuclear clusters, here only some of the metal-metal bonds are supported by the bridging ligands and the influence of the bridging ligands is often not so stringent for the properties of these bonds.

For this reason it is convenient to discuss separately the structural relationships in binuclear and cluster compounds, and their dependence on the presence of particular ligands.

B. BINUCLEAR COMPOUNDS, DESCRIPTION ACCORDING TO THE BRIDGING LIGAND

1. Non-bridged

This situation is rare for the Fe-Fe binuclear compounds. It has been established for the $[(CO)_4Fe\text{-}Fe(CO)_4]^{2-}$ anion

(Fe-Fe = 2.75 *(499)* and 2.787 Å *(144)*), and it is probably present in $(Cl_3Si)_2Fe_2(CO)_8$ (20) *(360)* and in $Fe_2(CO)_8I_2$ *(176)*. On the contrary this situation is common when iron is bonded to a transition metal of Groups VI and VII. For instance in $Cp(CO)_2Fe\text{-}Mo(CO)_3Cp$ *(378)*, $Cp(CO)_2Fe\text{-}Re(CO)_5$ *(448)*, $[(CO)_4Fe\text{-}Mn(CO)_5]^-$ *(485)*, and in $Cp(CO)_2Fe\text{-}Mn(CO)_5$ *(378)*. In this last compound the Fe-Mn distance amounts to 2.843 Å *(308)*, and considering the Mn-Mn distance which has been found in $Mn_2(CO)_{10}$ (2.923 Å) *(203)*, it agrees with a hypothetical Fe-Fe distance of 2.76 Å.

2. Bridged by 2-Electron-Donors

The most important ligand of this class is carbon monoxide itself, and representative examples of binuclear iron compounds bridged by carbonyl groups are collected in Table 4. The presence of bridging carbonyl groups is also common when iron is bonded to cobalt and nickel.

Comparison with the iron-iron distances of the previous section shows considerable shortening, while at the same time M-C-M angles of about 80° are observed; both these effects can be rationalized on the basis of three-center molecular orbitals *(9,13)*. They provide the first evidence that we are approaching situations in which it is not possible to distinguish between the metal-metal bond and the bond with the bridging ligands. The same concept of coordination number at the iron atoms often breaks down because of our inability to separate this system of bonds into individual components. For instance we are obliged to consider the presence of a stereochemically inert iron-iron bond in $Cp_2Fe_2(CO)_4$ (27) *(13)*.

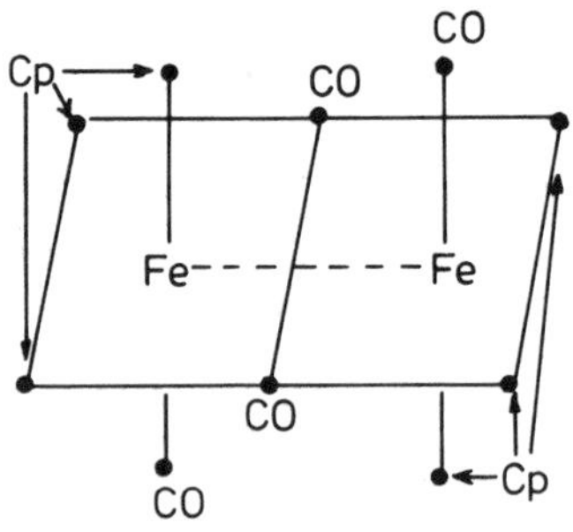

27

Table 4 shows that, when the metal-metal bond is supported only by carbonyl groups, the metal-metal distance is relatively independent of the different ligands on the two iron atoms. Only in the extreme cases of $(CO)_2(bipy)Fe(CO)_2Fe(CO)_3$ and $(CO)_3Fe(CO)(dppm)Fe(CO)_3$ where a strongly electron releasing group is present, or of Cp(CO)Fe(CO•$AlEt_3$)$_2$Fe(CO)Cp, where a Lewis acid has been added, a significant variation of the iron-iron distance (from 2.709 - 2.611 to 2.491 Å) is observed. Both the formation of adducts between bridging carbonyl groups and Lewis acids, such as $AlEt_3$, $AlBr_3$, BF_3, BCl_3, and BBr_3, and the lower carbonyl stretching frequencies (∿ 180 - 220 cm^{-1}) of such groups, is evidence that, compared with the terminal carbonyl groups, the bridging carbonyls pick up a larger amount of the available electron density. The formation of stable adducts is enhanced by electron releasing ligands such as Cp as shown by comparison of the behaviour of $Cp_2Fe_2(CO)_4$ *vs.* $Fe_2(CO)_9$ and $Cp_4Fe_4(CO)_4$ *vs.* $Fe_3(CO)_{12}$ *(398)*.

Careful examination of the IR *(274,418,419)* and ^{1}H-NMR *(124)* spectra of compounds such as Cp(CO)Fe(CO)$_2$Fe(CO)Cp and Cp(CO)Fe(CO)$_2$Co(CO)$_3$ indicates the presence of an equilibrium in which the unbridged isomer is also present. In all cases the unbridged form 29 is favoured on increasing the temperature, as would be expected from the higher degree of freedom

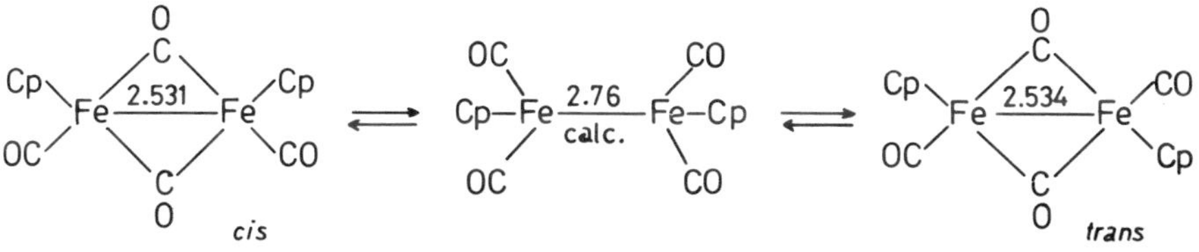

28 29 30

of such an isomer, whereas the bridged isomers 28 and 30 are enthalpy favoured (∿ -4 kcal mol^{-1}) *(274)* due to the presence of reinforced Fe-Fe bonds.

Examples of other types of 2-electron bridging ligands are summarized in Table 5. Here a meaningful comparison of the iron-iron distances is not possible, because the different bridging groups *di per se* introduce different electronic and steric requirements into the bridged system of bonds.

In the compounds bridged by GeR_2 and SO_2 groups, the bridging atoms have been found in approximately tetrahedral situation with the R groups (or the oxygen atoms) lying in a plane perpendicular to the iron-iron bond.

Introduction of a further carbonyl group in Cp(CO)Fe(CO)(SO_2)Fe(CO)Cp to give Cp(CO)$_2$Fe(SO_2)Fe(CO)$_2$Cp results in the breaking of the iron-iron bond (3.909 Å) *(157)* in agreement with the presence of two electrons in excess of the noble gas rule.

Finally, the existence of the remarkable hydride bridged cation [(triphos)Fe(H)$_3$Fe(triphos)]$^+$ (Fe-Fe = 2.34 Å) *(210)* must be mentioned; a simple electron counting would require a triple bond between the two iron atoms. Bridging hydrogen

Table 4: Representative examples of the binuclear compounds of iron bridged by carbonyl groups

Compound		d(Fe–M) [Å]	*(Ref.)*
$(CO)_3Fe(CO)_3Fe(CO)_3$		2.523(1)	*(186,474)*
$[(CO)_3Fe(CO)_2(H)Fe(CO)_3]^-$		2.521(1)	*(146)*
$(CO)_2(bipy)Fe(CO)_2Fe(CO)_3$		2.611	*(182)*
$(CO)_3Fe(CO)(dppm)Fe(CO)_3$		2.709(2)	*(185)*
$Cp(CO)Fe(CO)_2Fe(CO)Cp$	*cis* *trans*	2.531(2) 2.534(2)	*(122,123)*
$Cp(CO)Fe(CO)_2Fe(t\text{-}BuNC)Cp$		2.523(2)	*(50,363)*
$Cp(CO)Fe(CO{\cdot}AlEt_3)_2Fe(CO)Cp$		2.49(1)	*(376)*
$(\eta^5\text{-}C_{10}H_{11})(CO)Fe(CO)_2Fe(CO)(\eta^5\text{-}C_{10}H_{11})$		2.553(2)	*(179)*
$(\eta^5\text{-}C_5H_4)(CO)Fe(CO)_2Fe(CO)(\eta^5\text{-}C_5H_4)$ (C₅H₄ rings linked by $SiMe_2$)		2.512(3)	*(523)*
$(\eta^5\text{-}C_5H_4)(CO)Fe(CO)_2Fe(CO)(\eta^5\text{-}C_5H_4)$ (C₅H₄ rings linked by $(CH\text{-}NMe_2)_2$)		2.510(2)	*(500)*
$Cp(CO)Fe(CO)_2Fe[P(OPh)_3]Cp$		2.545(2)	*(301,187)*
$[(\eta\text{-}B_9)(CO)Fe(CO)_2Fe(CO)(\eta\text{-}B_9)]^{2-}$ $(\eta\text{-}B_9 = \eta\text{-}(3)\text{-}1,2\text{-}B_9C_2H_{11})$		2.591	*(293)*
$[(CO)_4Fe(CO)Co(CO)_3]^-$		2.585(3)	*(144,485)*
$Ind(CO)Fe(CO)_2Co(CO)_3$		2.552(2)	*(502)*
$Cp(CO)Fe(CO)_2Co(CO)_2(PPh_2Me)$		2.540(4)	*(212)*
$Cp(CO)Fe(CO)_2Co(CO)(C_4H_6)$		2.546(1)	*(132)*
$Cp(CO)Fe(CO)_2Co(CO)_3$		2.545	*(419,133)*
$Cp(CO)Fe(CO)_2NiCp$		-	*(432,536)*

Table 5: Types of 2-electron bridging ligands in binuclear compounds with iron-iron bonds

Bridging ligand	Examples of compounds	Fe-Fe distance [Å]	*(Ref.)*
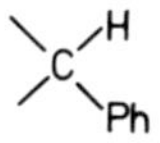	$(CO)_4Fe(CHPh)Fe(CO)_4$		*(272)*
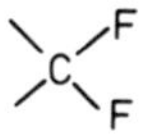	$(CO)_3Fe(CF_2)_2(CO)Fe(CO)_3$		*(495)*
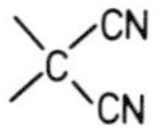	$Cp(CO)Fe(CO)(C_3N_2)Fe(CO)Cp$		*(390)*
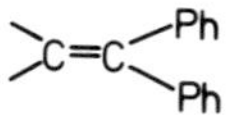	$(CO)_4Fe(C_2Ph_2)Fe(CO)_4$	2.635	*(440)*
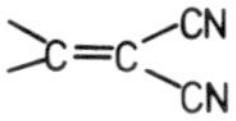	$Cp(CO)Fe(CO)(C_4N_2)Fe(CO)Cp$	2.511	*(390,391)*
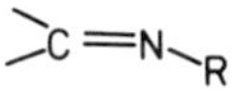	$Cp(CO)Fe(CNPh)(CO)Fe(CO)Cp$	2.53	*(370)*
	$Cp(CO)Fe(CNMe)_2Fe(CO)Cp$	2.538	*(181)*
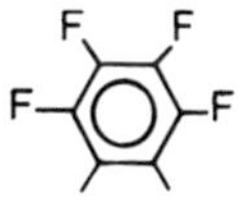	$(CO)_4Fe(C_6F_4)Fe(CO)_4$	2.797	*(96,480)*
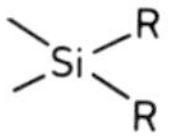	$(CO)_3Fe(CO)(SiMe_2)Fe(CO)_3$		*(405)*
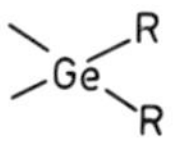	$(CO)_3Fe(GePh_2)_2(CO)Fe(CO)_3$	2.67	*(250)*
	$(CO)_3Fe(GeMe_2)_3Fe(CO)_3$	2.750	*(249)*
	$Cp(CO)Fe(GeMe_2)(CO)Fe(CO)Cp$	2.628	*(366,51)*
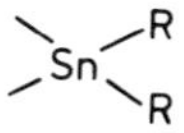	$Cp(CO)Fe(SnPh_2)(CO)Fe(CO)Cp$		*(162)*
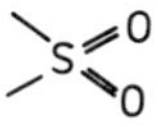	$(CO)_4Fe(SO_2)Fe(CO)_4$	2.717	*(438)*
	$Cp(CO)Fe(SO_2)(CO)Fe(CO)Cp$	2.597-2.584	*(158)*

atoms are rather common *(26,27)* and, for instance, this type of bridge is present in the $[(CO)_3Fe(CO)_2(H)Fe(CO)_3]^-$ anion *(146)*.

3. Bridged by 3-Electron Donors

Three-electron bridging ligands are very common, representative examples being summarized in Table 6.

Compounds having different 3-electron bridging ligands are well known, for instance $(CO)_3Fe(SPh)(PPh_2)Fe(CO)_3$ *(365)*, $(CO)_3Fe[P(CF_3)_2](I)Fe(CO)_3$ *(297)*, $(CO)_3Fe(TeC_6F_5)[As(C_6F_5)_2]$-$Fe(CO)_3$ *(397)*, and $(CO)_3Fe(Ph_2CN)(I)Fe(CO)_3$ *(375)*. Compounds in which there are two different transition metals are also known, for instance $(CO)_4Fe(PPh_2)Mn(CO)_4$ *(99,534)*, $(CO)_4Fe$-$(AsMe_2)Mn(CO)_4$ *(520)*, $(CO)_3Fe(PPh_2)(CO)Ru(CO)Cp$ *(305)*, $(CO)_3$-$Fe(PPh_2)Co(CO)_3$ *(99)*, $(CO)_3Fe(PPh_2)(CO)NiCp$ *(534)*, and $(CO)_4$-$Fe(PPh_2)Pd_2Cl_2(PPh_2)Fe(CO)_4$ *(99)*.

It seems important to point out that these compounds generally conform to the "magic number" (see Fig. 3 and Section III.*C*.1.) which in this case corresponds to the well known noble gas rule. The presence of two 3-electron ligands corresponds to a quite flexible situation, in which the iron-iron distance can change from a bonded to a non-bonded situation and *vice versa* and the dimeric unit can be held together just by the bridging ligands.

Some relevant data on the mercapto bridged compounds are compiled in Table 7. For instance the compound $Cp(CO)Fe$-$(SPh)_2Fe(CO)Cp$, which should have two excess electrons, in fact has an iron-iron distance of 3.39 Å *(264)* which corresponds to the absence of a metal-metal bond, and therefore effectively obeys the noble gas rule. Conversely, oxidation of the methyl compound with $AgSbF_6$ gives the corresponding cation $[Cp(CO)Fe(SMe)_2Fe(CO)Cp]^+$ *(385)* in which the iron-iron distance is decreased to 2.925 Å *(168)*.

Still more stringent is the case of the series *(227,509)*

$$cis\text{-}[Cp(CO)Fe(PPh_2)_2Fe(CO)Cp]^{n+} \begin{cases} n = 0 & \text{Fe-Fe} = 3.498(4) \\ n = 1 & \text{Fe-Fe} = 3.14(2) \\ n = 2 & \text{Fe-Fe} = 2.764(4) \end{cases}$$

Breaking the Fe-Fe bond in the species $(CO)_3Fe(L)_2Fe(CO)_3$ ($L = SMe$, PMe_2, and $AsMe_2$) is also in agreement with the results of electrochemical reduction *(227,228)*. Firstly there is formation of a paramagnetic anion $[Fe_2(CO)_6(L)_2]^-$ which does not conform to the noble gas rule, and then of the diamagnetic dianion $[Fe_2(CO)_6(L)_2]^{2-}$ in which probably there is no more Fe-Fe bonding. The same situation is expected in the dihydride derivative $H(CO)_3Fe(PPh_2)_2Fe(CO)_3H$ *(232)*.

All these changes can be simply rationalized assuming

Table 6: Types of 3-electron bridging ligands in binuclear compounds of iron

Bridging ligand	Examples of compounds	Fe-Fe distance [Å]	*(Ref.)*
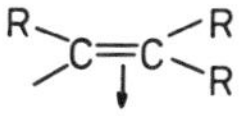	$Cp(CO)Fe(CH{=}CHCOCH_3)(CO)Fe(CO)_3$	2.556	*(65,66)*
	$(CO)_3Fe(CH{=}CHCOPh)(CO)W(CO)_2Cp$	2.81	*(68,452)*
	$(CO)_3Fe(CH{=}CHBr)(Br)Fe(CO)_3$	2.525	*(401)*
	$(CO)_3Fe(CH{=}CHCl)(Cl)Fe(CO)_3$		*(394)*
	$(CO)_3Fe(CH{=}CH_2)(SCH_3)Fe(CO)_3$		*(381,382)*
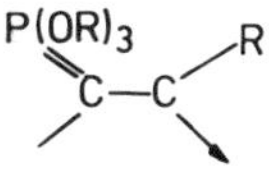	$(CO)_3Fe[(EtO)_3PC_2Ph](PPh_2)Fe(CO)_3$	2.671	*(532)*
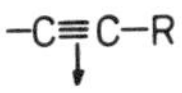	$(CO)_3Fe(C{\equiv}CPh)(PPh_2)Fe(CO)_3$		*(466)*
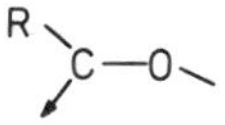	$(CO)_3Fe(PhCO)_2Fe(CO)_3$	2.568	*(413)*
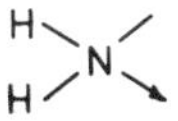	$(CO)_3Fe(NH_2)_2(CO)_3$	2.402	*(207)*
	$(PEt_3)(CO)_2Fe(NH_2)_2Fe(CO)_2(PEt_3)$		*(280)*
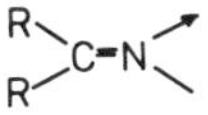	$(CO)_3Fe[(p\text{-}MeC_6H_4)_2CN]_2Fe(CO)_3$	2.403	*(113)*
	$(CO)_3Fe(Ph_2CN)_2Fe(CO)_3$		*(375)*
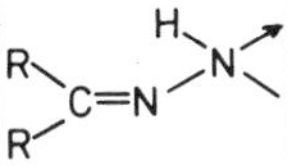	$(CO)_3Fe(R_2CN_2H)_2Fe(CO)_3$ ($R = p\text{-}CH_3C_6H_4$)	2.40	*(76)*
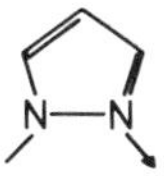	$(CO)_3Fe(N_2C_3H_3)_2Fe(CO)_3$		*(389)*
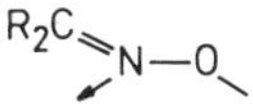	$(CO)_3Fe(Me_2CNO)(C_3H_7NH)Fe(CO)_3$	2.470	*(53)*
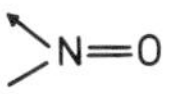	$CpFe(NO)_2FeCp$	2.326	*(121,131)*
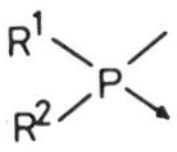	$(CO)_3Fe(PR^1R^2)_2Fe(CO)_3$		*(143,170, 296)*

Table 6: (continued)

	$Cp(CO)Fe(PPh_2)(H)Fe(CO)Cp$		*(311)*
	$Cp(CO)Fe(PMe_2)(CO)Fe(CO)_3$	2.626	*(302,521)*
	$(CO)_3Fe(PR_2)(OH)Fe(CO)_3$	2.511	*(514,515)*
	$(R = p\text{-}CH_3\text{-}C_6H_4)$		
	$[Cp(CO)Fe(PPh_2)_2Fe(CO)Cp]^+$	3.14	*(509)*
	$[Cp(CO)Fe(PPh_2)_2Fe(CO)Cp]^{2+}$	2.764	*(509)*
	$(NO)_2Fe[P(CF_3)_2]_2Fe(NO)_2$		*(232)*
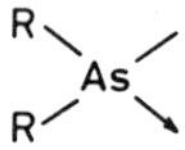	$(CO)_3Fe(AsR_2)_2Fe(CO)_3$		*(143,170)*
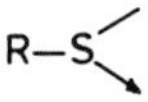	$(CO)_3Fe(SEt)_2Fe(CO)_3$	2.537	*(204)*
	$(PR_3)(CO)_2Fe(SR)_2Fe(CO)_2(PR_3)$		*(92,332)*
	$(NO)_2Fe(SEt)_2Fe(NO)_2$	2.720	*(511)*
	$(CO)_3Fe(SPh)_2Fe(CO)_3$	2.516	*(314)*
	$[Cp(CO)Fe(SR)_2Fe(CO)Cp]^+$ (R = Me;	2.925)	*(90,168, 281)*
	$[Cp(CO)Fe(SR)_2Fe(CO)Cp]^{2+}$		*(90,281)*
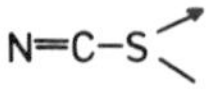	$(NO)_2Fe(NCS)_2Fe(NO)_2$		*(333)*
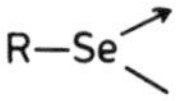	$(CO)_3Fe(SeEt)_2Fe(CO)_3$		*(326)*
	$(CO)_3Fe(SeCF_3)_2Fe(CO)_3$		*(482)*
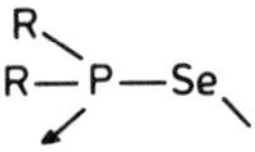	$(CO)_3Fe[P(CF_3)_2Se][P(CF_3)_2]Fe(CO)_3$		*(233)*
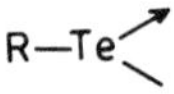	$(CO)_3Fe(TeC_6H_4\text{-}p\text{-}OMe)_2Fe(CO)_3$		*(328)*
	$(NO)_2Fe(I)_2Fe(NO)_2$	3.05	*(208)*
	$(NO)_2Fe(X)_2Fe(NO)_2$ (X = Cl, Br)		*(357)*
	$(CO)_3Fe(X)_2Fe(CO)_3$ (X = Br, I)		*(396)*

Table 7: Fe-Fe distances in mercapto bridged iron complexes

Compound	Excess electron over the n.g.r.	Fe-Fe [Å]	*(Ref.)*
$(CO)_3Fe(SPh)_2Fe(CO)_3$	0	2.516	*(314)*
$[(CO)_3Fe(SMe)_3Fe(CO)_3]^+$	1	3.062	*(492)*
$[Cp(CO)Fe(SMe)_2Fe(CO)Cp]^+$	1	2.925	*(168)*
$Cp(CO)Fe(SPh)_2Fe(CO)Cp$	2	3.39	*(264)*

that the frontier orbitals are of predominant metallic character, and some detailed M.O. calculations agree with this interpretation *(509)*. We shall see later (Section III.*F*.) that this conclusion is probably quite general.

Three electrons are probably also provided by the ethylthioxantate ligands in the compound $Fe_2(S_2CSEt)_4(SEt)_2$ (Fe-Fe = 2.618 Å) *(192)*.

4. Bridged by 4-Electron Donors

Four-electron donor bridging ligands are relatively rare, the most important classes being acetylenes, allenes, and tertiary diphosphines. Examples of acetylene bridged compounds are $(C_2R_2)Fe_2(CO)_6$ (R = CMe_3, Fe-Fe = 2.316(4)) *(189)* and $(C_2R_2)_2Fe_2(CO)_4$ (R = CMe_3) *(455)*, and allene bridged compounds are $(C_3H_4)Fe_2(CO)_6(PPh_3)$ *(213)* and $(C_{11}H_{18})Fe_2(CO)_7$ *(414)*.

A bridging ethoxy carbene also provides 4 electrons in $(CO)_4Fe(RC\text{-}OEt)Fe(CO)_3$ (R = 2,4,6-trimethoxyphenyl, Fe-Fe = 2.535 Å) *(353)*. A similar bridging situation is present in the $Fe_2(CO)_6$(maleic anhydride)(2,4-diphenylpyridazine) derivative *(465)* and probably in the phthalazine derivative $(CO)_3Fe(C_8H_8N_2)Fe(CO)_3$ *(59)*.

With tertiary diphosphines of the type $Ph_2P\text{-}Y\text{-}PPh_2$ (Y = CH_2, C_2H_2, C_2H_4, C_3H_6, NC_2H_5) a series of compounds of the type 31 has been obtained *(300)*, while the more rigid diphos-

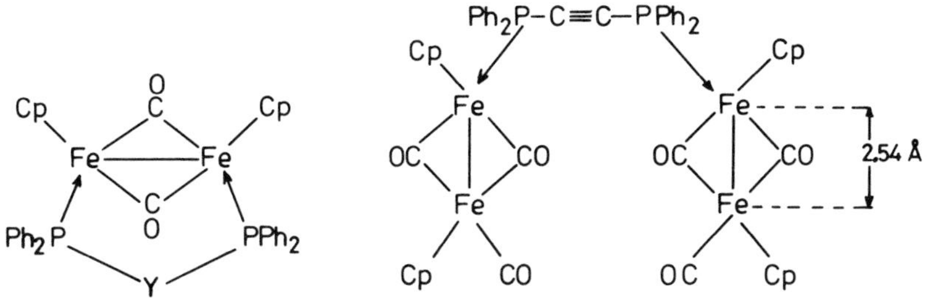

31 32

phine $Ph_2P-C\equiv C-PPh_2$ gives rise to the dimeric compound 32
(135). Here it is interesting to note that while 32 is oxidized by iodine with breaking of the iron-iron bonds, the
compounds 31 are oxidized not only by iodine and silver salts,
but also by very mild agents such as $SnCl_4$ and $HgCl_2$, giving
paramagnetic cations $\{Cp_2Fe_2(CO)_2[Ph_2P(CH_2)_n PPh_2]\}^+$ (300).
Recent work (266,303) has shown that these cations, which
have formal iron-iron bond orders of 1.5, are reasonably
stable in CH_2Cl_2, while in acetonitrile they easily disproportionate to give a 1:1 mixture of the neutral compound and the
dication $[Cp(CO)_2(CH_3CN)Fe(Ph_2P-R-PPh_2)Fe(CH_3CN)(CO)_2Cp]^{2+}$, in
which there is no iron-iron bond. In these cases it seems
probable that the diphosphine contributes to both the ease of
oxidation, owing to its high donor properties, and to the
stability of these unusual cations, because of its bridging
ability.

In the dithiolene substituted carbonylmetal complexes
these ligands generally provide 4 electrons, although when the
dithiolenes are in bridging positions they may provide up to
6 electrons (next section). This particular area has been
carefully reviewed (34) and we will just mention the existence
of electrochemically related series such as shown in equation
[68].

Finally, we can add that the interesting four-electron
donor nitrosobenzene gives a dimeric compound (33) which, in
agreement with the noble gas rule, does not contain iron-iron
bonds (86). The reaction between diphenylacetylene and $(CO)_3$-
$Fe(PPh_2)(CO)NiCp$ results in displacement of the bridging CO
group and insertion into the Ni-P bond to give the unusual
five-electron donor bridging ligand of structure 34 (85).

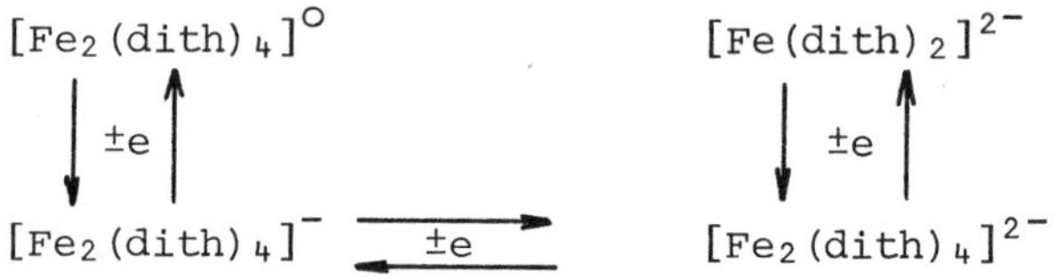

$$[Fe_2(dith)_4]^0 \qquad\qquad [Fe(dith)_2]^{2-}$$

$$\downarrow \pm e \uparrow \qquad\qquad\qquad \downarrow \pm e \uparrow$$

$$[Fe_2(dith)_4]^- \quad\underrightarrow{\pm e}\quad\qquad [Fe_2(dith)_4]^{2-} \qquad\qquad [68]$$

Fe-Fe = 2.767 Å   Fe-Fe = 3.02 Å
dith = $S_2C_2(CF_3)_2$ (492)   dith = $S_2C_2(CN)_2$ (307)

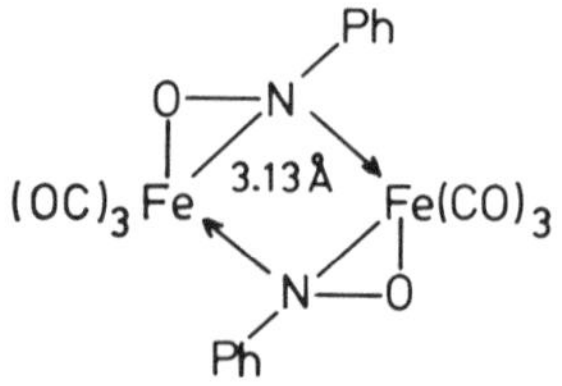

33

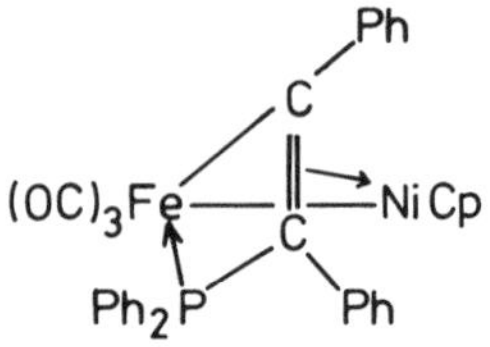

34

5. Bridged by 6-Electron Donors

Examination of Table 6, which summarizes the three-electron bridging ligands, shows that in the majority of the examples there are present two such ligands. The six-electron donor ligands of the present section mainly correspond to a similar bridging situation and, as the three-electron donors, they are generally situated in a plane approximately perpendicular to the iron-iron bond. Moreover, they are often strictly chemically related to the three-electron donors, as is apparent from comparison of Tables 6 and 8. The extremely complicated bonding situation which can arise with these 6-electron bridging ligands is well exemplified by the binuclear $(CO)_3Fe(L)Fe(CO)_3$ derivatives 35, 36, and 37.

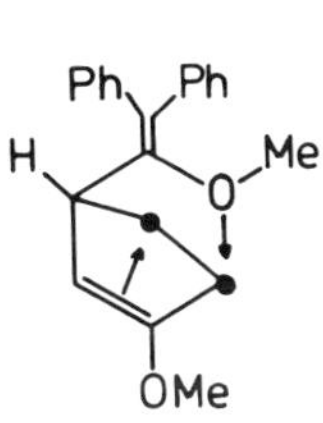

35

d(Fe-Fe) = 2.62 Å *(359)*

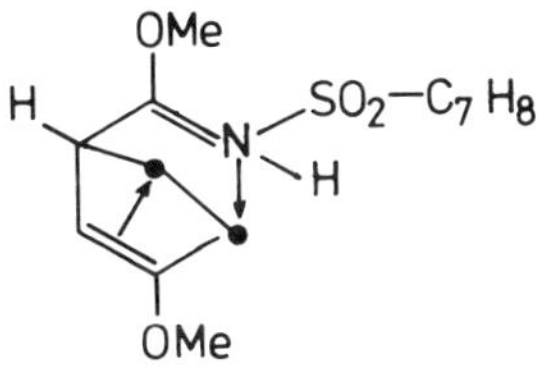

36

d(Fe-Fe) = 2.64 Å *(479)*

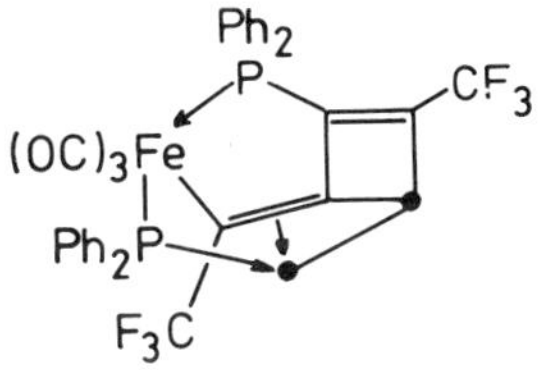

37

d(Fe-Fe) = 2.554 Å *(459)*

(the black circles indicate the iron atoms).

It seems possible that the cation $[Fe_2(CO)_6(N_3)_2]^{2+}$ contains two bent free radical ligands (38); this hypothesis is in agreement with the exceptionally high value of the paramagnetism of this ion (μ = 10.58 B.M.) *(408)* and with the noble gas rule.

38

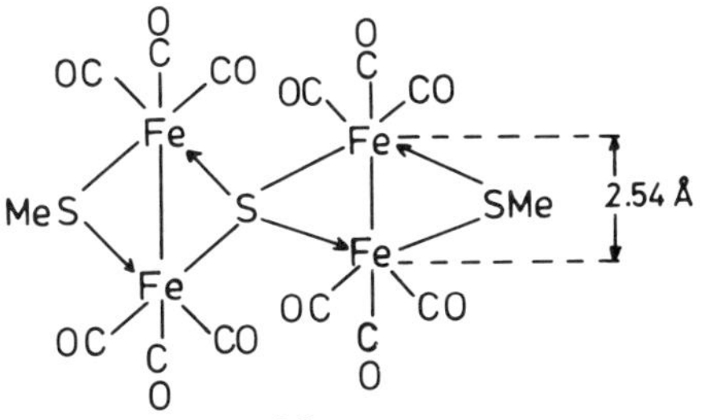

39

A peculiar six-electron donor is tetrahedral sulphur, because it is able to donate 3 electrons to each of the iron-iron systems present in perpendicular planes (39) *(165)*, a related structure is present in $Fe_4(CO)_{12}S(CSNMe_2)(CNMe_2)$

Table 8: Types of 6-electron bridging ligands (containing donor atoms different from carbon) in binuclear compounds of iron.

Bridging ligand	Examples of compounds	Distance Fe-Fe [Å]	*(Ref.)*
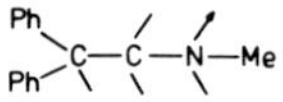	$(CO)_3Fe(Ph_2C_2NMe)Fe(CO)_3$	2.544	*(461)*
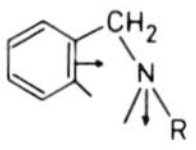	$(CO)_3Fe(C_{14}H_{13}N)Fe(CO)_3$ (R = *p*-Me-C_6H_4)	2.43	*(78)*
	$(CO)_3Fe(NC_{11}H_{13})Fe(CO)_3$		*(276)*
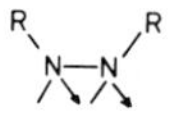	$(CO)_3Fe(N_2Me_2)Fe(CO)_3$	2.496	*(238)*
	$(CO)_3Fe(N_2C_{12}H_8)Fe(CO)_3$	2.508	*(239)*
	$(CO)_3Fe(N_2C_{16}H_{10})Fe(CO)_2(PPh_3)$	2.53	*(407)*
	$(CO)_3Fe(N_2C_6H_{10})Fe(CO)_3$	2.490	*(415)*
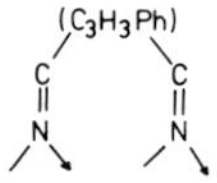	$(CO)_3Fe(N_2C_{23}H_{18})Fe(CO)_3$	2.392, 2.393	*(219)* *(136)*
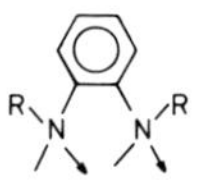	$(CO)_3Fe[N_2(H)(Ph)C_6H_4]Fe(CO)_3$	2.372	*(80)*
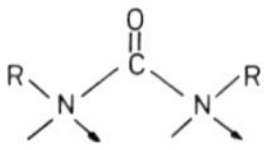	$(CO)_3Fe(N_2Ph_2CO)Fe(CO)_3$	2.402, 2.416	*(369, 470)*
	$(CO)_3Fe(N_2Me_2CO)Fe(CO)_3$	2.391	*(236)*
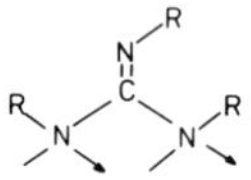	$(CO)_3Fe[N_3C(i\text{-}Pr)_3]Fe(CO)_3$		*(112)*
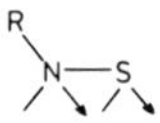	$(CO)_3Fe(PhNS)Fe(CO)_3$		*(462)*
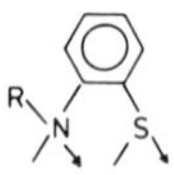	$(CO)_3Fe[N(H)SC_6H_4]Fe(CO)_3$	2.411	*(421, 409)*
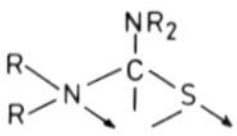	$(CO)_3Fe[N_2CS(CH_3)_4]Fe(CO)_3$		*(60)*
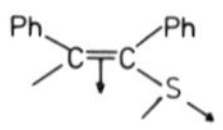	$(CO)_3Fe(Ph_2C_2S)Fe(CO)_3$	2.533	*(489,490)*
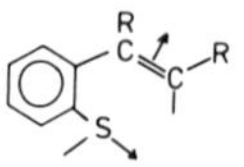	$(CO)_3Fe(C_8H_8S)Fe(CO)_3$		*(379)*

Table 8: (continued)

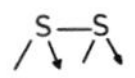	$(CO)_3Fe(S_2)Fe(CO)_3$	2.552	*(526)*
	$(PR_3)(CO)_2Fe(S_2)Fe(CO)_2(PR_3)$		*(93,332)*
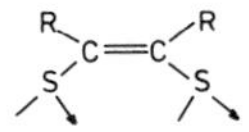	$(CO)_3Fe(S_2C_2Ph_2)Fe(CO)_3$	2.507	*(524)*
	$(NO)_2Fe[S_2C_2(CF_3)_2]Fe(NO)_2$		*(380)*
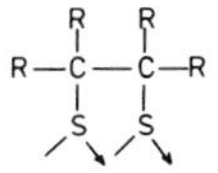	$(CO)_3Fe(S_2C_2H_4)Fe(CO)_3$		*(386)*
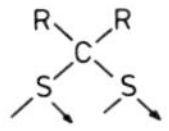	$(CO)_3Fe(S_2CR_2)Fe(CO)_3$		*(58)*
	(R = 2,4,6-trimethylphenyl)		
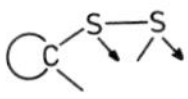	$(CO)_3Fe(S_2C_{16}H_{14})Fe(CO)_3$		*(58,61)*
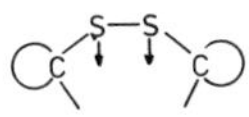	$(CO)_3Fe(S_2C_{32}H_{28})Fe(CO)_3$		*(61)*
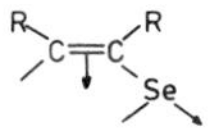	$(CO)_3Fe(SeC_{14}H_{10})Fe(CO)_3$		*(490)*
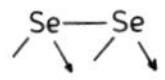	$(CO)_3Fe(Se_2)Fe(CO)_3$		*(326)*
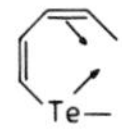	$(CO)_3Fe(TeC_4H_4)Fe(CO)_3$		*(460)*
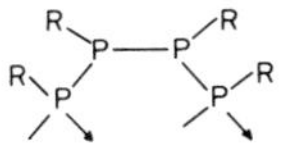	$(CO)_3Fe(P_4Ph_4)Fe(CO)_3$		*(71)*
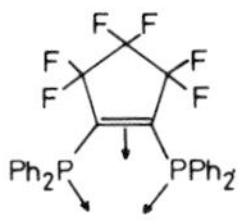	$(CO)_3Fe(f_6fos)Fe(CO)_3$		*(196)*
	$(CO)_3Fe(f_4fos)Fe(CO)_3$		*(196)*
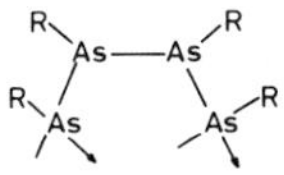	$(CO)_3Fe(As_4Me_4)Fe(CO)_3$	2.680	*(253,284)*
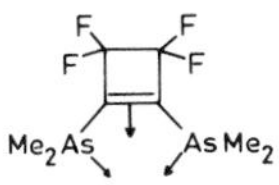	$(CO)_3Fe(f_4ars)Fe(CO)_3$	2.88	*(244)*
	$(CO)(f_4asp)Fe(f_4asp)Fe(CO)_3$	2.869	*(247)*

(217).

Organic six-electron bridging ligands which originate on condensation of acetylenes *(254,338,468,469,471)*, allenes *(98, 463)*, or cyclic unsaturated hydrocarbons *(177,178)* are very common. Bridging ligands which donate more than 6 electrons are also fairly common, for instance, azulene *(151)*, acenaphthylene *(153)*, cyclooctatetraene *(277)*, and diazulene *(152)*. A recent review on this aspect is available *(37)*, they are also discussed in other chapters of this book. We wish only to point out that iron substituted 5 and 6 membered rings which are π bonded to other iron atoms are usually present, and here we are often approximating the situation in which the iron-iron bond seems to be a condition imposed by the peculiar ligand.

C. *CLUSTER COMPOUNDS*

1. Geometrical Description

In Figure 3 we have reported the structural classification which will be used in this section. This geometrical classification is mainly useful for correlation of carbonyl complexes having different bridging ligands. In the same figure we have reported also the number of electrons formally present in the valence shell of each metal atom in these different structural situations, and, for better clarity, these "magic numbers" have been calculated without counting the metal-metal bonds. (They can be easily evaluated by adding together all the valence shell electrons of the metal atoms and all the electrons donated by the ligands, and then dividing by the number of metal atoms). With the exception of the octahedron *(149)* these numbers correspond to the simple noble gas rule.

A decrease of the space available for the ligands takes place on increasing the size of the metal polyhedron, and the limiting situation is reached in a flat surface where there is room for about one carbonyl per metal atom *(150)*. As a result, steric effects dominate the chemistry of the high nuclearity clusters, especially in the case of the early transition metals of Group VIII (Fe, Ru, Os) where a high number of ligands is required *(15)*. Experimentally no tetrahedral carbonyl cluster containing more than 13 carbonyls is known, and the tetranuclear carbonyl clusters $[Re_4(CO)_{16}]^{2-}$ and $[Re_4(CO)_{15}H_4]^{2-}$ present more open arrangements of metal atoms *(16)*. Steric compression of the carbonyls is already present in the short C••••C contacts (2.5-2.6 Å) found in the $[Fe_4(CO)_{13}]^{2-}$ dianion *(235)*, and the reversible opening of this tetrahedron into the "butterfly" $[Fe_4(CO)_{13}H]^{-}$ anion,

Class	Type of structure [a]	Typical Fe-Fe distances [Å]
a. linear	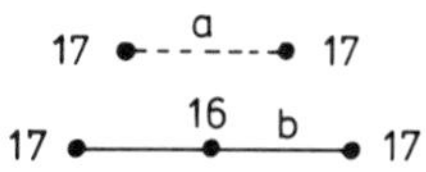	a = 2.75 (unbridged) b = 2.708 (calculated)
b. bent	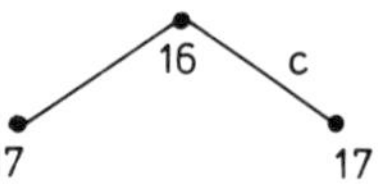	c = 2.43-2.65 (bridged)
c. triangular regular	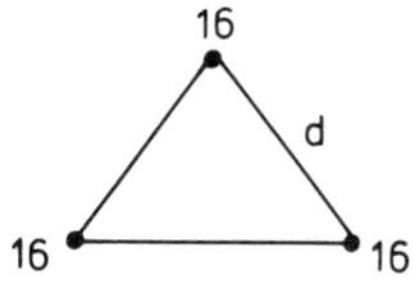	d = 2.53 (face bridged)
triangular isosceles	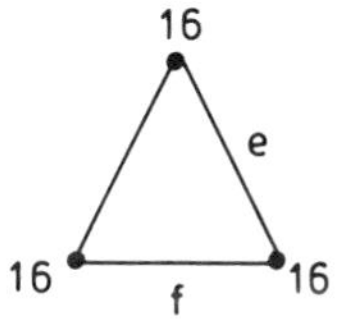	e = 2.65-2.71 f = 2.53-2.58 (bridged)
d. tetrahedral	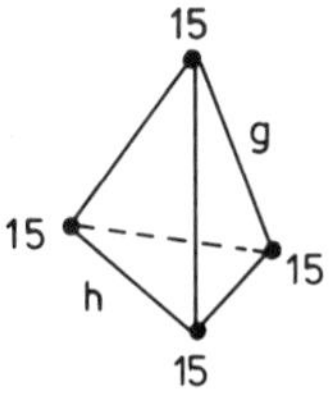	g = 2.58 h = 2.50 (bridged)
e. tetragonal pyramid and octahedron	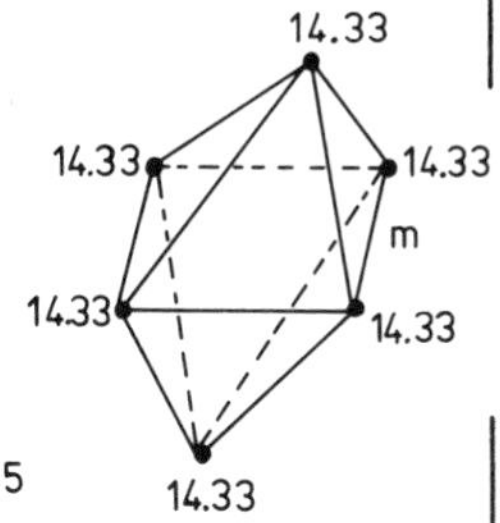 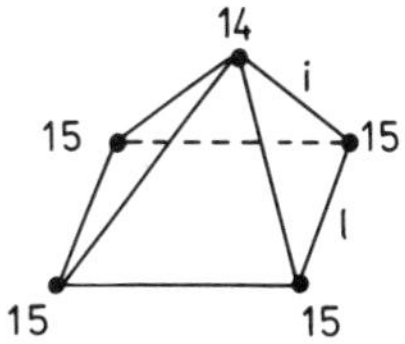	i = 2.63 l = 2.66 m = 2.69 (average)

[a] the numbers are the "magic numbers" (see text)

Fig. 3: Structural classification of the carbonyliron clusters.

Table 9: Bent cluster compounds of iron

Compound		Preparation *(Ref.)*	d(M-M), bonding [Å] *(Ref.)*
$Fe_3(CO)_{10}(S_2CH_2)$		*(316)*	
$Fe_3(CO)_{9-n}(L)_n(\mu_3\text{-}S)_2$	n = 0,1,2[a]	*(142,323)*	2.591(1) (n = 0) *(527)*
$Fe_3(CO)_{9-n}(L)_n(\mu_3\text{-}Se)_2$	n = 0,1,2[a]	*(142,323)*	2.655(8) (n = 0) *(205)*
$Fe_3(CO)_{9-n}(L)_n(\mu_3\text{-}Te)_2$	n = 0,1,2[a]	*(142,323)*	
$Fe_3(CO)_9(\mu_3\text{-}NR)_2$	R = Me,Pr,*t*-Bu	*(57,218)*	2.462(7) (R = Me) *(237)*
$Fe_3(CO)_{9-n}[P(OMe)_3]_n(\mu_3\text{-}PPh)_2$	n = 0,1,2	*(514)*	2.706 (n = 0) *(352)*
$Fe_3(CO)_9(\mu_3\text{-}AsPh)_2$		*(354,356)*	2.765 *(354)* 2.773 *(356)*
$Fe_3(CO)_9(\mu_2\text{-}S\text{-}t{\cdot}Bu)(\mu_3\text{-}S\text{-}t{\cdot}Bu)$		*(91)*	
$Fe_3(CO)_9(\mu_3\text{-}N_2CPh)_2$		*(79)*	2.44 *(79)*
$Fe_3(CO)_9(\mu_2\text{-}AsMe_2)(\mu_3\text{-}AsCH_2)$		*(197)*	
$Fe_3(CO)_9(\mu_2\text{-}AsMe_2)(\mu_3\text{-}C_4F_4AsMe_2)$		*(197)*	2.667(5) 2.917(5) *(245)*
$Fe_3(CO)_9(\mu_2\text{-}AsMe_2)(\mu_3\text{-}C_4F_4PPh_2)$		*(246)*	2.679(6) 2.863(6) *(246)*
$Fe_3(CO)_7(\mu_2\text{-}PPh_2)[\mu_3\text{-}C_4(CF_3)_2PPh_2]$		*(137)*	2.532(11) 2.665(8) *(137)*
$[Fe_2Rh(CO)_2(\eta^5\text{-}CH_3C_5H_4)_2(\mu_2\text{-}CO)_2(\mu_2\text{-}PPh_2)_2]^+$		*(304)*	2.659(2) 2.674(1) *(426)*
$Fe_3(CO)_2Cp_3(\mu_2\text{-}SR)(\mu_3\text{-}S)$[b]		*(306)*	
$[FeAu_2Cp(PPh_3)_2(\mu_3\text{-}C_5H_4)]^+$		*(70)*	2.818(9) *(70)*

[a] L = PBu_3, $P(OPh)_3$, $AsPh_3$; [b] R = Me, Et, *t*-Bu, Bz

which takes place on simple protonation, is probably a related steric effect *(417)*. Similarly 18 carbonyls have been found only in the larger ruthenium and osmium octahedra, and actually the radical anion $[Fe_3(CO)_{11}]^-$ as it is shown in eq. [132], does not dimerize to the expected $[Fe_6(CO)_{18}]^{2-}$ dianion. Moreover the central carbide present in $Fe_5(CO)_{15}C$ and $[Fe_6(CO)_{16}C]^{2-}$ is clearly able to decrease the number of carbonyls and to avoid excessive steric crowding.

Steric and conformational effects are also involved in the ligand fluxionality observed in several iron carbonyl clusters, the available data are summarized in recent reviews *(7,18)*.

2. Linear Structures

Linear structures of three metal atoms are known only in the mixed clusters containing elements of Group VII (compare with Section III.*B*.1.): $(CO)_5Mn\text{-}Fe(CO)_4\text{-}Mn(CO)_5$ *(259,491)*

(Mn-Fe = 2.815 Å *(52)*), $(CO)_5Mn-Fe(CO)_4-Re(CO)_5$ *(258,259)*, and $(CO)_5Re-Fe(CO)_4-Re(CO)_5$ *(259,491)*.

3. Bent Structures

Bent structures are much more common and are summarized in Table 9. The bent structures can often be ascribed to the presence of particular ligands which use three tetrahedral coordination positions, and therefore are able to compel the iron atoms to approximate the geometry of the basal plane of a tetrahedron, while, at the same time, they provide excess electrons for closing all the metal-metal bonds. For instance the presence of two such ligands and of three iron atoms gives rise to the distorted trigonal bipyramid 40 for compounds of the type $Fe_3(CO)_9(\mu_3-L)_2$. In this stereochemical situation

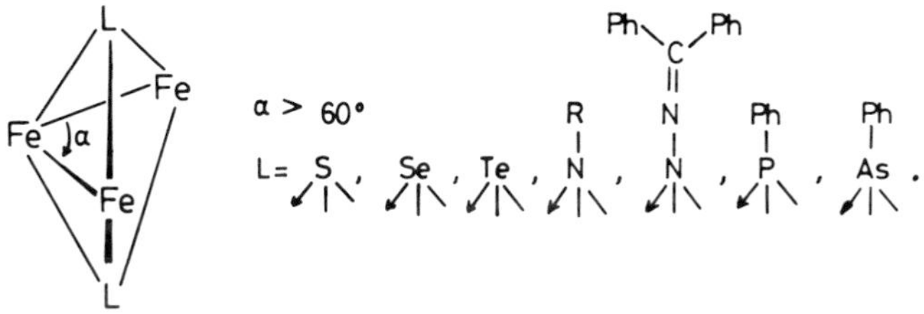

40

all these ligands are four-electron donors. In several cases the triply bridging ligand includes a carbon-iron σ-bond besides the bonds with the donor groups, as in the structures 58 and 59 which are shown later on.

4. Closed Triangular Clusters

The triangular clusters of iron, in which metal-metal bonding along all the three edges is present, are reported in Table 10. They can be divided in five subclasses:

a) unbridged (41); example $Fe_2Os(CO)_{12}$;
b) doubly bridged along one edge (42); example $Fe_3(CO)_{12}$;
c) triply bridged over one face (43); example $FeCo_2(CO)_9S$;
d) triply bridged over both faces (44); example $[Fe_3(CO)_{11}]^{2-}$;
e) miscellaneous (45, 46, and 47).

Iron is extremely reluctant to form the unbridged structure 41, and this structure appears only in the presence of second and third row transition metals as in $FeRu_2(CO)_{12}$, $Fe_2Os(CO)_{12}$, and $FeOs_2(CO)_{12}$. We may assume that formation of bridging carbonyl groups depends both on the ability of the metal atoms to donate further negative charge to these bridging ligands, and to distribute the ligands with minimum crowding *(13)*. The first effect being illustrated by the induction

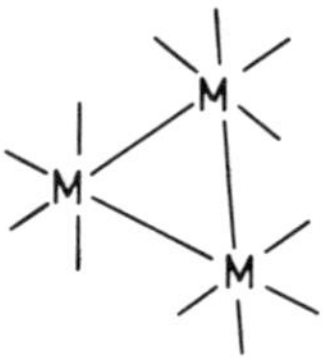

41

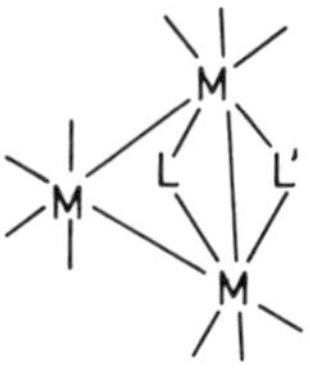

42

L = L′ = CO
L = CO, L′ = H, $CNMe_2$

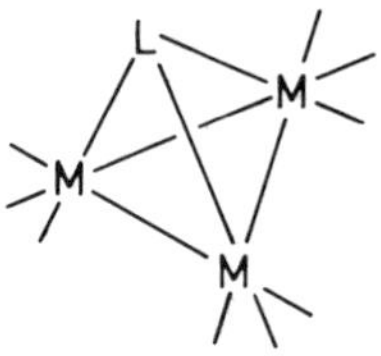

43

L = S, Se, Te, C_2Ph, C_2Ph_2, $C_3H_6N_2S$

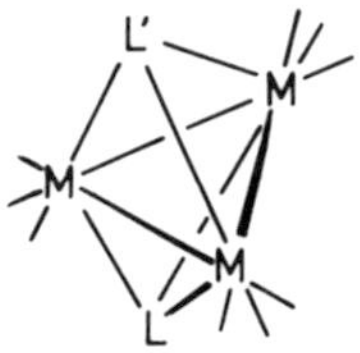

44

L = L′ = CO, As, C_2Ph_2
L = CO, L′ = $NSiMe_3$

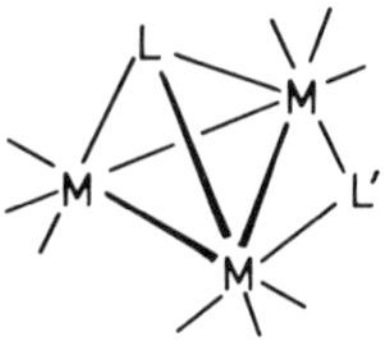

45

L = SR, L′ = H
L = C_2Ph, L′ = CO

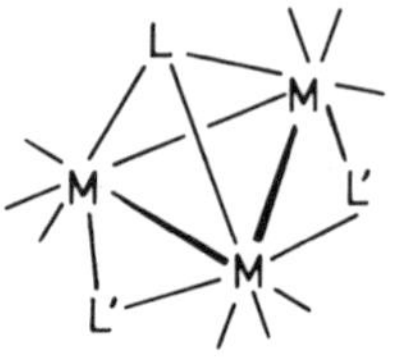

46

L = $NSiMe_3$
L′ = H

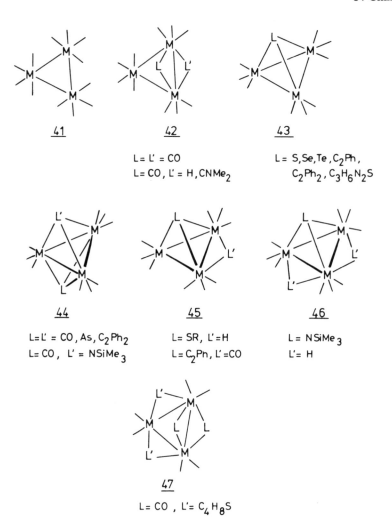

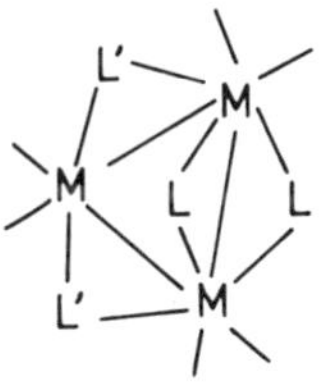

47

L = CO, L′ = C_4H_8S

of bridging carbonyl groups in $Ru_3(CO)_{12}$ on complexation with Lewis acids such as $AlBr_3$ *(398)*, while the second effect is illustrated by the different distribution of the carbonyl groups in structures 41 and 42 which correspond to the triangularly bicapped hexagon *(371)* and to the icosahedron *(175)*.

In comparison with $Fe_2Ru(CO)_{12}$, the presence of one atom of osmium in $Fe_2Os(CO)_{12}$ can hardly introduce any significant difference in steric factors, and we are inclined to consider the elements of the second and third transition row as increasingly efficient electron "sinks".

Structure 42 represents the usual type when 12 two-elec-

Table 10: Triangular clusters of iron

Compound	Preparation and IR data *(Ref.)*	d(M-M) [Å] *(Ref.)*
a) *unbridged* (41)		
$FeRu_2(CO)_{12}$	*(538)*	
$Fe_2Os(CO)_{12}$	*(442)*	
$FeOs_2(CO)_{12}$	*(442)*	
$(CO)_8Fe_2Pt(L)_n(CO)_{2-n}$ $n = 1^*$, $L = PPh_3^*$, $PMePh_2$, PMe_2Ph, $AsPh_3$ $n = 2$, $L = PMe_2Ph$, $PMePh_2$, $PPh(OMe)_2$, $P(OPh)_3$	*(119)*	{ Fe-Fe = 2.758(8) Fe^1-Pt = 2.530(5) Fe^2-Pt = 2.597(5) *(424,427)* } *
$(CO)_4FePt_2[P(OPh)_3]_3CO$	*(55)*	{ Fe-Pt^1 = 2.550(5) Fe-Pt^2 = 2.583(6) *(55)* }
b) *doubly bridged along one edge* (42)		
$Fe_3(CO)_{10}(\mu_2\text{-}CO)_2$	*(434,472)*	2.558(1)[a], 2.680(2) *(184,529)*
$Fe_3(CO)_9(L)(\mu_2\text{-}CO)_2$ $L = PPh_3$, $P(OMe)_3$	*(72,473)*	2.568(8)[a], 2.684(8) ($L = PPh_3$) *(209)*
$Fe_3(CO)_8(L_2)(\mu_2\text{-}CO)_2$ L_2 = $f_4$ars, 2 $P(OMe)_3$	*(197,473)*	2.527(6)[a], 2.651(7) (L = $f_4$ars) *(477)*
$Fe_3(CO)_7(L)_3(\mu_2\text{-}CO)_2$ $L = PPhMe_2$, $P(OMe)_3$	*(433,473)*	2.540(7)[a], 2.688(7) ($L = PPhMe_2$) *(475)*
$Fe_2Ru(CO)_{10}(\mu_2\text{-}CO)_2$	*(538)*	
$[Fe_2Mn(CO)_{10}(\mu_2\text{-}CO)_2]^-$	*(63,411)*	
$[Fe_2Mn(CO)_9(L)(\mu_2\text{-}CO)_2]^-$ $L = PPh_2Me$, PPh_3, $P(OPr)_3$	*(171)*	
$[Fe_2Tc(CO)_{10}(\mu_2\text{-}CO)_2]^-$	*(411)*	
$[Fe_2Re(CO)_{10}(\mu_2\text{-}CO)_2]^-$	*(257,411)*	
$(CO)_7Fe_2CoCp(\mu_2\text{-}CO)_2$	*(393)*	
$(CO)_6Fe_2RhCp(CO)(\mu_2\text{-}CO)_2$	*(393)*	
$(CO)_3FeRh_2Cp_2(CO)(\mu_2\text{-}CO)_2$	*(393)*	
$[Fe_3(CO)_{10}(\mu_2\text{-}CO)(\mu_2\text{-}H)]^-$	*(240,262, 317,434)*	2.577(3)[a], 2.690(3) *(206)*
$Fe_3(CO)_{10}(\mu_2\text{-}CNMe_2)(\mu_2\text{-}H)$	*(292)*	
c) *triply bridged over one side* (43)		
$Fe_3(CO)_9(\mu_3\text{-}C_2Ph_2)$	*(105)*	2.480(10), 2.501(9), 2.579(11) *(105)*
$Fe_3(CO)_9(\mu_3\text{-}SN_2C_3H_6)$[b]	*(60)*	
$Fe_3(CO)_9(\mu_3\text{-}N_2C_{14}H_{14})$[c]	*(77)*	
$Fe_3(CO)_9(\mu_3\text{-}NCR)$ (R = Me,Pr)	*(64)*	*(64)*
$FeCo_2(CO)_9(\mu_3\text{-}S)$	*(373)*	2.554(3) *(503)*
$FeCo_2(CO)_9(\mu_3\text{-}Se)$	*(505)*	2.577(1) *(505)*
$FeCo_2(CO)_9(\mu_3\text{-}Te)$	*(505)*	2.598(2) *(505)*
$FeCo_2(CO)_{9-n}(L)_n(\mu_3\text{-}S)$; $L = PPh_3$, n = 1,2; L = CN-t-Bu, n = 2,3	*(125,454)*	
$(CO)_6Fe_2RuCp(PPh_3)(\mu_3\text{-}C_2Ph)$	*(48,120)*	

Table 10: (continued)

$CpFeCo_2(CO)_6(\mu_3\text{-}C_2Ph)$	*(535)*	
$(CO)_3FeNi_2Cp_2(C_2PhR)$; R = H, Ph	*(513)*	
d) ***triply bridged over both sides*** (44)		
$[Fe_3(CO)_9(\mu_3\text{-}CO)_2]^{2-}$	*(240,262,320)*	2.592, 2.599 *(416)*
$Fe_3(CO)_{9-n}(PR_3)_n(\mu_3\text{-}NSiMe_3)(\mu_3\text{-}CO)$; n = 0,1,2,3 R = OCH_3, Bu	*(275,395)*	2.535(2) (n = 0) *(84)*
$Fe_3(CO)_9(\mu_3\text{-}As)_2$	*(220)*	2.623(7) *(220)*
$Fe_3(CO)_9[\mu_3\text{-}SnFe(CO)_2Cp]_2$	*(435)*	2.792(6) *(435)*
$[Fe_3(CO)_8(\mu_3\text{-}C_2Ph_2)_2]^{n-}$; n = 0,1	*(226,350)*	2.469(5), 2.457(5), 2.592(5) (n = 0) *(234)*
$(CO)_3FeNi_2Cp_2(\mu_3\text{-}CO)_2$	*(345)*	Fe-Ni = 2.436 *(510)*
e) *Miscellaneous*		
$[Fe_3(CO)_{10}(\mu_2\text{-}COCH_3)]^-$	*(497)*	
$Fe_3(CO)_9(\mu_2\text{-}H)(\mu_3\text{-}MeC{=}NH)$	*(64)*	*(64)*
$Fe_3(CO)_9(\mu_2\text{-}H)(\mu_3\text{-}MeCH{=}N)$	*(64)*	
$Fe_3(CO)_9(\mu_2\text{-}H)_2(\mu_3\text{-}NR)$; (R=Et,Ph)	*(64)*	
$Fe_3(CO)_9(\mu_2\text{-}H)(\mu_3\text{-}SR)$ R = *i*-Pr, *s*-Bu, *t*-Bu	*(91)*	2.640(2), 2.653(2), 2.678(2) (R = *i*-Pr) *(88)*
$Fe_3(CO)_6Cp(\mu_2\text{-}CO)(\mu_3\text{-}C_2Ph)$	*(535)*	2.524(1), 2.634(1) *(537)*
$Fe_3(CO)_{9-n}(PR_3)_n(\mu_2\text{-}H)_2(\mu_3\text{-}NSiMe_3)$; n = 0,1,2 R = OCH_3, etc.	*(275)*	
$Fe_3(CO)_6(\mu_2\text{-}SC_4H_8)_2(\mu_2\text{-}CO)_2$[d]	*(184)*	2.611(2), 2.645(2) *(188)*

[a]Bridged; [b]ethylenethiourea; [c]from diazotoluene; [d]tetrahydrothiophene.

tron ligands are present; it is common not only to $Fe_3(CO)_{12}$ and its tertiary phosphine derivatives, but also to the isoelectronic anions $[Fe_2M(CO)_{12}]^-$ (M = Mn, Tc, Re).

Structure 43 is clearly one imposed by the presence of a triply bridging ligand, while structure 44 is common not only in the presence of triply bridging ligands such as As and C_2Ph_2, but also when only 11 two-electron ligands are present on the cluster as in the case of the $[Fe_3(CO)_{11}]^{2-}$ dianion and of the compound $FeNi_2Cp_2(CO)_5$ (1 Cp $\hat{=}$ 3 L). In the $[Fe_3(CO)_{11}]^{2-}$ dianion the two triply bridging carbonyls are both distorted, one may be represented as a distorted terminal (Fe-C = 1.876 and 2.217 (x2) Å) and the other as a distorted edge bridging (Fe-C = 1.949 (x2) and 2.713 Å), and both lay in the unique plane of symmetry *(416)*.

Stepwise reduction of acetonitrile on a Fe_3-face, μ_3-($CH_3C{\equiv}N \rightarrow CH_3C{=}NH \rightarrow CH_3CH{=}N \rightarrow CH_3\text{-}CH_2\text{-}N$), has been recently observed *(64)*. In the presence of the typical triply bridging ligands, such as S and As, oxidation and reduction are expected to interconvert bent clusters into closed triangular ones and *vice versa*. Finally structure 45 can be considered as having the features of both structures, 42 and 43, while structures 46 and 47 indicate that other different arrangements of ligands are possible.

5. Tetrahedral Clusters

Here we may usefully distinguish (Table 11) between a) tetrahedra (48) and b) cubane-tetrahedra (49). The simple unbridged tetrahedron is present only in the anions $[Fe_2M_2(CO)_{10}Cp_2]^{2-}$ (M = Mo,W), where Fe(CO)$_3$ and M(CO)$_2$Cp

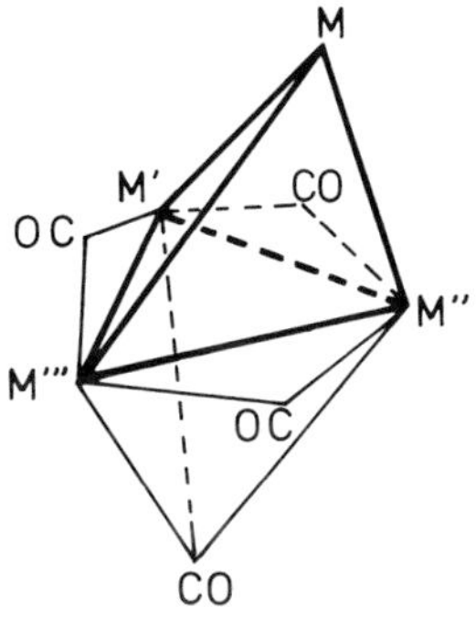

groups alternate at the apices of the tetrahedron *(348)*. The other tetrahedra can be described on the basis of structure 48 in which the basal plane can be free or occupied by a triply bridging carbonyl group, as shown in Table 12. The cubane-tetrahedron 49 is characterized by the presence of four ligands on the four faces of the tetrahedron to give a cubane-like structure in which half of the vertices are occupied by the triply bridging ligands. The presence of these triply bridging ligands is expected to give rise to a robust frame and therefore to allow subtraction or addition of electrons from the frontier metal orbitals. In fact, Cp$_4$Fe$_4$(CO)$_4$ can withstand cyclic voltammetry to give two cations and one anion (equation [69]) *(265,267)*. The dication is unstable and rapidly disproportionates. The shortening of all the iron-

$$[Cp_4 Fe_4 (CO)_4]^{2+} \xrightarrow{1.08V} [Cp_4 Fe_4 (CO)_4]^+ \xrightarrow{0.32V} [Cp_4 Fe_4(CO)_4] \xrightarrow{-1.30 V} [Cp_4 Fe_4(CO)_4]^-$$

$\tilde{\nu}$ (CO) [cm^{-1}]	1696	1623	1576	[69]
d (Fe-Fe)[Å]	2.484 (2)	2.520 (1)	—	
μ_{eff} [B.M.]	2.46	diamagnetic	paramagnetic	

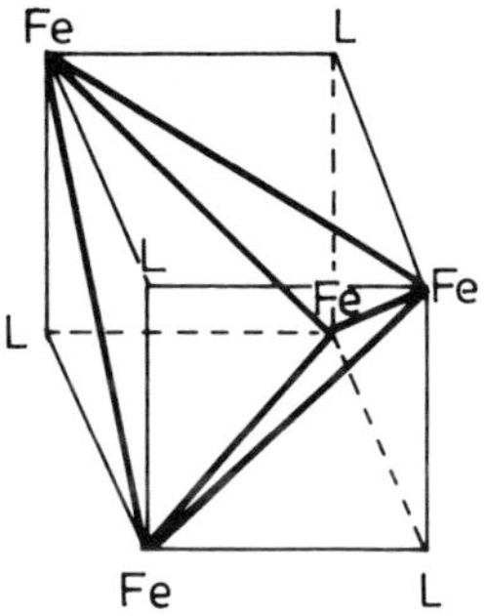

Table 11: Tetrahedral cluster compounds of iron

Compound	Preparation and IR data (*Ref.*)	Average Fe-Fe distance in Å and X-ray data (*Ref.*)
a) *tetrahedra*		
$[Fe_4(CO)_{13}]^{2-}$	*(240,262,318,324)*	2.545 *(235)*
$[Fe_2Mo_2Cp_2(CO)_{10}]^{2-}$	*(348)*	
$[Fe_2W_2Cp_2(CO)_{10}]^{2-}$	*(348)*	
$FeRu_3(CO)_{13}H_2$	*(538)*	*(288)*
$FeRu_2Os(CO)_{13}H_2$	*(285)*	
$FeRuOs_2(CO)_{13}H_2$	*(285)*	
$FeOs_3(CO)_{13}H_2$	*(442)*	
$FeCo_3(CO)_{12}H$	*(148,429)*	
$FeCo_3(CO)_{12-n}(PR_3)_nH$ n = 1,2,3 R = Et,Ph,OMe,OPr,OPh,*etc.*	*(172)*	*(351)* n = 3 R = OMe
$[FeCo_3(CO)_{12}]^-$	*(148)*	
$[FeCo_3(CO)_{10}(L)]^-$ L = $PPh_2Me,PPh_3,P(OPr)_3$	*(171)*	
$[FeCo_3(CO)_{10}(L)_2]^-$ L_2 = dppe,C_2Ph_2	*(171)*	
$Fe_3Rh(CO)_{11}Cp$	*(393)*	2.577 *(156)*
$Fe_2Rh_2(CO)_8Cp_2$	*(393)*	2.539 *(155)*
b) *cubane-tetrahedra*		
$[Fe_4Cp_4(\mu_3\text{-}CO)_4]^{n+}$ n = 2,1,0,-1	*(265,384)*	2.520 (n=0) *(453)* 2.484 (n=1) *(517)*
$Fe_4Cp_4(\mu_3\text{-}CO{\cdot}AlEt_3)_4$	*(447)*	
$Fe_4(NO)_4(\mu_3\text{-}S)_4$	*(282)*	2.634 *(282)*
$Fe_4(NO)_4(\mu_3\text{-}S)_2(\mu_3\text{-}N\text{-}t\text{-}Bu)_2$	*(282)*	2.564 *(282)*

Table 11: (continued)

$[Fe_4Cp_4(\mu_3\text{-}S)_4]^{n+}$; n = 3,2,1,0,-1	*(262,518)*	(n = 0)* *(493,528)*; (n = 1,2)* *(518)*
$[Fe_4(SR)_4(\mu_3\text{-}S)_4]^{2-}$; R = Me, Et, *t*-Bu, Bz, Ph, CH_2Cy, *p*-$C_6H_5NO_2$, C_6F_5, C_6Cl_5	*(74)*	2.746 R=Bz *(74)*
$\{Fe_4[S_2C_2(CF_3)_2]_4(\mu_3\text{-}S)_4\}^{2-}$	*(83)*	2.73 *(100)*
c) *miscellaneous*		
$[Fe_4(CO)_{12}(\mu_4\text{-}CO)H]^-$	*(262,318)*	2.63 *(417)*
$Fe_4(CO)_8(\mu_2\text{-}CO)_3(\mu_4\text{-}NEt)(\mu_4\text{-}ONEt)$	*(286)*	2.514 *(286)*
$Fe_4(CO)_8(\mu_2\text{-}CO)_3(\mu_4\text{-}EtC_2H)_2$	*(487)*	2.600 *(487)*
$[Fe_4(NO)_7(\mu_3\text{-}S)_3]^-$		2.70 *(367)*
$Fe_2Ni_2(CO)_8Cp_2(C_2RPh)$ R = H,Ph	*(513)*	

*see eq. [70]

iron distances in the monocation indicates that one electron has been removed from a fully delocalized frontier molecular orbital of predominant metallic character *(517)* and agrees with the equivalence of the iron atoms as shown by Mössbauer spectroscopy *(291)*.

Similar structures have been recently found in the two isoelectronic nitrosyl derivatives $Fe_4(NO)_4(\mu_3\text{-}S)_4$ and $Fe_4(NO)_4(\mu_3\text{-}S)_2(\mu_3\text{-}N\text{-}t\text{-}Bu)_2$ *(282)*. A related structure has also been found in the $[Fe_4(SR)_4(\mu_3\text{-}S)_4]^{2-}$ dianions (R = Me, Et, *t*-Bu, Ph, Bz, CH_2Cy, *p*-$C_6H_4NO_2$, C_6F_5, and C_6Cl_5); Fe-Fe = 2.746 Å *(74)* when R = Bz. Here the ligands provide only 20+2 electrons in place of 28, and the noble gas rule clearly breaks down *(282)*. The relevance of these compounds as models for the active site of certain iron-sulphur proteins has been discussed *(343,425)*. The related $Cp_4Fe_4(\mu_3\text{-}S)_4$ cluster is well known, but here the ligands are able to provide 36 electrons and, in agreement with the noble gas rule, only two of the six tetrahedral iron-iron edges are bonding (Fe-Fe = 2.650 Å) *(493,528)*. Again this cluster can withstand cyclic voltammetry (equation [70]) *(265)*, and a continuous increase in the number of iron-iron bonds is observed in the cationic species *(518)*.

Table 12: Structural description of the tetrahedral carbonyl clusters of iron

Compound	Apical group M	Equatorial groups M', M", M"	Doubly bridging equatorial CO ligands	Triply bridging basal CO
$[Fe_4(CO)_{13}]^{2-}$	$Fe(CO)_3$	$Fe(CO)_3$	3	present
$[FeCo_3(CO)_{12}]^{-}$	$Fe(CO)_3$	$Co(CO)_3$	3	absent
$Fe_3Rh(CO)_{11}Cp$	$Fe(CO)_3$	2 $Fe(CO)_3$ + RhCp	2 (Rh-Fe)	absent
$Fe_2Rh_2(CO)_8Cp_2$	$Fe(CO)_3$	$Fe(CO)_2$ + 2 RhCp	3	absent
$FeM_3(CO)_{13}H_2$ M = Ru	$M(CO)_3$	$Fe(CO)_2$ + 2 $M(CO)_3$	2 (M-Fe)[a]	absent[b]

[a] asymmetric; [b] the two hydrogen atoms are present along M-M' and M-M" of 48 (M = M' = M" ≠ Fe).

Fe-Fe distances [Å] bonding	intermediate	nonbonding		
			$[Cp_4Fe_4S_4]^{3+}$	
			↓ 1.41 V	
4 2.834 (3)	—	2 3.254 (3)	$[Cp_4Fe_4S_4]^{2+}$	
			↓ 0.88 V	[70]
2 2.652 (4)	2 3.188 (3)	2 3.319 (3)	$[Cp_4Fe_4S_4]^{+}$	
			↓ 0.33 V	
2 2.650 (6)	—	4 3.363 (10)	$[Cp_4Fe_4S_4]$	
			↓ -0.33 V	
			$[Cp_4Fe_4S_4]^{-}$	

The tetrahedral structure with only two iron-iron bonds has been found in the bis(trifluoromethyl)dithiolenesulphur dianion $\{Fe_4[S_2C_2(CF_3)_2]_4(\mu_3\text{-}S)_4\}^{2-}$ (Fe-Fe = 2.73 Å) *(100)*. Finally, the presence of three iron-iron bonds in the anion of Roussins' salt $[Fe_4(NO)_7(\mu_3\text{-}S)_3]^-$ (Fe-Fe = 2.70 Å) *(367)* must be mentioned, while in the mixed cluster $Cp_2Ni_2Fe_2(CO)_6$-$(\mu_2\text{-}C_2R_2)$ the acetylene ligand is believed to bridge the two iron atoms and only four metal-metal bonds are assumed to result in some sort of butterfly structure *(513)*. The unexpected "butterfly" structure found in the $[Fe_4(CO)_{13}H]^-$ anion *(417)*, in which one carbonyl is bonded both from oxygen and carbon and provides four electrons, is schematically shown in Table 13. A square tetrahedrally distorted arrangement of iron atoms has been found both in $Fe_4(CO)_{11}(\mu_4\text{-}NEt)(\mu_4\text{-}ONEt)$ *(286)* and in $Fe_4(CO)_{11}(\mu_4\text{-}EtC_2H)_2$ *(487)*.

6. Clusters with 5 and 6 Metal Atoms

In the three-dimensional clusters a hole is formed between the metal atoms, and the dimension of the hole depends on the geometry of the cluster: for a tetrahedron of iron atoms the radius of the hole can be approximated in about 0.29 Å assuming a metallic radius of 1.3 Å for iron, while in the case of a square pyramid and an octahedron it amounts to *ca.* 0.53 Å. There is abundant evidence that the hole is real for the higher polynuclear clusters, and the presence of a central carbide in the penta- and hexanuclear clusters of iron provides one example of such evidence.

At first sight the dimension of the carbon atom (covalent radius 0.77 Å) seems to be large for this hole, but recent ^{13}C-NMR measurements on a related rhodium species provides evidence that this carbon atom can carry some positive charge and give rise to a reduced radius *(15,56)*. The bonds between the central carbon and the iron atoms are expected to add a considerable contribution to the energy required for the thermodynamic stability of the clusters.

The pentanuclear $Fe_5(CO)_{15}C$ and the hexanuclear dianion $[Fe_6(CO)_{16}C]^{2-}$ have the structures <u>50</u> *(111)* and <u>51</u> *(154,159)*. The structure of the related pentanuclear dianion $[Fe_5(CO)_{14}C]^{2-}$ is believed to be derived from <u>50</u> by loss of one of the apical carbonyl groups *(347)*.

These structures also illustrate how the concept of coordination number at a metallic centre becomes complicated in these high clusters. In fact, apparent coordination numbers of 8, having a very odd distribution of bonds in favour of the interior of the clusters, such as 3 ext. + 5 int., are present in both structures. This is because of the increased tendency to become like the pure metals, and results in mixing the "usual" coordination numbers with the "high" coordination numbers of the metals.

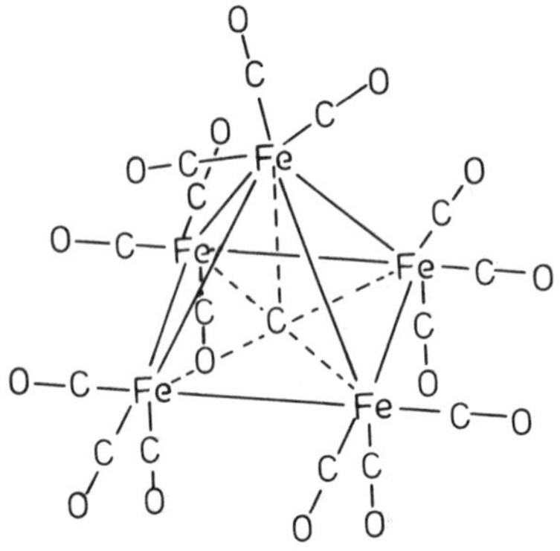

50

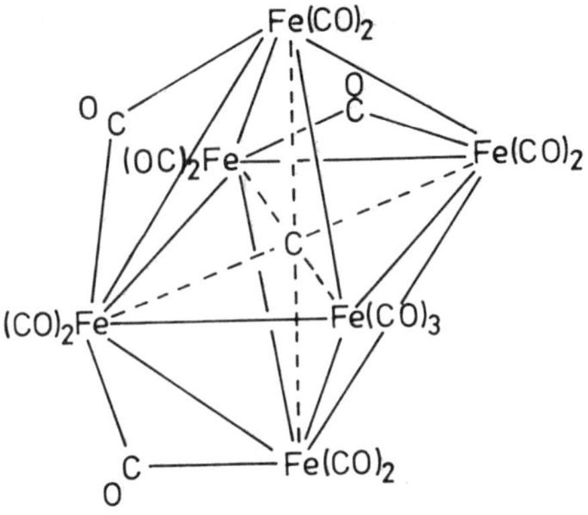

51

D. CARBONYLFERRATES

The two series of carbonylferrates are shown in Table 13. These strictly interrelated species are mainly polynuclear and have many peculiar properties. It seems therefore appropriate, at this point, to discuss briefly their properties and reactions, and at the same time we can also partially cover the more polar situations of the metal-iron bonds, which have not been considered in Section II.

The spectroscopic properties of these anions are very important for identification purposes. As shown in Table 13, the UV-vis spectrum provides a first method, which was once largely used. The energy of light absorbed becomes lower on increasing the number of metal atoms, and a darkening in colours is observed, a trend which is also found in the neutral carbonyls.

Of major significance at the moment are the infrared spectra; here the wavenumber of the principal absorption increases in each of the two series on increasing the number of metal atoms, due to the decrease in average negative charge and associated back-donation. Unfortunately these values are rather sensitive to the particular cation and to the particular solvent. Often, owing to different ion pairing and associated different symmetries, slightly different band positions and shapes are observed in the IR spectra.

As a first approximation each monoanion-dianion pair is related by a simple hydrolysis equilibrium of the type [71]

$$[Fe(CO)_4]^{2-} + H_2O \rightleftharpoons [Fe(CO)_4H]^- + OH^- \qquad [71]$$

Table 13: The carbonylferrates

Anion	$[Fe(CO)_4]^{2-}$	$[Fe_2(CO)_8]^{2-}$	$[Fe_3(CO)_{11}]^{2-}$	$[Fe_4(CO)_{13}]^{2-}$
Structure	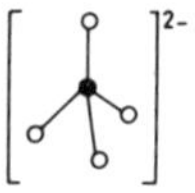	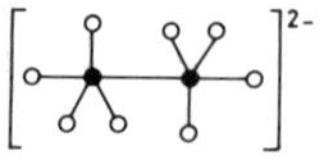	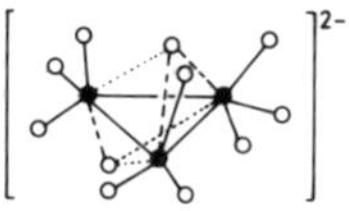	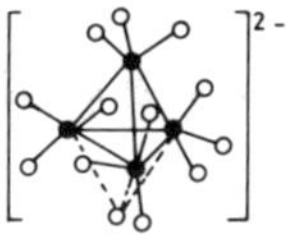
(Ref.)	*(145,508)*	*(144)*	*(416)*	*(235)*
Colour	colourless	orange	red	brown
λ max [nm] *(Ref.)*	300 *(329)*	347 *(329)*	485 *(329)*	500 *(329)*
$\tilde{\nu}$(CO) [cm^{-1}][a] *(Ref.)*	1730 *(240)*	1866 *(240)*	1941 *(240)*	1967 *(240)*[b]
Preparation *(Ref.)*	*(166,169)*	*(262,317,320)*	*(262,320)*	*(318)*
Anion	$[Fe(CO)_4H]^-$	$[Fe_2(CO)_8H]^-$	$[Fe_3(CO)_{11}H]^-$	$[Fe_4(CO)_{13}H]^-$
Structure	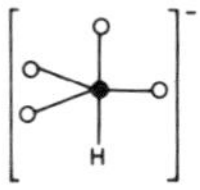	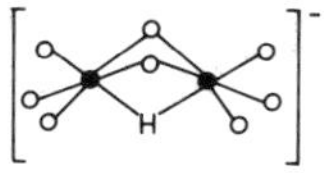	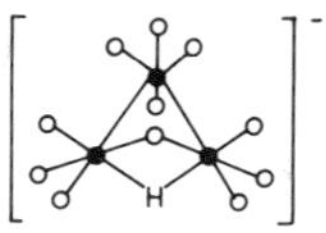	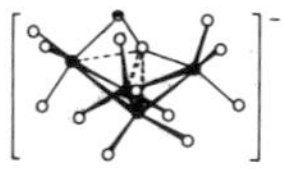
(Ref.)	*(498)*	*(146,261)*	*(206)*	*(417)*
Colour	colourless (?)	yellow-brown	dark red	brown
λ max [nm] *(Ref.)*	-	-	540 *(138)*	527 *(334)*
$\tilde{\nu}$(CO) [cm^{-1}][a] *(Ref.)*	1880 *(240)*[c]	1930 *(240)*	2004 *(240)*[d]	1980-2020 *(262)*[e]
Preparation *(Ref.)*	*(262,318,522)*	*(167)*	*(138)*	*(318)*

[a] Strongest absorption band in DMF; [b] 1955 in MeOH *(533)*, 1949 in acetone *(262)*; [c] 1878 in CH_2Cl_2 *(362)*, 1880 in MeOH *(522)*; [d] 2009 in MeOH *(262)*, 1995 in MeOH *(522)*, 2000 in THF *(63)*; [e] in acetone, 1990-2018 in MeOH *(533)*.

which is displaced to the right due to the weak character of all the corresponding monoacids. Practically it is relatively easy to transform the dianions into monoanions, although slow condensation by elimination of hydrogen, such as decomposition of the thermally unstable $[Fe_2(CO)_8H]^-$ *(167)*, can complicate this simple transformation (compare with Section III.*E*.5). On the contrary, the reverse reaction from the monoanions to the dianions, which also takes place with structural rearrangements (see Table 13), is often a slow process and requires excess OH^- ions. In these conditions nucleophilic fragmentation of the polynuclear anion has been frequently observed (compare with Section III.*F*.4) *(262,320)*.

Probably some of these complications can be avoided using non-aqueous solvents, as it is the case for $[Fe_4(CO)_{13}H]^-$ in DMF *(417)*. The preparative situation is not yet completely clear, further systematic study of the preparation and of the infrared spectra of these salts will be an important contribution to the field.

The relative acid strengths of the monohydride anions, $[HFe_x(CO)_y]^-$, is expected to increase regularly on increasing the number of iron atoms *(320)*, but the $[Fe_3(CO)_{11}H]^-$ is found to be particularly resistent to deprotonation *(416)*. A related behaviour is expected for the dihydrido neutral compounds $H_2Fe_x(CO)_y$, although here the variation in their ability to undergo the necessary structural rearrangements can be more stringent. A meaningful comparison is difficult because of the paucity of the data reported in the literature and of the additional influence of different solvents on the dissociation process. It is well established, for instance, that $Fe(CO)_4H_2$ is a very weak acid ($K_1 \sim 10^{-5}$, $K_2 \sim 10^{-14}$) *(402)*. On the contrary, recent results suggest that $[H]^+[Fe_4(CO)_{13}H]^-$ in THF is a very strong acid *(416)* which does not transform easily into the covalent hydride derivative $Fe_4(CO)_{13}H_2$ *(320)*; the same behaviour has been found for $[H]^+[Fe_3(CO)_{11}H]^-$ *(341)*. At -80°C in CH_2Cl_2 the ^1H-NMR spectrum is consistent with the structure of $Fe_3(CO)_{11}H_2$ beeing $[Fe_3(CO)_{10}(\mu\text{-}H)(\mu\text{-}COH)]$ (δ = - 18,4 and + 15 ppm) *(341)*.

The reducing power should decrease on increasing the number of metal atoms, corresponding to a decrease in average negative charge, but comparative data are lacking.

Several reactions of preparative significance for the polynuclear anions will be pointed out in the next sections; see, *e.g.*, reactions [97], [98], [126], [128], and [133].

E. *SYNTHESES*

1. Condensation of Neutral Species

A convenient starting point is the fact that fragmentation of $Fe_2(CO)_9$ and $Fe_3(CO)_{12}$ by carbon monoxide takes place under very mild conditions (eqs. [72] *(229)* and [73] *(533)*),

$$Fe_2(CO)_9 + CO \xrightarrow[1\ \text{atm}]{25°C} 2\ Fe(CO)_5 \qquad [72]$$

$$Fe_3(CO)_{12} + 2\ CO \xrightarrow[\text{THF, 1 atm}]{25°C,\ \text{slow}} 3\ Fe(CO)_5 \qquad [73]$$

and that conversely it has not been possible to obtain $Fe_2(CO)_9$ or $Fe_3(CO)_{12}$ by simple thermal decomposition of $Fe(CO)_5$. Therefore, both these polynuclear species must be considered to be metastable compounds. Moreover, reaction [74] *(183,202)* shows that $Fe_2(CO)_9$ is the less stable species,

$$3\ Fe_2(CO)_9 \xrightarrow[\text{organic solvents}]{25\text{-}70°C} Fe_3(CO)_{12} + 3\ Fe(CO)_5 \qquad [74]$$

and it is not surprising that it is the more reactive neutral iron carbonyl.

This dinuclear $Fe_2(CO)_9$ seems to be the best reagent for the thermal synthesis of polynuclear compounds, a type of reaction which should be carried out at the lowest temperature compatible with a reasonable reaction rate, otherwise thermal decomposition of the desired reaction product can severely compete. Some examples of the use of $Fe_2(CO)_9$ are the reactions with $Cp_2Ni_2(CO)_2$ *(345)*, $Pt(PR_3)_4$ derivatives *(119)*, and the reactions [75] *(393)* and [76] *(442)*. These mild con-

$$Fe_2(CO)_9 + CpM(CO)_2 \xrightarrow[\text{petroleum ether}]{25°C,\ 20\ \text{min}} Fe_2M(CO)_8Cp + 2\ CO \quad (M = Co,\ Rh) \qquad [75]$$

$$Fe_2(CO)_9 + Os(CO)_4H_2 \xrightarrow[\text{heptane}]{25°C,\ 3\ \text{h}} \underset{(70\ \%)}{Fe_2Os(CO)_{12}} + \underset{(3\ \%)}{FeOs_3(CO)_{13}H_2} + Fe(CO)_5 \qquad [76]$$

ditions should be compared with the temperature required for the thermal synthesis of $Cp_4Fe_4(CO)_4$ (eq. [77]) *(384)*. Re-

$$2\ Cp_2Fe_2(CO)_4 \xrightarrow[\text{days}]{130\text{-}140°C} Cp_4Fe_4(CO)_4 + 4\ CO \qquad [77]$$

placement of $Fe_2(CO)_9$ by $Fe_3(CO)_{12}$ in such reactions generally also requires severe conditions *(393)*. In every case it is important to realize that removal of the evolved carbon monoxide can be a decisive component of the reaction driving force.

It is also possible to use pentacarbonyliron under ultraviolet irradiation. In this case there is probably a primary reaction (eq. [78]) which leads to a very reactive coordinately unsaturated species. The further condensation of this

$$Fe(CO)_5 \xrightarrow{h\nu/25°C} [Fe(CO)_4] + CO \qquad [78]$$

species with $Fe(CO)_5$ itself provides the well-known basic method for preparing $Fe_2(CO)_9$ *(383)*. A similar reaction has been used for obtaining a cluster compound containing three different transition metals (eq. [79]) *(258)*.

$$(CO)_5Mn\text{-}Re(CO)_5 + Fe(CO)_5 \xrightarrow[\text{hexane}]{h\nu/25\,°C} (CO)_5Mn\text{-}Fe(CO)_4\text{-}Re(CO)_5 + CO \qquad [79]$$

2. Condensation of Neutral Species in the Presence of Particular Ligands

The formation of metal-metal bonds in the presence of bridging ligands, which are able to stabilize a polynuclear species, is a very common process. Some typical examples of this broad class of reactions are shown in equations [80] *(365)*, [81] *(112)*, [82] *(327)*, and [83] *(383)*. In the last

$$2\ Fe(CO)_5 + Ph_2P\text{-}SPh \xrightarrow[24\ h]{150\,°C} (CO)_3Fe(PPh_2)(SPh)Fe(CO)_3 + 4\ CO \quad (75\ \%) \qquad [80]$$

$$2\ Fe(CO)_5 + 2\ R\text{-}N{=}C{=}N\text{-}R \xrightarrow[\text{heptane}]{\text{reflux, 72 h}} (CO)_3Fe[(RN)_2C{=}NR]Fe(CO)_3 + 4\ CO + RNC \qquad [81]$$

$$(R = C_6H_{11}, i\text{-}Pr)$$

$$2\ Fe(CO)_2(NO)_2 + Ph_2Se_2 \xrightarrow[\text{petroleum ether}]{25\,°C,\ \text{days}} (NO)_2Fe(SePh)_2Fe(NO)_2 + 4\ CO \qquad [82]$$

$$2\ Fe(CO)_5 + C_{10}H_{12} \xrightarrow{140\,°C} Cp(CO)Fe(CO)_2Fe(CO)Cp + \tfrac{1}{2}\ C_5H_{10}(?) + 6\ CO \qquad [83]$$

case the ligand is not a bridging one, it has a more subtle function giving rise probably to the intermediate species $CpFe(CO)_2H$, which then dimerizes by some sort of hydrogen elimination.

An example of cluster synthesis is given by equation [84] *(373)*. Here it is interesting to note that the analogous reactions between $Fe_2(CO)_9$ and $Co_2(CO)_6(SR)_2$ ($R = C_6Cl_5$, C_6F_5), or between $Fe_2(CO)_6(SR)_2$ (R = Et, C_6Cl_5) and $Co_2(CO)_8$, failed

$$Fe_2(CO)_6S_2 + 2\ Co_2(CO)_8 \xrightarrow[\text{hexane}]{25\,°C} 2\ FeCo_2(CO)_9S + 6\ CO \qquad [84]$$

to give the desired isoelectronic $Fe_2Co(CO)_9(SR)$ *(446)*, and it seems that the bridging ligand RS is reluctant to donate the required 5 electrons.

3. Halide Elimination

Double exchange reactions of the type shown in equations [85] *(419)* and [86] *(337)* have been already discussed in Sec-

$$[Co(CO)_4]^- + CpFe(CO)_2Cl \xrightarrow[THF]{25°C} Cp(CO)Fe(CO)_2Co(CO)_3 + Cl^- \quad [85]$$

$$[Mn(CO)_5]^- + Fe(NO)_2(PPh_3)Br \xrightarrow[THF]{-190°C} (CO)_5Mn\text{-}Fe(NO)_2(PPh_3) + Br^- \quad [86]$$

tion II.*B*.1. In this particular case which involves only transition metal derivatives they have been little used, and they can be probably extended to several other types of halides.

Direct dehalogenation through one of the species which takes part in the reaction has also found little application, for instance, equation [87] *(538)*. Recently there has been

$$Ru_2(CO)_6Cl_2 + Fe(CO)_5 \xrightarrow{90\text{-}100°C} Fe_2Ru(CO)_{12} + FeRu_2(CO)_{12} + Ru_3(CO)_{12} + Fe_3(CO)_{12} + FeRu_3(CO)_{13}H_2 + FeCl_2 \quad [87]$$

much interest in building the metal-metal system through step by step condensation of species in which a bridging PR_2 group is already present (eqs. [88] *(243,302)*, [89] *(534)*, and [90] *(99)*).

$$(CO)_4Fe(PMe_2Cl) + [CpFe(CO)_2]^- \longrightarrow (CO)_4Fe(PMe_2)Fe(CO)_2Cp + Cl^- \xrightarrow{h\nu} (CO)_3Fe(PMe_2)(CO)Fe(CO)Cp + CO \quad [88]$$

$$Fe(CO)_4(PPh_2H) + CpNi(CO)I \xrightarrow{NEt_2H} (CO)_3Fe(PPh_2)(CO)NiCp + CO + [NEt_2H_2]I \quad [89]$$

$$Fe(CO)_4(PPh_2H) + [(\eta^3\text{-}C_3H_5)PdCl]_2 \longrightarrow$$

$$[90]$$

$$(CO)_4Fe(PPh_2)Pd_2Cl_2(PPh_2)Fe(CO)_4 + 2\ C_3H_6$$

A related sequence of reactions which could have been discussed in the previous Section is shown in equation [91] *(298)*.

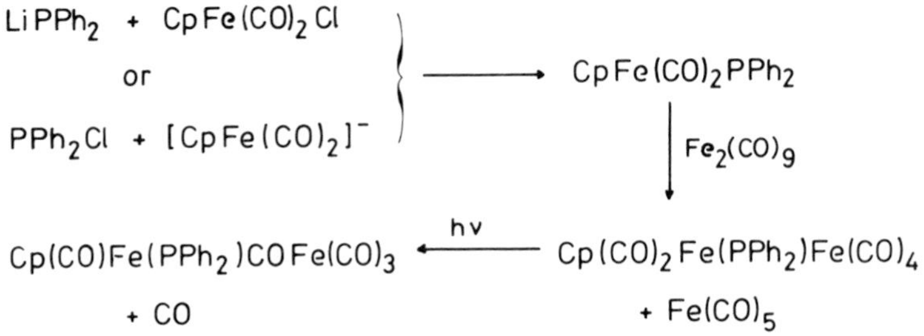

$$LiPPh_2 + CpFe(CO)_2Cl$$

or

$$PPh_2Cl + [CpFe(CO)_2]^-$$

$$\left.\vphantom{\begin{array}{c}a\\b\\c\end{array}}\right\} \longrightarrow CpFe(CO)_2PPh_2$$

$$\downarrow Fe_2(CO)_9$$

$$Cp(CO)Fe(PPh_2)COFe(CO)_3 \xleftarrow{\ h\nu\ } Cp(CO)_2Fe(PPh_2)Fe(CO)_4 \quad [91]$$

$$+ CO \hspace{5cm} + Fe(CO)_5$$

4. Condensation of Anionic with Neutral Species

It seems probable that condensation of a carbonylmetallate and a neutral metal carbonyl, when the average distribution of metal-metal and metal-carbon bonds remains constant, is

$$Fe_3(CO)_{12} + [Mn(CO)_5]^- \xrightarrow[\text{diglyme}]{}$$

$$[92]$$

$$[MnFe_2(CO)_{12}]^- + Fe(CO)_5$$

[93]

$$2\ Fe_3(CO)_{12} + [Co(CO)_4]^- \xrightarrow[\text{THF}]{25\,°C}$$

$$[Fe_3Co(CO)_{13}]^- + 3\ Fe(CO)_5$$

[94]

just dominated by the tendency to redistribute the negative charge over a higher number of metal atoms and to increase the average back-donation. Examples are reactions [93] *(63)* and [94] *(533)*.

More often the relative distribution of bonds changes and there is formation of new metal-metal bonds at the expense of metal-carbon bonds, *e.g.*, reactions [95] *(485)* and [96] *(335)*. In these cases, owing to the unfavourable balance of

$$Fe(CO)_5 + [Mn(CO)_5]^- \xrightarrow[\text{THF}]{h\nu} [MnFe(CO)_9]^- + CO \qquad [95]$$

$$Fe(CO)_5 + [Fe_3(CO)_{11}]^{2-} \xrightarrow[\text{py}]{85\,°C} [Fe_4(CO)_{13}]^{2-} + 3\ CO\ [96]$$

the metal-metal and metal-(carbon monoxide) energies (Section III.*A*), elimination of carbon monoxide may be crucial. Addition of $Na_2[Fe(CO)_4]$ to $Fe_2Ru(CO)_{12}$, $Ru_2Os(CO)_{12}$ and $RuOs_2(CO)_{12}$, followed by acidification, gives the mixed clusters $H_2Fe_2Ru_2(CO)_{13}$, $H_2FeRu_2Os(CO)_{13}$, and $H_2FeRuOs_2(CO)_{13}$ *(285)*.

Reactions of this type are particularly important for the synthesis of the polynuclear anions of iron (eqs. [97] *(138)*, and [98] *(318)*). In both cases the compounds are easily iso-

$$3\ Fe(CO)_5 + Et_3N + H_2O \xrightarrow[10\ h]{80°C} \underset{(95\ \%)}{[NEt_3H]^+[Fe_3(CO)_{11}H]^-} + CO_2 + 3\ CO \qquad [97]$$

$$5\ Fe(CO)_5 + 6\ py \xrightarrow[py]{h\nu/85°C} [Fepy_6][Fe_4(CO)_{13}] + 12\ CO \qquad [98]$$

lated in crystalline form and it seems probable that both syntheses involve a chain of condensations similar to [96], which starts from the primary anions $[Fe(CO)_4H]^-$ or $[Fe(CO)_4]^{2-}$ (compare with Section III.*F*.4, reactions [127] and [129]).

Finally, at 150-160°C in diglyme the reaction of $Fe(CO)_5$ with a carbonylmetallate such as $[Co(CO)_4]^-$, $[Fe(CO)_4]^{2-}$, $[Mn(CO)_5]^-$, and $[V(CO)_6]^-$ gives the hexanuclear carbide anion $[Fe_6(CO)_{16}C]^{2-}$ *(504)*, while isolation of the intermediate $[Fe_5(CO)_{14}C]^{2-}$ pentanuclear dianion has been claimed when using $[CpMo(CO)_3]^-$ as reducing agent *(347)*. The extreme dependence of these syntheses on the experimental conditions is well exemplified by the different results obtained with $Fe(CO)_5$ and $[Mn(CO)_5]^-$ (eq. [99]) *(63,347)*. It seems probable

$$Fe(CO)_5 + [Mn(CO)_5]^- \begin{cases} \xrightarrow[heating]{THF} [Fe(CO)_4]^{2-} + Mn_2(CO)_{10} \\ \xrightarrow[reflux,\ 5\ min]{diglyme} [Fe_2Mn(CO)_{12}]^- \\ \xrightarrow[reflux,\ 1\ h]{diglyme} [Fe_6(CO)_{16}C]^{2-} \end{cases} \qquad [99]$$

that at high temperature nucleophilic condensation takes place together with disproportionation of carbon monoxide to carbon and carbon dioxide, which is a well-known thermodynamically possible process.

5. Oxidation of Anionic Species

Oxidation of carbonylmetallates is expected to involve substitution of negative charges with metal-metal bonds. Generally this is a very delicate reaction, because the excess of oxidizing agent can destroy the desired intermediate cluster compound. A mild oxidizing agent is trityl chloride, and an interesting recent example of its use is reaction [100] *(273)*. The anion $[Fe_2Re(CO)_{12}]^-$ has been reported to undergo

$$2\ Fe(CO)_5 + 2\ PhLi \xrightarrow[Et_2O]{-60°C} 2\ Li[Fe(CO)_4(COPh)]$$

$$\xrightarrow[2Ph_3CCl]{-60°C} \qquad [100]$$

$$(CO)_3Fe(PhCO)_2Fe(CO)_3 + C_2Ph_6 + 2\ LiCl + 2\ CO$$

a similar oxidation *(279)*, although the claimed formation of $[Fe_2Re(CO)_{12}]_2$ is not apparently consistent with the fact that this neutral species is reported to absorb in the same infrared region as the starting anion.

Another case of particular interest is the oxidation of the anion $[Fe(CO)_4H]^-$ with hydroxylamine, a reaction which in aqueous ammonia occurs in several steps. Firstly there is a simple oxidation according to equation [101] *(329)*. Further

$$2\ [Fe(CO)_4H]^- + NH_2OH \xrightarrow[NH_3(aq.)]{} \qquad [101]$$

$$[Fe_2(CO)_8]^{2-} + NH_3 + H_2O$$

$$[Fe_2(CO)_8]^{2-} + NH_2OH \longrightarrow$$

$$\underset{(\sim 30\ \%)}{[Fe(CO)_3NO]^-} + \underset{(\sim 10\ \%)}{(CO)_3Fe(NH_2)_2Fe(CO)_3} + Fe(OH)_3 \qquad [102]$$

addition of oxidizing agent gives a more complex result (eq. [102]) *(330)*. Alkaline oxidation of tetracarbonylferrate dianion with EO_3^{2-} species (E = S, Se, Te), followed by acidification at 0°C, has been reported to follow a formal stoichiometry such as shown in equation [103] *(323)*, while a similar

$$3\ [Fe(CO)_4]^{2-} + 2\ SO_3^{2-} + 10\ H^+ \xrightarrow[H_2O]{} \qquad [103]$$

$$Fe_3(CO)_9S_2 + 2\ CO + CO_2 + 5\ H_2O$$

$$2\ [Fe(CO)_4]^{2-} + S_5^{2-} + 6\ H^+ \xrightarrow[H_2O]{} Fe_2(CO)_6S_2 + 3\ H_2S + 2\ CO \qquad [104]$$

reaction ([104]) in the presence of a polysulphide gives a different result *(323)*. In both cases the mechanism is obscure, and formation of iron hydride derivatives by acidification is probably significant.

A formally similar case, in which oxidation is combined with acidification, represents one of the better syntheses for $Fe_3(CO)_{12}$. This process is based on an oxidative condensation of $[Fe(CO)_4H]^-$ with MnO_2 similar to reaction [101], which is followed by further oxidation (eq. [105]), and the neutral

$$[Fe(CO)_4H]^- + [Fe_2(CO)_8]^{2-} + MnO_2 + 2\ H_2O \xrightarrow[H_2O]{} [Fe_3(CO)_{11}H]^- + Mn(OH)_2 + 2\ OH^- + CO \qquad [105]$$

carbonyl is then generated by addition of 50 % H_2SO_4 through decomposition of the unstable $Fe_3(CO)_{11}H_2$ *(381,382)*.

The equivalence between oxidation and elimination of hydrogen from intermediate hydrido derivatives is clearly evident in the synthesis of $Fe_5(CO)_{15}C$ both by reaction of the anion $[Fe_6(CO)_{16}C]^{2-}$ with strong mineral acids and by oxidation with reagents such as $AgBF_4$ *(504)*.

Carbon dioxide has been reported to oxidize $Na[CpFe(CO)_2]$ to $Cp_2Fe_2(CO)_4$ with contemporary disproportionation into CO and Na_2CO_3 *(260)*.

F. *REACTIVITY*

1. General Discussion

A main characteristic of the iron-transition metal bonds is their high reactivity towards very different types of chemical agents. Here we will consider the following three broad classes of reactions:

(a) electrophilic attack and oxidation,
(b) nucleophilic substitution,
(c) nucleophilic attack and reduction.

In reactions of class (a) there is formation of a transition state associated with decreased electron density of the compound, and in class (c) the opposite is true. Class (b) is formed by reactions of both types (S_N1 and S_N2), which unfortunately we can not yet often distinguish *(5,45)*.

We have already seen that, both for binuclear (Section III.*B*.3.) and for polynuclear derivatives (Section III.*E*.5.), there is good evidence that the frontier orbitals have predominantly metal-metal character. Simple transfer of electrons is therefore expected to be mainly associated with weakening or strengthening of such bonds *(168,267,509,518)*.

The situation is much more complicated with other reagents which cannot tunnel so easily through the barrier of ligands which is present around the cluster. It seems very probable that in most cases the reagent should initially interact with the external polarizable ligands; for instance an electrophile is expected to interact with the oxygen atoms of the carbonyl groups *(398)* while a nucleophile should attack the carbon atoms of the same groups *(73,211,399)*. The next step of the reaction is then expected to depend on the available possibilities of forming strong bonds, and for instance a soft nucleophilic agent such as a phosphine seems generally able to move toward the metallic frame, while a harder nucleophile such as an aliphatic primary or secondary amine can eliminate a proton and give stable carbimido derivatives *(73)*.

A characteristic behaviour of the iron-metal bonds is the general possibility of oxidative addition reactions. This type of reaction results in the breaking of metal-metal bonds, and it is shown in each of the three classes (see eqs. [106] - [108]).

$$\text{(a)}\quad \text{M-M} + \text{X}^{+} \longrightarrow [\text{M-X-M}]^{+} \quad (\text{X} = \text{H, I, } etc.) \qquad [106]$$

$$\text{(b)}\quad \underset{\text{M}}{\text{M-M}} + 3\ \text{L} \longrightarrow 3\ \text{ML} \quad (\text{L} = \text{CO, PR}_3\text{, } etc.) \qquad [107]$$

$$\text{(c)}\quad \text{M-M} + 2\ \text{e}^{-} \longrightarrow 2\ \text{M}^{-} \qquad [108]$$

Sometimes the ease of this oxidative addition can be taken as a rough comparative index of the relative strengths of iron-metal bonds.

2. Electrophilic Attack and Oxidations

There are only a few examples of electrophilic attack on compounds with iron-metal bonds, although it has been shown that such an attack can be used for synthetic purposes in order to labilize the carbonyl groups in simple compounds such as $Fe(CO)_5$ and $Fe(CO)_4(PPh_3)$ *(87)*. In fact the lowering of electron density at the metallic centre is reflected in less back-donation from metal to carbon monoxide and hence in less bonding energy and increased lability.

The binuclear compound $Cp_2Fe_2(CO)_4$ is protonated in concentrated sulphuric acid and in liquid hydrochloric acid *(507)* to give a stable cation, which has been isolated as the hexafluorophosphate salt (eq. [109]) *(214)*. Iodine gives a related result (eq. [110]) *(116)*.

$$Cp_2Fe_2(CO)_4 + H^+ \xrightarrow[98\ \%\ H_2SO_4]{} [Cp_2Fe_2(CO)_4H]^+ \qquad [109]$$

($\tilde{\nu}$(CO) = 2005,1958,1781 cm^{-1}
in CCl_4) ($\tilde{\nu}$(CO) = 2068,2045,2022 cm^{-1})

$$Cp_2Fe_2(CO)_4 + I_2 \xrightarrow[CHCl_3]{0°C} [Cp_2Fe_2(CO)_4I]^+ + X^- \qquad [110]$$

(X^- = I^- or I_3^-) ($\tilde{\nu}$(CO) = 2062,2044,2011 cm^{-1} in $CHCl_3$)

The IR spectra of the reaction products are very similar. The structure of $[Cp_2Fe_2(CO)_4I]^+$ has an Fe-I-Fe bond angle of 110.8° and does not contain an iron-iron bond *(180)*; the same structure is believed to be found for all such cationic species *(200,271)*.

The unusual $[Fe_3(CO)_{10}(\mu\text{-}COCH_3)]^-$ anion, in which the bridge may be considered a carbyne which provides 3 electrons, has been obtained by the following reaction *(497)*:

$$[Fe_3(CO)_{11}]^{2-} + CH_3SO_3F \xrightarrow{CH_3CN} [Fe_3(CO)_{10}(COCH_3)]^- + SO_3F^- \qquad [111]$$

Currently there is much interest in oxidation reactions of clusters as it has already been mentioned in Sections III.*B*.3, III.*B*.4, and III.*E*.5. Both electrochemical and chemical methods can be used for these oxidations. Common oxidizing agents include I_2, Br_2, $AgClO_4$, and $AgPF_6$; less use has been made of O_2/H^+ *(90)*, $NOPF_6$ *(90)*, anhydrous iron(III) perchlorate *(368)*, and $[Cp_4Fe_4(CO)_4]^+$ itself *(110)*. For instance, oxidation to the cations $[Cp(CO)Fe(SR)_2(CO)FeCp]^{n+}$ (n = 1,2) has been carefully studied, as examplified in eq. [112] *(90,281)*.

$$Cp(CO)Fe(SR)_2Fe(CO)Cp + ox_1 \rightleftharpoons [Cp(CO)Fe(SR)_2Fe(CO)Cp]^+ + red_1 \qquad [112]$$

ox_1 = Br_2, I_2, O_2/H^+, $AgSbF_6$, $NOPF_6$

Cyclic voltammetry has shown that the ease of these oxidations decreases in the series Me ∿ Et ∿ Bu > *t*-Bu > Ph, probably due to the presence of both electronic and steric effects *(90)*.

Both the mono- and the dication have been isolated in pure form, although it has not been generally possible to decide between five possible isomers (52 - 56). Oxidation with silver salts is very dependent on the anion present in the salt. For instance *(530)*, the formation of the cation $[CpFe(CO)_2(Me_2CO)]^+$ (eq. [113]) can be contrasted with the formation of the neutral derivatives Y-Fe$(CO)_2$Cp (eq. [114]).

Sometimes, careful reaction with halogens results in stepwise breaking of the metal-metal bonds *(176,336)*, *e.g.*, reaction

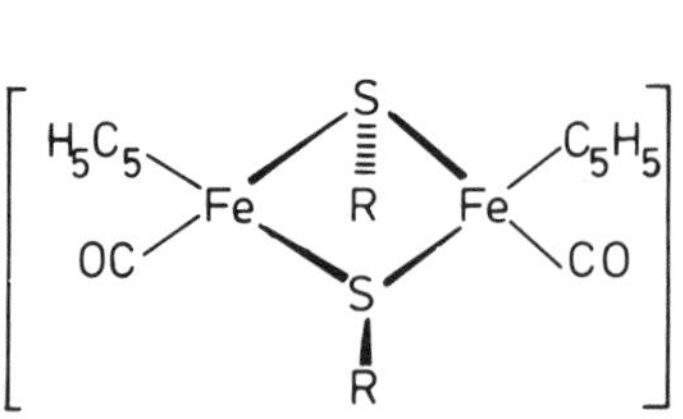

52

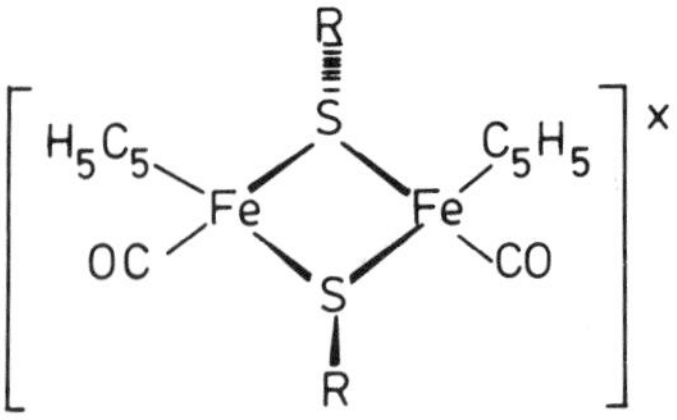

53

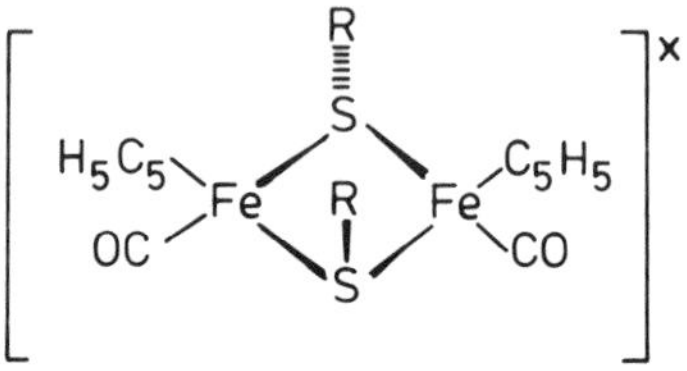

54

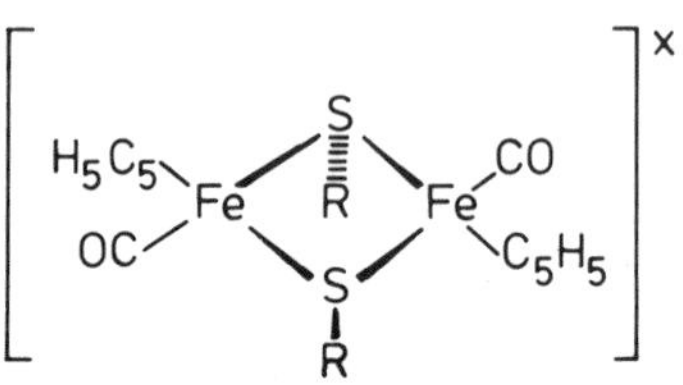

55

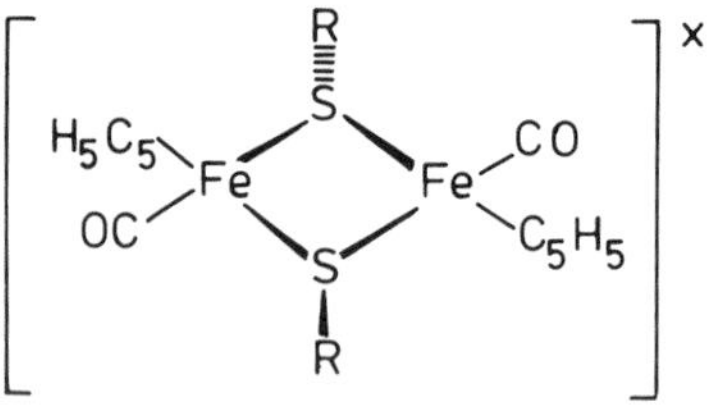

56

$$Cp_2Fe_2(CO)_4 + 2\ AgX \xrightarrow[\text{room temp.}]{Me_2CO} 2\ [CpFe(CO)_2(Me_2CO)]^+X^- + 2\ Ag \qquad [113]$$

($X = ClO_4$, PF_6, SbF_6)

$$Cp_2Fe_2(CO)_4 + 2\ AgY \xrightarrow[\text{room temp.}]{Me_2CO} 2\ CpFe(CO)_2Y + 2\ Ag \qquad [114]$$

($Y = NO_3$, CF_3CO_2, SCN, OCN, $PhCO_2$, *p*-$MeC_6H_4SO_2$, $OP(O)(OPh)_2$)

[115]. Reaction with excess iodine in pyridine at 80°C has been used for the determination of the carbon monoxide content *(148)*.

All the carbonyl derivatives of iron(O) are thermodynami-

cally unstable towards air. However, they are frequently moderately stable in air in the solid state because of kinetic

$$2\ Fe_3(CO)_9S_2\ +\ 5\ Br_2\ \longrightarrow\ 2\left[(OC)_3Fe\overset{\displaystyle S\diagdown\ S}{\diagup\diagdown}Fe(CO)_3\atop\underset{\displaystyle Fe}{\underset{Br\diagdown\ \diagdown Br}{\overset{|\qquad|}{Br\qquad Br}}}\right]^+\ +\ 2\ Br^-$$

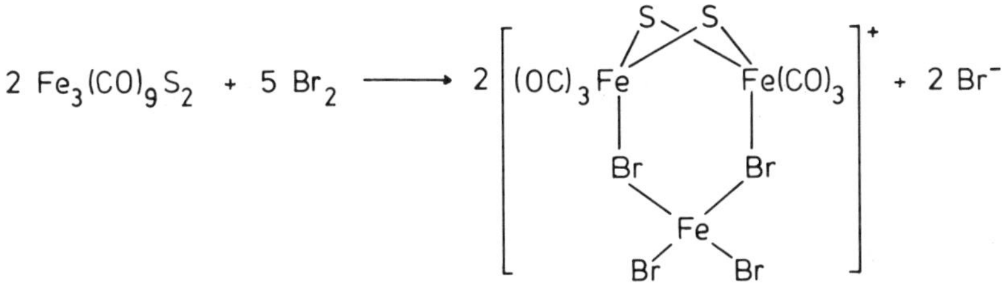

[115]

factors, but in solution they are generally much more sensitive. This should be contrasted with the behaviour of the an-
ionic derivatives (formal oxidation numbers less than zero)
which must generally be handled in a rigorously inert atmosphere. In every case it is important to remember that *every
solid compound of this type is a potentially dangerous pyro-
phoric material,* especially when the presence of finely divided metallic iron is suspected.

3. Nucleophilic Substitution

The two extreme situations in which we are interested are
a) simple substitution by different ligands, and b) simple
oxidative addition with breaking of metal-metal bonds. These
two different results are exemplified by reactions [116] *(332)*
and [117] *(333)* which can be related to the different lengths
and strengths of the iron-iron bonds. This different behaviour agrees also with the recent preparation of

$$(CO)_3Fe(SEt)_2Fe(CO)_3\ +\ PPh_3\ \xrightarrow[petroleum\ ether]{20°C,\ PPh_3}$$

(Fe-Fe = 2.537 Å *(204)*)   [116]

$$(CO)_3Fe(SEt)_2Fe(CO)_2PPh_3\ +\ CO$$

$$(NO)_2Fe(SPh)_2Fe(NO)_2\ +\ 4\ PPh_3\ \xrightarrow[benzene]{20°C}$$

(Fe-Fe = 2.72 Å *(511)*; R = Et)   [117]

$$2\ Fe(NO)_2(PPh_3)_2\ +\ PhS\text{-}SPh$$

$(CO)_2(PPh_3)Fe(SPh)_2Fe(PPh_3)(CO)_2$ and $(CO)_2(L)Fe(SPh)_2Fe(L)_2(CO)$
(L = $P(OPh)_3$) when working in boiling xylene *(89)*. The extreme
stability of the mercapto bridge in the carbonyl derivative is
also confirmed by the rather unusual bridge-exchange reaction
[118] *(374)*. Generally these two extreme cases are not so
clearly separated and often the same compound can lead to different reaction products on changing the experimental conditions (*e.g.,* temperature, time, solvent, molar ratio of rea-

$$(CO)_3Fe(COPh)_2Fe(CO)_3 + 2\ REH \xrightarrow[25^\circ C]{hexane} (CO)_3Fe(ER)_2Fe(CO)_3 + 2\ PhCHO \quad [118]$$

$$(E = S,\ R = Et;\ E = Se,\ R = Ph)$$

gents). Reaction [119] between $Fe_3(CO)_{12}$ and PPh_3 is a typical example. Clearly this is a case of concurrent reactions,

$$Fe_3(CO)_{11}PPh_3\ (+\ Fe(CO)_4PPh_3 + Fe(CO)_3(PPh_3)_2)\ \textit{(72)}$$

$$\uparrow\ 45^\circ C,\ CHCl_3$$

$$Fe_3(CO)_{12} + PPh_3 \quad [119]$$

$$\downarrow\ 70^\circ C,\ THF$$

$$Fe(CO)_4PPh_3 + Fe(CO)_3(PPh_3)_2\ \textit{(163)}$$

and the products obtained are extremely dependent on the reaction conditions. It is worth noting that when using ligands of small steric requirements, such as $PPhMe_2$ *(433)* and $P(OMe)_3$ *(473)*, it has been possible to isolate both the bis- and tris-substituted derivatives, $Fe_3(CO)_{10}(L)_2$ and $Fe_3(CO)_9(L)_3$, respectively (see Table 10).

The clusters $Fe_3(CO)_9E_2$ (E = S, Se, Te), in which additional bridging ligands are present, are much less prone towards oxidative addition. They can be easily transformed into mono-substituted derivatives $Fe_3(CO)_8(L)E_2$ and, in more severe conditions, into bis-substituted $Fe_3(CO)_7(L)_2E_2$ (L = PBu_3, $P(OPh)_3$, $AsPh_3$) *(142)*. Also the carbide $Fe_5(CO)_{15}C$ is readily substituted at room temperature to give $Fe_5(CO)_{15-n}(C)(L)_n$ derivatives (n = 1,2,3; L = PPh_3, PMe_2Ph, $P(OPr)_3$, $P(OPh)_3$ *(173)*.

It might be expected that polynuclear anions, in which the negative charge may be regarded as being spread over a number of atoms, would undergo nucleophilic substitution more readily than mononuclear anions. In fact, substitution reactions have recently been carried out on the anions $[FeCo_3(CO)_{12}]^-$ and $[Fe_2Mn(CO)_{12}]^-$ to give, for instance, the substituted anions $[FeCo_3(CO)_{11}(L)]^-$ and $[Fe_2Mn(CO)_{11}(L)]^-$ (L = PPh_3, PPh_2Me, $P(OPr)_3$) *(171)*.

A most interesting example in this class of substitution reactions concerns the bidentate ligand 1,2-bis(dimethylarsine)tetrafluorocyclobutene ($f_4$ars) *(197)*. A simple substitution reaction under UV irradiation in acetone (eq. [120]) *(477)* gives <u>57</u>. By heating at 80°C this reaction is followed by an internal oxidative addition, which requires breaking of an As-C bond and oxidative addition of the two fragments leading to 58 *(245)*. Structures 57 and 58 have been established by X-ray analysis *(245,477)*. Further heating at 80°C gives a new product which has spectroscopic data in agreement with structure 59 *(202)*. In fact, this is a good example for pointing out that an organic ligand in a cluster can be considered as a true excited ligand due to minor energetic differences between several possible bonding situations and to moderate entropy requirements. Unusual behaviours of other

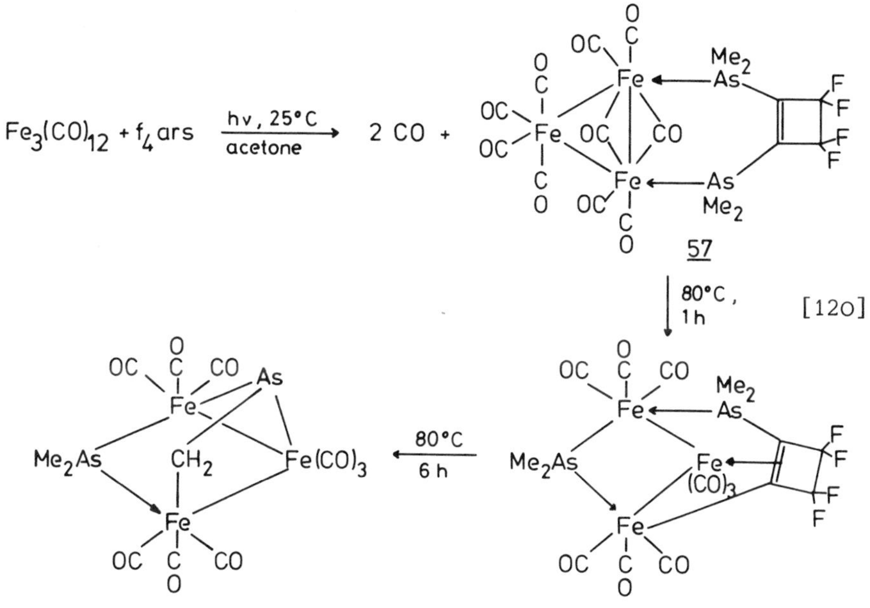

[120]

59 58

organic ligands such as acetylenes, allenes, dienes, *etc.*, have been discussed in other chapters of this book and are clearly related to similar effects.

An interesting example of substitution which confirms the higher reactivity of $Fe_2(CO)_9$ is shown in equation [121] *(269)*.

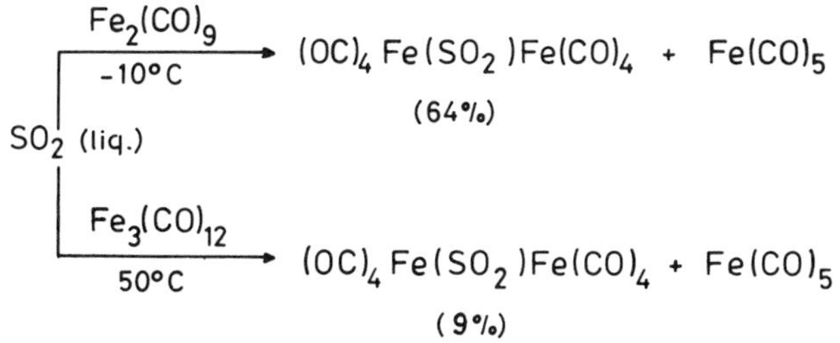

[121]

Finally, a particular type of nucleophilic reagent is carbon monoxide itself, because, in this case, considerable electrophilic assistance through back-donation imparts an ambiguous character. Examples of such reactions are [122] *(142)* and [123] *(318)*. The related addition of carbon monoxide to the sodium salt $Na_2[Fe_4(CO)_{13}]$ is reported to take place only under high pressure (300 atm) and elevated tempe-

$$Fe_3(CO)_9Te_2 + CO \xrightarrow[\text{heptane}]{30°C,\ 70\text{-}80\ atm} Fe_3CO)_{10}Te_2 \qquad [122]$$

$$[Fepy_6][Fe_4(CO)_{13}] + 12\ CO \xrightarrow[1\ atm]{60°C} 5\ Fe(CO)_5 + 6\ py \qquad [123]$$

rature (200°C) *(329)*. Such severe conditions lend credance to the relevance of internal electron redistribution processes when the cation is a transition metal *(13)*.

4. Nucleophilic Attack and Reduction

Attack by anions which are able to give strong bonds with sp^2 carbon atoms, such as H^-, R^-, OH^-, OR^-, SR^-, and NR_2^-, is conveniently represented as a nucleophilic attack which takes place on a carbonyl group itself. In fact, such groups are expected to be polarized, or polarizable, with a positive charge on the carbon atom $\left(\overset{\delta+}{M}\ \overset{\delta-}{C=O}\right)$ owing to insufficient back-donation. A typical reaction of this type is [124] *(320)* where the stepwise oxidation of carbon monoxide to

$$Fe_3(CO)_{12} + OH^- \xrightarrow[MeOH]{25°C} [Fe_3(CO)_{11}(COOH)]^- \ (\underline{60}) \xrightarrow{OH^-} [Fe_3(CO)_{11}]^{2-} + CO_2 + H_2O \qquad [124]$$

carbon dioxide through the unstable intermediate <u>60</u> is in agreement with such direct nucleophilic attack, was originally proposed by Kruck *(399)*. Further fragmentation is easy, and generally, using reactions such as [124], the synthesis of pure products requires carefully controlled conditions *(138, 262,320)*.

When using amines in the presence of water, a similar mechanism has been found by Edgell in the case of pentacarbonyliron (equation [125]) *(240)*, and a related mechanism seems to initiate the reaction of $Fe_3(CO)_{12}$ with triethylamine in the presence of water (reaction [97]).

In other cases the picture is much less clear, and simultaneous oxidation of iron(0) to iron(II) takes place according to a disproportionation scheme such as [126]. Some representative results of such a disproportionation are summarized in

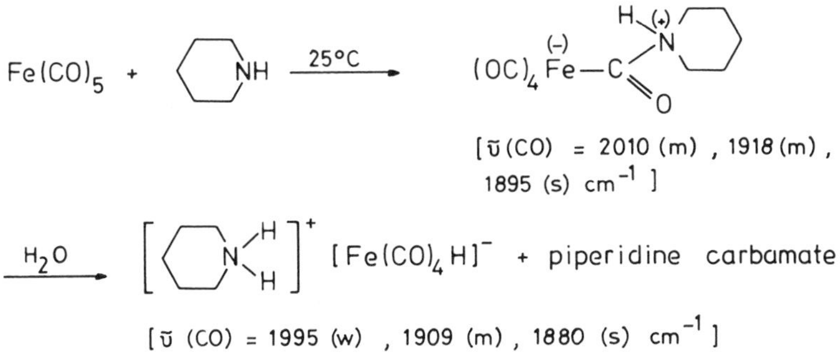

$$Fe(CO)_5 + \text{(piperidine NH)} \xrightarrow{25°C} (OC)_4 \overset{(-)}{Fe}-C \underset{O}{\overset{H}{\diagdown}} \text{(piperidine N}^{(+)})$$

$$[\bar{\nu}(CO) = 2010\,(m)\,,\,1918\,(m)\,,\,1895\,(s)\;cm^{-1}]$$ [125]

$$\xrightarrow{H_2O} \left[\text{(piperidine N}^{+}\text{HH)} \right]^{+} [Fe(CO)_4H]^{-} + \text{piperidine carbamate}$$

$$[\bar{\nu}(CO) = 1995\,(w)\,,\,1909\,(m)\,,\,1880\,(s)\;cm^{-1}]$$

$$4\,Fe_3(CO)_{12} \xrightarrow{B} 3\,[FeB_6][Fe_3(CO)_{11}] + Y$$ [126]

Table 14. The stoichiometric evolution of carbon monoxide is

Table 14: Representative examples of $Fe_3(CO)_{12}$ disproportionation according to equation [126].

Base B	Reaction temp.	Product Y	*Ref.*
NH_3	-33°C	$CO(NH_2)_2$	*(94,319,335)*
py	30°C	CO	*(335)*
C_2H_5OH [a]	30°C	$Fe(CO)_5$	*(321,335)*
CH_3COCH_3 [a]	60°C	$Fe(CO)_5$	*(335)*

[a] Formation of $[Fe_3(CO)_{11}H]^{-}$ owing to hydrolysis.

rarely observed, and more usually there is addition to the amine to give formamides and ureas *(94,322)*, or formation of $Fe(CO)_5$.
An oversimplified scheme involving initial oxidative addition to $Fe_3(CO)_{12}$ ([127]) agrees with the recently discovered behaviour of $Fe_2(CO)_9$ (reaction [128]) *(183)*. The

$$Fe_3(CO)_{12} + 3\,B \longrightarrow 3\,Fe(CO)_4(B)$$ [127]

$$Fe_2(CO)_9 + B \xrightarrow[0°C]{THF} Fe(CO)_4(B) + Fe(CO)_5$$ [128]

(B = THF, py, pyrazine)

$$6\,Fe(CO)_4(B) \rightleftharpoons [Fe(B)_6]^{2+} + [Fe(CO)_4]^{2-} + 4\,Fe(CO)_5$$ [129]

$$[Fe(CO)_4]^{2-} + Fe_3(CO)_{12} \rightleftharpoons [Fe_3(CO)_{11}]^{2-} + Fe(CO)_5$$ [130]

formation of $Fe(CO)_4B$ could be followed by a disproportionation, such as [129], and then by a reduction of the starting carbonyl species (eq. [130]). This speculative sequence of reactions would account for equation [126] when Y = $Fe(CO)_5$, although, as it is shown by the formation of free carbon monoxide when B is pyridine, the complete picture is clearly complicated by other competing reactions and presently not well understood.

Possibly in several cases there is competition between disproportionation of iron(O) and carbon monoxide oxidation. This ambiguity is well illustrated by the reaction with ethylenediamine hydrate which, in spite of the presence of water and the associated by-production of carbonates *(317)*, has often been oversimplified using the scheme [131] which points

$$4\ Fe_3(CO)_{12} + 9\ en \xrightarrow[en]{40^\circ C} 3\ [Fe(en)_3][Fe_3(CO)_{11}] \xrightarrow[en]{90^\circ C}$$

$$4\ [Fe(en)_3][Fe_2(CO)_8] \xrightarrow[en]{145^\circ C} 6\ [Fe(en)_3][Fe(CO)_4] \qquad [131]$$

out the step by step fragmentation which takes place on increasing the temperature.

The first step in alkali metals or electrochemical reduction of iron carbonyls, and in the oxidation of the corresponding dianions, is an one electron transfer which has been reported to give rise to several radical anions *(109,439, 467)*. Only the radical anions $[Fe_2(CO)_8]^-$ (g = 2.0385), $[Fe_3(CO)_{11}]^-$ (g = 2.0497), $[Fe_3(CO)_{12}]^-$ (g = 2.0016), and $[Fe_4(CO)_{13}]^-$ (g = 2.0134) have been satisfactorily characterized by ESR studies on isotopically labelled species *(403)*. The violet radical anion $[Fe_3(CO)_{11}]^-$ has a long life at room temperature and can be even isolated in the solid state. It can be readily obtained according, for instance, to eq. [132] *(416)*.

$$Fe_3(CO)_{12} + [Fe_3(CO)_{11}]^{2-} \xrightarrow[THF]{25^\circ} 2\ [Fe_3(CO)_{11}]^- + CO \qquad [132]$$

Quantitative formation of pure $Na_2[Fe(CO)_4]$ from $Fe(CO)_5$ is possible in THF either using sodium amalgam *(169)* or benzophenone ketyl *(166)* as an electron carrier. Other reduction reactions of iron clusters have received little attention, and for instance, $Fe_3(CO)_{12}$ has been reduced using $NaBH_4$ to give $Na[Fe_3(CO)_{11}H]$ *(130)*.

Acknowledgements: It is a pleasure to acknowledge Dr. C. Wright and Dr. B. Heaton for correction of the English, Prof. S. Cenini and Dr. G. Longoni for critical reading of the first manuscript in 1972.

LIST OF REVIEWS

1. Abel, E.W., and Crosse, B.C., *Sulphur-containing Metal Carbonyls, Organometal. Chem. Rev., 2,* 443 (1967).
2. Abel, E.W., and Stone, F.G.A., *The Chemistry of Transition Metal Carbonyls, Structural Considerations, Quart. Rev. (London), 23,* 325 (1969).
3. Abel, E.W., and Tyfield, S.P., *Metal Carbonyl Cations, Advan. Organometal. Chem., 8,* 117 (1970).
4. Abel, E.W., and Stone, F.G.A., *The Chemistry of Transition Metal Carbonyls, Synthesis and Reactivity, Quart. Rev. (London), 24,* 498 (1970).
5. Angelici, R.J., *Kinetics and Mechanisms of Substitution Reactions of Metal Carbonyl Compounds, Organometal. Chem. Rev. A, 3,* 173 (1968).
6. Baird, M.C., *Metal-metal Bonds in Transition Metal Compounds, Progr. Inorg. Chem., 9,* 1 (1968).
7. Band, E., and Muetterties, E.L., *Mechanistic Features of Metal Cluster Rearrangements, Chem. Rev., 78,* 639 (1978).
8. Biryukov, B.P., and Struchkov, Yu.T., *Metal-metal Bonds and Covalent Atomic Radii of Transition Metals in Their π-Complexes and Polynuclear Carbonyls, Russ. Chem. Rev., 39,* 789 (1970).
9. Braterman, P.S., *Spectra and Bonding in Metal Carbonyls, Part A: Bonding, Struct. Bonding (Berlin), 10,* 57 (1971).
10. Brooks, E.H., and Cross, R.J., *Group IV B Metal Derivatives of Transition Elements, Organometal. Chem. Rev. A, 6,* 227 (1970).
11. Calderazzo, F., *Halogeno Metal Carbonyls and Related Compounds,* in V. Gutmann (Ed.), *Halogen Chemistry, Vol. 3,* Academic Press, London - New York 1967, p. 383.
12. Calderazzo, F., Ercoli, R., and Natta, G., *Metal Carbonyls: Preparation, Structure and Properties,* in I. Wender and P. Pino (Eds.), *Organic Synthesis Via Metal Carbonyls, Vol. 1,* Interscience, New York 1968, p. 1.
13. Chini, P., *The Closed Metal Carbonyl Clusters, Inorg. Chim. Acta Rev., 2,* 31 (1968).
14. Chini, P., *Some Aspects of the Chemistry of Polynuclear Metal Carbonyl Compounds, Pure Appl. Chem., 23,* 489 (1970).
15. Chini, P., Longoni, G., and Albano, V.G., *High Nuclearity Metal Carbonyl Clusters, Advan. Organometal. Chem., 14,* 285 (1976).
16. Chini, P., and Heaton, B.T., *Tetranuclear Carbonyl Clusters, Top. Curr. Chem., 71,* 1 (1977).
17. Connor, J.A., *Thermochemical Studies of Organo-Transi-*

tion Metal Carbonyls and Related Compounds, Top. Curr. Chem., *71*, 71 (1977).

18. Evans, J., *Molecular Rearrangements in Polynuclear Transition Metal Complexes, Advan. Organometal. Chem.*, *16*, 319 (1977).

19. Glockling, F., and Stobart, S.R., *Organometallic Complexes Containing Group III (B to Tl) and Group IV (Si to Pb) Ligands, Int. Rev. Sci., Inorg. Chem., Ser. I*, *6*, 63 (1972).

20. Hsieh, A.T.T., and Mays, M.J., *Complexes Containing Transition Metal-Group II B Metal Bonds, Int. Rev. Sci., Inorg. Chem., Ser. I*, *6*, 43 (1972).

21. Hsieh, A.T.T., *Organometallic Complexes Containing Bonds Between Transition Metals and Group III B Metals, Inorg. Chim. Acta*, *14*, 87 (1975).

22. Hübel, W., *Organometallic Derivatives from Metal Carbonyls and Acetylenic Compounds*, in I. Wender and P. Pino (Eds.), *Organic Synthesis Via Metal Carbonyls*, Interscience, New York 1968, p. 273.

23. Humphries, A.P., and Kaesz, H.D., *The Hydrido Transition-Metal Cluster Complexes, Progr. Inorg. Chem.*, in the press.

24. Johnson, B.F.G., and McCleverty, J.A., *Nitric Oxide Compounds of Transition Metals, Progr. Inorg. Chem.*, *7*, 277 (1966).

25. Johnston, R.D., *Transition Metal Clusters with π-Acidic Ligands, Advan. Inorg. Chem. Radiochem.*, *13*, 471 (1970).

26. Kaesz, H.D., and Saillant, R.B., *Hydride Complexes of the Transitions Metals, Chem. Rev.*, *72*, 231 (1972).

27. Kaesz, H.D., *Hydrido Transition-Metal Cluster Complexes, Chem. Brit.*, *9*, 344 (1973).

28. King, R.B., *Reactions of Alkali Derivatives of Metal Carbonyls and Related Compounds, Advan. Organometal. Chem.*, *2*, 157 (1964).

29. King, R.B., *Transition Metal Cluster Compounds, Progr. Inorg. Chem.*, *15*, 287 (1972).

30. Kolobova, N.E., Antonova, A.B., and Anisimov, K.N., *Derivatives of Metal Carbonyls Containing a Bond Between Atoms of Transition Metals and Group IV B Elements, Russ. Chem. Rev.*, *38*, 822 (1969).

31. Kotz, J.C., and Pedrotty, D.G., *The Lewis Basicity of Transition Metals, Organometal. Chem. Rev. A*, *4*, 479 (1969).

32. Manuel, A.T., *Lewis Bases Metal-Carbonyl Compounds, Advan. Organometal. Chem.*, *3*, 181 (1965).

33. Maslowski, E., Jr., *Vibrational Spectra of Intra- and Inter-metal and Semimetal Bonds, Chem. Rev.*, *71*, 507

(1971).
34. McCleverty, J.A., *Metal 1,2-Dithiolene and Related Complexes*, *Progr. Inorg. Chem.*, *10*, 49 (1968).
35. Meyer, T.J., *Oxidation-Reduction and Related Reactions of Metal-Metal Bonds*, *Progr. Inorg. Chem.*, *19*, 1 (1975).
36. Muetterties, E.L., Rhodin, T.L., Band, E., Brucker, C.F., and Pretzer, W.R., *Clusters and Surfaces*, *Chem. Rev.*, in the press.
37. Nesmeyanov, A.N., Rybinskaya, M.I., Rybin, L.V., and Kaganovich, V.S., *Binuclear Complexes of Transition Metals with a Common Unsaturated Ligand*, *J. Organometal. Chem.*, *47*, 1 (1973).
38. Penfold, B.R., *Stereochemistry of Metal Cluster Compounds*, *Perspect. Struct. Chem.*, *2*, 71 (1968).
39. Schmid, G., *Metal-Boron Compounds*, *Angew. Chem.*, *82*, 920 (1970); *Angew. Chem. Int. Ed. Engl.*, *9*, 819 (1970).
40. Smith, A.K., and Basset, J.M., *Transition Metal Complexes as Catalysts. A Review.*, *J. Mol. Catal.*, *2*, 229 (1977).
41. Spiro, T.G., *Vibrational Spectra and Metal-metal Bonds*, *Progr. Inorg. Chem.*, *11*, 1 (1969).
42. Todd, L.J., *Transition Metal-Carborane Complexes*, *Advan. Organometal. Chem.*, *8*, 87 (1970).
43. Vyazankin, N.S., Razuvaev, G.A., and Kruglaya, O.A., *Organometallic Compounds with Metal-metal Bonds Between Different Metals*, *Organometal. Chem. Rev. A*, *3*, 323 (1968).
44. Watters, K.L., and Risen, W.M., Jr., *Spectroscopic Studies of Metal-metal Bonded Compounds*, *Inorg. Chim. Acta Rev.*, *3*, 129 (1969).
45. Werner, H., *Kinetic Studies on Substitution Reactions of Carbonyl Metal Complexes*, *Angew. Chem.*, *80*, 1017 (1968); *Angew. Chem. Int. Ed. Engl.*, *7*, 930 (1968).
46. Young, J.F., *Transition Metal Complexes with Group IV B Elements*, *Advan. Inorg. Chem. Radiochem.*, *11*, 91 (1968).

REFERENCES

47. Abel, E.W., and Moorhouse, S., *J. Organometal. Chem.*, *24*, 687 (1970).
48. Abu Salah, O.M., and Bruce, M.I., *J. Chem. Soc. Dalton Trans.*, *1975*, 2311.
49. Adams, D.M., Cook, D.J., and Kemmitt, R.D.W., *J. Chem. Soc. A*, *1968*, 1067.
50. Adams, R.D., Cotton, F.A., and Troup, J.M., *Inorg. Chem.*, *13*, 257 (1974).

51. Adams, R.D., Brice, M.D., and Cotton, F.A., *Inorg. Chem.*, *13*, 1080 (1974).
52. Agron, P.A., Ellison, R.D., and Levy, H.A., *Acta Crystallogr.*, *23*, 1079 (1967).
53. Aime, S., Gervasio, G., Milone, L., Rossetti, R., and Stanghellini, P.L., *J. Chem. Soc. Chem. Commun.*, *1976*, 370.
54. van den Akker, M., and Jellinek, J., *J. Organometal. Chem.*, *10*, P 37 (1967).
55. Albano, V.G., and Ciani, G., *J. Organometal. Chem.*, *66*, 311 (1974).
56. Albano, V.G., Chini, P., Martinengo, S., McCaffrey, D.J.A., Strumolo, D., and Heaton, B.T., *J. Amer. Chem. Soc.*, *96*, 8106 (1974).
57. Alper, H., *Inorg. Chem.*, *11*, 976 (1972).
58. Alper, H., and Chan, A.S.K., *J. Amer. Chem. Soc.*, *95*, 4905 (1973).
59. Alper, H., *J. Organometal. Chem.*, *50*, 209 (1973).
60. Alper, H., and Chan, A.S.K., *Inorg. Chem.*, *13*, 225 (1974).
61. Alper, H., and Chan, A.S.K., *Inorg. Chem.*, *13*, 232 (1974).
62. Amberger, E., Mühlhofer, E., and Stern, H., *J. Organometal. Chem.*, *17*, P 5 (1969).
63. Anders, U., and Graham, W.A.G., *Chem. Commun.*, *1966*, 291.
64. Andrews, M.A., and Kaesz, H.D., *J. Amer. Chem. Soc.*, *99*, 6764 (1977).
65. Andrianov, V.G., and Struchkov, Yu.T., *Chem. Commun.*, *1968*, 1590.
66. Andrianov, V.G., and Struchkov, Yu.T., *Zh. Strukt. Khim.*, *9*, 845 (1968); *J. Struct. Chem.*, *9*, 737 (1968).
67. Andrianov, V.G., Martynov, V.P., Anisimov, K.N., Kolobova, N.E., and Skripkin, V.V., *J. Chem. Soc. D, Chem. Commun.*, *1970*, 1252.
68. Andrianov, V.G., and Struchkov, Yu.T., *Zh. Strukt. Khim.*, *12*, 336 (1971); *J. Struct. Chem.*, *12*, 312 (1971).
69. Andrianov, V.G., Martynov, V.P., and Struchkov, Yu.T., *Zh. Strukt. Khim.*, *12*, 866 (1971); *J. Struct. Chem.*, *12*, 793 (1971).
70. Andrianov, V.G., Struchkov, Yu.T., and Rossinskaja, E.R., *J. Chem. Soc., Chem. Commun.*, *1973*, 338.
71. Ang, H.G., and West, B.O., *Aust. J. Chem.*, *20*, 1133 (1967).
72. Angelici, R.J., and Siefert, E.E., *Inorg. Chem.*, *5*, 1457 (1966).
73. Angelici, R.J., and Blacik, L.J., *Inorg. Chem.*, *11*,

1754 (1972).
74. Averill, B.A., Herskovitz, T., Holm, R.H., and Ibers, J.A., *J. Amer. Chem. Soc.*, *95*, 3523 (1973).
75. Aylett, B.J., Campbell, J.M., and Walton, A., *J. Chem. Soc. A*, *1969*, 2110.
76. Bagga, M.M., Baikie, P.E., Mills, O.S., and Pauson, P.L., *Chem. Commun.*, *1967*, 1106.
77. Bagga, M.M., Flannigan, W.T., Knox, G.R., and Pauson, P.L., *J. Chem. Soc. C*, *1969*, 1534.
78. Baikie, P.E., and Mills, O.S., *Chem. Commun.*, *1966*, 707.
79. Baikie, P.E., and Mills, O.S., *Chem. Commun.*, *1967*, 1228.
80. Baikie, P.E., and Mills, O.S., *Inorg. Chim. Acta*, *1*, 55 (1967).
81. Baird, W., and Dahl, L.F., *J. Organometal. Chem.*, *7*, 503 (1967).
82. Baker, R.W., and Pauling, P., *J. Chem. Soc. D, Chem. Commun.*, *1970*, 573.
83. Balch, A.L., *J. Amer. Chem. Soc.*, *91*, 6962 (1969).
84. Barnett, B.L., and Krüger, C., *Angew. Chem.*, *83*, 969 (1971); *Angew. Chem. Int. Ed. Engl.*, *10*, 910 (1971).
85. Barnett, B.L., and Krüger, C., *Cryst. Struct. Commun.*, *2*, 347 (1973).
86. Barrow, M.J., and Mills, O.S., *Angew. Chem.*, *81*, 898 (1969); *Angew. Chem. Int. Ed. Engl.*, *8*, 879 (1969).
87. Basolo, F., Brault, A.T., and Poë, A.J., *J. Chem. Soc.*, *1964*, 676.
88. Bau, R., Don, B., Greatrex, R., Haines, R.J., Love, R.A., and Wilson, R.D., *Inorg. Chem.*, *14*, 3021 (1975).
89. de Beer, J.A., and Haines, R.J., *J. Organometal. Chem.*, *36*, 297 (1972).
90. de Beer, J.A., Haines, R.J., Greatrex, R., and van Wyk, J.A., *J. Chem. Soc. Dalton Trans.*, *1973*, 2341.
91. de Beer, J.A., and Haines, R.J., *J. Organometal. Chem.*, *24*, 757 (1970).
92. de Beer, J.A., Haines, R.J., Greatrex, R., and Greenwood, N.N., *J. Chem. Soc. A*, *1971*, 3271.
93. de Beer, J.A., Haines, R.J., Greatrex, R., and Greenwood, N.N., *J. Organometal. Chem.*, *27*, C 33 (1971).
94. Behrens, H., and Wakamatsu, H., *Z. Anorg. Allg. Chem.*, *320*, 30 (1963).
95. Behrens, H., Feilner, H.-D., Lindner, E., and Uhlig, D., *Z. Naturforsch.*, *B 26*, 990 (1971).
96. Bennett, M.J., Graham, W.A.G., Stewart, R.P., Jr., and Tuggle, R.M., *Inorg. Chem.*, *12*, 2944 (1973).
97. Bennett, M.J., Graham, W.A.G., Smith, R.A., and

Stewart, Jr., R.P., *J. Amer. Chem. Soc.*, *95*, 1684 (1973).

98. Ben-Shoshan, R., and Pettit, R., *Chem. Commun.*, *1968*, 247.

99. Benson, B.C., Jackson, R., Joshi, K.K., and Thompson, D.T., *Chem. Commun.*, *1968*, 1506.

100. Bernal, I., Davis, B.R., Good, M.R., and Chandra, S., *J. Coord. Chem.*, *2*, 61 (1972).

101. Bichler, R.E.J., and Clark, H.C., *J. Organometal. Chem.*, *23*, 427 (1970).

102. Bichler, R.E.J., Booth, M.R., and Clark, H.C., *J. Organometal. Chem.*, *24*, 145 (1970).

103. Bird, S.R.A., Donaldson, J.D., Holding, A.F.LeC., Senior, B.J., and Tricker, M.J., *J. Chem. Soc. A*, *1971*, 1616.

104. Bir'yukov, B.P., Struchkov, Yu.T., Anisimov, K.N., Kolobova, N.E., and Skripkin, V.V.. *Chem. Commun.*, *1968*, 159.

105. Blount, J.F., Dahl, L.F., Hoogzand, C., and Hübel, W., *J. Amer. Chem. Soc.*, *88*, 292 (1966).

106. Bonati, F., and Wilkinson, G., *J. Chem. Soc.*, *1964*, 179.

107. Bonati, F., Cenini, S., Morelli, D., and Ugo, R., *Inorg. Nucl. Chem. Lett.*, *1*, 107 (1965).

108. Bonati, F., and Minghetti, G., *J. Organometal. Chem.*, *16*, 332 (1969).

109. Bond, A.M., Dawson, P.A., Peake, B.M., Robinson, B.H., and Simpson, J., *Inorg. Chem.*, *16*, 2199 (1977).

110. Braddock, J.N., and Meyer, T.J., *Inorg. Chem.*, *12*, 723 (1973).

111. Braye, E.H., Dahl, L.F., Hübel, W., and Wampler, D.L., *J. Amer. Chem. Soc.*, *84*, 4633 (1962).

112. Bremer, N.J., Cutcliffe, A.B., Farona, M.F., and Kofron, W.G., *J. Chem. Soc. A*, *1971*, 3264.

113. Bright, D., and Mills, O.S., *Bull. Soc. Chim. Belg.*, *76*, 245 (1967).

114. Brooks, E.H., Elder, M., Graham, W.A.G., and Hall, D., *J. Amer. Chem. Soc.*, *90*, 3587 (1968).

115. Brotherton, P.D., Kepert, D.L., White, A.H., and Wild, S.B., *J. Chem. Soc. Dalton Trans.*, *1976*, 1870.

116. Brown, D.A., Manning, A.R., and Thornhill, D.J., *J. Chem. Soc. D, Chem. Commun.*, *1969*, 338.

117. Brown, M.P., and Webster, D.E., *J. Phys. Chem.*, *64*, 698 (1960).

118. Brown, T.L., and Morgan, G.L., *Inorg. Chem.*, *2*, 736 (1963).

119. Bruce, M.I., Shaw, G., and Stone, F.G.A., *J. Chem. Soc. D, Chem. Commun.*, *1971*, 1288.

120. Bruce, M.I., Abu Salah, O.M., Davis, R.E., and Raghavan, N.V., *J. Organometal. Chem.*, *64*, C 48 (1974).
121. Brunner, H., *J. Organometal. Chem.* *14*, 173 (1968).
122. Bryan, R.F., and Greene, P.T., *J. Chem. Soc. A*, *1970*, 3064.
123. Bryan, R.F., Greene, P.T., Newlands, M.J., and Field, D.S., *J. Chem. Soc. A*, *1970*, 3068.
124. Bullitt, J.G., Cotton, F.A., and Marks, T.J., *J. Amer. Chem. Soc.*, *92*, 2155 (1970).
125. Burger, K., Korecz, L., and Bor, G., *J. Inorg. Nucl. Chem.*, *31*, 1527 (1969).
126. Burlitch, J.M., *J. Amer. Chem. Soc.*, *91*, 4562 (1969).
127. Burlitch, J.M., and Ulmer, S.W., *J. Organometal. Chem.*, *19*, P 21 (1969).
128. Burlitch, J.M., and Ferrari, A., *Inorg. Chem.*, *9*, 563 (1970).
129. Burlitch, J.M., and Hayes, S.E., *J. Organometal. Chem.*, *42*, C 13 (1972).
130. Burton, R., Pratt, L., and Wilkinson, G., *J. Chem. Soc.*, *1961*, 594.
131. Calderón, J.L., Fontana, S., Frauendorfer, E., Day, V.W., and Iske, S.D.A., *J. Organometal. Chem.*, *64*, C 16 (1974).
132. Campbell, I.L.C., and Stephens, F.S., *J. Chem. Soc. Dalton Trans.*, *1974*, 923.
133. Campbell, I.L.C., and Stephens, F.S., *J. Chem. Soc. Dalton Trans.*, *1975*, 22.
134. Cane, D.J., Forbes, E.J., and Hamor, T.A., *J. Organometal. Chem.*, *117*, C 101 (1976).
135. Carty, A.J., Efraty, A., Ng, T.W., and Birchall, T., *Inorg. Chem.*, *9*, 1263 (1970).
136. Carty, A.J., Madden, D.P., Mathew, M., Palenik, G.J., and Birchall, T., *J. Chem. Soc. D, Chem. Commun.*, *1970*, 1664.
137. Carty, A.J., Ferguson, G., Paik, H.N., and Restivo, R., *J. Organometal. Chem.*, *74*, C 14 (1974).
138. Case, J.R., and Whiting, M.C., *J. Chem. Soc.*, *1960*, 4632.
139. Casey, M., and Manning, A.R., *J. Chem. Soc. A*, *1970*, 2258.
140. Casey, M., and Manning, A.R., *J. Chem. Soc. A*, *1971*, 256.
141. Casey, M., and Manning, A.R., *J. Chem. Soc. A*, *1971*, 2989.
142. Cetini, G., Stanghellini, P.M., Rossetti, R., and Gambino, O., *J. Organometal. Chem.*, *15*, 373 (1968).
143. Chatt, J., and Thornton, D.A., *J. Chem. Soc.*, *1964*, 1005.

144. Chin, H.B., Smith, M.B., Wilson, R.D., and Bau, R., *J. Amer. Chem. Soc., 96,* 5285 (1974).
145. Chin, H.B., and Bau, R., *J. Amer. Chem. Soc., 98,* 2434 (1976).
146. Chin, H.B., and Bau, R., *Inorg. Chem.,* in the press.
147. Chini, P., *XVII IUPAC Congress,* Munich 1959, Proceedings p. 23.
148. Chini, P., Colli, L., and Peralde, M., *Gazz. Chim. Ital., 90,* 1005 (1960).
149. Chini, P., Cavalieri, A., and Martinengo, S., *Coord. Chem. Rev., 8,* 3 (1972).
150. Chini, P., *Gazz. Chim. Ital.,* in the press.
151. Churchill, M.R., *Inorg. Chem., 6,* 190 (1967).
152. Churchill, M.R., and Bird, P.H., *Inorg. Chem., 8,* 1941 (1969).
153. Churchill, M.R., and Wormald, J., *Inorg. Chem., 9,* 2239 (1970).
154. Churchill, M.R., Wormald, J., Knight, J., and Mays, M.J., *J. Amer. Chem. Soc., 93,* 3073 (1971).
155. Churchill, M.R., and Veidis, M.V., *J. Chem. Soc. A, 1971,* 2170.
156. Churchill, M.R., and Veidis, M.V., *J. Chem. Soc. A, 1971,* 2995.
157. Churchill, M.R., de Boer, B.G., and Kalra, K.L., *Inorg. Chem., 12,* 1646 (1973).
158. Churchill, M.R., and Kalra, K.L., *Inorg. Chem., 12,* 1650 (1973).
159. Churchill, M.R., and Wormald, J., *J. Chem. Soc. Dalton Trans., 1974,* 2410.
160. Clark, H.C., and Rake, A.T., *J. Organometal. Chem., 74,* 29 (1974).
161. Cleland, A.J., Fieldhouse, S.A., Freeland, B.H., Mann, C.D.M., and O'Brien, R.J., *J. Chem. Soc. A, 1971,* 736.
162. Cleland, A.J., Fieldhouse, S.A., Freeland, B.H., and O'Brien, R.J., *J. Organometal. Chem., 32,* C 15 (1971).
163. Clifford, A.F., and Mukherjee, A.K., *Inorg. Chem., 2,* 151 (1963).
164. Cohen, S.C., Sage, S.H., Baker, Jr., W.A., Burlitch, J.M., and Petersen, R.B., *J. Organometal. Chem., 27,* C 44 (1971).
165. Coleman, J.M., Wojcicki, A., Pollick, P.J., and Dahl, L.F., *Inorg. Chem., 6,* 1236 (1967).
166. Collman, J.P., Winter, S.R., and Clark, D.R., *J. Amer. Chem. Soc., 94,* 1788 (1972).
167. Collmann, J.P., Finke, R.G., Matlock, P.L., Wahren, R., Komoto, R.G., and Brauman, J.I., *J. Amer. Chem. Soc., 100,* 1119 (1978).
168. Connelly, N.G., and Dahl, L.F., *J. Amer. Chem. Soc.,*

92, 7472 (1970).
169. Cooke, M.P., *J. Amer. Chem. Soc.*, *92*, 6080 (1970).
170. Cooke, M., Green, M., and Kirkpatrick, D., *J. Chem. Soc. A*, *1968*, 1507.
171. Cooke, C.G., and Mays, M.J., *J. Organometal. Chem.*, *74*, 449 (1974).
172. Cooke, C.G., and Mays, M.J., *J. Chem. Soc. Dalton Trans.*, *1975*, 455.
173. Cooke, C.G., and Mays, M.J., *J. Organometal. Chem.*, *88*, 231 (1975).
174. Connor, J.A., Skinner, H.A., and Virmani, Y., *Faraday Symp. Chem. Soc.*, *8*, 18 (1974).
175. Corradini, P., and Paiaro, G., *Ric. Sci.*, *36*, 365 (1966).
176. Cotton, F.A., and Johnson, B.F.G., *Inorg. Chem.*, *6*, 2113 (1967).
177. Cotton, F.A., and Edwards, W.T., *J. Amer. Chem. Soc.*, *91*, 843 (1969).
178. Cotton, F.A., de Boer, B.G., and Marks, T.J., *J. Amer. Chem. Soc.*, *93*, 5069 (1971).
179. Cotton, F.A., Frenz, B.A., Troup, J.M., and Deganello, G., *J. Organometal. Chem.*, *59*, 317 (1973).
180. Cotton, F.A., Frenz, B.A., and White, A.J., *J. Organometal. Chem.*, *60*, 147 (1973).
181. Cotton, F.A., and Frenz, B.A., *Inorg. Chem.*, *13*, 253 (1974).
182. Cotton, F.A., and Troup, J.M., *J. Amer. Chem. Soc.*, *96*, 1233 (1974).
183. Cotton, F.A., and Troup, J.M., *J. Amer. Chem. Soc.*, *96*, 3438 (1974).
184. Cotton, F.A., and Troup, J.M., *J. Amer. Chem. Soc.*, *96*, 4155 (1974).
185. Cotton, F.A., and Troup, J.M., *J. Amer. Chem. Soc.*, *96*, 4422 (1974).
186. Cotton, F.A., and Troup, J.M., *J. Chem. Soc. Dalton Trans.*, *1974*, 800.
187. Cotton, F.A., Frenz, B.A., and White, A.J., *Inorg. Chem.*, *13*, 1407 (1974).
188. Cotton, F.A., and Troup, J.M., *J. Amer. Chem. Soc.*, *96*, 5070 (1974).
189. Cotton, F.A., Jamerson, J.D., and Stults, B.R., *J. Organometal. Chem.*, *94*, C 53 (1975).
190. Cotton, J.D., Knox, S.A.R., Paul, I., and Stone, F.G.A., *J. Chem. Soc. A*, *1967*, 264.
191. Cotton, J.D., and Peachey, R.M., *Inorg. Nucl. Chem. Lett.*, *6*, 727 (1970).
192. Coucouvanis, D., Lippard, S.J., and Zubieta, J.A., *Inorg. Chem.*, *9*, 2775 (1970).

193. Cox, J.D., and Pilcher, G., *Thermochemistry of Organic and Organometallic Compounds*, Academic Press, London-New York 1970.
194. Cross, R.J., and Glockling, F., *J. Organometal. Chem.*, *3*, 253 (1965).
195. Cullen, W.R., and Hayter, R.G., *J. Amer. Chem. Soc.*, *86*, 1030 (1964).
196. Cullen, W.R., Harbourne, D.A., Liengme, B.V., and Sams, J.R., *Inorg. Chem.*, *8*, 95 (1969).
197. Cullen, W.R., Harbourne, D.A., Liengme, B.V., and Sams, J.R., *Inorg. Chem.*, *9*, 702 (1970).
198. Cullen, W.R., Sams, J.R., and Thompson, J.A.J., *Inorg. Chem.*, *10*, 843 (1971).
199. Cullen, W.R., Patmore, D.J., Sams, J.R., Newlands, M.J., and Thompson, L.K., *J. Chem. Soc. D, Chem. Commun.*, *1971*, 952.
200. Cullen, W.R., Patmore, D.J., and Sams, J.R., *Inorg. Chem.*, *12*, 867 (1973).
201. Curtis, M.D., and Job, R.C., *J. Amer. Chem. Soc.*, *94*, 2153 (1972).
202. Cutforth, H.G., and Selwood, P.W., *J. Amer. Chem. Soc.*, *65*, 2414 (1943).
203. Dahl, L.F., and Rundle, R.E., *Acta Crystallogr.*, *16*, 419 (1963).
204. Dahl, L.F., and Wei, C.-H., *Inorg. Chem.*, *2*, 328 (1963).
205. Dahl, L.F., and Sutton, P.W., *Inorg. Chem.*, *2*, 1067 (1963).
206. Dahl, L.F., and Blount, J.F., *Inorg. Chem.*, *4*, 1373 (1965).
207. Dahl, L.F., Costello, W.R., and King, R.B., *J. Amer. Chem. Soc.*, *90*, 5422 (1968).
208. Dahl, L.F., de Gil, E.R., and Feltham, R., *J. Amer. Chem. Soc.*, *91*, 1653 (1969).
209. Dahm, D.J., and Jacobson, R.A., *J. Amer. Chem. Soc.*, *90*, 5106 (1968).
210. Dapporto, P., Fallani, G., Midollini, S., and Sacconi, L., *J. Amer. Chem. Soc.*, *95*, 2021 (1973).
211. Darensbourg, M.Y., Conder, H.L., Darensbourg, D.J., and Hasday, C., *J. Amer. Chem. Soc.*, *95*, 5919 (1973).
212. Davey, G., and Stephens, F.S., *J. Chem. Soc. Dalton Trans.*, *1974*, 698.
213. Davis, R.E., *Chem. Commun.*, *1968*, 248.
214. Davison, A., McFarlane, W., Pratt, L., and Wilkinson, G., *J. Chem. Soc.*, *1962*, 3653.
215. Davison, A., McCleverty, J.A., and Wilkinson, G., *J. Chem. Soc.*, *1963*, 1133.
216. Dean, P.A.W., Ibbott, D.G., and Bancroft, G.M., *J.*

Chem. Soc. Chem. Commun., 1976, 901.
217. Dean, W.K., and Vanderveer, D.G., *J. Organometal. Chem., 146*, 143 (1978).
218. Dekker, M., and Knox, G.R., *Chem. Commun., 1967*, 1243.
219. Delbaere, L.T.J., McBride, D.W., and Ferguson, R.B., *Acta Crystallogr., B 26*, 515 (1970).
220. Delbaere, L.T.J., Kruczynski, L.J., and McBride, D.W., *J. Chem. Soc. Dalton Trans., 1973*, 307.
221. Demerseman, B., Bouquet, G., and Bigorgne, M., *J. Organometal. Chem., 35*, 341 (1972).
222. Dessy, R.E., Weissman, P.M., and Pohl, R.L., *J. Amer. Chem. Soc., 88*, 5117 (1966).
223. Dessy, R.E., Pohl, R.L., and King, R.B., *J. Amer. Chem. Soc., 88*, 5121 (1966).
224. Dessy, R.E., and Weissman, P.M., *J. Amer. Chem. Soc., 88*, 5124 (1966).
225. Dessy, R.E., and Weissman, P.M., *J. Amer. Chem. Soc., 88*, 5129 (1966).
226. Dessy, R.E., and Pohl, R.L., *J. Amer. Chem. Soc., 90*, 1995 (1968).
227. Dessy, R.E., Kornmann, R., Smith, C., and Haytor, R., *J. Amer. Chem. Soc., 90*, 2001 (1968).
228. Dessy, R.E., and Wieczorek, L., *J. Amer. Chem. Soc., 91*, 4963 (1969).
229. Dewar, J., and Jones, H.O., *Proc. Roy. Soc. (London), A 79*, 66 (1906).
230. Dighe, S.V., and Orchin, M., *J. Amer. Chem. Soc., 86*, 3895 (1964).
231. Dighe, S.V., and Orchin, M., *J. Amer. Chem. Soc., 87*, 1146 (1965).
232. Dobbie, R.C., Hopkinson, M.J., and Whittaker, D., *J. Chem. Soc. Dalton Trans., 1972*, 1030.
233. Dobbie, R.C., and Hopkinson, M.J., *J. Chem. Soc. Dalton Trans., 1974*, 1290.
234. Dodge, R.P., and Shomaker, V., *J. Organometal. Chem., 3*, 274 (1965).
235. Doedens, R.J., and Dahl, L.F., *J. Amer. Chem. Soc., 88*, 4847 (1966).
236. Doedens, R.J., *Inorg. Chem., 7*, 2323 (1968).
237. Doedens, R.J., *Inorg. Chem., 8*, 570 (1969).
238. Doedens, R.J., and Ibers, J.A., *Inorg. Chem., 8*, 2709 (1969).
239. Doedens, R.J., *Inorg. Chem., 9*, 429 (1970).
240. Edgell, W.F., Yang, M.T., Bulkin, B.J., Bayer, R., and Koizumi, N., *J. Amer. Chem. Soc., 87*, 3080 (1965).
241. Edmonson, R.C., and Newlands, M.J., *Chem. Commun., 1968*, 1219.
242. Edmonson, R.C., Eisner, E., Newlands, M.J., and

Thompson, L.K., *J. Organometal. Chem., 35,* 119 (1972).

243. Ehrl, W., and Vahrenkamp, H., *J. Organometal. Chem., 63,* 389 (1973).

244. Einstein, F.W.B., and Trotter, J., *J. Chem. Soc. A, 1967,* 824.

245. Einstein, F.W.B., and Svensson, A.-M., *J. Amer. Chem. Soc., 91,* 3663 (1969).

246. Einstein, F.W.B., and Jones, R.D.G., *J. Chem. Soc. Dalton Trans., 1972,* 2563.

247. Einstein, F.W.B., and Jones, R.D.G., *Inorg. Chem., 12,* 255 (1973).

248. Einstein, F.W.B., and Jones, R.D.G., *Inorg. Chem., 12,* 1690 (1973).

249. Elder, M., and Hall, D., *Inorg. Chem., 8,* 1424 (1969).

250. Elder, M., *Inorg. Chem., 8,* 2703 (1969).

251. Elder, M., and Hutcheon, W., *J. Chem. Soc. Dalton Trans., 1972,* 175.

252. Elmes, P.S., Leverett, P., and West, B.O., *J. Chem. Soc. D, Chem. Commun., 1971,* 747.

253. Elmes, P.S., and West, B.O., *J. Organometal. Chem., 32,* 365 (1971).

254. Epstein, E.F., and Dahl, L.F., *J. Amer. Chem. Soc., 92,* 502 (1970).

255. Ernst, R.D., Marks, T.J., and Ibers, J.A., *J. Amer. Chem. Soc., 99,* 2090 (1977).

256. Ernst, R.D., Marks, T.J., and Ibers, J.A., *J. Amer. Chem. Soc., 99,* 2098 (1977).

257. Evans, G.O., Hargaden, J.P., and Sheline, R.K., *Chem. Commun., 1967,* 186.

258. Evans, G.O., and Sheline, R.K., *J. Inorg. Nucl. Chem., 30,* 2862 (1968).

259. Evans, G.O., and Sheline, R.K., *Inorg. Chem., 10,* 1598 (1971).

260. Evans, G.O., Walter, W.F., Mills, D.R., and Streit, C.A., *J. Organometal. Chem., 144,* C 34 (1978).

261. Farmery, K., Kilner, M., Greatrex, R., and Greenwood, N.N., *Chem. Commun., 1968,* 593.

262. Farmery, K., Kilner, M., Greatrex, R., and Greenwood, N.N., *J. Chem. Soc. A, 1969,* 2339, and references therein.

263. Felkin, H., Knowles, P.J., and Meunier, B., *J. Organometal. Chem., 146,* 151 (1978).

264. Ferguson, G., Hannaway, G., and Islam, K.M.S., *Chem. Commun., 1968,* 1165.

265. Ferguson, J.A., and Meyer, T.J., *J. Chem. Soc. D, Chem. Commun., 1971,* 623.

266. Ferguson, J.A., and Meyer, T.J., *Inorg. Chem., 11,* 631 (1972).

267. Ferguson, J.A., and Meyer, T.J., *J. Amer. Chem. Soc.*, *94*, 3409 (1972).

268. Field, D.S., and Newlands, M.J., *J. Organometal. Chem.*, *27*, 213 (1971).

269. Field, D.S., and Newlands, M.J., *J. Organometal. Chem.*, *27*, 221 (1971).

270. Fischer, E.O., and Böttcher, R., *Z. Naturforsch.*, *B 10*, 600 (1955).

271. Fischer, E.O., and Moser, E., *Z. Anorg. Allg. Chem.*, *342*, 156 (1966).

272. Fischer, E.O., Kiener, V., and Fischer, R.D., *J. Organometal. Chem.*, *16*, P 60 (1969).

273. Fischer, E.O., and Kiener, V., *J. Organometal. Chem.*, *23*, 215 (1970).

274. Fischer, R.D., Vogler, A., and Noak, K., *J. Organometal. Chem.*, *7*, 135 (1967).

275. Fischler, I., Wagner, R., and Koerner von Gustorf, E.A., *J. Organometal. Chem.*, *112*, 155 (1976).

276. Flannigan, W.T., Knox, G.R., and Pauson, P.L., *J. Chem. Soc. C*, *1969*, 2077.

277. Fleischer, E.B., Stone, A.L., Dewar, R.B.K., Wright, J.D., Keller, C.E., and Pettit, R., *J. Amer. Chem. Soc.*, *88*, 3158 (1966).

278. Flitcroft, N., Harbourne, D.A, Paul, I., Tucker, P.M., and Stone, F.G.A., *J. Chem. Soc. A*, *1966*, 1130.

279. Flitcroft, N., and Leach, J.M., *J. Organometal. Chem.*, *18*, 367 (1969).

280. Frey, V., Hieber, W., and Mills, O.S., *Z. Naturforsch.*, *B 23*, 105 (1968).

281. Frisch, P.D., Lloyd, M.K., McCleverty, J.A., and Seddon, D., *J. Chem. Soc. Dalton Trans.*, *1973*, 2268.

282. Gall, R.S., Ting-Wah Chu, C., and Dahl, L.F., *J. Amer. Chem. Soc.*, *96*, 4019 (1974).

283. Garner, C.D., and Roger, Sen., G., *Inorg. Nucl. Chem. Lett.*, *10*, 609 (1974).

284. Gatehouse, B.M., *J. Chem. Soc. D, Chem. Commun.*, *1969*, 948.

285. Geoffrey, G.L., and Gladfelter, W.L., *J. Amer. Chem. Soc.*, *99*, 7565 (1977).

286. Gervasio, G., Rossetti, R., and Stanghellini, P.L., *J. Chem. Soc. Chem. Sommun.*, *1977*, 387.

287. Cerveau, G., Colomer, E., Corriu, R., and Douglas, W.E., *J. Chem. Soc. Chem. Commun.*, *1975*, 410.

288. Gilmore, C.J., and Woodward, P., *J. Chem. Soc. A*, *1971*, 3453.

289. Gilmore, C.J., and Woodward, P., *J. Chem. Soc. Dalton Trans.*, *1972*, 1387.

290. Gorsich, R.D., *J. Amer. Chem. Soc.*, *84*, 2486 (1962).

291. Greatrex, R., and Greenwood, N.N., *Discuss. Faraday Soc.*, *47*, 126 (1969).
292. Greatrex, R., Greenwood, N.N., Rhee, I., Ryang, M., and Tsutsumi, S., *J. Chem. Soc. D, Chem. Commun.*, *1970*, 1193.
293. Greene, P.T., and Bryan, R.F., *Inorg. Chem.*, *9*, 1464 (1970).
294. Greene, P.T., and Bryan, R.F., *J. Chem. Soc. A*, *1970*, 1696.
295. Greene, P.T., and Bryan, R.F., *J. Chem. Soc. A*, *1970*, 2261.
296. Grobe, J., *Z. Anorg. Allg. Chem.*, *361*, 32 (1968).
297. Grobe, J., *Z. Anorg. Allg. Chem.*, *361*, 47 (1968).
298. Grynkewich, G.W., Ho, B.Y.K., Marks, T.J., Tomaja, D.L., and Zuckerman, J.J., *Inorg. Chem.*, *12*, 2522 (1972).
299. Hackett, P., and Manning, A.R., *J. Organometal. Chem.*, *34*, C 15 (1972).
300. Haines, R.J., du Preez, A.L., and Wittmann, G.T.W., *Chem. Commun.*, *1968*, 611.
301. Haines, R.J., and du Preez, A.L., *Inorg. Chem.*, *8*, 1459 (1969).
302. Haines, R.J., Nolte, C.R., Greatrex, R., and Greenwood, N.N., *J. Organometal. Chem.*, *26*, C 45 (1971).
303. Haines, R.J., and du Preez, A.L., *Inorg. Chem.*, *11*, 330 (1972).
304. Haines, R.J., Mason, R., Zubieta, J.A., and Nolte, C.R., *J. Chem. Soc. Chem. Commun.*, *1972*, 990.
305. Haines, R.J., du Preez, A.L., and Nolte, C.R., *J. Organometal. Chem.*, *55*, 199 (1973).
306. Haines, R.J., de Beer, J.A., and Greatrex, R., *J. Organometal. Chem.*, *55*, C 30 (1973).
307. Hamilton, W.C., and Bernal, I., *Inorg. Chem.*, *6*, 2003 (1967).
308. Hansen, P.J., and Jacobson, R.A., *J. Organometal. Chem.*, *6*, 389 (1966).
309. Harrison, P.G., King, T.J., and Richards, J.A., *J. Chem. Soc. Dalton Trans.*, *1975*, 2097.
310. Housecroft, C.E., Wade, K., and Smith, B.C., *J. Chem. Soc. Chem. Commun.*, *1978*, 765.
311. Hayter, R.G., *J. Amer. Chem. Soc.*, *85*, 3120 (1963).
312. Hein, F., and Heuser, E., *Z. Anorg. Allg. Chem.*, *249*, 293 (1942).
313. Hein, F., and Jehn, W., *Justus Liebigs Ann. Chem.*, *684*, 4 (1965) and references therein.
314. Henslee, W., and Davis, R.E., *Cryst. Struct. Commun.*, *1*, 403 (1972).
315. Hieber, W., and Fack, E., *Z. Anorg. Allg. Chem.*, *236*,

83 (1938).
316. Hieber, W., and Scharfenberg, C., *Chem. Ber.*, *73*, 1012 (1940).
317. Hieber, W., Sedlmeier, J., and Werner, R., *Chem. Ber.*, *90*, 278 (1957).
318. Hieber, W., and Werner, R., *Chem. Ber.*, *90*, 286 (1957).
319. Hieber, W., and Werner, R., *Chem. Ber.*, *90*, 1116 (1957).
320. Hieber, W., and Brendel, G., *Z. Anorg. Allg. Chem.*, *289*, 324 (1957).
321. Hieber, W., and Brendel, G., *Z. Anorg. Allg. Chem.*, *289*, 338 (1957).
322. Hieber, W., and Kahlen, N., *Chem. Ber.*, *91*, 2234 (1958).
323. Hieber, W., and Gruber, J., *Z. Anorg. Allg. Chem.*, *296*, 91 (1958).
324. Hieber, W., and Lipp, A., *Chem. Ber.*, *92*, 2085 (1959).
325. Hieber, W., Gruber, J., and Lux, F., *Z. Anorg. Allg. Chem.*, *300*, 275 (1959).
326. Hieber, W., and Beck, W., *Z. Anorg. Allg. Chem.*, *305*, 265 (1960).
327. Hieber, W., and Beck, W., *Z. Anorg. Allg. Chem.*, *305*, 274 (1960).
328. Hieber, W., and Kruck, T., *Chem. Ber.*, *95*, 2027 (1962).
329. Hieber, W., and Beutner, H., *Z. Naturforsch.*, *B 17*, 211 (1962).
330. Hieber, W., and Beutner, H., *Z. Anorg. Allg. Chem.*, *319*, 285 (1963).
331. Hieber, W., and Beutner, H., *Z. Anorg. Allg. Chem.*, *320*, 101 (1963).
332. Hieber, W., and Zeidler, A., *Z. Anorg. Allg. Chem.*, *329*, 92 (1964).
333. Hieber, W., Bauer, I., and Neumair, G., *Z. Anorg. Allg. Chem.*, *335*, 250 (1965).
334. Hieber, W., and Schubert, E.H., *Z. Anorg. Allg. Chem.*, *338*, 32 (1965).
335. Hieber, W., and Schubert, E.H., *Z. Anorg. Allg. Chem.*, *338*, 37 (1965).
336. Hieber, W., and Kaiser, K., *Chem. Ber.*, *102*, 4043 (1969).
337. Hieber, W., and Führling, H., *Z. Naturforsch.*, *B 25*, 663 (1970).
338. Hock, A.A., and Mills, O.S., *Acta Crystallogr.*, *14*, 139 (1961).
339. Hock, H., and Stuhlmann, H., *Chem. Ber.*, *B 61*, 2097 (1928).
340. Hock, H., and Stuhlmann, H., *Chem. Ber.*, *B 62*, 431 (1929).

341. Hodali, H.A. Shriver, D.F., and Ammlung, C.A., *J. Amer. Chem. Soc.*, *100*, 5239 (1978).
342. Höfler, M., and Scheuren, J., *J. Organometal. Chem.*, *55*, 177 (1973).
343. Holm, R.H., Averill, B.A., Herskovitz, T., Frankel, R.B., Gray, H.B., Siiman, O., and Grunthaner, F.J., *J. Amer. Chem. Soc.*, *96*, 2644 (1974).
344. Hsieh, A.T.T., Mays, M.J., and Platt, R.H., *J. Chem. Soc. A*, *1971*, 3296.
345. Hsieh, A.T.T., and Knight, J., *J. Organometal. Chem.*, *26*, 125 (1971).
346. Hsieh, A.T.T., and Mays, M.J., *J. Organometal. Chem.*, *37*, 9 (1972).
347. Hsieh, A.T.T., and Mays, M.J., *J. Organometal. Chem.*, *37*, C 53 (1972).
348. Hsieh, A.T.T., and Mays, M.J., *J. Organometal. Chem.*, *39*, 157 (1972).
349. Hsieh, A.T.T., and Wilkinson, G., *J. Chem. Soc. Dalton Trans.*, *1973*, 867.
350. Hübel, W., and Braye, E.H., *J. Inorg. Nucl. Chem.*, *10*, 250 (1959).
351. Huie, B.T., Knobler, C.B., and Kaesz, H.D., *J. Chem. Soc. Chem. Commun.*, *1975*, 684.
352. Huntsmann, J.R., and Dahl, L.F., to be published.
353. Huttner, G., and Regler, D., *Chem. Ber.*, *105*, 2726 (1972).
354. Huttner, G., Mohr, G., Frank, A., and Schubert, U., *J. Organometal. Chem.*, *118*, C 73 (1976).
355. Ibekwe, S.D., and Newlands, M.J., *J. Chem. Soc. A*, *1967*, 1783.
356. Jacob, M., and Weiss, E., *J. Organometal. Chem.*, *131*, 263 (1977).
357. Jahn, A., *Z. Anorg. Allg. Chem.*, *301*, 301 (1959).
358. Jansen, P.R., Oskam, A., and Olic, K., *Cryst. Struct. Commun.*, *4*, 667 (1975).
359. Jeffreys, J.A.D., Willis, C.M., Robertson, I.C., Ferguson, G., and Sime, J.G., *J. Chem. Soc. Dalton Trans.*, *1973*, 749.
360. Jetz, W., and Graham, W.A.G., *Inorg. Chem.*, *10*, 4 (1971).
361. Jetz, W., and Graham, W.A.G., *Inorg. Chem.*, *10*, 1159 (1971).
362. Jetz, W., and Graham, W.A.G., *Inorg. Chem.*, *10*, 1647 (1971).
363. Jetz, W., and Angelici, R.J., *J. Organometal. Chem.*, *35*, C 37 (1972) and references therein.
364. Jetz, W., and Graham, W.A.G., *J. Organometal. Chem.*, *69*, 383 (1974).

365. Job, B.E., McLean, R.A.N., and Thompson, D.T., *Chem. Commun., 1966,* 895.
366. Job, R.C., and Curtis, M.D., *Inorg. Chem., 12,* 2514 (1973).
367. Johansson, G., and Lipscomb, W.N., *Acta Crystallogr., 11,* 594 (1958).
368. Johnson, E.C., Meyer, T.J., and Winterton, N., *Inorg. Chem., 10,* 1673 (1971).
369. Jarvis, J.A.J., Job, B.E., Kilbourn, B.T., Mais, R.H.B., Owsten, P.G., and Todd, P.F., *Chem. Commun., 1967,* 1149.
370. Joshi, K.K., Mills, O.S., Pauson, P.L., Shaw, B.W., and Stubbs, W.H., *Chem. Commun., 1965,* 181.
371. Kaesz, H.D., Fontal, B., Bau, R., Kirtley, S.W., and Churchill, M.R., *J. Amer. Chem. Soc., 91,* 1021 (1969).
372. Kahn, O., and Bigorgne, M., *J. Organometal. Chem., 10,* 137 (1967).
373. Khattab, S., Markó, L., Bor, G., and Markó, B., *J. Organometal. Chem., 1,* 373 (1964).
374. Kiener, V., and Fischer, E.O., *J. Organometal. Chem., 42,* 447 (1972).
375. Kilner, M., and Midcalf, C., *J. Chem. Soc. Dalton Trans., 1974,* 1620.
376. Kim, N.E., Nelson, N.J., and Shriver, D.F., *Inorg. Chim. Acta, 7,* 393 (1973).
377. King, R.B., and Stone, F.G.A., *J. Amer. Chem. Soc., 82,* 3833 (1960).
378. King, R.B., Treichel, P.M., and Stone, F.G.A., *Chem. Ind. (London), 1961,* 747.
379. King, R.B., Treichel, P.M., and Stone, F.G.A., *J. Amer. Chem. Soc., 83,* 3600 (1961).
380. King, R.B., *Inorg. Chem., 2,* 1275 (1963).
381. King, R.B., and Stone, F.G.A., *Inorg. Synth., 7,* 193 (1963).
382. King, R.B., and Stone, F.G.A., *Inorg. Synth., 7,* 196 (1963).
383. King, R.B., in J. Eisch, and R.B. King (Eds.), *Organometallic Syntheses, Vol. 1,* Academic Press, New York 1965, p. 114.
384. King, R.B., *Inorg. Chem., 5,* 2227 (1966).
385. King, R.B., and Bisnette, M.B., *Inorg. Chem., 6,* 469 (1967).
386. King, R.B., and Eggers, C.A., *Inorg. Chem., 7,* 1214 (1968).
387. King, R.B., and Pannell, K.H., *Inorg. Chem., 7,* 1510 (1968).
388. King, R.B., Pannell, K.H., Bennett, C.R., and Ishaq, M., *J. Organometal. Chem., 19,* 327 (1969).

389. King, R.B., and Bond, A., *J. Organometal. Chem.*, *46*, C 53 (1972).
390. King, R.B., and Saran, M.S., *J. Amer. Chem. Soc.*, *95*, 1811 (1973).
391. Kirchner, R.M., and Ibers, J.A., *J. Organometal. Chem.*, *82*, 243 (1974).
392. Kläui, W., and Werner, H., *J. Organometal. Chem.*, *54*, 331 (1973).
393. Knight, J., and Mays, M.J., *J. Chem. Soc. A*, *1970*, 654.
394. Koerner von Gustorf, E., Grevels, F.-W., and Hogan, J.C., *Angew. Chem.*, *81*, 918 (1969); *Angew. Chem. Int. Ed. Engl.*, *8*, 899 (1969).
395. Koerner von Gustorf, E., and Wagner, R., *Angew. Chem.*, *83*, 968 (1971); *Angew. Chem. Int. Ed. Engl.*, *10*, 910 (1971).
396. Koerner von Gustorf, E., Hogan, J.C., and Wagner, R., *Z. Naturforsch.*, *B 27*, 140 (1972).
397. Kostiner, E., Reddy, M.L.N., Urch, D.S., and Massey, A.G., *J. Organometal. Chem.*, *15*, 383 (1968).
398. Kristoff, J.S., and Shriver, D.F., *Inorg. Chem.*, *13*, 499 (1974).
399. Kruck, T., Hofler, M., and Noack, M., *Chem. Ber.*, *99*, 1153 (1966).
400. Kruck, T., Job, E., and Klose, U., *Angew. Chem.*, *80*, 360 (1968); *Angew. Chem. Int. Ed. Engl.*, *7*, 374 (1968).
401. Krüger, C., Tsay, Y.H., Grevels, F.-W., and Koerner von Gustorf, E., *Israel J. Chem.*, *10*, 201 (1972).
402. Krumholz, P., and Stettiner, H.M.A., *J. Amer. Chem. Soc.*, *71*, 3035 (1949).
403. Krusic, P.J., San Filippo, J., Hutchinson, B., Hance, R.L., and Daniels, L.M., submitted to *J. Amer. Chem. Soc.*
404. Kubas, G.J., and Spiro, T.G., *Inorg. Chem.*, *12*, 1797 (1973).
405. Kummer, D., and Furrer, J., *Z. Naturforsch.*, *B 26*, 162 (1971).
406. Kummer, R., and Graham, W.A.G., *Inorg. Chem.*, *7*, 1208 (1968).
407. Kuz'mina, L.G., Bokii, N.G., Struchkov, Yu.T., Arutyunyan, A.V., Rybin, L.V., and Rybinskaya, M.I., *Zh. Strukt. Khim.*, *12*, 875 (1971); *J. Struct. Chem.*, *12*, 801 (1971).
408. Lange, G., and Dehnicke, K., *Z. Anorg. Allg. Chem.*, *344*, 167 (1966).
409. Le Borgne, G., and Grandjean, D., *Acta Crystallogr.*, *B 29*, 1040 (1973).
410. Lewis, J., and Wild, S.B., *J. Chem. Soc. A*, *1966*, 69.
411. Lindauer, M.W., Evans, G.O., and Sheline, R.K., *Inorg.

Chem., *7*, 1249 (1968).
412. Lindley, P.F., and Woodward, P., *J. Chem. Soc. A*, *1967*, 382.
413. Lindley, P.F., and Mills, O.S., *J. Chem. Soc. A*, *1969*, 1279.
414. Lindley, P.F., and Mills, O.S., *J. Chem. Soc. A*, *1970*, 38.
415. Little, R.G., and Doedens, R.J., *Inorg. Chem.*, *11*, 1392 (1972).
416. Longoni, G., Dahl, L.F., and Chini, P., unpublished results.
417. Manassero, M., Sansoni, M., and Longoni, G., *J. Chem. Soc. Chem. Commun.*, *1976*, 919.
418. Manning, A.R., *J. Chem. Soc. A*, *1968*, 1319.
419. Manning, A.R., *J. Chem. Soc. A*, *1971*, 2321.
420. Manojlović-Muir, L., Muir, K.W., and Ibers, J.A., *Inorg. Chem.*, *9*, 447 (1970).
421. Manuel, T.A., and Meyer, T.J., *Inorg. Chem.*, *3*, 1049 (1964).
422. Marks, T.J., and Seyam, A.M., *J. Organometal. Chem.*, *31*, C 62 (1971).
423. Marks, T.J., and Newman, A.R., *J. Amer. Chem. Soc.*, *95*, 769 (1973).
424. Mason, R., Zubieta, J., Hsieh, A.T.T., Knight, J., and Mays, M.J., *J. Chem. Soc. Chem. Commun.*, *1972*, 200.
425. Mason, R., and Zubieta, J.A., *Angew. Chem.*, *85*, 390 (1973); *Angew. Chem. Int. Ed. Engl.*, *12*, 390 (1973).
426. Mason, R., and Zubieta, J.A., *J. Organometal. Chem.*, *66*, 279 (1974).
427. Mason, R., and Zubieta, A., *J. Organometal. Chem.*, *66*, 289 (1974).
428. Mays, M.J., and Robb, J.D., *J. Chem. Soc. A*, *1968*, 329.
429. Mays, M.J., and Simpson, R.N.F., *J. Chem. Soc. A*, *1968*, 1444.
430. Mays, M.J., and Prater, B.E., *J. Chem. Soc. A*, *1969*, 2525.
431. Mays, M.J., and Sears, P.L., *J. Chem. Soc. Dalton Trans.*, *1973*, 1873.
432. McArdle, P., and Manning, A.R., *J. Chem. Soc. A*, *1971*, 717.
433. McDonald, W.S., Moss, J.R., Raper, G., Shaw, B.L., Greatrex, R., and Greenwood, N.N., *J. Chem. Soc. D, Chem. Commun.*, *1969*, 1295.
434. McFarlane, W., and Wilkinson, G., *Inorg. Synth.*, *8*, 181 (1966).
435. McNeese, T.J., Wreford, S.S., Tipton, D.L., and Bau, R., *J. Chem. Soc. Chem. Commun.*, *1977*, 390.

436. McVicker, G.B., and Matyas, R.S., *J. Chem. Soc. Chem. Commun.*, *1972*, 972.

437. McVicker, G.B., *Inorg. Synth.*, *16*, 56 (1976).

438. Meunier-Piret, J., Piret, P., and Van Meerssche, M., *Bull. Soc. Chim. Belg.*, *76*, 374 (1967).

439. Miholova, D., Klima, J., and Vlcek, A.A., *Inorg. Chim. Acta*, *27*, L 67 (1978).

440. Mills, O.S., and Redhouse, A.D., *J. Chem. Soc. A*, *1968*, 1282.

441. Mond, L., and Langer, C., *J. Chem. Soc.*, *59*, 1090 (1891).

442. Moss, J.R., and Graham, W.A.G., *J. Organometal. Chem.*, *23*, C 23 (1970).

443. Murahashi, S.-I., Mizoguchi, T., Hosokawa, T., Moritani, I., Kai, Y., Kohara, M., Yasuoka, N., and Kasai, N., *J. Chem. Soc. Chem. Commun.*, *1974*, 563.

444. Nasta, M.A., and Mac Diarmid, A.G., *J. Organometal. Chem.*, *18*, P 11 (1969).

445. Nasta, M., and Mac Diarmid, A.G., *J. Amer. Chem. Soc.*, *93*, 2813 (1971).

446. Natile, G., and Bor, G., *J. Organometal. Chem.*, *35*, 185 (1972).

447. Nelson, N.J., Kime, N.E., and Shriver, D.F., *J. Amer. Chem. Soc.*, *91*, 5173 (1969).

448. Nesmeyanov, A.N., Anisimov, K.N., Kolobova, N.E., and Khandozhko, V.N., *Dokl. Akad. Nauk SSSR*, *156*, 383 (1964); *Dokl. Chem.*, *156*, 502 (1964).

449. Nesmeyanov, A.N., Anisimov, K.N., Kolobova, N.E., and Skripkin, V.V., *Izv. Akad. Nauk SSSR, Ser. Khim*, *1966*, 1292; *Bull. Acad. Sci. USSR, Div. Chem. Sci.*, *1966*, 1248.

450. Nesmeyanov, A.N., Anisimov, K.N., Kolobova, N.E., and Denisov, F.S., *Izv. Akad. Nauk SSSR, Ser. Khim.*, *1968*, 1419; *Bull. Acad. Sci. USSR, Div. Chem. Sci.*, *1968*, 1348.

451. Nesmeyanov, A.N., Makarova, L.G., and Vinogradova, V.N., *Izv. Akad. Nauk SSSR, Ser. Khim.*, *1971*, 1984; *Bull. Acad. Sci. USSR, Div. Chem. Sci.*, *1971*, 1869.

452. Nesmeyanov, A.N., Rybinskaya, M.I., Rybin, L.V., Kaganovich, V.S., and Petrovskii, P.V., *J. Organometal. Chem.*, *31*, 257 (1971).

453. Neuman, M.A., Dahl, L.F., and King, R.B., unpublished results quoted by Doedens R.J., and Dahl, L.F., *J. Amer. Chem. Soc.*, *88*, 4847 (1966).

454. Newman, J., and Manning, A.R., *J. Chem. Soc. Dalton Trans.*, *1974*, 2549.

455. Nicholas, K., Bray, L.S., Davis, R.E., and Pettit, R., *J. Chem. Soc. D, Chem. Commun.*, *1971*, 608.

456. Nowell, I.M., and Russel, D.R., *Chem. Commun., 1967,* 817.
457. O'Connor, J.E., and Corey, E.R., *Inorg. Chem., 6,* 968 (1967).
458. O'Connor, J.E., and Corey, E.R., *J. Amer. Chem. Soc., 89,* 3930 (1967).
459. O'Connor, T., Carty, A.J., Mathew, M., and Palenik, G.J.,*J. Organometal. Chem., 38,* C 15 (1972).
460. Öfele, K., and Dotzauer, E., *J. Organometal. Chem., 42, C 87* (1972).
461. Ogawa, K., Torii, A., Kobayashi-Tamura, H., Watanabê, T., Yoshida, T., and Otsuka, S., *J. Chem. Soc. D, Chem. Commun., 1971,* 991.
462. Otsuka, S., Yoshida, T., and Nakamura, A., *Inorg. Chem., 7,* 1833 (1968).
463. Otsuka, S., Nakamura, A., and Tani, K., *J. Chem. Soc. A, 1968,* 2248.
464. Pardue, J.E., and Dobson, G.R., *J. Organometal. Chem., 132,* 121 (1977).
465. Patel, H.A., Carty, A.J., Mathew, M., and Palenik, G.J., *J. Chem. Soc. Chem. Commun., 1972,* 810.
466. Patel, H.A., Fischer, R.G., Carty, A.J., Naik, D.V., and Palenik, G.J., *J. Organometal. Chem., 60,* C 49 (1973).
467. Peake, B.M., Robinson, B.H., Simpson, J., and Watson, D.J., *J. Chem. Soc. Chem. Commun., 1974,* 945.
468. Piret, P., Meunier-Piret, J., van Meerssche, M., and King, G.S.D., *Acta Crystallogr., 19,* 78 (1965).
469. Piret, P., Meunier-Piret, J., and van Meerssche, M., *Acta Crystallogr., 19,* 85 (1965).
470. Piron, J., Piret, P., and van Meerssche, M., *Bull. Soc. Chim. Belg., 76,* 505 (1967).
471. Piron, J., Piret, P., Meunier-Piret, J., and van Meerssche, M., *Bull. Soc. Chim. Belg., 78,* 121 (1969).
472. Poliakoff, N., and Turner, J.J., *J. Chem. Soc. A, 1971,* 654.
473. Pollick, P.J., and Wojcicki, A., *J. Organometal. Chem., 14,* 469 (1968).
474. Powell, H.M., and Ewens, R.V.G., *J. Chem. Soc., 1939,* 286.
475. Raper, G., and McDonald, W.S., *J. Chem. Soc. A, 1971,* 3430.
476. Restivo, R., and Bryan, R.F., *J. Chem. Soc. A, 1971,* 3364 and references therein.
477. Roberts, P.J., Penfold, B.R., and Trotter, J., *Inorg. Chem., 9,* 2137 (1970).
478. Roberts, R.M.G., *J. Organometal. Chem., 47,* 359 (1973).
479. Rodrique, L., van Meerssche, M., and Piret, P., *Acta*

Crystallogr., *B 25*, 519 (1969).
480. Roe, D.M., and Massey, A.G., *J. Organometal. Chem.*, *23*, 547 (1970).
481. Reger, D.L., and Coleman, C., *J. Organometal. Chem.*, *131*, 153 (1977).
482. Rosenbuch, P., and Welcman, N., *J. Chem. Soc. Dalton Trans.*, *1972*, 1963.
483. Ruff, J.K., *Inorg. Chem.*, *6*, 1502 (1967).
484. Ruff, J.K., *Inorg. Chem.*, *7*, 1499 (1968).
485. Ruff, J.K., *Inorg. Chem.*, *7*, 1818 (1968).
486. San Filippo, Jr., J., and Sniadoch, H.J., *Inorg. Chem.*, *12*, 2326 (1973).
487. Sappa, E., Tiripicchio, A., and Tiripicchio Camellini, M., *J. Chem. Soc. Dalton Trans.*, *1978*, 419.
488. Schmid, G., and Balk, H.J., *J. Organometal. Chem.*, *80*, 257 (1974).
489. Schrauzer, G.N., Rabinowitz, H.N., Frank, J.A.K., and Paul, I.C., *J. Amer. Chem. Soc.*, *92*, 212 (1970).
490. Schrauzer, G.N., and Kisch, H., *J. Amer. Chem. Soc.*, *95*, 2501 (1973).
491. Schubert, E.H., and Sheline, R.K., *Z. Naturforsch.*, *B 20*, 1306 (1965).
492. Schultz, A.J., and Eisenberg, R., *Inorg. Chem.*, *12*, 518 (1973).
493. Schunn, R.A., Fritchie, Jr., C.J., and Prewitt, C.T., *Inorg. Chem.*, *5*, 892 (1966).
494. Scovell, W.M., and Spiro, T.G., *Inorg. Chem.*, *13*, 304 (1974).
495. Seel, F., and Röschenthaler, G.V., *Angew. Chem.*, *82*, 182 (1970); *Angew. Chem. Int. Ed. Engl.*, *9*, 166 (1970).
496. Shklober, V.E., Skripkin, V.V., Gusev, A.I., and Struchkov, Yu.T., *Zh. Strukt. Khim.*, *13*, 744 (1972); *J. Struct. Chem.*, *13*, 698 (1972).
497. Shriver, D.F., Lehman, D., and Strope, D., *J. Amer. Chem. Soc.*, *97*, 1594 (1975).
498. Smith, M.B., and Bau, R., *J. Amer. Chem. Soc.*, *95*, 2388 (1973).
499. Stephens, F.S., and Mills, O.S., quoted in ref. 56 by Handy, L.B., Ruff, J.K., and Dahl, L.F., *J. Amer. Chem. Soc.*, *92*, 7312 (1970).
500. Stephens, F.S., *J. Chem. Soc. A*, *1970*, 1722.
501. Stephens, F.S., *J. Chem. Soc. Dalton Trans.*, *1972*, 2257.
502. Stephens, F.S., *J. Chem. Soc. Dalton Trans.*, *1974*, 13.
503. Stevenson, D.L., Wei, C.H., and Dahl, L.F., *J. Amer. Chem. Soc.*, *93*, 6027 (1971).
504. Stewart, R.P., Anders, U., and Graham, W.A.G., *J. Organometal. Chem.*, *32*, C 49 (1971).

505. Strouse, C.E., and Dahl, L.F., *J. Amer. Chem. Soc., 93,* 6032 (1971).
506. Sweet, R.M., Fritchie, Jr., C.J., and Schunn, R.A., *Inorg. Chem., 6,* 749 (1967).
507. Symon, D.A., and Waddington, T.C., *J. Chem. Soc. A, 1971,* 953.
508. Teller, R.G., Finke, R.G., Collman, J.P., Chin, H.B., and Bau, R., *J. Amer. Chem. Soc., 99,* 1104 (1977).
509. Teo, B.K., Hall, M.B., Fenske, R.F., and Dahl, L.F., *J. Organometal. Chem., 70,* 413 (1974) and references therein.
510. Teo, B.K., Hsieh, A.T.T., Knight, J., and Dahl, L.F., submitted for publication.
511. Thomas, J.T., Robertson, J.H., and Cox, E.G., *Acta Crystallogr., 11,* 599 (1958).
512. Thompson, L.K., Eisner, E., and Newlands, M.J., *J. Organometal. Chem., 56,* 327 (1973).
513. Tilney-Bassett, J.F., *J. Chem. Soc., 1963,* 4784.
514. Treichel, P.M., Dean, W.K., and Douglas, W.M., *Inorg. Chem., 11,* 1609 (1972).
515. Treichel, P.M., Dean, W.K., and Calabrese, J.C., *Inorg. Chem., 12,* 2908 (1973).
516. Trinh-Toan, and Dahl, L.F., *J. Amer. Chem. Soc., 93,* 2654 (1971).
517. Trinh-Toan, Felhammer, W.P., and Dahl, L.F., *J. Amer. Chem. Soc., 94,* 3389 (1972).
518. Trinh-Toan, Teo, B.K., Ferguson, J.A., Meyer, T.A., and Dahl, L.F., *J. Amer. Chem. Soc., 99,* 408 (1977).
519. Ugo, R., Cenini, S., and Bonati, F., *Inorg. Chim. Acta, 1,* 451 (1967).
520. Vahrenkamp, H., *Chem. Ber., 106,* 2570 (1973).
521. Vahrenkamp, H., *J. Organometal. Chem., 63,* 399 (1973).
522. Wada, F., and Matsuda, T., *J. Organometal. Chem., 61,* 365 (1973).
523. Weaver, J., and Woodward, P., *J. Chem. Soc. Dalton Trans., 1973,* 1439.
524. Weber, H.P., and Bryan, R.F., *J. Chem. Soc. A, 1967,* 182.
525. Bryan, R.F., *J. Chem. Soc. A, 1967,* 192.
526. Wei, C.H., and Dahl, L.F., *Inorg. Chem., 4,* 1 (1965).
527. Wei, C.H., and Dahl, L.F., *Inorg. Chem., 4,* 493 (1965).
528. Wei, C.H., Wilkes, G.R., Treichel, P.M., and Dahl, L.F., *Inorg. Chem., 5,* 900 (1966).
529. Wei, C.H., and Dahl, L.F., *J. Amer. Chem. Soc., 91,* 1351 (1969).
530. Williams, W.E., and Lalor, F.J., *J. Chem. Soc. Dalton Trans., 1973,* 1329.
531. Windus, C., Sujishi, S., and Giering, W.P., *J. Amer.*

Chem. Soc., *96*, 1951 (1974).

532. Wong, Y.S., Paik, H.N., Chieh, P.C., and Carty, A.J., *J. Chem. Soc. Chem. Commun.*, *1975*, 309.

533. Wright, C., and Chini, P., unpublished work.

534. Yasufuku, K., and Yamazaki, H., *J. Organometal. Chem.*, *28*, 415 (1971).

535. Yasufuku, K., and Yamazaki, H., *Bull. Chem. Soc. Jap.*, *45*, 2664 (1972).

536. Yasufuku, K., and Yamazaki, H., *J. Organometal. Chem.*, *38*, 367 (1972).

537. Yasufuku, K., Aoki, K., and Yamazaki, H., *Bull. Chem. Soc. Jap.*, *48*, 1616 (1975).

538. Yawney, D.B.W., and Stone, F.G.A., *J. Chem. Soc. A*, *1969*, 502.

539. Zimmer, J.-C., and Huber, M., *C.R. Acad. Sci., Ser. C*, *267*, 1685 (1968).

COMPLEXES WITH SULPHUR-CONTAINING LIGANDS

BY LÁSZLO MARKÓ

Department of Organic Chemistry
University of Chemical Engineering
Veszprém, Hungary

and

BERNADETT MARKÓ-MONOSTORY

Hungarian Oil and Gas Research Institute,
Veszprém, Hungary

TABLE OF CONTENTS

ISBN 0-12-417102-8

I. INTRODUCTION

The first sulphur-containing organoiron compounds to be prepared were the ethylmercapto derivative $[Fe(CO)_3(SEt)]_2$, reported by Reihlen and co-workers and (although only in solution) bis(L-cysteinato)dicarbonyliron, observed by Cremer, both in 1928. This review covers the literature till early 1979 and deals with nearly 1000 compounds with iron-carbon bonds and containing sulphur, selenium or tellurium, most of which - in accordance with the rapid expansion of organometallic chemistry - were synthesized in the last few years. The great majority of these complexes belongs to the carbonyliron derivatives and - since ferrocene type compounds have been omitted - the number of complexes not containing carbon monoxide as a ligand is unexpectedly small.

Research work done in this field of chemistry has been dominated by preparative chemists. A large part of the knowledge regarding these compounds is limited to their methods of synthesis and a few physical (mostly spectroscopic) data useful for characterization and identification. Usually little is known about the chemistry of these complexes, perhaps with the exception of some rather simple substitution reactions.

In addition to organoiron sulphur compounds, this chapter also covers some related areas. The analogous selenium and tellurium derivatives are included too, and one section deals with the use of carbonyliron derivatives in the chemistry of organic sulphur compounds. For those who are interested in some further problems related to the subject of this chapter a few review articles have been compiled *(472-484)*.

II. COMPLEXES WITH SULPHUR AND INORGANIC SULPHUR CONTAINING LIGANDS

Two carbonyliron sulphides are known, $Fe_2(CO)_6S_2$ *(217, 219,334,400,401)* and $Fe_3(CO)_9S_2$ *(90,211,218,219,334,411)*, their structures (<u>1</u> and <u>2</u>) have been established by X-ray diffraction *(119,448,449)*. They probably contain bent iron-iron bonds *(121)*. One to three of their CO ligands may be replaced by phosphines and similar Lewis bases *(86,87,91,226)*. Unexpectedly, $Fe_3(CO)_9S_2$ exists in two modifications in the solid state *(260,448)*, differing only in the orientation of the carbonyl ligands surrounding the seven-coordinate iron atom. This trinuclear carbonyliron sulphide may play some role as an intermediate in the synthesis of $Fe(CO)_5$ from iron in the presence of sulphur *(218)*. Its selenium and tellurium containing derivatives $Fe_3(CO)_9SX$ (X = Se, Te) have been prepared recently *(382,383)*. $Fe_3(CO)_9S_2$ often appears as a by-product in

1

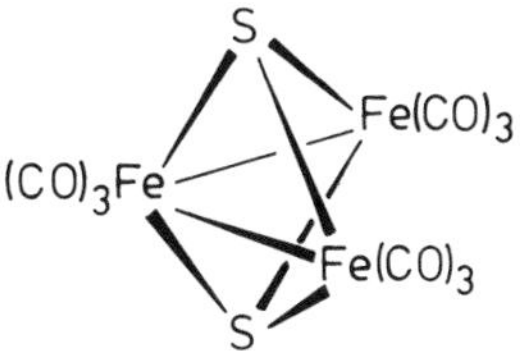

2

reactions between iron carbonyls and sulphur compounds, its formation was observed even in a gas pipeline used for methane and carbon monoxide *(106)*. $Fe_2(CO)_6S_2$ is transformed to $Fe_3(CO)_9S_2$ by $Fe_3(CO)_{12}$ *(332)*. Some other, less well defined carbonyliron sulphides have been reported *(160,283)*. $Fe_3(CO)_{10}(S_2CH_2)$ *(224)* was later found to be identical with $Fe_3(CO)_9S_2$ *(211)*.

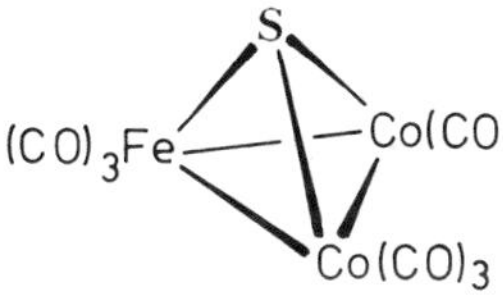

3

$FeCo_2(CO)_9S$ (3) *(415)* was first prepared by reacting thiophene with $Fe(CO)_5$ and $Co_2(CO)_8$ under "hydroformylation conditions" (200°C, 300 bar CO/H_2) *(255)*. It is a rather stable compound and is formed under such conditions even in the absence of added $Fe(CO)_5$, the iron furnished in this latter case by the corrosion of the autoclave walls. It may thus contaminate the products obtained by the hydroformylation of sulphur-containing olefins *(254)*. The yield of $FeCo_2(CO)_9S$ is almost quantitative if $Fe(CO)_5$, $Co_2(CO)_8$, and EtSH are reacted in stoichiometric ratio at 160°C and 200 bar CO/H_2 *(314)*; its formation has been observed also in other reaction mixtures containing iron and cobalt carbonyls *(326)*. In the substituted derivatives $FeCo_2(CO)_8L$ and $FeCo_2(CO)_7L_2$ (L = PPh_3, CN-*t*-Bu) the ligands are attached to the cobalt atoms *(55,73, 279,346)*. $[(\eta^5\text{-}C_5H_5)(CO)_2FeSMn(CO)_4]_2$ is a further example of a carbonyliron sulphide containing another transition metal, too *(438)*. Starting from $FeCo_2(CO)_9S$ two carbonyliron sulphides containing three different transition metals could be prepared: $FeCo_2(CO)_9SCr(CO)_5$ (4) *(375)* and $(\eta^5\text{-}C_5H_5)$-$MoFeCo(CO)_8S$ (5) *(376)*. Latter complex is the first chiral tetrahedral cluster compound.

A tetrahedral Fe_3S cluster structure has been proposed for $(\eta^5\text{-}C_5H_5)_3Fe_3(CO)_2(S)(SR)$ *(205,206)*. $(\eta^5\text{-}C_5H_5)_2Fe_2(S_2)$-$(SEt)_2$ (6) contains an Fe-S-S-Fe brigde with no direct inter-

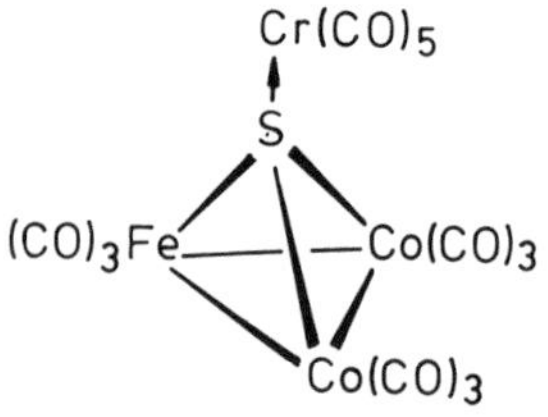

4

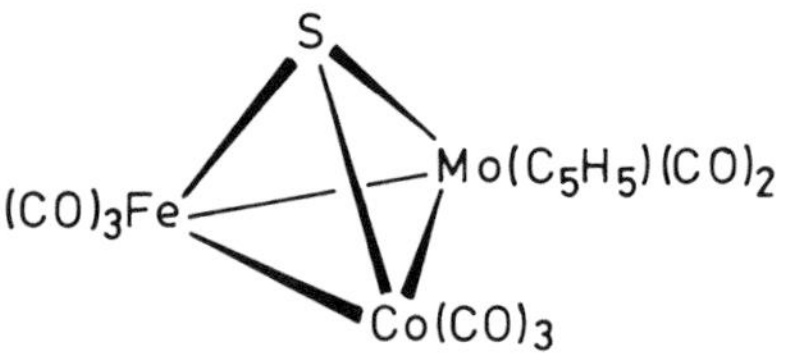

5

action between the iron atoms *(284,425)* but is nevertheless diamagnetic due to a coupling of the two odd electrons on the two iron atoms over the S_2-bridge *(442)*. It can be oxidized electrochemically to a paramagnetic monocation *(285)* in which the Fe•••Fe distance is reduced to 306 pm indicative of a one-electron bond *(442)*. In $[(\eta^5\text{-}CH_3C_5H_4)FeS_2]_2CO$ (7) the two S_2 ligands form a core of four sulphur atoms between the two iron atoms *(193)*.

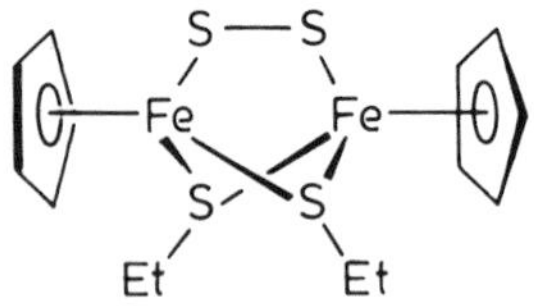

6

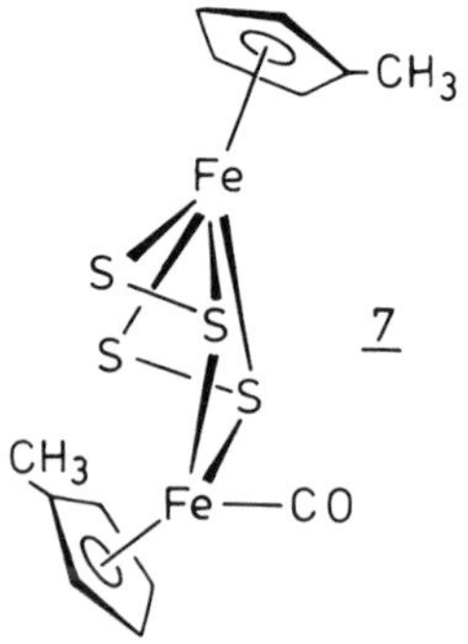

7

Sulphur *(399)* or cyclohexene sulphide *(450)* convert $[(\eta^5\text{-}C_5H_5)Fe(CO)_2]_2$ to $[(\eta^5\text{-}C_5H_5)FeS]_4$ (8), the structure of which has been verified by X-ray investigations *(399,450)*. $[(\eta^5\text{-}C_5H_5)Fe(CO)S]_2$ is formed as a by-product *(105,399)* in

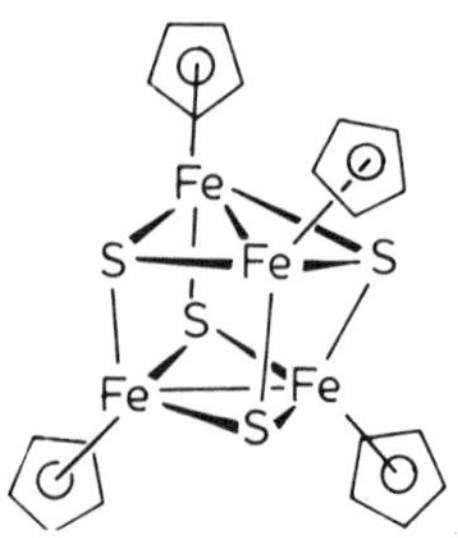

8

these preparation; the complex reported as [$(\eta^5$-$C_5H_5)$-$Fe(CO)_2]S_3$ *(430)* is probably identical with this. [$(\eta^5$-$C_5H_5)$-$FeS]_4$ can be oxidized electrochemically to a paramagnetic monovalent and a diamagnetic divalent cation *(182)* in which two of the long non-bonding Fe-Fe distances present in the neutral complex are significantly shortened indicating the removal of the electrons from an antibonding iron cluster orbital *(432,433)*.

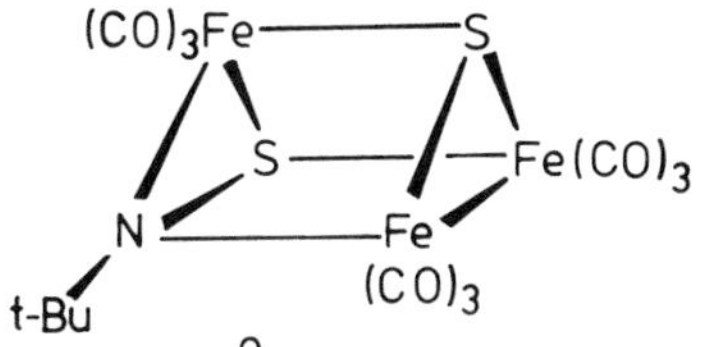

9

$Fe_3(CO)_9(S)(SN$-*t*-Bu) (9) *(319)* is a further example of an iron-sulphur cluster in which the sulphur atom as a four-electron donor is bonded to three iron atoms.

The complexes $[Fe_2(CO)_6(SR)]_2S$ (R = Me *(107)*, *t*-Bu *(132)*, CF_3 *(125)*, and C_6F_5 *(53)*) and 28 constitute examples in which sulphur is coordinated to four iron atoms. They appear as by-products formed in rather minor amounts, no efficient method for their synthesis is yet known.

A number of organoiron compounds containing the thiocyanato or isothiocyanato ligand or group have been prepared. The two isomers, $(\eta^5$-$C_5H_5)Fe(CO)_2(SCN)$ and $(\eta^5$-$C_5H_5)Fe(CO)_2(NCS)$ are the best investigated representatives *(31,410)*. The sulphur-bonded isomer transforms in the solid state or in nujol suspension to the nitrogen-bonded isomer, this isomerization does not take place in solution *(410)*. In some cases such as $Fe(CO)_2(PEt_3)_2(NCS)_2$ *(1)* and $[(PPh_3)_2N]^+[Fe(CO)_4(NCS)]^-$ *(388)* the bonding of the (NCS) ligand is unclear. The pentacarbonyliron derivative $Fe(CO)_5(SCN)_2$ contains seven-coordinated iron *(177,178,459)*, in $(\eta^5$-$C_5H_5)Fe(CO)(PPh_3)I\cdot(SCN)_2$ the sulphur containing ligand is probably NCS-I-SCN bonded through nitrogen to iron *(286)*, and $Fe(CO)_2(SCN)_2$ is polymeric *(198)*. In several complexes the SCN or NCS ligand is bonded to a heteroatom such as Sn *(32,50,336)*, P *(157)*, Hg *(317)*, or Ge *(337)* and in this way coordinated to iron.

The first iron complexes described to contain the thiocarbonyl group as a ligand were the salts of the $[(\eta^5$-$C_5H_5)$-$Fe(CO)_2(CS)]^+$ cation *(77,78)*. IR *(77)*, Mössbauer *(74)*, and ^{13}C-NMR spectra showed that the CS ligand is even a better π-acceptor than the CO ligand. This is in accordance with theoretical considerations *(373)*. Reacting this cation with hydrazine or the azide ion, $(\eta^5$-$C_5H_5)Fe(CO)_2(NCS)$ is formed *(79)*, proving the reactivity of the thiocarbonyl ligand against the attack of nucleophiles. In contrast to this, the

CO ligand is always substituted by simple two-electron donor ligands such as phosphines *(81)*. Sodium hydride reduction of $[\eta^5-C_5H_5)Fe(CO)_2(CS)]^+$ gives *cis-* and *trans-* $[(\eta^5-C_5H_5)Fe(CO)(CS)]_2$ both containing the CS groups in bridging position *(161)*. In the related complex $(\eta^5-C_5H_5)_2Fe_2(CO)_3(CS)$ with only one CS group *(361,444)* this ligand occupies again the bridging position; thus, these data show a clear preference of the thiocarbonyl group over the carbonyl group to act as a bridging ligand. The CS group may be easily alkylated to form stable S-alkyl derivatives, the structure of the ethyl derivative is shown in 10 *(444)*. The unusual $CSEt^+$ bridging group is partly similar to a bridging carbyne ligand.

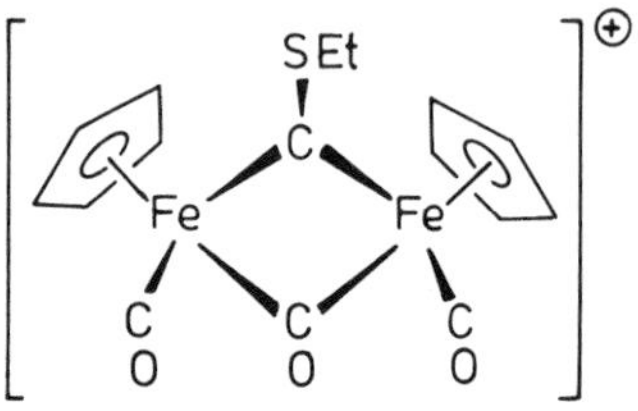

10

Reductive cleavage of $(\eta^5-C_5H_5)_2Fe_2(CO)_3(CS)$ with Na(Hg) gives the thiocarbonyl anion $[(\eta^5-C_5H_5)Fe(CO)(CS)]^-$ which reacts with Ph_3SnCl to form $(\eta^5-C_5H_5)Fe(CO)(CS)SnPh_3$ *(361)*.

Recently, $Fe(CO)_4(CS)$ could be prepared from $Na_2Fe(CO)_4$ and thiophosgene *(357)*, and thiocarbonyl complexes of iron-(II) porphyrins have been described *(306)*.

The complex $[(\eta^5-C_5H_5)Fe(CO)_2]_2SnS$ *(336)* contains the unusual SnS *(336)* ligand.

Organoiron compounds containing sulphur dioxide are obtained from the appropriate carbonyl *(58,75,98,99,183,368)* or phosphite *(202)* complexes. X-ray data *(321)* indicate structures with more or less distorted tetrahedral units for the dinuclear complexes $Fe_2(CO)_8(SO_2)$ *(321)*, $[(\eta^5-C_5H_5)Fe(CO)_2]_2$-$SO_2$ *(97,98)*, and $(\eta^5-C_5H_5)_2Fe_2(CO)_3(SO_2)$ *(99)*. If SO_2 is present in large excess, complex 11 containing a dithionite

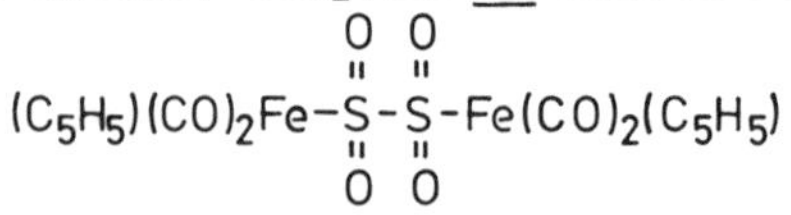

11

ligand is formed from $Na[(\eta^5-C_5H_5)Fe(CO)_2]$ *(422)*, whereas from the corresponding potassium salt the sulfinate anion $K^+[(\eta^5-C_5H_5)Fe(CO)_2(SO_2)]^-$ is produced *(234)*. The latter may be alkylated at sulphur or oxygen.

In the case of mononuclear $Fe-SO_2$ derivatives, a direct Fe-S bond can be assumed on the basis of IR and NMR data *(75)*.

The iron(II) fluorosulphate $Fe(CO)_4(SO_3F)_2$ is stable only in the solid state *(63)*. The structure of $Fe(CO)_3O(SO_2)_2$, formed in the reaction between $Fe(CO)_5$ and SO_3 *(453)*, is unclear. The same is true for $Fe(CO)(N_4S_4)$ *(62)*, which may be a polymer, and for $Fe_2(CO)_6S_2(GeCl_2)$ and $[Fe_2(CO)_6S_2]_2Ge$ both formed from $Fe_2(CO)_6S_2$ and adducts of $HGeCl_3$ *(333)*. $[(\eta^5\text{-}C_5H_5)Fe(CO)_2]_2\text{-}SnS_4$ probably contains a five-membered SnS_4 ring *(436)*.

Hydrolysis of $(\eta^5\text{-}C_5H_5)(CO)_2FeS(O)_2OMe$ yields the strong acid $(\eta^5\text{-}C_5H_5)(CO)_2FeSO_3H$, the first organometallic sulphonic acid known *(359)*. $(\eta^5\text{-}C_5H_5)Fe(CO)_2SP(S)F_2$ *(303)* contains the dithiodifluorophosphate group, $Fe(CO)_4[PMe_2P(S)Me_2]$ *(431)* tetramethyldiphosphine sulphide and $Fe(CO)_{5-n}(P_4S_3I_2)_n$ (n = 1,2) *(38)* diiodo-α-tetraphosphorous trisulphide as a monodentate ligand.

III. DERIVATIVES OF MERCAPTANS

The organoiron mercaptides with monothiols belong mainly to the following three classes of complexes: the mononuclear $(\eta^5\text{-}C_5H_5)Fe(CO)_2SR$, and the dinuclear $[Fe(CO)_3(SR)]_2$ and $[(\eta^5\text{-}C_5H_5)Fe(CO)(SR)]_2$ derivatives.

As mentioned before, $[Fe(CO)_3(SEt)]_2$ was described already in 1928 *(369)*, its first preparation was achieved by reacting an alkaline suspension of $Fe(OH)_2$ with EtSH and CO. Subsequent investigations have shown that this complex and the analogous alkyl- and arylthio-tricarbonyliron derivatives may be prepared by a number of ways, starting from $Fe_3(CO)_{12}$ *(42, 225,265,330)*, $Fe(CO)_5$ *(269,330,370)*, $Fe_2(CO)_9$ *(330)*, $Fe(CO)_4H_2$ *(176)* or $[HFe_3(CO)_{11}]^-$ *(330)* and mercaptans, organic disulphides or thioethers; some of them are even commercially available. This points to their increased stability and, in accordance with this, they are often observed as by-products in quite different reactions *(147,272,307)*. They frequently serve as model compounds for different spectroscopic investigations such as Mössbauer *(194,200,213,273)*, Raman *(401)*, ^{13}C-NMR *(288)*, ^{19}F-NMR *(170)*, mass *(162,264,354)* and combination IR spectra *(57)*.

The structure of $[Fe(CO)_3(SEt)]_2$ as determined by X-ray diffraction is shown in 12 *(120)*. As first shown in the case of the analogous methyl derivative *(195)*, most (but not all) of the $[Fe(CO)_3(SR)]_2$ complexes exist in two isomeric forms which are designed as *syn* and *anti* isomers and can be identified by their IR and NMR spectra. The structure of $[Fe(CO)_3(SEt)]_2$ in the solid state is that of the *anti* isomer (12, with axial and equatorial ethyl groups), the *syn* configuration (with both ethyl groups in the equatorial positions) is shown in 13. The two forms are stable enough for chromatographic separation *(54,259,312,313)* but isomerization proceeds slowly even at room temperature *(54,312,313)*, and in the case of the *t*-butyl derivative becomes fast in the NMR time scale above 90°C *(327)*. Little information is available on the mechanism of this process *(327)*. In general, the *anti* isomer seems to be the thermodynamically more stable form of these compounds *(54,212,312,313)*. More recently the somewhat similar

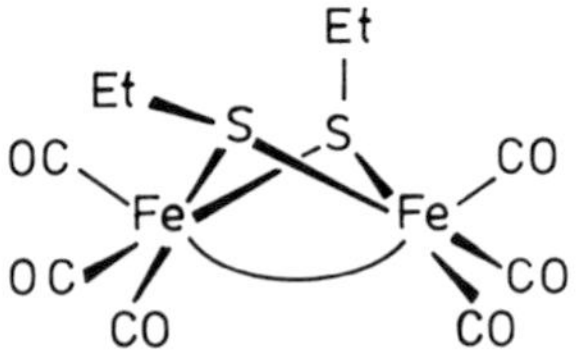

12

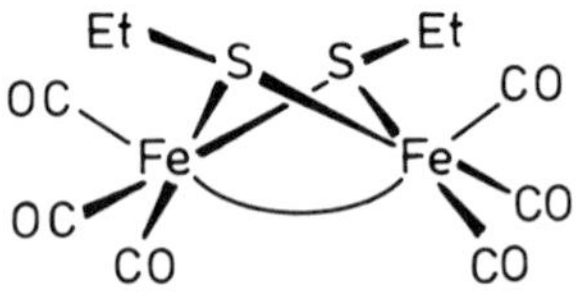

13

complexes $[Fe(CO)_3(SR)]_2C_4F_6$ (14) have been prepared which have both R groups in an axial position *(124,127,315)*. This configuration is probably made possible by the absence of the Fe-Fe bond. The cation $[Fe_2(CO)_6(SMe)_3]^+$ (15), contains three equatorial CH_3 groups *(398)*. The structure of the rather labile $Fe_2(CO)_5(S\text{-}i\text{-}Pr)_3I$ may be similar *(175)*.

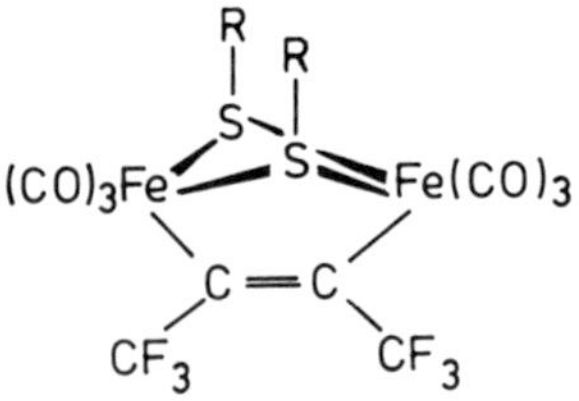

14

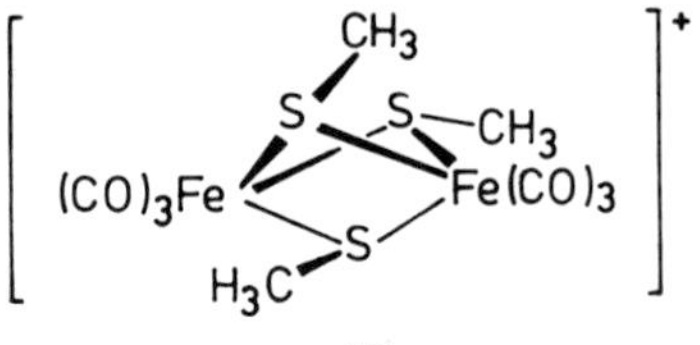

15

In the presence of free thiol, the SR-group is exchanged; this reaction takes place through the rupture of the Fe-S bonds *(287,331)*.

A few "mixed" complexes of type $Fe_2(CO)_6(SR)(SR')$ *(53,281, 365)* are also known. Structurally *(107)*, the complexes of type $[Fe_2(CO)_6(SR)]_2S$ *(53,107,132,133)* may be also regarded as members of this class.

The reaction of $[Fe(CO)_3(SR)]_2$ complexes with different phosphines and similar Lewis-base ligands leads to at least five different types of mono-, di-, and trisubstituted derivatives *(134,135,137,138,226)*. The kinetics of this substitution has been studied *(39)*. The *syn-anti* isomerism described in the case of the unsubstituted complexes may be observed with such derivatives, too *(2,125,126,291)*. The arsenic or phosphorous atom may be used to form a bridge to other transition metals as in $Fe_2(CO)_5(SMe)_2\text{-}AsMe_2\text{-}Mn(CO)_5$ and similar complexes *(374)*.

Reaction with halogens leads to the rupture of the "bent"

iron-iron bond - which is the HOMO orbital of such complexes *(423)* - and affords several types of halogen containing carbonyliron mercaptides, for example $[Fe(CO)_3(SEt)X]_2$ *(220)* or $[Fe(CO)_2(PPh_3)(SPh)]_2I^+$ *(207)*. Accordingly, $HgCl_2$ also reacts with this iron-iron bond forming adducts with an Fe-Fe-Hg three-centre bond such as $[Fe(CO)_2(PMe_3)(SMe)]_2 \cdot HgCl_2$ *(33)*. Oxidants generally lead to the rupture of the iron-iron bond and subsequently many different types of ligands may be inserted, *e.g.* SR in the presence of RSSR (leading to complexes of type 15 *(247)*), or F if $AgPF_6$ is used as an oxidant *(316)* (see 16).

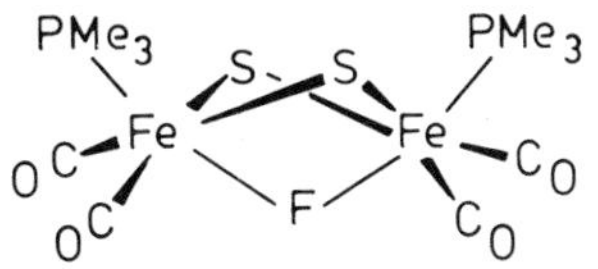

16

If two CO ligands are substituted by a more basic phosphine the iron-iron bond may even be protonated *(180)*. In $[Fe(CO)_2(SCF_3)]_2S$ *(203)* the Fe-Fe bond is presumably replaced by the bridging sulphur ligand. The electrochemistry of $[Fe(CO)_3(SR)]_2$ and similar complexes has been studied *(141, 143,145)*.

The mononuclear $(\eta^5\text{-}C_5H_5)Fe(CO)_2SR$ complexes are best prepared from $(\eta^5\text{-}C_5H_5)Fe(CO)_2Br$ and RSNa *(4)*. Alkynes R'C≡CR' may be inserted into the iron-sulphur bond to form $(\eta^5\text{-}C_5H_5)$-$Fe(CO)_2$-C(CR')=C(CR')(SR) derivatives *(127)*. The alkylthio derivatives readily dimerize - simultaneously loosing carbon monoxide - to dinuclear $[(\eta^5\text{-}C_5H_5)Fe(CO)(SR)]_2$ *(267)*; this reaction may be reversed by CO in tetrahydrofuran *(445)*. The tendency of dimerization is much less, however, with R = Ph and could not be achieved in the case of the C_6F_5 *(110,278)* and C_6Cl_5 *(253)* compounds. This trend has been ascribed to the decreased nucleophilic character of the sulphur atom in the latter compounds. The nucleophilic character of the sulphur atom enables the formation of the unusual sulphonium salt $\{[(\eta^5\text{-}C_5H_5)Fe(CO)_2]_2SMe\}^+Cl^-$ *(169)* and the carbonyliron derivatives $(\eta^5\text{-}C_5H_5)Fe(CO)_2S(R)Fe(CO)_4$ *(208,209)*. Two isomeric

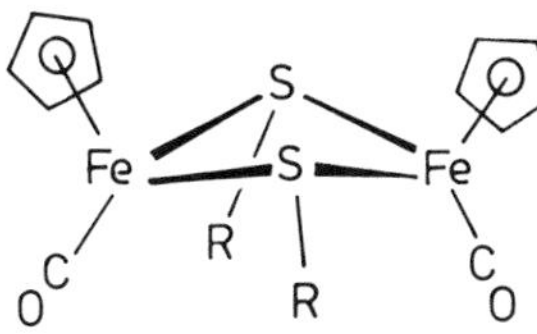

17

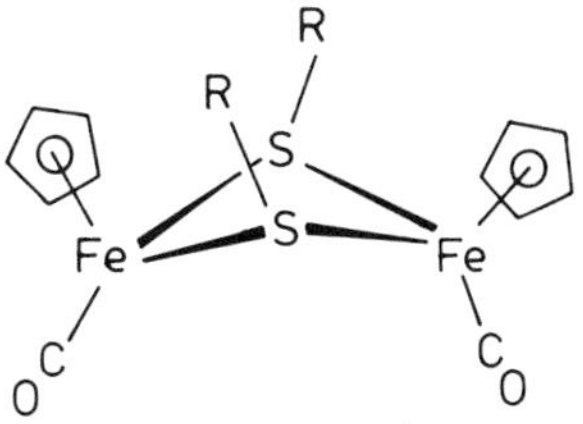

18

forms ("stable" and "unstable") of the dimers have been observed *(3,4,126,140,389)*, their structure is shown in 17 and 18. The structure of the stable isomer (R = Ph) has been proved by X-ray diffraction *(181)*, that of the unstable one is supported by its NMR spectrum *(140)*. The isomerization of the unstable form into the stable form is reversible *(445)* and has a rather complicated mechanism: it apparently proceeds in part by the inversion of the sulphur atom and in part through a dissociative process *(140)*.

The analogous cyclohexadienyl and cycloheptadienyl complexes and $(\eta^3\text{-}C_3H_5)Fe(CO)_3SCF_3$ *(277)* have also been prepared, they are much less stable than the cyclopentadienyl derivatives treated above.

$[(\eta^5\text{-}C_5H_5)Fe(CO)(SMe)]_2$ can be oxidized both electrochemically *(142,144)* and chemically *(108,109,136,139)* to the $\{[(\eta^5\text{-}C_5H_5)Fe(CO)(SMe)]_2\}^+$ cation, in which there is a formally one-electron bond between the two iron atoms. The structure of this species - which is of interest also with respect to the structures of oxidized and reduced ferredoxins - is shown in 19 *(100)*. The Fe•••Fe distance in the uncharged analogue 17 is 339 pm and points to no interaction between the two metal atoms; this distance is decreased to 293 pm in 19. This oxidation was shown to be reversible and rather general for $[(\eta^5\text{-}C_5H_5)Fe(CO)(SR)]_2$ complexes *(139,191,441)*. Stronger oxidants or higher anodic potentials lead to divalent cations of type $[(\eta^5\text{-}C_5H_5)Fe(CO)(SR)]^{2+}$ *(136,139,191)*. The monocation $[(\eta^5\text{-}C_5H_5)_2Fe_2(CO)_3(SEt)]^+$ has *cis* configuration and possesses a two-electron iron-iron bond *(173,174)*.

Using secondary or tertiary mercaptans trinuclear intermediates of the types $Fe_3(CO)_9(SR)_2$ and $Fe_3(CO)_9H(SR)$ could be isolated in the reaction with $Fe_3(CO)_{12}$ *(132,133)*; their structure is shown in 20 *(40)*. These are transformed by an excess of mercaptan to $[Fe(CO)_3(SR)]_2$, no such intermediates were observed with primary thiols, however.

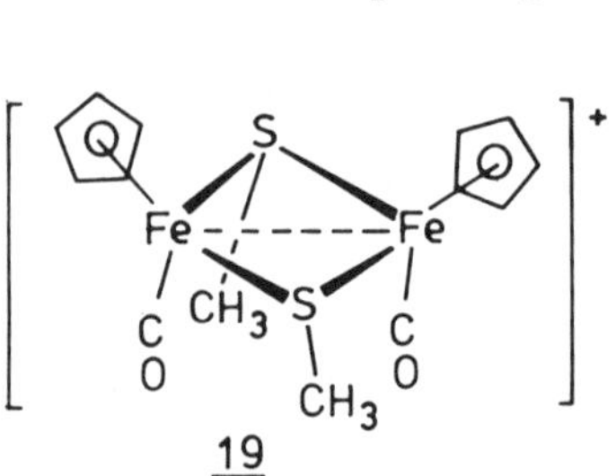

19

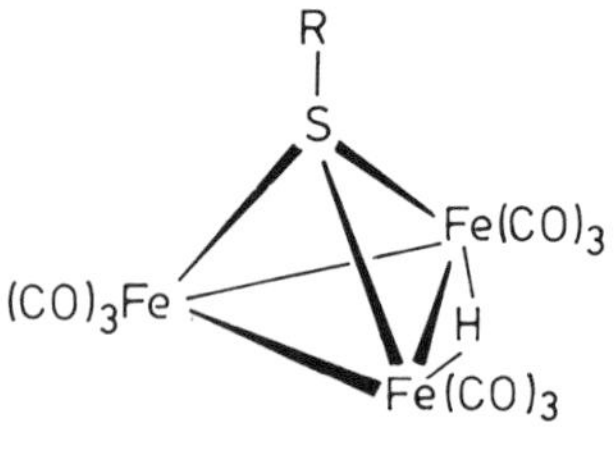

20

Organoiron complexes with saturated dithiols are represented by $[Fe(CO)_3(SCH_2)]_2$ *(171,404)* and $[Fe(CO)_3(SCF_2)]_2$ *(261)*, those of dithiophenols by $[Fe(CO)_3(1,2,4\text{-}S_2C_6H_3Me)]_2$ *(171)* and the tetrathionaphthalene derivative $C_{10}H_4S_4$-

$[Fe_2(CO)_6]_2$ *(424)*. $[(CO)_3FeS]_2CH_2$ may be regarded as the organoiron derivative of methane dithiol, $HSCH_2SH$, non-existent as a free molecule *(404)*. Most compounds of such types, however, are 1,2-dithiolene derivatives *(89)* with the -S-C=C-S grouping and possess, therefore, a 5-membered chelate ring with strongly delocalized π-electrons. Most of these are dinuclear complexes with the general formula $Fe_2(CO)_6(S_2C_2RR')$ (with Fe-Fe bond) or $[(\eta^5\text{-}C_5H_5)Fe(CO)_2]_2(S_2C_2RR')$ (without Fe-Fe bond) where R and/or R' may be H *(49,266,271,394)*, CH_3 *(394)*, aryl *(49,71,261,392,394,395,446)*, CF_3 *(245,246,261,270,324)*, and cyanide *(301)*; only a few mononuclear representatives such as $Fe(CO)_nL_m[S_2C_2(CF_3)_2]$ *(246,324,387)*, $Fe(CO)[S_2C_2(CF_3)_2]_2$ *(323)*, and $(\eta^5\text{-}C_5H_5)Fe[S_2C_2(CF_3)_2]$ *(323)* have been described. The mixed dithiolene-mercapto carbonyliron compounds $[Fe(CO)(SR)\text{-}S_2C_2(CF_3)_2]_n$ *(245)* are di- or tetranuclear complexes.

If *o*-aminothiophenol reacts with $Fe_3(CO)_{12}$, $Fe_2(CO)_6\text{-}(SNHC_6H_4)$ is formed *(308)* with the probable structure 21. 2-Mercaptopyridine reacts with $Fe_3(CO)_{12}$ to form 22 *(289)*, and the ligand benzene-1-thiophenyl-2-thiol transforms

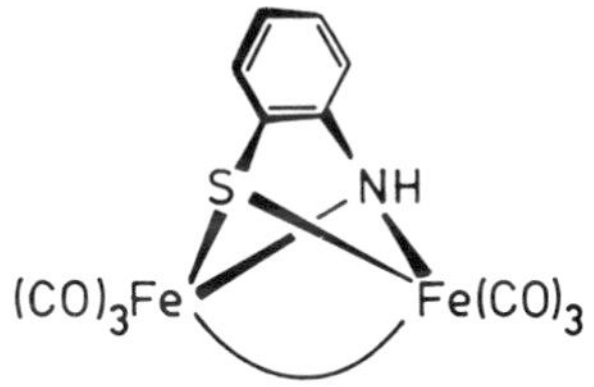

21

$[(\eta^5\text{-}C_5H_5)Fe(CO)_2]_2$ into $(\eta^5\text{-}C_5H_5)Fe(CO)SC_6H_4SC_6H_5\text{-}o$, a thiolate stabilized by a chelating thioether group *(257)*. A further example for a dinuclear organoiron mercaptide with another bridging ligand is $Fe_2(CO)_6(SPh)(PPh_2)$ *(242,468)* (23). Mixed organoiron-transition metal mercaptides such as $(\eta^5\text{-}C_5H_5)Fe(CO)_2(SR)W(CO)_5$ have been described *(168,169,170, 439)*.

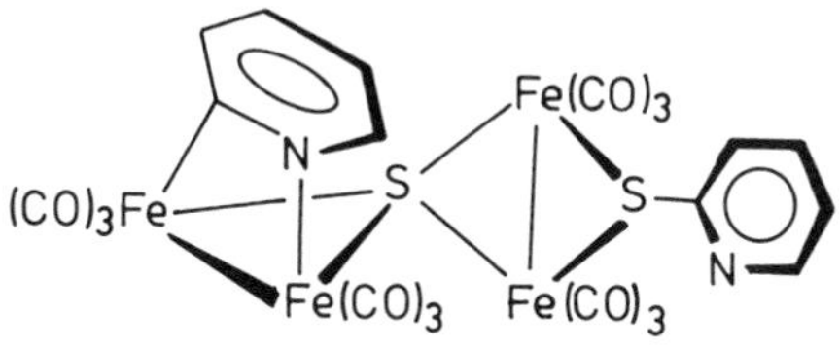

22

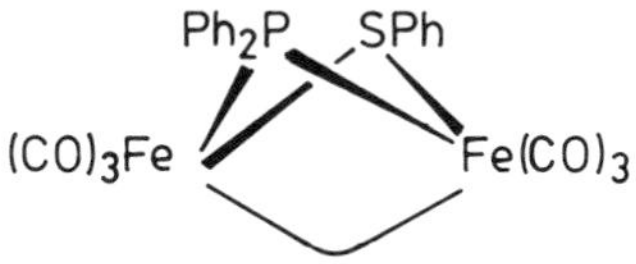

23

Tris(dimethylamino)borthiine reacts with $Fe_2(CO)_9$ to form $Me_2NBS_2Fe_2(CO)_6$ in which Me_2NBS_2 probably functions like a bridging dithiolato ligand *(349)*.

IV. SULPHINATES AND RELATED COMPOUNDS

The great majority of these complexes belongs to the
class of cyclopentadienyl-dicarbonyliron sulphinates and their
substituted derivatives, readily accessible by the insertion
of sulphur dioxide into the Fe-C bond of the corresponding
iron alkyls *(45,46,199,238,334,419,458)* (eq.[1]). The kinetics
of this reaction have been studied *(239)*. IR and X-ray *(102)*
work has shown that these organoiron sulphinates are S-sul-
phinates with the structure Fe-S(O)$_2$-R, but O-sulphinates with
the grouping Fe-O-S(O)-R were observed as intermediates of the
SO$_2$ insertion *(235,236)*. The reaction is best interpreted as
an electrophilic attack of the SO$_2$ molecule *(199,237,238)*
which leads through an iron-η^2-alkenesulphinate zwitterion
(93,94,159,381). It is stereospecific at carbon *(6,8,34,51,*

$$(\eta^5\text{-}C_5H_5)Fe(CO)(L)R + SO_2 \xrightarrow[(liq.)]{SO_2} (\eta^5\text{-}C_5H_5)Fe(CO)(L)SO_2R \qquad [1]$$

$$(L = CO, PR_3, P(OR)_3)$$

185,187, but see also *414)* but, in contrast to CO insertion
(454), proceeds mainly through an inversion at the carbon atom
(51,156,455). The insertion is also stereospecific at iron
(186,367), where it proceeds with retention of configuration
(95,322). It seems to be an irreversible process *(7,9,66,457)*.
    The insertion of sulphur dioxide is often accompanied by
a rearrangement of the σ-bonded organic group. Some of these
are usual allyl-rearrangements *(158,159,320,381)*, e.g., reaction [2], but in the case of alkinyl *(101,103,296,384,457)* or
cyclopropyl *(117,196)* groups the reaction takes a different
course (eqs. [3] - [5]). The organoiron sulphinate is not an
intermediate of this cyclization *(426)*.

$$(\eta^5\text{-}C_5H_5)Fe(CO)_2\text{-}CH_2\text{-}CH=CH\text{-}CH_3 \xrightarrow{SO_2}$$

$$(\eta^5\text{-}C_5H_5)Fe(CO)_2SO_2\text{-}CH(CH_3)\text{-}CH=CH_2 \qquad [2]$$

$$(\eta^5\text{-}C_5H_5)Fe(CO)_2\text{-}CH_2\text{-}C\equiv C\text{-}CH_3 \xrightarrow{SO_2} (\eta^5\text{-}C_5H_5)(CO)_2Fe\text{-}C\underset{CH_2\text{-}O}{\overset{C\text{-}S=O}{\diagup\diagdown}} \qquad [3]$$

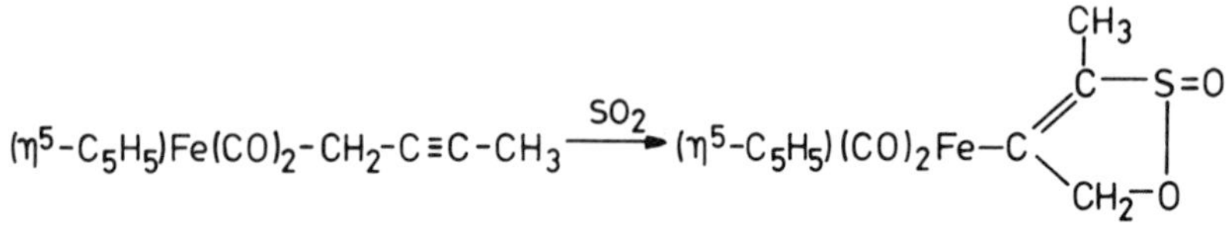

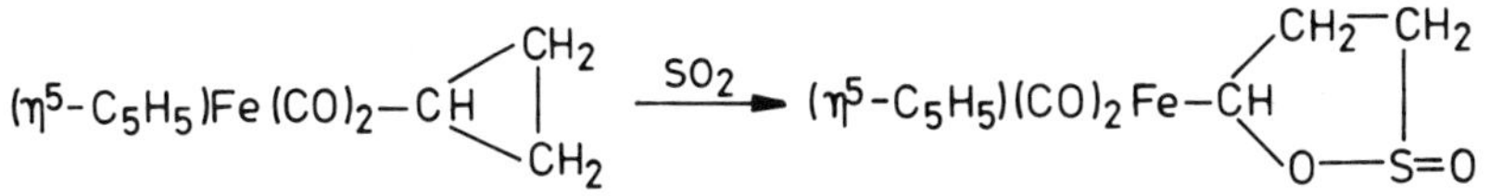 [4]

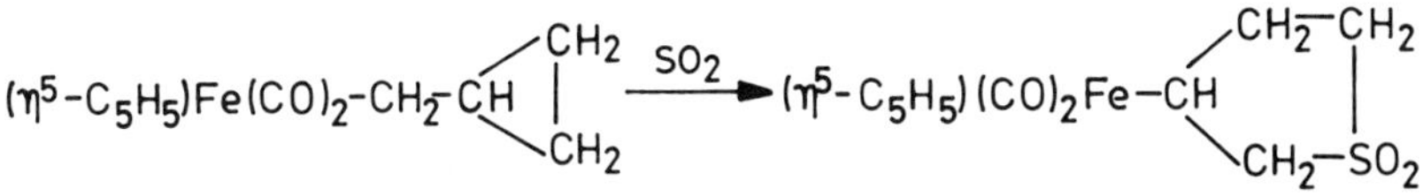 [5]

These reactions are not confined to sulphur dioxide, analogous compounds were obtained by using *N*-sulphinyl-aniline *(377)*, *N*-sulphinylsulphonamides *(403)*, bis(methylsulphonyl)-sulphur diimide *(403)*, or sulphur trioxide *(295,385)*.

If the organoiron compound contains an iron-tin-carbon bond, SO_2 is inserted between the tin and carbon atoms *(47,70, 163,164,371)* and *O*-sulphinates are formed.

In addition to SO_2 insertion, the cyclopentadienyl-dicarbonyliron sulphinates may be prepared by several other methods, too *(45,46,66,298,299,366,458)*. These may be useful if the parent alkyl (aryl) iron compounds do not react with sulphur dioxide, *e.g.*, reaction [6]. Mössbauer *(233)*, ^{19}F-NMR *(65)*, and ^{1}H-NMR *(263)* spectra of $(\eta^5\text{-}C_5H_5)Fe(CO)_2S(O)_2R$ complexes have been reported.

Only one rather labile organoiron sulphinate, $Fe(CO)_4Cl(SO_2CF_3)$ has been described *(23,299)*, which does not contain a cyclopentadienyl ligand.

Cyclopentadienyl-dicarbonyliron alkylsulphite complexes $(\eta^5\text{-}C_5H_5)(CO)_2FeS(O)_2OR$ have been prepared recently *(359)*.

$$Na[(\eta^5\text{-}C_5H_5)Fe(CO)_2] + C_6F_5\text{-}SO_2Cl \longrightarrow (\eta^5\text{-}C_5H_5)Fe(CO)_2\text{-}SO_2\text{-}C_6F_5 + NaCl \qquad [6]$$

V. MISCELLANEOUS SULPHUR CONTAINING LIGANDS

A. CARBON DISULPHIDE AND ITS ORGANIC DERIVATIVES

$Fe_2(CO)_9$ and (benzylideneacetone)tricarbonyliron readily react with carbon disulphide; if this reaction is carried out in the presence of tertiary phosphines, complexes of type $Fe(CO)_2L_2(CS_2)$ can be isolated *(37,60)*. The bonding of the CS_2 ligand is shown in <u>24</u> *(292)*. The C=S group in these complexes is a strong nucleophile and can be coordinated to another transition metal atom forming binuclear derivatives with a bridging CS_2 ligand *(352)* such as 25.

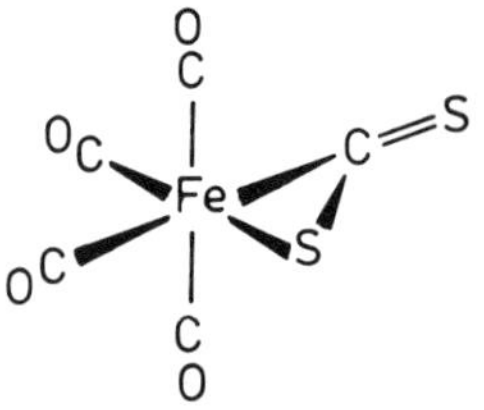

24

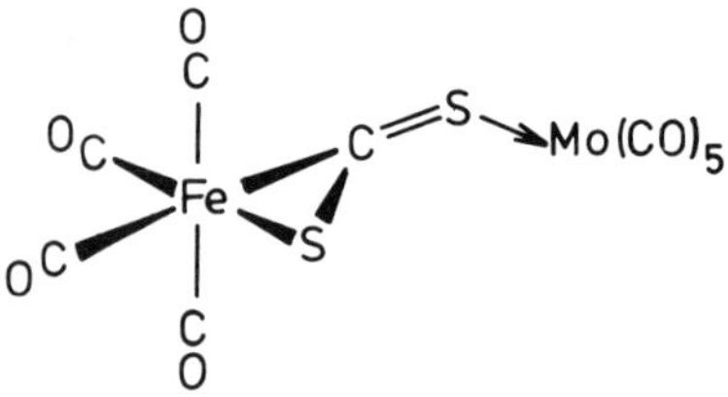

25

A different type of a bridging CS_2 ligand can be found in $[(\eta^5\text{-}C_5H_5)Fe(CO)_2]_2CS_2$ (26) which may be regarded as a monodentate metaldithiocarboxylate *(172)*.

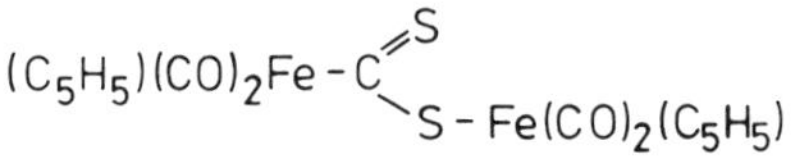

26

Carbene complexes $Fe(CO)_2L_2(CS_2C_2R_2)$ are formed from $Fe(CO)_2L_2(CS_2)$ [$L = P(OMe)_3$] and activated alkynes *(293)*, whereas the CS_2 ligand is selectively eliminated by $NO^+PF_6^-$ *(429)*.

Complex 27, formally a derivative of dithioformic acid, has been obtained from an irondicarbollide and CS_2 *(189)* (eq. [7]).

$$K[(1,2\text{-}B_9C_2H_{10})_2Fe] + CS_2 \xrightarrow[CS_2 \text{ solvent}]{HCl/AlCl_3} [(1,2\text{-}B_9C_2H_{10})_2S_2CH]Fe \qquad [7]$$

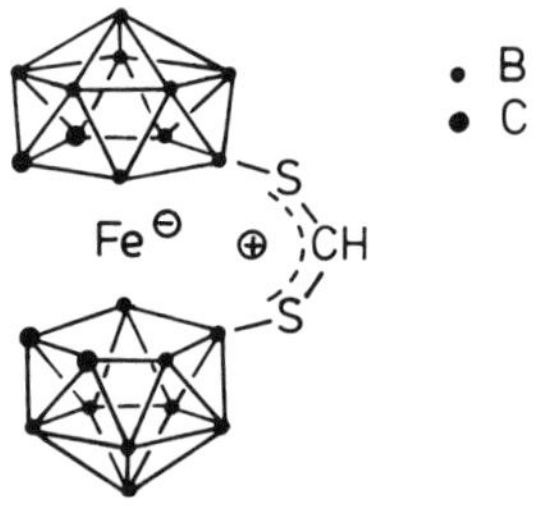

27

Dialkyldithiocarbamate groups may function as one-electron unidentate and as three-electron bidentate ligands, as exemplified by $(\eta^5\text{-}C_5H_5)Fe(CO)_2(SCSNMe_2)$ *(351)* or $[Fe(SCSNEt_2)(CO)_2(SMe)]_2$ *(83)* and $(\eta^5\text{-}C_5H_5)Fe(CO)(SSCNR_2)$ *(112,*

460) or $Fe(SSCNR_2)_2(CO)_2$ *(83,372,460)*, respectively. Alkyl- or aryltrithiocarbonate (RSCSS) *(67)* and dithiocarboxylate (RCSS) *(80,82)* groups show a similar behaviour. The chiral $(\eta^5\text{-}C_5H_5)Fe(CO)(S_2CNH\text{-}i\text{-}Pr)$ rapidly racemizes at room temperature as a consequence of the rotation around the $S_2C\text{-}NH\text{-}i\text{-}Pr$ bond *(68)*.

The monodentate, sulphur bonded dialkyl thiocarbamate $(\eta^5\text{-}C_5H_5)(CO)_2Fe(SCONC_4H_8)$ could not be decarbonylated to a derivative containing a bidentate thiocarbamate ligand *(412)*.

Dimethylthiocarbamoyl chloride, Me_2NCSCl, reacts with several carbonyliron complexes *(128)* to yield compounds containing the thiocarboxamide ligand, as in 28 *(131)* and 29 *(129)* or the chelating dicarbene ligand $(Me_2NC)_2S$ as in 30 *(130)*.

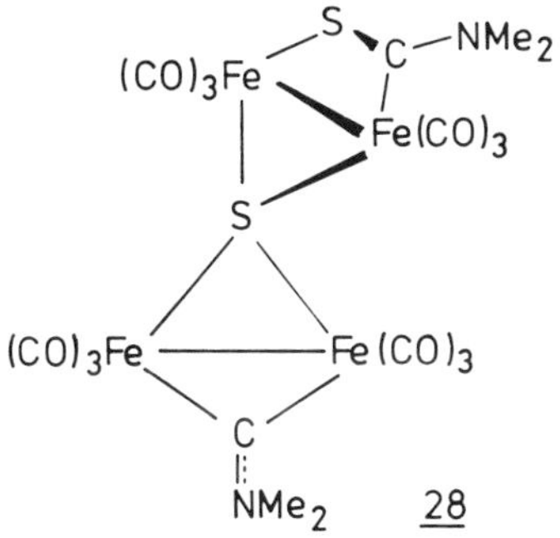

28

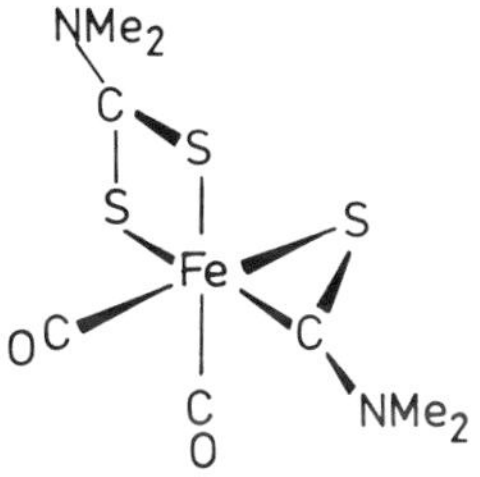

29

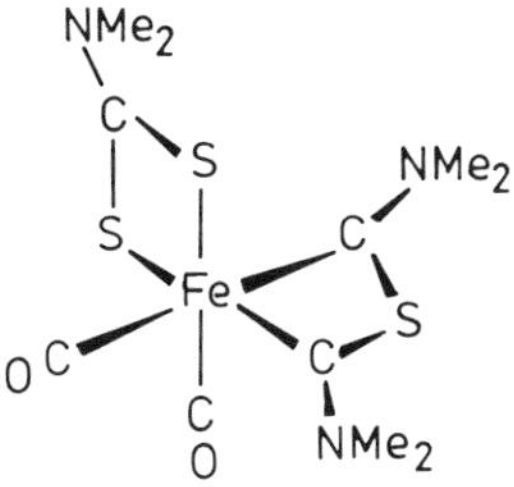

30

Thioamides $R\text{-}C(S)\text{-}NR_2$ and thioureas $R_2N\text{-}C(S)\text{-}NR_2$ form several types of carbonyliron derivatives in which these molecules are bonded either through the sulphur atom, or through both the sulphur and nitrogen atoms to iron *(17,69)*.

The carbon disulphide ligand can be converted into the thiocarbonyl ligand by the reaction sequence [8] *(78,155)*.

$$[(\eta^5\text{-}C_5H_5)Fe(CO)_2]^- \xrightarrow{CS_2} [(\eta^5\text{-}C_5H_5)Fe(CO)_2(CS_2)]^- \xrightarrow{MeI}$$

$$(\eta^5\text{-}C_5H_5)Fe(CO)_2(CSSMe) \xrightarrow{H^+} [(\eta^5\text{-}C_5H_5)Fe(CO)_2(CS)]^+ + CH_3SH \qquad [8]$$

The last step may be reversible, since the CS ligand
reacts with different nucleophiles (e.g., RO^-, RNH_2) to give
thioesters and thiocarboxamides (79) (eq. [9]).

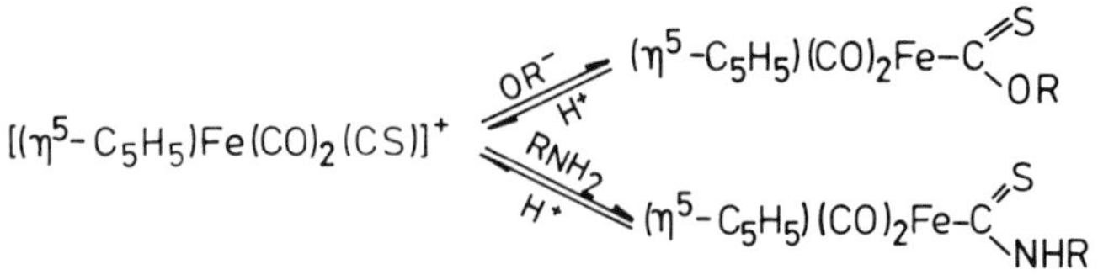

$$[(\eta^5-C_5H_5)Fe(CO)_2(CS)]^+ \xrightarrow[\substack{RNH_2 \\ H^+}]{\substack{OR^- \\ H^+}} \begin{cases} (\eta^5-C_5H_5)(CO)_2Fe-C\overset{\displaystyle S}{\underset{\displaystyle OR}{<}} \\ \\ (\eta^5-C_5H_5)(CO)_2Fe-C\overset{\displaystyle S}{\underset{\displaystyle NHR}{<}} \end{cases} \qquad [9]$$

These additions may be reversed by acidification. Since
the thioesters are also accessible starting from
$[(\eta^5-C_5H_5)Fe(CO)_2]^-$ and chlorothioformates [ClC(S)OR], this
opens another route for the formation of the thiocarbonyl ligand (77).
     The complexes $(\eta^5-C_5H_5)Fe(CO)_2(CSOPh)$ (262) and
$\{Fe(CO)_3[SC(NH)C_6H_4O-p-Me]\}_2$ (20) are examples for thiocarboxylic and thiocarboxamide groups bonded through their sulphur atoms to iron.

B.   COMPOUNDS WITH IRON-SULPHUR BONDS

     Examples with simple thioethers are cations of type
$[(\eta^5-C_5H_5)Fe(CO)_2(SR_2)]^+$ (154,244,456), $(\eta^5-C_5H_5)Fe(CO)-$
$(SEt_2)CH_3$ (188), several (1,3-dithiane)Fe(CO)$_4$ derivatives
(84,111) and $Fe(CO)_4[S(CH_2)_4]$ (417) (the latter two identified
only in solution). Bidentate ligands are more suitable for the
formation of stable organoiron complexes with coordinative
sulphur-iron bonds. Carbonyliron derivatives containing 2,5-
dithiahexane (61,64,215), diphenyl disulphide (221),
$(MeS)_2C_2Ph_2$ (392), or the $MeSC_2H_4CO$ group (268) have been
described. Dimethyl sulphoxide also forms an S-bonded compound
$Fe(CO)_4(SOMe_2)$ if it reacts with $Fe(CO)_5$ under the influence
of irradiation (416,417); in the thermal reaction an ionic
complex with O-bonded DMSO, $[Fe(OSMe_2)_6]$ $[Fe_4(CO)_{13}]$, is
formed (223). The complexes $(\eta^5-C_5H_5)Fe(CO)(COCH_2C_6H_{11})$(DMSO)
(454), [Fe(CO)(phthalocyanin)(DMSO)] (192) and $[(\eta^5-C_5H_5)_2-$
$Fe_2(CO)_3(DMSO)]$ (10) have been prepared and characterized by
IR or NMR only in solution.
     The observation that carbon monoxide inhibits the oxida-
tion of cysteine catalysed by ferrous ions (113) has led to
the discovery of bis(L-cysteinato)dicarbonyliron(II),
$Fe(CO)_2(SCH_2-CH-NH_2-CO_2H)_2$ (114,115), later prepared also in
a crystalline form (397). It has a high molecular rotation
value (21,000°) suggesting that its optical activity is due
not only to the asymmetry of the cysteine ligands, but also
to a configurational effect (428). The complex is light sensi-

tive and looses both of the CO ligands on irradiation *(114, 115,427)*.

Compounds with similar properties have been synthesized by using benzene-1-thiomethyl-2-thiol *(402)* or tetradentate N_2S_2 ligands *(252)* instead of cysteine.

Reacting $Ni(S_2Me_2C_2Ph_2)_2$ with $Fe(CO)_5$ under the influence of light, a thioketocarbene intermediate is stabilized in the form of a hexacarbonyldiiron complex with the structure 31 *(396)*. The analogous parent compound with hydrogen atoms in the place of the phenyl groups has been prepared by a different route *(392)*.

The thioacrolein complex $(\eta^4\text{-}C_3H_4S)Fe(CO)_2(PPh_3)$ (32) *(148,210,420)* is another example for an unstable sulphur containing organic molecule stabilized by complexation. It may be dimerized to a compound with Fe_2S_2 array and oxidized to an S-oxide *(148)*.

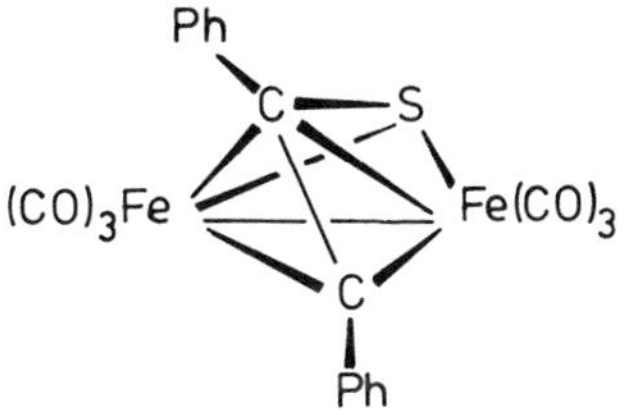

31

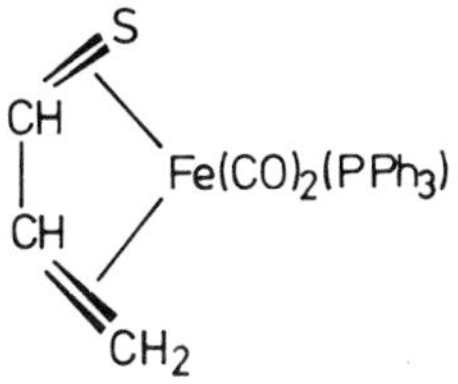

32

The dinuclear structural unit 33 *(41)* is frequently formed by reacting iron carbonyls with quite different sulphur compounds such as vinyl sulphides *(276)*, thianaphthene *(275)*, and dithienyl *(290,308)*, *etc.*. For example, thiomaleic anhydride and $Fe(CO)_5$ react to form 34 *(227)*. $CH_3C_4H_3SFe_2(CO)_6$ (35) is formed (as a by-product) from 2-methylthiophene *(231)* and $(CH_3)_4C_3SOFe_2(CO)_6$ (36) from tetramethyl thietanone *(118)*.

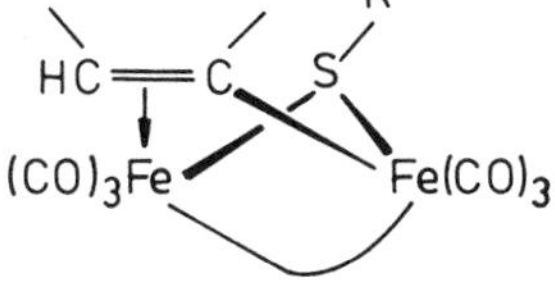

33

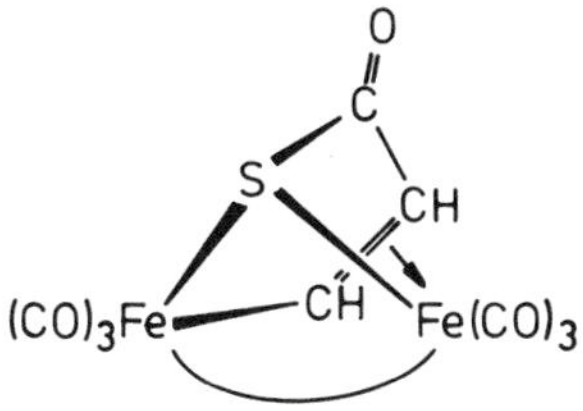

34

Reacting thioketones with $Fe_2(CO)_9$ several dinuclear carbonyliron derivatives are formed *(16,18,28)*. These include in the case of aromatic thioketones *ortho*-metallated species as

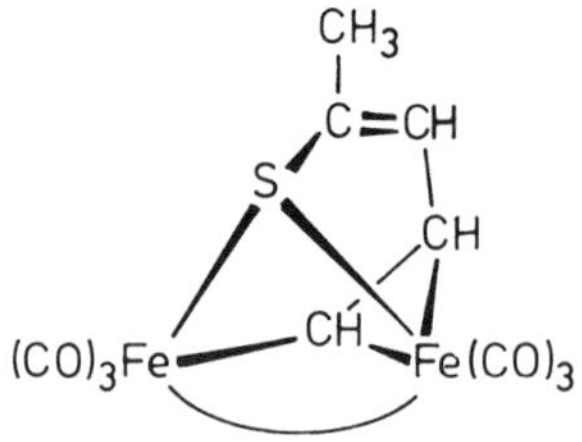

35

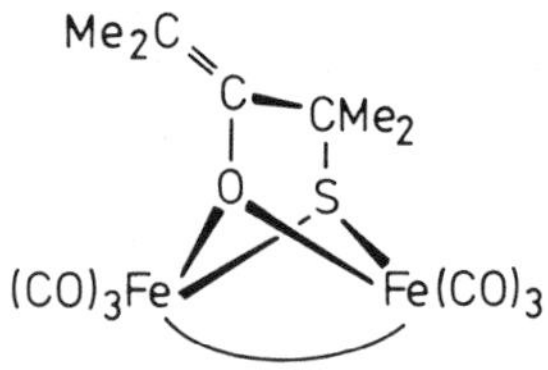

36

shown by 37 *(13,14,16,19,88)*. Diaryl sulphines yield the same complexes *(12)*. Carbon monoxide insertion may be observed, too *(28)*. These complexes may be transformed into sulphur-free

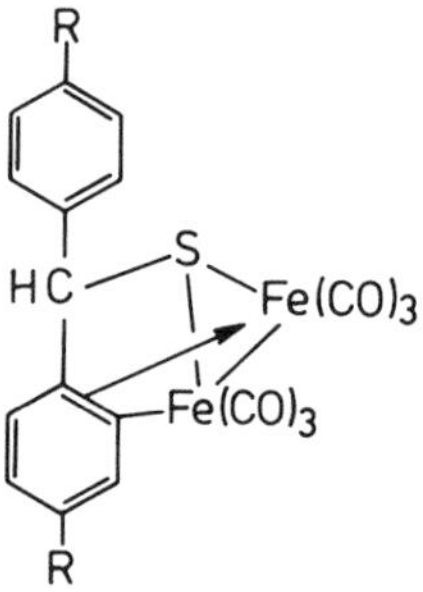

37

lactones *(25)* and are cleaved by mercuric acetate to ethers and esters *(26)*. A similar *ortho*-metallation has been observed in the reaction between thiocarboxylic acid esters and $Fe_2(CO)_9$ *(15,21)*.

In contrast to the above thioketones, diphenylcyclopropenethion, $Ph_2C_2C{=}S$, reacts with $Fe_2(CO)_9$ as a Lewis base to yield the simple monosubstituted tetracarbonyliron derivative $Fe(CO)_4(SCC_2Ph_2)$ *(146)*. A thioketone complex has been described, too *(43)*. The dithioacetylacetonate complexes $Fe(CO)(L)(S_2C_5H_7)$ (L = CO, pyridine) may be regarded as non-heme iron protein models *(60)*.

The preparation of the neutral thiophene complex $Fe(CO)_2(C_4H_4S)$ *(76)* has not been confirmed, but a few cationic organoiron thiophene derivatives have been described *(36,56)*.

The dinuclear compounds 38 *(121,319)* and 39 *(319)* are formed from $Fe_2(CO)_9$ and the corresponding diimido sulphur derivative.

In the 1,2,5-thiadiborolene (L) complexes the planar heterocyclic ring functions either as a four-electron ligand like in $Fe(CO)_3L$ compounds *(5,405,406)*, or it may be part of a tripeldecker sandwich *(407)*.

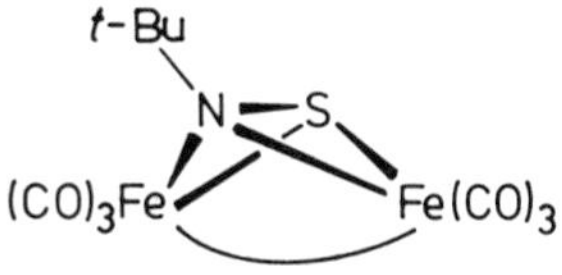

38

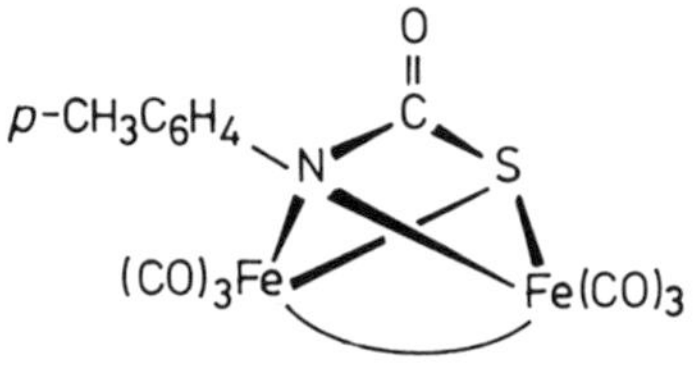

39

C. *COMPOUNDS WITHOUT IRON-SULPHUR BONDS*

(η^5-C_5H_5)Fe(CO)$_2$R compounds, in which R contains a thioether group, have been described *(110,266,268,294,348)*. One of these, (η^5-C_5H_5)Fe(CO)$_2$-CH_2-CH_2SMe, if irradiated is transformed into the acyl derivative 40, a reaction which is an example for carbon monoxide insertion as a result of intramolecular nucleophilic attack *(266,268)*. In an analogous reaction (η^5-C_5H_5)Fe(CO)$_2$C(CF_3)=C(CF_3)(SC_6F_5) is transformed to 41 *(127)* which reacts with hexafluorobut-2-yne to give (η^5-C_5H_5)-Fe[(η^4-C_4(CF_3)$_4$S(C_6F_5)]. Tetramethyl thiacycloheptyne, another ligand with a thioether group, is bonded as an alkyne to the iron atoms in ($C_{10}H_{16}S_2$)$_2$$Fe_2(CO)_4$ (which complex contains an

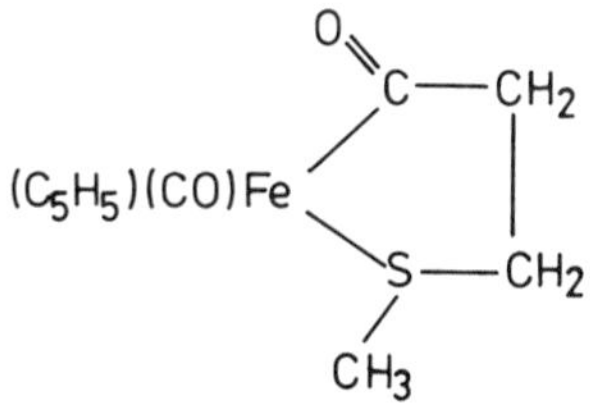

40

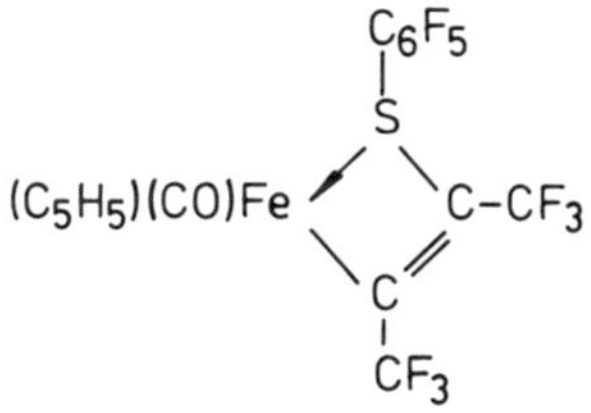

41

iron-iron double bond *(391)*) and dimercaptoheptafulvenes act as η^4-diolefin ligands to form LFe(CO)$_3$ complexes *(197)*.

Carbonyliron porphyrins with a thioether group in a side chain may serve as models for cytochrome C *(72)*. A thioanisolyl ligand is σ-bonded to iron in Na[PhSCH_2Fe(phthalocyanin)]•4 THF *(421)*.

The reaction of (η^4-butadiene)Fe(CO)$_3$ with liquid sulphur dioxide in the presence of boron trifluoride leads to 42

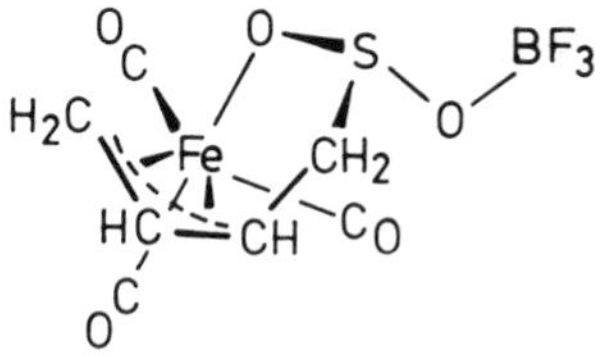

42

(100,104). This unusual complex may be regarded as an intermediate of a Friedel-Crafts sulphination.

Sulphonated derivatives of (η^4-butadiene)-, (η^4-cyclobutadiene)-, and (η^4-cyclohexadiene)tricarbonyliron are known *(48,309,310,311)*. The *O*-tosylate of (7-norbornadienyl)tricarbonyliron *(232)* and the *N*-tosylates of several carbonyliron derivatives with *N*-heterocycles *(59,85,379)* may be mentioned here.

Organoiron compounds with sulphur-containing ligands π-bonded to the iron atom are represented by (η^4-thiophene sulphone)$Fe(CO)_3$ complexes *(96,165,166,204,228,393,437,451, 464,465)*, $(C_3H_4SO_2)Fe(CO)_4$ (43) *(318)*, the η^2-thiophene sulphone derivative 44 *(204)*, the dihydrothiophene oxide complex 45 *(166,167,440)*, $Fe(CO)_4(R\text{-}CO\text{-}CH{=}CH\text{-}SO_2R')$ *(29,342)*, $[(\eta^5\text{-}PhS\text{-}C_5H_4)Fe(\eta^6\text{-}C_6H_6)]^+(PF_6)^-$ *(344)*, and derivatives of the $[(\eta^5\text{-}C_5H_5)Fe(\eta^6\text{-}C_6H_6)]^+$ cation with thioether groups on

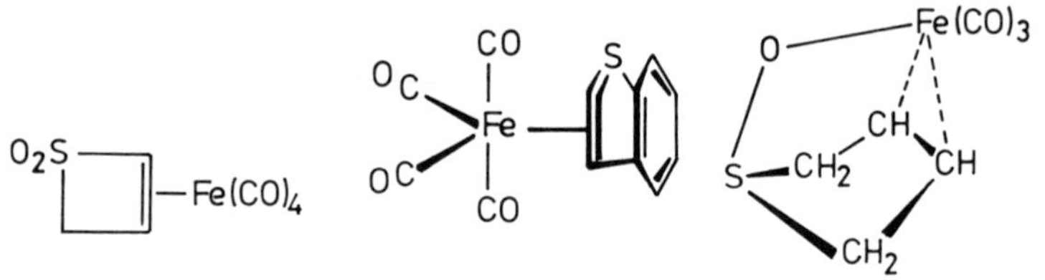

43 44 45

the benzene ring *(343,345,408,409)*. One of these, $(C_4H_4SO_2)Fe(CO)_3$ (46), contains a ligand existing as a free molecule only in solution.

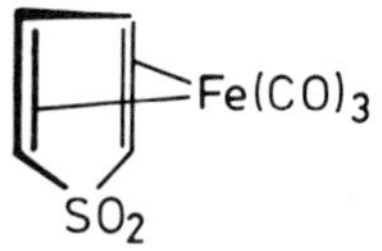

46

Some complexes contain a thiophene ring σ- *(30,338,339, 363)* or π-bonded *(184,443)* to the iron atom, in others these two are linked by a methylene group *(274)*, a carbonyl group, *(338)* or an η^4-cyclohexadiene unit *(243,251,305)*.

$Fe(CO)_4L$ type complexes with sulphur containing carbenes as ligands have been synthesized *(190,201,325)* and sulphur ylides may also function as ligands *(304,447)*.

Recently a fairly large number of γ-lactames with the general formula 47 starting from $(\eta^5\text{-}C_5H_5)Fe(CO)_2$(alkyl) complexes and isocyanates (XNCO) have been prepared in which X is a sulphur containing group such as SO_2Cl, SO_2NR_2, tosyl, and others *(116,195,362,461,462,463)*. Analogous reactions *(116, 195)* lead to γ-sultams or sulphonic acids, 48 and 49. Benzothiazole *(340,341)* and a 1,3,4-thiadiazoline derivative *(214)*

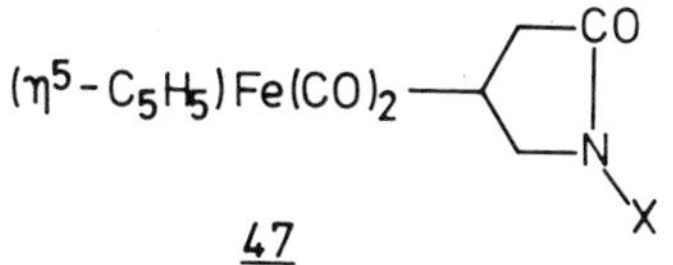

47

form organoiron complexes in which these ligands are bonded through nitrogen to iron.

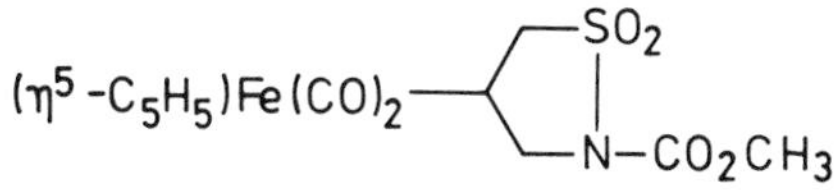

48

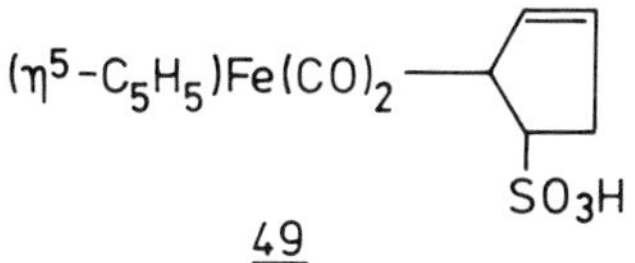

49

Several cyclopentadienyl-dicarbonyliron derivatives containing iron-tin, iron-germanium or iron-phosphorous bonds and with sulphur-containing groups bonded to Sn,Ge or P are known. Those which have not yet been mentioned include $[(\eta^5\text{-}C_5H_5)\text{-}Fe(CO)_2]_2[Sn(SPh)_2]$ *(336,360)*, $[(\eta^5\text{-}C_5H_5)Fe(CO)_2]_2[Ge(SEt)_2]$ *(335)*, and $(\eta^5\text{-}C_5H_5)Fe(CO)_2P(S)R_2$ [R = CH_3 *(358)*, CF_3 *(151, 152)*].

VI. SELENIUM AND TELLURIUM CONTAINING LIGANDS

Several organoiron complexes containing Se or Te have been prepared and characterized; these compounds are compiled in Table 1. Since all of these Se or Te compounds are carbonyliron derivatives and for all of them the sulphur-containing analogues are known too, discussion in this section will primarily deal with the comparison of the properties of these chalcogenides as ligands in carbonyliron complexes.

The trend of ν(CO) values in $Fe_3(CO)_9X_2$ (X = S, Se, Te) complexes show a decreasing π-acceptor ability of the chalcogenides in this order *(87)*. Accordingly, the CO ligands most easily dissociate in the case of the sulphur containing compound and are substituted by an S_N1 mechanism. On the other hand, the "soft" character of the Te atom leads to an S_N2 mechanism characterized by the formation of adducts *(86,87)* according to reaction [10].

$$Fe_3(CO)_9Te_2 + L \rightleftharpoons Fe_3(CO)_9(L)Te_2 \rightleftharpoons Fe_3(CO)_8(L)Te_2 + CO \qquad [10]$$

The ligand L (CO, PBu_3, *etc.*) is presumed to coordinate initially to the Te atom.

The same trend of the π-acceptor properties has been observed in the case of $[Fe(CO)_3XR]_2$ (X = S, Se, Te) complexes *(281)*.

The competitive reaction of $(C_6F_5)_2S_2$ and $(C_6F_5)_2Se_2$ with $Fe_3(CO)_{12}$ leads to the predominant formation of $[Fe(CO)_3\text{-}(SC_6F_5)]_2$ *(281)*, and the tendency for dimerization of the $(\eta^5\text{-}C_5H_5)Fe(CO)_2XPh$ complexes decreases in the order S>Se>Te *(389)*. Both results indicate the diminishing of σ-donor and π-acceptor ability.

Selenium- and nitrogen-bonded organoiron-selenocyanate isomers have been prepared by reaction [11] *(240)*.

$$(\eta^5\text{-}C_5H_5)Fe(CO)(PPh_3)(\text{-}CH_2Ph) \xrightarrow{Se(SeCN)_2}$$
$$\underset{(48\%)}{(\eta^5\text{-}C_5H_5)Fe(CO)(PPh_3)NCSe} + \underset{(17\%)}{(\eta^5\text{-}C_5H_5)Fe(CO)(PPh_3)SeCN} \qquad [11]$$

As observed with the analogous thiocyanate complexes *(410)*, the stability of the *N*-bonded isomer is enhanced by an increase of the electron density at the central metal atom *(179)*. Accordingly, the *N*-bonded isomer of the unsubstituted derivative $(\eta^5\text{-}C_5H_5)Fe(CO)_2SeCN$ is not known *(241)*.

The insertion of SeO_2 into the iron-carbon bond leads to $(\eta^5\text{-}C_5H_5)Fe(CO)_2Se(O)_2CH_3$ *(302)*.

VII. APPLICATION OF CARBONYLIRON REAGENTS IN THE CHEMISTRY OF ORGANIC SULPHUR COMPOUNDS

Iron carbonyls have been successfully applied for the elimination of sulphur *(20,24,122,123,248,328,329,355,356,386,*

434,435,467,471), halogen *(11,230,282,297,300,437)*, or oxygen *(22,122,123,282)* from organic sulphur compounds, and some carbonyliron derivatives could be used as starting materials for the synthesis of sulphur (or selenium) containing heterocycles *(16,28,230,282,470)*. These diverse uses of iron carbonyls in organic sulphur chemistry are compiled in Table 2.

Raney nickel is widely used for the desulphurization of organic sulphur compounds, and due to the principal similarities of finely divided metals and metal carbonyls the use of iron carbonyls in the organic chemistry of sulphur seems to deserve more attention.

It should be mentioned here, that sulphur containing carbonyliron derivatives are claimed to be useful as polymerization initiators *(466,468,469)* and to improve the selectivity of the reaction between thiols and olefins *(249,250)*.

Table 1: Selenium or Tellurium Containing Organoiron Compounds

Compound	Preparation	IR	NMR	Other physical data[a]	Chemical properties[a]
$FeCo_2(CO)_9Se$	418			418 (X)	
$Fe(CO)(PPh_3)_2(NO)(SeCN)$	153	153			
$Fe(CO)(NO)_2[P(CF_3)_2SeP(CF_3)_2]$	149	149	149		
$(C_5H_5)Fe(CO)_2(SePh)$	389	389	389		
$(C_5H_5)Fe(CO)_2(SeCF_3)$	452				
$(C_5H_5)Fe(CO)_2(SeCN)$	241	241			240
$[(C_5H_5)Fe(CO)_2(PCy_3)](SeCN)$	240	240	240		
$(C_5H_5)Fe(CO)(PPh_3)(SeCN)$	240	240	240		240
$(C_5H_5)Fe(CO)(PPh_3)(NCSe)$	240	240	240		240
$(C_5H_5)Fe(CO)[P(OPh)_3](SeCN)$	240	240	240		
$(C_5H_5)Fe(CO)_2[SeP(CF_3)_2]$	150	150	150		
$(C_5H_5)Fe(CO)_2[P(Se)(CF_3)_2]$	149,150	149,150	149,150		150
$Fe_2(CO)_6(SPh)(SePh)$	281	280,364		280 (M)	
$Fe_2(CO)_6(SC_6F_5)(SePh)$	281	280,281,364		280 (M)	
$Fe_2(CO)_6(SeC_2Ph_2)$	392	392	392	392 (v)	
$Fe_2(CO)_6(SPh)(SeC_6F_5)$	281	281		280 (M)	
$Fe_2(CO)_6(SC_6F_5)(SeC_6F_5)$	281,365	364		280 (M),281 (M)	
$Fe_2(CO)_6[P(CF_3)_2]_2Se$	149	149	149		
$[Fe(CO)_3Se]_2$				89 (m),216 (D)	
$[Fe(CO)_3(SeEt)]_2$	216,380	216,364,380		216 (D)	
$Fe_2(CO)_5(PPh_3)(SeEt)_2$	226				
$[Fe(CO)_3(SeR)]_2$ (R = Me, *i*-Pr, C_2F_5)	380	380			

$[Fe(CO)_3(SePh)]_2$	*256,281*	*256,281,364*	*256*	*168* (D),*280* (M)	
$Fe_2(CO)_5L(SePh)_2$ ($L = PPh_3, PCy_3, P(OPh)_3$)	*226*			*226* (D)	
$Fe_2(CO)_6(SePh)(SeC_6F_5)$	*281*	*281,364*		*280* (M)	
$[Fe(CO)_3(SeC_6F_5)]_2$	*281,365*	*280,281,364*		*280* (M)	
$[Fe(CO)_3(SeCF_3)]_2$	*92,380,452*	*92,380*	*92*	*92* (m)	
$[(C_5H_5)Fe(CO)(SeR)]_2$ ($R = Et, Pr, CF_3, C_2F_5, C_3F_7$)	*380*	*380*			
$[(C_5H_5)Fe(CO)(SePh)]_2$	*389,390*	*389,390*	*389*	*390* (m)	*389* (i)
$Fe_3(CO)_9SSe$	*382,383*	*382*			*383*
$Fe_3(CO)_7[P(OPh)_3]_2SSe$	*383*	*383*			
$Fe_3(CO)_8(L)SSe$ ($L = AsPh_3, P(OPh)_3$)	*383*	*383*			
$Fe_3(CO)_9Se_2$	*218,219*	*219*		*89* (m),*119* (X)	*87*
$Fe_2(CO)_8(L)Se_2$ ($L = AsPh_3, P(OPh)_3, P\text{-}n\text{-}Bu_3$)	*86,87*	*86*			
$Fe_3(CO)_7(L)_2Se_2$ ($L = P(OPh)_3, P\text{-}n\text{-}Bu_3$)	*86*	*86*			
$[Fe(CO)_2(Se_2Ph_2)]_3$	*221*	*221*			
$\{[Fe(CO)(SePh)][S_2C_2(CF_3)_2]\}_n$ (n = 2 or 4)	*245*	*245*			*245* (e)
$FeCo_2(CO)_9Te$	*418*			*418* (X)	
$Fe(CO)_4(TePh_2)$	*222*	*222*		*222* (D)	
$Fe(CO)_3(TePh_2)Br_2$	*222*				
$Fe(CO)_3(TePh_2)I_2$	*222*	*222*			
$Fe(CO)(NO)_2(TePh_2)$	*222*	*222*			
$(C_5H_5)Fe(CO)_2(TePh)$	*389*	*389*	*389*		

Table 1: (continued)

Compound	Preparation	IR	NMR	Other physical data	Chemical properties [a]
$Fe_2(CO)_6(TeC_4H_4)$	*353*	*353*	*353*		*353*
$Fe_3(CO)_9STe$	*382,383*	*382*			*383*
$Fe_3(CO)_8(L)STe$ ($L = AsPh_3$, $P(OPh)_3$)	*383*	*383*			
$Fe_3(CO)_7(L)_2STe$ ($L = AsPh_3$, $P(OPh)_3$)	*383*	*383*			
$[Fe(CO)_3(TePh)]_2$	*281*	*281,364*			
$[Fe(CO)_3(TeC_6H_4\text{-}p\text{-}OMe)]_2$	*222*	*222,364*			
$[Fe(CO)_3(TeC_6F_5)]_2$	*281,365*	*281,365*		*281* (m)	
$[(C_5H_5)Fe(CO)(TePh)]_2$	*389,390*	*389,390*	*389*	*390* (m)	*389* (i)
$Fe_3(CO)_{10}Te_2$	*86*	*86*			
$Fe_3(CO)_9(L)Te_2$ ($L = AsPh_3$, $P(OPh)_3$, $P\text{-}n\text{-}Bu_3$)	*86,87*	*86*			
$Fe_3(CO)_9Te_2$	*219,353*	*219*		*89* (m)	*87,413*
$Fe_3(CO)_8(L)Te_2$ ($L = AsPh_3$, $P(OPh)_3$, $P\text{-}n\text{-}Bu_3$)	*86*	*86*			
$Fe_3(CO)_7(L)_2Te_2$ ($L = AsPh_3$, $P(OPh)_3$, $P\text{-}n\text{-}Bu_3$)	*86*	*86*			
$Fe_3(CO)_9SeTe$	*382,383*	*382*			*383*
$Fe_3(CO)_8(L)SeTe$ ($L = AsPh_3$, $P(OPh)_3$)	*383*	*383*			
$Fe_3(CO)_7[P(OPh)_3]_2SeTe$	*383*	*383*			

[a] (M) = Mössbauer spectrum; (m) = Mass spectrum; (X) = X-ray data; (D) = Dipole moment; (i) = Isomers or isomerization; (e) = Electrochemistry; (v) = visual and UV spectrum.

Table 2: Application of Carbonyliron Complexes in the Chemistry of Organic Sulphur Compounds

Reactants	Reaction conditions	Products	Yield [%]	*Refs.*
$RSO_2Cl + Fe(CO)_5 + BF_3{\cdot}Et_2O$ (R = alkyl, aryl)	0 - 75°C	$RSSO_2R$	36-71	*11,23*
$Ar\text{-}C(S)\text{-}NH_2 + Fe(CO)_5$	reflux/Bu_2O	ArCN	60-64	*20*
$Ph\text{-}C(S)\text{-}NHPh + Fe(CO)_5$	reflux/Bu_2O	PhCHNPh	47	*20*
$R_2SO + Fe(CO)_5$	130°C	R_2S	48-96	*22*
$Ar_2CS + Fe_2(CO)_9$	80°C/PPh_3 or amines	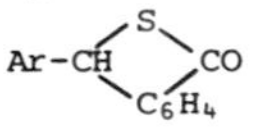	40-80	*27,28*
	20°C/$Hg(OOCCF_3)_2$	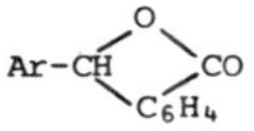		*27*
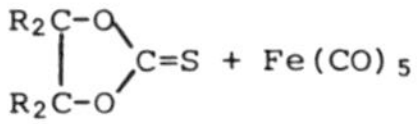	100°C	$R_2C{=}CR_2$	10-80	*122,123*
$(Ph_4C_4)Fe_2(CO)_6 + S$		Ph_4C_4S	80	*230,470*
$(Ph_4C_4)Fe_2(CO)_6 + Se$		Ph_4C_4Se	60	*230,470*
$(Ph_4C_4)Fe_2(CO)_6 + p\text{-}MeC_6H_4SO_2NCl_2$	hν	$(p\text{-}MeC_6H_4SO_2NHCPh{=}CPh)_2$		*230,470*
$(OCPh_4C_4)Fe_2(CO)_6 + Na_2S_x$ (or S)	hν	$OCPh_4C_4S$	24 (or 19)	*230*
$(OCPh_4C_4)Fe_2(CO)_6 + K_2Se_x$	hν	$OCPh_4C_4Se$	3	*230*
$RSCl + Fe(CO)_5$	< 0°C	RSSR	> 90	*297*
$C_2Cl_5SCl + Fe(CO)_5$	-80°C	$(C_2Cl_5S)_2 + (C_2Cl_3S)_2$		*297,299*
$(MeCHBr)_2CO + C_4H_4S + Fe_2(CO)_9$	40°C	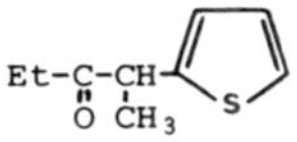	37	*350*

Table 2: (continued)

Reactants	Reaction conditions	Products	Yield [%]	*Refs.*
$(PhN)_2S$ + $Fe(CO)_5$ (or $Fe_2(CO)_9$)	80°C (or 40°C)	PhNNPh	10	*356*
$\beta\text{-}C_{10}H_7SSCN$ + $Fe(CO)_5$	-70°C	$(\beta\text{-}C_{10}H_7S)_2$	88	*386*
$\begin{array}{c} R_2C \\ \quad \diagdown \\ \qquad S \\ \quad \diagup \\ R_2C \end{array}$ + $Fe_2(CO)_9$ (or $Fe_3(CO)_{12}$)	reflux/benzene	$R_2C{=}CR_2$	80	*434*

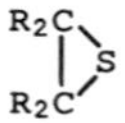

REFERENCES

1. Adams, D.M., *J. Chem. Soc.*, *1964*, 1771.
2. Adams, R.D., Cotton, F.A., Cullen, W.R., Hunter, D.L., and Mihichuk, L., *Inorg. Chem.*, *14*, 1395 (1975).
3. Ahmad, M., Bruce, R., and Knox, G.R., *Z. Naturforsch.*, *B 21*, 289 (1966).
4. Ahmad, M., Bruce, R., and Knox, G.R., *J. Organometal. Chem.*, *6*, 1 (1966).
5. Albright, T.A., and Hoffmann, R., *Chem. Ber.*, *111*, 1578 (1978).
6. Alexander, J.J., and Wojcicki, A., *Progress in Coordination Chemistry*, Elsevier, New York, 1968, p. 383.
7. Alexander, J.J., and Wojcicki, A., *J. Organometal. Chem.*, *15*, P 23 (1968).
8. Alexander, J.J., and Wojcicki, A., *Inorg. Chim. Acta*, *5*, 655 (1971).
9. Alexander, J.J., and Wojcicki, A., *Inorg. Chem.*, *12*, 74 (1973).
10. Allen, D.M., Cox, A., and Kemp, T.J., *J. Chem. Soc. Dalton Trans.*, *1973*, 1899.
11. Alper, H., *Tetrahedron Lett.*, *1969*, 1239.
12. Alper, H., *Organometal. Chem.*, *84*, 347 (1975).
13. Alper, H., *Org. Magn. Reson.*, *8*, 587 (1976).
14. Alper, H., and Chan, A.S.K., *J. Chem. Soc. D, Chem. Commun.*, *1971*, 1203.
15. Alper, H., and Chan, A.S.K., *J. Chem. Soc. Chem. Commun.*, *1973*, 724.
16. Alper, H., and Chan, A.S.K., *J. Amer. Chem. Soc.*, *95*, 4905 (1973).
17. Alper, H., and Chan, A.S.K., *Inorg. Chem.*, *13*, 225 (1974).
18. Alper, H., and Chan, A.S.K., *Inorg. Chem.*, *13*, 232 (1974).
19. Alper, H., and Des Roches, D., *J. Organometal. Chem.*, *117*, C 44 (1976).
20. Alper, H., and Edward, J.T., *Can. J. Chem.*, *46*, 3112 (1968).
21. Alper, H., and Foo, C.K., *Inorg. Chem.*, *14*, 2928 (1975).
22. Alper, H., and Keung, E.C.H., *Tetrahedron Lett.*, *1970*, 53.
23. Alper, H., and Keung, E.C.H., *J. Org. Chem.*, *37*, 2566 (1972).
24. Alper, H., and Paik, H.N., *J. Org. Chem.*, *42*, 3522 (1977).
25. Alper, H., and Root, W.G., *J. Chem. Soc. Chem. Commun.*, *1974*, 956.

26. Alper, H., and Root, W.G., *Tetrahedron Lett.*, *1974*, 1611.
27. Alper, H., and Root, W.G., *J. Amer. Chem. Soc.*, *97*, 4251 (1975).
28. Alper, H., Root, W.G., and Chan, A.S.K., *J. Organometal. Chem.*, *71*, C 14 (1974).
29. Andrianov, V.G., Korotkevich, A.F., and Struchkov, Yu.T., *Zh. Strukt. Khim.*, *9*, 712 (1968); *J. Struct. Chem.*, *9*, 622 (1968).
30. Andrianov, V.G., Sergeeva, G.N., Struchkov, Yu.T., Anisimov, K.N., Kolobova, N.E., and Beschastnov, A.S., *Zh. Strukt. Khim.*, *11*, 168 (1970); *J. Struct. Chem.*, *11*, 163 (1970).
31. Angelici, R.J., and Busetto, L., *J. Amer. Chem. Soc.*, *91*, 3197 (1969).
32. Anisimov, K.N., Lokshin, B.V., Kolobova, N.E., and Skripkin, V.V., *Izv. Akad. Nauk SSSR, Ser. Khim.*, *1968*, 1024; *Bull. Acad. Sci. USSR, Div. Chem. Sci.*, *1968*, 978.
33. Arabi, M.S., Mathieu, R., and Poilblanc, R., *Inorg. Chim. Acta*, *23*, L 17 (1977).
34. Attig, T.G., and Wojcicki, A., *J. Amer. Chem. Soc.*, *96*, 262 (1974).
35. Averill, B.A., Herskowitz, T., Holm, R.H., and Ibers, J.A., *J. Amer. Chem. Soc.*, *95*, 3523 (1973).
36. Bachmann, P., and Singer, H., *Z. Naturforsch.*, *B 31*, 525 (1976).
37. Baird, M.C., Hartwell, G., Jr., and Wilkinson, G., *J. Chem. Soc. A*, *1967*, 2037.
38. Baudler, M., and Mozaffar-Zanganeh, H., *Z.anorg. allg. Chem.*, *423*, 193 (1976).
39. Basato, M., *J. Chem. Soc. Dalton Trans.*, *1975*, 911.
40. Bau, R., Don, B., Greatrex, R., Haines, R.J., Love, R.A., and Wilson, R.D., *Inorg. Chem.*, *14*, 3021 (1975).
41. Bayer, E., Breitmaier, E., and Schurig, V., *Chem. Ber.*, *101*, 1594 (1968).
42. Beck, W., Stetter, K.H., Tadros, S., and Schwarzhans, K.E., *Chem. Ber.*, *100*, 3944 (1967).
43. Behrens, U., and Edelmann, F., *J. Organometal. Chem.*, *118*, C 41 (1976).
44. Best, S.A., Brant, P., Feltham, R.D., Rauchfuss, T.B., Roundhill, D.M., and Walton, R.A., *Inorg. Chem.*, *16*, 1976 (1977).
45. Bibler, J.P., and Wojcicki, A., *J. Amer. Chem. Soc.*, *86*, 5051 (1964).
46. Bibler, J.P., and Wojcicki, A., *J. Amer. Chem. Soc.*, *88*, 4862 (1966).
47. Bichler, R.E.J., and Clark, H.C., *J. Organometal.*

Chem., *23*, *427* (1970).
48. Birch, A.J., Jenkins, I.D., and Liepa, A.J., *Tetrahedron Lett.*, *1975*, 1723.
49. Bird, C.W., and Hollins, E.M., *J. Organometal. Chem.*, *4*, 245 (1965).
50. Bird, S.R.A., Donaldson, J.D., Le, A.F., Holding, C., Senior, B.J., and Tricker, M.J., *J. Chem. Soc. A*, *1971*, 1616.
51. Bock, P.L., Boschetto, D.J., Rasmussen, J.R., Demers, J.P., and Whitesides, G.M., *J. Amer. Chem. Soc.*, *96*, 2814 (1974).
52. Bodner, G.M., *Inorg. Chem.*, *13*, 2563 (1974).
53. Booth, G., and Chatt, J., *J. Chem. Soc.*, *1962*, 2099.
54. Bor, G., *J. Organometal. Chem.*, *11*, 195 (1968).
55. Bor, G., *Proc. Symp. Coord. Chem.*, *Tihany, Hungary*, *1964*, 361.
56. Braitsch, D.M., and Kumarappan, R., *J. Organometal. Chem.*, *84*, C 37 (1975).
57. Bratermann P.S., and Thompson, D.T., *J. Chem. Soc. A*, *1968*, 1454.
58. Braye, E.H., and Hübel, W., *Angew. Chem.*, *75*, 345 (1963); *Angew. Chem. Int. Ed. Engl.*, *2*, 217 (1963).
59. Braye, E.H., and Hübel, W., *J. Organometal. Chem.*, *9*, 370 (1967).
60. Broitman, M.O., Borodko, Yu.G., Stolyarova, T.A., and Shilov, A.E., *Izv. Akad. Nauk SSSR, Ser. Khim.*, *1970*, 937; *Bull. Acad. Sci. USSR, Div. Chem. Sci.*, *1970*, 889.
61. Brown, M.L., Cramer, J.L., Ferguson, J.A., Meyer, T.J., and Winterton, N., *J. Amer. Chem. Soc.*, *94*, 8707 (1972).
62. Brown, D.A., and Frimmel, F., *J. Chem. Soc. D, Chem. Commun.*, *1971*, 579.
63. Brown, S.D., and Gard, G.L., *Inorg. Chem.*, *17*, 1363 (1978).
64. Brown, M.L., Meyer, T.J., and Winterton, N., *J. Chem. Soc. D, Chem. Commun.*, *1971*, 309.
65. Bruce, M.I., *J. Chem. Soc. A*, *1968*, 1459.
66. Bruce, M.I., and Redhouse, A.D., *J. Organometal. Chem.*, *30*, C 78 (1971).
67. Bruce, R., and Knox, G.R., *J. Organometal. Chem.*, *6*, 67 (1966).
68. Brunner, H., Burgemeister, T., and Wachter, J., *Chem. Ber.*, *108*, 3349 (1975).
69. Brunner, H., and Wachter, J., *J. Organometal. Chem.*, *142*, 133 (1977).
70. Bryan, R.F., and Manning, A.R., *Chem. Commun.*, *1968*, 1220.
71. Bryan, R.F., and Weber, H.P., *Chem. Commun.*, *1966*, 329.

72. Buckingham, D.A., and Rauchfuss, T.B., *J. Chem. Soc. Chem. Commun., 1978,* 705.
73. Burger, K., Korecz, L., and Bor, G., Magy. Kém. *Lapja, 74,* 542 (1968); *J. Inorg. Nucl. Chem., 31,* 1527 (1969).
74. Burger, K., Korecz, L., Mag, P., Belluco, U., and Busetto, L., *Inorg. Chim. Acta, 5,* 362 (1971).
75. Burt, R., Cooke, M., and Green, M., *J. Chem. Soc. A, 1969,* 2645.
76. Burton, R., Green, M.L.H., Abel, E.W., and Wilkinson, G., *Chem. Ind. (London), 1958,* 1592.
77. Busetto, L., and Angelici, R.J., *J. Amer. Chem. Soc., 90,* 3283 (1968).
78. Busetto, L., Belluco, U., and Angelici, R.J., *J. Organometal. Chem., 18,* 213 (1969).
79. Busetto, L., Graziani, M., and Belluco, U., *Inorg. Chem., 10,* 78 (1971).
80. Busetto, L., and Palazzi, A., *Chim. Ind. (Milan), 58,* 804 (1976).
81. Busetto, L., and Palazzi, A., *Inorg. Chim. Acta, 19,* 233 (1976).
82. Busetto, L., Palazzi, A., Serantoni, E.F., and Di Sanseverino, L.R., *J. Organometal. Chem., 129,* C 55 (1977).
83. Büttner, H., and Feltham, R.D., *Inorg. Chem., 11,* 971 (1972).
84. Cane, D.J., Graham, W.A.G., and Vancea, L., *Can. J. Chem., 56,* 1538 (1978).
85. Carty, A.J., Kan, G., Madden, D.P., Snieckus, V., Stanton, M., and Birchall, T., *J. Organometal. Chem., 32,* 241 (1971).
86. Cetini, G., Stanghellini, P.L., Rossetti, R., and Gambino, O., *J. Organometal. Chem., 15,* 373 (1968).
87. Cetini, G., Stanghellini, P.L., Rossetti, R., and Gambino, O., *Inorg. Chim. Acta, 2,* 433 (1968).
88. Chan, A.S.-K., *Diss. Abstr., B 34,* 3698 (1974).
89. Chaudhuri, M.K., Haas, A., Rosenberg, M., Velicescu, M., and Welcman, N., *J. Organometal. Chem., 124,* 37 (1977).
90. Chaudhuri, M.K., Haas, A., and Welcman, N., *J. Organometal. Chem., 85,* 85 (1975).
91. Chaudhuri, M.K., Haas, A., and Welcman, N., *J. Organometal. Chem., 91,* 81 (1975).
92. Chaudhuri, M.K., Haas, A., and Wensky, A., *J. Organometal. Chem., 116,* 323 (1976).
93. Chen, L.S., Su, S.R., and Wojcicki, A., *J. Amer. Chem. Soc., 96,* 5655 (1974).
94. Chen, L.S., Su, S.R., and Wojcicki, A., *Inorg. Chim. Acta, 27,* 79, (1978).

95. Chou, C.-K., Miles. D.L., Bau, R., and Flood, T.C., *J. Amer. Chem. Soc., 100,* 7271 (1978).
96. Chow, Y.L., Fossey, J., and Perry, R.A., *J. Chem. Soc. Chem. Commun., 1972,* 501.
97. Churchill, M.R., DeBoer, B.G., and Kalra, K.L., *Inorg. Chem., 12,* 1646 (1973).
98. Churchill, M.R., DeBoer, B.G., Kalra, K.L., Reich-Rohrwig, P., and Wojcicki, A., *J. Chem. Soc. Chem. Commun., 1972,* 981.
99. Churchill, M.R., and Kalra, K.L., *Inorg. Chem., 12,* 1650 (1973).
100. Churchill, M.R., and Wormald, J., *Inorg. Chem., 9,* 2430 (1970).
101. Churchill, M.R., and Wormald, J., *J. Amer. Chem. Soc., 93,* 354 (1971).
102. Churchill, M.R., and Wormald, J., *Inorg. Chem., 10,* 572 (1971).
103. Churchill, M.R., Wormald, J., Ross, D.A., Thomasson, J.E., and Wojcicki, A., *J. Amer. Chem. Soc., 92,* 1795 (1970).
104. Churchill, M.R., Wormald, J., Young, D.A.T., and Kaesz, H.D., *J. Amer. Chem. Soc., 91,* 7201 (1969).
105. Clare, M., Hill, H.A.O., Johnson, C.E., and Richards, R., *J. Chem. Soc. D, Chem. Commun., 1970,* 1376.
106. Cole-Hamilton, D.J., Stephenson, T.A., Whan, D.A., and Talbot, P., *Chem. Ind. (London), 1975,* 649.
107. Coleman, J.M., Wojcicki, A., Pollick, P.J., and Dahl, L.F., *Inorg. Chem., 6,* 1236 (1967).
108. Connelly, N.G., and Dahl, L.F., *J. Amer. Chem. Soc., 92,* 7472 (1970).
109. Connelly, N.G., and Davies, J.D., *J. Organometal. Chem., 38,* 385 (1972).
110. Cooke, J., Green, M., and Stone, F.G.A., *J. Chem. Soc. A, 1968,* 170.
111. Cotton, F.A., Kolb, J.R., and Stults, B.R., *Inorg. Chim. Acta, 15,* 239 (1975).
112. Cotton, F.A., and McCleverty, J.A., *Inorg. Chem., 3,* 1398 (1964).
113. Cremer, W., *Biochem. Z., 201,* 490 (1928).
114. Cremer, W., *Biochem. Z., 194,* 231 (1928).
115. Cremer, W., *Biochem. Z., 206,* 228 (1929).
116. Cutler, A., Entholt, D., Giering, W.P., Lennon, P., Raghu, S., Rosan, A., Rosenblum, M., Tancrede, J., and Wells, D., *J. Amer. Chem. Soc., 98,* 3495 (1976).
117. Cutler, A., Fish, R.W., Giering, W.P., and Rosenblum, M., *J. Amer. Chem. Soc., 94,* 4354 (1972).
118. Czauderna, B., Jogun, K.H., Stezowski, J.J., and Föhlisch, B., *J. Amer. Chem. Soc., 98,* 6696 (1976).

119. Dahl, L.F., and Sutton, P.W., *Inorg. Chem.*, *2*, 1067 (1963).
120. Dahl, L.F., and Wei, C.-H., *Inorg. Chem.*, *2*, 328 (1963).
121. van Dam, H., Stufkens, D.J., and Oskam, A., *Inorg. Chim. Acta*, *31*, L 377 (1978).
122. Daub, J., Trautz, V., and Erhardt, U., *Tetrahedron Lett.*, *1972*, 4435.
123. Daub, J., Erhardt, U., Kappler, J., and Trautz, V., *J. Organometal. Chem.*, *69*, 423 (1974).
124. Davidson, J.L., Harrison, W., Sharp, D.W.A., and Sim, G.A., *J. Organometal. Chem.*, *46*, C 47 (1972).
125. Davidson, J.L., and Sharp, D.W.A., *J. Chem. Soc. Dalton Trans.*, *1972*, 107.
126. Davidson, J.L., and Sharp, D.W.A., *J. Chem. Soc. Dalton Trans.*, *1973*, 1957.
127. Davidson, J.L., and Sharp, D.W.A., *J. Chem. Soc. Dalton Trans.*, *1975*, 2283.
128. Dean, W.K., *J. Organometal. Chem.*, *135*, 195 (1977).
129. Dean, W.K., and Vanderveer, D.G., *J. Organometal. Chem.*, *144*, 65 (1978).
130. Dean, W.K., and Vanderveer, D.G., *J. Organometal. Chem.*, *145*, 49 (1978).
131. Dean, W.K., and Vanderveer, D.G., *J. Organometal. Chem.*, *146*, 143 (1978).
132. de Beer, J.A., and Haines, R.J., *J. Chem. Soc. D, Chem. Commun.*, *1970*, 288.
133. de Beer, J.A., and Haines, R.J., *J. Organometal. Chem.*, *24*, 757 (1970).
134. de Beer, J.A., and Haines, R.J., *J. Organometal. Chem.*, *36*, 297 (1972).
135. de Beer, J.A., and Haines, R.J., *J. Organometal. Chem.*, *37*, 173 (1972).
136. de Beer, J.A., Haines, R.J., and Greatrex, R., *J. Chem. Soc. Chem. Commun.*, *1972*, 1094.
137. de Beer, J.A., Haines, R.J., Greatrex, R., and Greenwood, N.N., *J. Chem. Soc. A*, *1971*, 3271.
138. de Beer, J.A., Haines, R.J., Greatrex, R., and Greenwood, N.N., *J. Organometal. Chem.*, *27*, C 33 (1971); *J. Organometal. Chem.*, *29*, C 22 (1971).
139. de Beer, J.A., Haines, R.J., Greatrex, R., and van Wyk, J.A., *J. Chem. Soc. Dalton Trans.*, *1973*, 2341.
140. Dekker, M., Knox, G.R., and Robertson, C.G., *J. Organometal. Chem.*, *18*, 161 (1969).
141. Dessy, R.E., King, R.B., and Waldrop, M., *J. Amer. Chem. Soc.*, *88*, 5112 (1966).
142. Dessy, R.E., Kornmann, R., Smith, C., and Haytor, R., *J. Amer. Chem. Soc.*, *90*, 2001 (1968).

143. Dessy, R.E., and Pohl, R.L., *J. Amer. Chem. Soc.*, *90*, 1995 (1968).

144. Dessy, R.E., Stary, F.E., King, R.B., and Waldrop, M., *J. Amer. Chem. Soc.*, *88*, 471 (1966).

145. Dessy, R.E., and Wieczorek, L., *J. Amer. Chem. Soc.*, *91*, 4963 (1969).

146. Dettlaf, G., Behrens, U., and Weiss, E., *J. Organometal. Chem.*, *152*, 95 (1978).

147. Dias, A.R., and Green, M.L.H., *Rev. Port. Quim.*, *11*, 61 (1969); *Chem. Abstr.*, *73*, 4011 j (1970).

148. Dittmer, D.C., Takahashi, K., Iwanami, M., Tsai, A.I., Chang, P.L., Blidner, B.B., and Stamos, I.K., *J. Amer. Chem. Soc.*, *98*, 2795 (1976).

149. Dobbie, R.C., and Hopkinson, M.J., *J. Chem. Soc. Dalton Trans.*, *1974*, 1290.

150. Dobbie, R.C., and Mason, P.R., *J. Chem. Soc. Dalton Trans.*, *1973*, 1124.

151. Dobbie, R.C., and Mason, P.R., *J. Chem. Soc. Dalton Trans.*, *1974*, 2439.

152. Dobbie, R.C., Mason, P.R., and Porter, R.J., *J. Chem. Soc. Chem. Commun.*, *1972*, 612.

153. Dolcetti, G., Busetto, L., and Palazzi, A., *Inorg. Chem.*, *13*, 222 (1974).

154. Dombek, B.D., and Angelici, R.J., *Inorg. Chim. Acta*, *7*, 345 (1973).

155. Dombek, B.D., and Angelici, R.J., *Inorg. Synth.*, *17*, 100 (1977).

156. Dong, D., Slack , D.A., and Baird, M.C., *J. Organometal. Chem.*, *153*, 219 (1978).

157. Douglas, W.M., and Ruff, J.K., *J. Chem. Soc. A*, *1971*, 3558.

158. Downs, R.L., as quoted by Kitching, W., and Fong, C.W., *Organometal. Chem. Rev. A*, *5*, 281 (1970).

159. Downs, R.L., and Wojcicki, A., *Inorg. Chim. Acta*, *27*, 91 (1978).

160. Downs, R.L., Wojcicki, A., and Pollick, P.J., *148th Meeting Amer. Chem. Soc.*, Chicago, 1964, Abstracts, p. 31.

161. Dunker, J.W., Finer, J.S., Clardy, J., and Angelici, R.J., *J. Organometal. Chem.*, *114*, C 49 (1976).

162. Edgar, K., Johnson, B.F.G., Lewis, J., Williams, I.G., and Wilson, J.M., *J. Chem. Soc. A*, *1967*, 379.

163. Edmondson, R.C., Field, D.S., and Newlands, M.J., *Can. J. Chem.*, *49*, 618 (1971).

164. Edmondson, R.C., and Newlands, M.J., *Chem. Commun.*, *1968*, 1219.

165. Eekhof, J.H., Hogeveen, H., and Kellogg, R.M., *J. Chem. Soc. Chem. Commun.*, *1976*, 657.

166. Eekhof, J.H., Hogeveen, H., Kellogg, R.M., and Sawatzky, G.A., *J. Organometal. Chem.*, *111*, 349 (1976).
167. Eekhof, J.H., Hogeveen, H., Kellogg, R.M., and Schudde, E.P., *J. Organometal. Chem.*, *105*, C 35 (1976).
168. Ehrl, W., and Vahrenkamp, H., *Chem. Ber.*, *103*, 3563 (1970).
169. Ehrl, W., and Vahrenkamp, H., *Chem. Ber.*, *105*, 1471 (1972).
170. Ehrl, W., and Vahrenkamp, H., *Z. Naturforsch.*, *B 28*, 365 (1973).
171. Ellgen, P.C., and Gerlach, J.N., *Inorg. Chem.*, *12*, 2526 (1973).
172. Ellis, J.E., Fennell, R.W., and Flom, E.A., *Inorg. Chem.*, *15*, 2031 (1976).
173. English, R.B., Haines, R.J., and Nolte, C.R., *J. Chem. Soc. Dalton Trans.*, *1975*, 1030.
174. English, R.B., Nassimbeni, L.R., and Haines, R.J., *J. Organometal. Chem.*, *135*, 351 (1977).
175. Farmery, K., and Kilner, M., *J. Organometal. Chem.*, *17*, 127 (1969).
176. Farmery, K., and Kilner, M., *J. Chem. Soc. A*, *1970*, 634.
177. Farona, M.F., and Camp, G.R., *New Aspects Chem. Metal Carbonyls Deriv. Int. Symp., Proc.*, *1st 1968*, A 4; *Chem. Abstr.*, *71*, 118024 e (1969).
178. Farona, M.F., and Camp, G.R., *Inorg. Chim. Acta*, *3*, 395 (1969).
179. Farona, M.F., and Wojcicki, A., *Inorg. Chem.*, *4*, 1402 (1965).
180. Fauvel, K., Mathieu, R., and Poilblanc, R., *Inorg. Chem.*, *15*, 976 (1976).
181. Ferguson, G., Hannaway, C., and Islam, K.M.S., *Chem. Commun.*, *1968*, 1165.
182. Ferguson, J.A., and Meyer, T.J., *J. Chem. Soc. D, Chem. Commun.*, *1971*, 623.
183. Field, D.S., and Newlands, M.J., *J. Organometal. Chem.*, *27*, 221 (1971).
184. Flannigan, W.T., Knox, G.R., and Pauson, P.L., *J. Chem. Soc. C*, *1969*, 2077.
185. Flood, T.C., DiSanti, F.J., and Miles, D.L., *J. Chem. Soc. Chem. Commun.*, *1975*, 336.
186. Flood, T.C., DiSanti, F.J., and Miles, D.L., *Inorg. Chem.*, *15*, 1910 (1976).
187. Flood, T.C., and Miles, D.L., *J. Amer. Chem. Soc.*, *95*, 6460 (1973).
188. Folkes, C.R., and Rest, A.J., *J. Organometal. Chem.*, *136*, 355 (1977).
189. Francis, J.N., and Hawthorne, M.F., *Inorg. Chem.*, *10*,

594 (1971).
190. Fraser, P.J., Roper, W.R., and Stone, F.G.A., *J. Chem. Soc. Dalton Trans., 1974,* 760.
191. Frisch, P.D., Lloyd, M.K., McCleverty, J.A., and Seddon, D., *J. Chem. Soc. Dalton Trans., 1973,* 2268.
192. Gaspard, S., Viovy, R., Brégeault, J.-M., Jarjour, C., and Yolou, S., *C.R. Acad. Sci. Ser. C, 281,* 925 (1975).
193. Giannotti, C. Ducourant, A.M., Chanaud, H., Chiaroni, A., and Riche, C., *J. Organometal. Chem., 140,* 289 (1977).
194. Gibb, T.C., Greatrex, R., Greenwood, N.N., and Thompson, D.T., *J. Chem. Soc. A, 1967,* 1663.
195. Giering, W.P., Raghu, S., Rosenblum, M., Cutler, A., Ehntholt, D., and Fish, R.W., *J. Amer. Chem. Soc., 94,* 8251 (1972).
196. Giering, W.P., and Rosenblum, M., *J. Amer. Chem. Soc., 93,* 5299 (1971).
197. Gompper, R., and Reiser, W., *Tetrahedron Lett., 1976,* 1263.
198. Granifo, J., and Müller, H., *J. Chem. Soc. Dalton Trans., 1973,* 1891.
199. Graziani, M., and Wojcicki, A., *Inorg. Chim. Acta, 4,* 347 (1970).
200. Greatrex, R., and Greenwood, N.N., *Discuss. Faraday Soc., 47,* 126 (1969).
201. Green, M., Stone, F.G.A., and Underhill, M., *J. Chem. Soc. Dalton, Trans., 1975,* 939.
202. Green, M.L.H., and Whiteley, R.N., *J. Chem. Soc. A, 1971,* 1943.
203. Grobe, J., and Kober, F., *Z. Naturforsch., B 24,* 1346 (1969).
204. Guilard, R., and Dusausoy, Y., *J. Organometal. Chem., 77,* 393 (1974).
205. Haines, R.J., de Beer, J.A., and Greatrex, R., *J. Organometal. Chem., 55,* C 30 (1973).
206. Haines, R.J., De Beer, J.A., and Greatrex, R., *J. Organometal. Chem., 85,* 89 (1975).
207. Haines, R.J., De Beer, J.A., and Greatrex, R., *J. Chem. Soc. Dalton Trans., 1976,* 1749.
208. Haines, R.J., and Nolte, C.R., *J. Organometal. Chem., 36,* 163 (1972).
209. Haines, R.J., Nolte, C.R., Greatrex, R., and Greenwood, N.N., *J. Organometal. Chem., 26,* C 45 (1971).
210. Harlow, R.L., and Pfluger, C.E., *Acta Crystallogr., B 29,* 2633 (1973).
211. Havlin, R., and Knox, G.R., *J. Organometal. Chem., 4,* 247 (1965).

212. Henslee, W., and Davis, R.E., *Cryst. Struct. Commun.*, *1*, 403 (1972).
213. Herber, R.H., King, R.B., and Wertheim, G.K., *Inorg. Chem.*, *3*, 101 (1964).
214. Herberhold, M., and Leonhard, K., *J. Organometal. Chem.*, *78*, 253 (1974).
215. Hieber, W., Bader, G., and Ries, K., *Z. anorg. allg. Chem.*, *201*, 329 (1931).
216. Hieber, W., and Beck, W., *Z. anorg. allg. Chem.*, *305*, 265 (1960).
217. Hieber, W., and Beutner, H., *Z. Naturforsch.*, *B 15*, 324 (1960).
218. Hieber, W., and Geisenberger, O., *Z. anorg. allg. Chem.*, *262*, 15 (1950).
219. Hieber W., and Gruber, J., *Z. anorg. allg. Chem.*, *296*, 91 (1958).
220. Hieber, W., and Kaiser, K., *Chem. Ber.*, *102*, 4043 (1969).
221. Hieber, W., and Kaiser, K., *Z. Naturforsch.*, *B 24*, 778 (1969).
222. Hieber, W., and Kruck, J., *Chem. Ber.*, *95*, 2027 (1962).
223. Hieber, W., and Lipp, A., *Chem. Ber.*, *92*, 2085 (1959).
224. Hieber, W., and Scharfenberg, C., *Ber. Deutsch. Chem. Ges.*, *73*, 1012 (1940).
225. Hieber, W., and Spacu, P., *Z. anorg. allg. Chem.*, *233*, 353 (1937).
226. Hieber, W., and Zeidler, A., *Z. anorg. allg. Chem.*, *329*, 92 (1964).
227. Hoffmann, K., and Weiss, E., *J. Organometal. Chem.*, *128*, 225 (1977).
228. Hoffmann, K., and Weiss, E., *J. Organometal. Chem.*, *128*, 389 (1977).
229. Hogben, M.G., and Graham, W.A.G., *J. Amer. Chem. Soc.*, *91*, 283 (1969).
230. Hübel, W., in I. Wender and P. Pino (Eds.), *Organic Syntheses via Metal Carbonyls, Vol 1*, Interscience, New York, 1967, p. 273.
231. Hübener, P., and Weiss, E., *J. Organometal. Chem.*, *129*, 105 (1977).
232. Hunt, D.F., Lillya, C.P., and Rausch, M.D., *J. Amer. Chem. Soc.*, *90*, 2561 (1968).
233. Ingletto, G., Tondello, E., Di Sipio, L., Carturan, G., and Graziani, M., *J. Organometal. Chem.*, *56*, 335 (1973).
234. Jablonski, C.R., *J. Organometal. Chem.*, *142*, C 25 (1977).
235. Jacobson, S.E., Reich-Rohrwig, P., and Wojcicki, A., *J. Chem. Soc. D, Chem. Commun.*, *1971*, 1526.

236. Jacobson, S.E., Reich-Rohrwig, P., and Wojcicki, A., *Inorg. Chem., 12,* 717 (1973).
237. Jacobson, S.E., and Wojcicki, A., *J. Amer. Chem. Soc., 93,* 2535 (1971).
238. Jacobson, S.E., and Wojcicki, A., *J. Amer. Chem. Soc., 95,* 6962 (1973).
239. Jacobsen, S.E., and Wojcicki, A., *Inorg. Chim. Acta, 10,* 229 (1974).
240. Jennings, M.A., and Wojcicki, A., *Inorg. Chim. Acta, 3,* 335 (1969).
241. Jennings, M.A., and Wojcicki, A., *J. Organometal. Chem., 14,* 231 (1968).
242. Job, B.E., McLean, R.A.N., and Thompson, D.T., *Chem. Commun., 1966,* 895.
243. John, G.R., Kane-Maguire, L.A.P., and Eaborn, C., *J. Chem. Soc. Chem. Commun., 1975,* 481.
244. Johnson, E.C., Meyer, T.J., and Winterton, N., *J. Chem. Soc. D, Chem. Commun., 1970,* 934.
245. Jones, C.J., McCleverty, J.A., and Orchard, D.G., *J. Organometal. Chem., 26,* C 19 (1971).
246. Jones, C.J., McCleverty, J.A., and Orchard, D.G., *J. Chem. Soc. Dalton Trans., 1972,* 1109.
247. Jones, C.J., and McCleverty, J.A., *J. Chem. Soc. Dalton Trans., 1975,* 701.
248. Kaesz, H.D., King, R.B., Manuel, T.A., Nichols, L.D., and Stone, F.G.A., *J. Amer. Chem. Soc., 82,* 4749 (1960).
249. Kandror, I.I., Petrova, R.G., Petrovskii, P.V., and Freidlina, R.Kh., *Izv. Akad. Nauk SSSR, Ser. Khim., 1969,* 1621; *Bull. Acad. Sci. USSR, Div. Chem. Sci., 1969,* 1508.
250. Kandror, I.I., Petrovskii, P.V., and Petrova, R.G., *Izv. Akad. Nauk SSSR, Ser. Khim., 1970,* 1329; *Bull. Acad. Sci. USSR, Div. Chem. Sci., 1970,* 1256.
251. Kane-Maguire, L.A.P., and Mansfield, C.A., *J. Chem. Soc. Chem. Commun., 1973,* 540.
252. Karlin, K.D., and Lippard, S.J., *J. Amer. Chem. Soc., 98,* 6951 (1976).
253. Kemmitt, R.D.W., and Rimmer, G.D., *J. Inorg. Nucl. Chem., 35,* 3155 (1973).
254. Khattab, S.A., and Markó, L., *Acta Chim. Acad. Sci. Hung., 40,* 471 (1964).
255. Khattab, S.A., Markó, L., Bor, G., and Markó, B., *J. Organometal. Chem., 1,* 373 (1964).
256. Kiener, V., and Fischer, E.O., *J. Organometal. Chem., 42,* 447 (1972).
257. Killops, S.D., Knox, S.A.R., Riding, G.H., and Welch, A.J., *J. Chem. Soc. Chem. Commun., 1978,* 486.

258. Kimura, T., Nagata, Y., and Tsurugi, J., *J. Biol. Chem.*, *246*, 5140 (1971).
259. King, R.B., *J. Amer. Chem. Soc.*, *84*, 2460 (1962).
260. King, R.B., *Inorg. Chem.*, *2*, 326 (1963).
261. King, R.B., *J. Amer. Chem. Soc.*, *85*, 1584 (1963).
262. King, R.B., *J. Amer. Chem. Soc.*, *85*, 1918 (1963).
263. King, R.B., *Inorg. Chim. Acta*, *2*, 454 (1968).
264. King, R.B., *J. Amer. Chem. Soc.*, *90*, 1429 (1968).
265. King, R.B., *Organometallic Syntheses, Vol. 1, Transition-Metal Compounds*, Academic Press, New York, 1965, p. 180.
266. King, R.B., and Bisnette, M.B., *J. Amer. Chem. Soc.*, *86*, 1267 (1964).
267. King, R.B., and Bisnette, M.B., *Inorg. Chem.*, *4*, 482 (1965).
268. King, R.B., and Bisnette, M.B., *Inorg. Chem.*, *4*, 486 (1965).
269. King, R.B., and Bisnette, M.B., *Inorg. Chem.*, *4*, 1663 (1965).
270. King, R.B., and Bisnette, M.B., *Inorg. Chem.*, *6*, 469 (1967).
271. King, R.B., and Eggers, C.A., *Inorg. Chem.*, *7*, 340 (1968).
272. King, R.B., and Eggers, C.A., *Inorg. Chem.*, *7*, 1214 (1968).
273. King, R.B., Epstein, L.M., and Gowling, E.W., *J. Inorg. Nucl. Chem.*, *32*, 441 (1970).
274. King, R.B., and Kapoor, R.N., *Inorg. Chem.*, *8*, 2535 (1969).
275. King, R.B., and Stone, F.G.A., *J. Amer. Chem. Soc.*, *82*, 4557 (1960).
276. King, R.B., Treichel, P.M., and Stone, F.G.A., *J. Amer. Chem. Soc.*, *83*, 3600 (1961).
277. King, R.B., and Welcman, N., *Inorg. Chem.*, *8*, 2540 (1969).
278. Knox, R.G., and Pryde, A., *J. Organometal. Chem.*, *18*, 169 (1969).
279. Korecz, L., and Burger, K., *J. Inorg. Nucl. Chem.*, *30*, 781 (1968).
280. Kostiner, E., and Massey, A.G., *J. Organometal. Chem.*, *19*, 233 (1969).
281. Kostiner, E., Reddy, M.L.N., Urch, D.S., and Massey, A.G., *J. Organometal. Chem.*, *15*, 383 (1968).
282. Koyanagi, T., Hayami, J., and Kaji, A., *Chem. Lett.*, *1976*, 971.
283. Krause, R.A., and Ruggles, C.R., *Inorg. Nucl. Chem. Lett.*, *4*, 555 (1968).
284. Kubas, G.T., Spiro, T.G., and Terzis, A., *J. Amer.

Chem. Soc., *95*, 273 (1973).

285. Kubas, G.J., Vergamini, P.J., Eastman, M.P., and Prater, K.B., *J. Organometal. Chem.*, *117*, 71 (1976).
286. Kuhnhen, H.B., *Z. Naturforsch.*, *B 32*, 718 (1977).
287. Kukina, M.A., Lys, Ya.I, Tyurin, V.D., Fedoseev, V.M., and Nametkin, N.S., *Izv. Akad. Nauk SSSR, Ser. Khim.*, *1975*, 2383; *Bull. Acad. Sci. USSR, Div. Chem. Sci.*, *1975*, 2275.
288. Lauterbur, P.C., and King, R.B., *J. Amer. Chem. Soc.*, *87*, 3266 (1965).
289. Le Borgne, G., and Grandjean, D., *J. Organometal. Chem.*, *92*, 381 (1975).
290. Le Borgne, G., and Grandjean, D., *Acta Crystallogr.*, *B 33*, 344 (1977).
291. Le Borgne, G., Grandjean, D., Mathieu, R., and Poilblanc, R., *J. Organometal. Chem.*, *131*, 429 (1977).
292. Le Bozec, H., Dixneuf, P., Taylor, N.J., and Carty, A.J., *J. Organometal. Chem.*, *135*, C 29 (1977).
293. Le Bozec, H., Gorgues, A., and Dixneuf, P.H., *J. Amer. Chem. Soc.*, *100*, 3946 (1978).
294. Lennon, P., Madhavarao, M., Rosan, A., and Rosenblum, M., *J. Organometal. Chem.*, *108*, 93 (1976).
295. Lichtenberg, D.W., and Wojcicki, A., *J. Organometal. Chem.*, *33*, C 77 (1971).
296. Lichtenberg, D.W., and Wojcicki, A., *Inorg. Chim. Acta*, *7*, 311 (1973).
297. Lindner, E., and Vitzthum, G., *Angew. Chem.*, *81*, 532 (1969); *Angew. Chem. Int. Ed. Engl.*, *8*, 518 (1969).
298. Lindner, E., and Weber, H., *Z. Naturforsch.*, *B 22*, 1243 (1967).
299. Lindner, E., Weber, H., and Vitzthum, G., *J. Organometal. Chem.*, *13*, 431 (1968).
300. Lindner, E., Weber, H., Vitzthum, G., and Karman, H.-G., *Progress in Coordination Chemistry*, Elsevier, New York, 1968, p. 400.
301. Locke, J., and McCleverty, J.A., *Inorg. Chem.*, *5*, 1157 (1966).
302. Lorenz I.-P., *Angew. Chem.*, *90*, 60 (1978); *Angew. Chem. Int. Ed. Engl.*, *17*, 53 (1978).
303. Lustig, M., and Houk, L.W., *Inorg. Nucl. Chem. Lett.*, *5*, 851 (1969).
304. Malisch, W., *J. Organometal. Chem.*, *61*, C 15 (1973).
305. Mansfield, C.A., Al-Kathumi, K.M., and Kane-Maguire, L.A.P., *J. Organometal. Chem.*, *71*, C 11 (1974).
306. Mansuy, D., Battioni, J.P., and Chottard, J.C., *J. Amer. Chem. Soc.*, *100*, 4311 (1978).
307. Manuel, T.A., *Inorg. Chem.*, *3*, 1794 (1964).
308. Manuel, T.A., and Meyer, T.J., *Inorg. Chem.*, *3*, 1049

(1964).
309. Marcincal, P., and Cuingnet, E., *Tetrahedron Lett.*, *1975*, 1223.
310. Marcincal, P., and Cuingnet, E., *Tetrahedron Lett.*, *1975*, 3827.
311. Marcincal, P., and Cuingnet, E., *Bull. Soc. Chim. Fr.*, *1977*, 489.
312. Maresca, L., Bor, G., Greggio, F., and Sbrignadello, G., *Proc. Inorg. Chim. Acta, Third Internat. Symp.*, Venice, 1970, p. E 3.
313. Maresca, L., Greggio, F., Sbrignadello, G., and Bor, G., *Inorg. Chim. Acta*, *5*, 667 (1971).
314. Markó, L., unpublished result.
315. Mathieu, R., and Poilblanc, R., *J. Organometal. Chem.*, *142*, 351 (1977).
316. Mathieu, R., Poilblanc, R., Lemoine, P., and Gross, M., *J. Organometal. Chem.*, *165*, 243 (1979).
317. Mays, M.J., and Robb, J.D., *J. Chem. Soc. A*, *1968*, 329.
318. McCaskie, J.E., Chang, P.L., Nelsen, T.R., and Dittmer, D.C., *J. Org. Chem.*, *38*, 3963 (1973).
319. Meij, R., v.d.Helm, J., Stufkens, D.J., and Vrieze, K., *J. Chem. Soc. Chem. Commun.*, *1978*, 506.
320. Merour, J.Y., *C.R. Acad. Sci., Ser. C*, *271*, 1397 (1970).
321. Meunier-Piret, J., Piret, P., and Van Meerssche, M., *Bull. Soc. Chim. Belg.*, *76*, 374 (1967).
322. Miles, S.L., Miles, D.L., Bau, R., and Flood, T.C., *J. Amer. Chem. Soc.*, *100*, 7278 (1978).
323. Miller, J.S., *Inorg. Chem.*, *14*, 2011 (1975).
324. Miller, J., and Balch, A.L., *Inorg. Chem.*, *10*, 1410 (1971).
325. Müller, J., Öfele, K., and Krebs, G., *J. Organometal. Chem.*, *82*, 383 (1974).
326. Natile, G., and Bor, G., *J. Organometal. Chem.*, *35*, 185 (1972).
327. Natile, G., Maresca, L., and Bor, G., *Inorg. Chim. Acta*, *23*, 37 (1977).
328. Nametkin, N.S., Gubin, S.P., Tyurin, V.D., Podolskaya, I.P. Kolesnikova, L.P., and Zelikov, M.I., *Izv. Akad. Nauk SSSR, Ser. Khim.*, *1976*, 1678; *Bull. Acad. Sci. USSR, Div. Chem. Sci.*, *1976*, 1595.
329. Nametkin, N.S., Gubin, S.P., Tyurin, V.D., Podolskaya, I.P., Kolesnikova, L.P., and Zelikov, M.I., *Neftekhimiya*, *16*, 480 (1976).
330. Nametkin, N.S., Tyurin, V.D., Gubin, S.P., and Kukina, M.A., *Neftekhimiya*, *15*, 767 (1975).
331. Nametkin, N.S., Tyurin, V.D., and Kukina, M.A., *J. Organometal. Chem.*, *149*, 355 (1978).

332. Nametkin, N.S., Tyurin, V.D., Kukina, M.A., Nekhaew, A.I., Mavlonov, M., and Alekseeva, S.D., *Izv. Akad. Nauk SSSR, Ser. Khim., 1977,* 2384; *Bull. Acad. Sci. USSR, Div. Chem. Sci., 1977,* 2217.

333. Nametkin, N.S., Tyurin, V.D., Kuz'min, O.V., Nekhaev, A.I., and Mavlonov, M., *Izv. Akad. Nauk SSSR, Ser. Khim., 1976,* 2143; *Bull. Acad. Sci. USSR, Div. Chem. Sci., 1976,* 2006.

334. Nametkin, N.S., Tyurin, V.D., Nekhaev, A.I., Mavlonov, M., and Kukina, M.A., *Izv. Akad. Nauk SSSR, Ser. Khim., 1975,* 2846; *Bull. Acad. Sci. USSR, Div. Chem. Sci., 1975,* 2739.

335. Nesmeyanov, A.N., Anisimov, K.N., Kolobova, N.E., and Denisov, F.S., *Izv. Akad. Nauk SSSR, Ser. Khim., 1968,* 142; *Bull. Acad. Sci. USSR, Div. Chem. Sci., 1968,* 133.

336. Nesmeyanov, A.N., Anisimov, K.N., Kolobova, N.E., and Skripkin, V.V., *Izv. Akad. Nauk SSSR, Ser. Khim., 1966,* 1292; *Bull. Acad. Sci. USSR, Div. Chem. Sci., 1966,* 1248.

337. Nesmeyanov, A.N., Anisimov, K.N., Lokshin, B.V., Kolobova, N.E., and Denisov, F.S., *Izv. Akad. Nauk SSSR, Ser. Khim., 1969,* 758; *Bull. Acad. Sci. USSR, Div. Chem. Sci., 1969,* 690.

338. Nesmeyanov, A.N., Kolobova, N.E., Goncharenko, L.V., and Anisimov, K.N., *Izv. Akad. Nauk SSSR, Ser. Khim., 1976,* 153; *Bull. Acad. Sci. USSR, Div. Chem. Sci., 1976,* 142.

339. Nesmeyanov, A.N., Kolobova, N.E., Goncharenko, L.V., and Anisimov, K.N., *Izv. Akad. Nauk SSSR, Ser. Khim., 1976,* 1179; *Bull. Acad. Sci. USSR, Div. Chem. Sci., 1976,* 1149.

340. Nesmeyanov, A.N., Kolobova, N.E., Zlotina, I.B., Ivanova, L.V., and Anisimov, K.N., *Izv. Akad. Nauk SSSR, Ser. Khim., 1977,* 707; *Bull. Acad. Sci. USSR, Div. Chem. Sci., 1977,* 644.

341. Nesmeyanov, A.N., Kolobova, N.E., Zlotina, I.B., Solodova, M.Ya., and Anisimov, K., *Thezisy Dokl. Vses. Chugaevskoe Soveshch. Khim. Kompleksn. Soedin., 12th, 3,* 474 (1975).

342. Nesmeyanov, A.N., Rybin, L.V., Rybinskaya, M.I., Gubenko, N.T., Leshcheva, I.F., and Ustynyuk, Yu.A., *Zh. Obsh. Khim., 38,* 1476 (1968); *J. Gen. Chem. USSR, 38,* 1428 (1968).

343. Nesmeyanov, A.N., Vol'kenau, N.A., and Bolesova, I.N., *Dokl. Akad. Nauk SSSR, 175,* 606 (1967); *Dokl. Chem., 175,* 661 (1967).

344. Nesmeyanov, A.N., Vol'kenau, N.A., and Isaeva, L.S., *Dokl. Akad. Nauk SSSR, 176,* 106 (1967); *Dokl. Chem.,*

176, 772 (1967).
345. Nesmeyanov, A.N., Vol'kenau, N.A., Sirotkina, E.I., and Deryabin, V.V., *Dokl. Akad. Nauk SSSR*, *177*, 1110 (1967); *Dokl. Chem.*, *177*, 1170 (1967).
346. Newman, J., and Manning, A.R., *J. Chem. Soc. Dalton Trans.*, *1974*, 2549.
347. Nicholas, K., Raghu, S., and Rosenblum, M., *J. Organometal. Chem.*, *78*, 133 (1974).
348. Nicholas, K.M., and Rosan, A.M., *J. Organometal. Chem.*, *84*, 351 (1975).
349. Nöth, H., and Schuchardt, U., *Z. anorg. allg. Chem.*, *418*, 97 (1975).
350. Noyori, R., Baba, Y., Makino, S., and Takaya, H., *Tetrahedron Lett.*, *1973*, 1741.
351. O'Connor, C., Gilbert, J.D., and Wilkinson, G., *J. Chem. Soc. A*, *1969*, 84.
352. Oehmichen, U., Southern, T.G., Le Bozec, H., and Dixneuf, P., *J. Organometal. Chem.*, *156*, C 29 (1978).
353. Öfele, K., and Dotzauer, E., *J. Organometal. Chem.*, *42*, C 87 (1972).
354. Otsuka, S., Nakamura, A., and Yoshida, T., *Bull. Chem. Soc. Jap.*, *40*, 1266 (1967).
355. Otsuka, S., Nakamura, A., and Yoshida, T., *Justus Liebigs Ann. Chem.*, *719*, 54 (1969).
356. Otsuka, S., Yoshida, T., and Nakamura, A., *Inorg. Chem.*, *7*, 1833 (1968).
357. Petz, W., *J. Organometal. Chem.*, *146*, C 23 (1978).
358. Piraino, P., Faraone, F., and Aversa, M.C., *J. Chem. Soc. Dalton Trans.*, *1976*, 610.
359. Poffenberger, C.A., and Wojcicki, A., *J. Organometal. Chem.*, *165*, C 5 (1979).
360. Powell, P., *Inorg. Chem.*, *7*, 2458 (1968).
361. Quick, M.H., and Angelici, R.J., *J. Organometal. Chem.* *160*, 231 (1978).
362. Raghu, S., and Rosenblum, M., *J. Amer. Chem. Soc.*, *95*, 3060 (1973).
363. Rausch, M.D., Criswell. T.R., and Ignatowicz, A.K., *J. Organometal. Chem.*, *13*, 419 (1968).
364. Reddy, M.L.N., and Urch, D.S., *Discuss. Faraday Soc.*, *47*, 53 (1969).
365. Reddy, M.L.N., Wiles, M.R., and Massey, A.G., *Nature*, *217*, 740 (1968).
366. Redhouse, A.D., *J. Chem. Soc. Dalton Trans.*, *1974*, 1106.
367. Reich-Rohrwig, P., and Wojcicki, A., *Inorg. Chem.*, *13*, 2457 (1974).
368. Reich-Rohrwig, P., Clark, A.C., Down, R.L., and Wojcicki, A., *J. Organometal. Chem.*, *145*, 57 (1978).

369. Reihlen, H., Friedolsheim, A., and Oswald, W., *Justus Liebigs Ann. Chem., 465,* 72 (1928).
370. Reihlen, H., Gruhl, A., and Hessling, G., *Justus Liebigs Ann. Chem., 472,* 268 (1929).
371. Restivo, R., and Bryan, R.F., *J. Chem. Soc. A, 1971,* 3364.
372. Ricci, J.S., Jr., Eggers, C.A., and Bernal, I., *Inorg. Chim. Acta, 6,* 97 (1972).
373. Richards, W.G., *Trans. Faraday Soc., 63,* 257 (1967).
374. Richter , F., and Vahrenkamp, H., *J. Chem. Res. S, 1977,* 155.
375. Richter, F., and Vahrenkamp, H., *Angew. Chem., 90,* 474 (1978); *Angew. Chem. Int. Ed. Engl., 17,* 444 (1978).
376. Richter, F., and Vahrenkamp, H., *Angew. Chem., 90,* 916 (1978); *Angew. Chem. Int. Ed. Engl., 17,* 864 (1978).
377. Robinson, P.W., and Wojcicki, A., *J. Chem. Soc. D, Chem. Commun., 1970,* 951.
378. Röder, A., and Bayer, E., *Angew. Chem., 79,* 274 (1967); *Angew. Chem. Int. Ed. Engl., 6,* 263 (1967).
379. Rodrique, L., Van Meerssche, M., and Piret, P., *Acta Crystallogr., B 25,* 519 (1969).
380. Rosenbuch, P., and Welcman, N., *J. Chem. Soc. Dalton Trans., 1972,* 1963.
381. Ross, D.A., and Wojcicki, A., *Inorg. Chim. Acta, 28,* 59 (1978).
382. Rossetti, R., Cetini, G., Gambino, O., and Stanghellini, P.L., *Atti. Accad. Sci. Torino, 104,* 127 (1969-70).
383. Rossetti, R., Stanghellini, P.L., Gambino, O., and Cetini, G., *Inorg. Chim. Acta, 6,* 205 (1972).
384. Roustan, J.-L., and Charrier, C., *C.R. Acad. Sci., Ser. C, 268,* 2113 (1969).
385. Roustan, J.-L, Merour, J.Y., Benaim, J., and Charrier, C., *C.R. Acad. Sci., Ser. C, 274,* 537 (1972).
386. Roy, J., *Z. Naturforsch., B 25,* 1063 (1970).
387. Rubinson, K.A., and Palmer, G., *J. Amer. Chem. Soc., 94,* 8375 (1972).
388. Ruff, J.K., *Inorg. Chem., 8,* 86 (1969).
389. Schermer, E.D., and Baddley, W.H., *J. Organometal. Chem., 27,* 83 (1971).
390. Schermer, E.D., and Baddley, W.H., *J. Organometal. Chem., 30,* 67 (1971).
391. Schmitt, H.J., and Ziegler, M.L., *Z. Naturforsch., B 28,* 508 (1973).
392. Schrauzer, G.N., and Kisch, H., *J. Amer. Chem. Soc., 95,* 2501 (1973).
393. Schrauzer, G.N., and Kratel, G., *J. Organometal. Chem., 2,* 336 (1964).

394. Schrauzer, G.N., Mayweg, V.P., Finck, H.W., and Heinrich, W., *J. Amer. Chem. Soc.*, *88*, 4604 (1966).

395. Schrauzer, G.N., Mayweg, V.P., and Heinrich, W., *Inorg. Chem.*, *4*, 1615 (1965).

396. Schrauzer, G.N., Rabinowitz, H.N., Frank, J.A.K., and Paul, I.C., *J. Amer. Chem. Soc.*, *92*, 212 (1970).

397. Schubert, M.P., *J. Amer. Chem. Soc.*, *55*, 4563 (1933).

398. Schultz, A.J., and Eisenberg, R., *Inorg. Chem.*, *12*, 518 (1973).

399. Schunn, R.A., Fritchie, C.J., Jr., and Prewitt, C.T., *Inorg. Chem.*, *5*, 892 (1966).

400. Scovell, W.M., and Spiro, T.G., *163rd National Meeting Amer. Chem. Soc.*, Boston, 1972, Abstracts, p. 113.

401. Scovell, W.M., and Spiro, T.G., *Inorg. Chem.*, *13*, 304 (1974).

402. Sellmann, D., Kreutzer, P., and Unger, E., *Z. Naturforsch.*, *B 33*, 190 (1978).

403. Severson, R.G., and Wojcicki, A., *J. Organometal. Chem.*, *149*, C 66 (1978).

404. Shaver, A., Fitzpatrick, P.J., Steliou, K., and Butler, I.S., *J. Amer. Chem. Soc.*, *101*, 1313 (1979).

405. Siebert, W., Augustin, G., Full, R., Krüger, C., and Tsay, Y.-H., *Angew. Chem.*, *87*, 286 (1975); *Angew. Chem. Int. Ed. Engl.*, *14*, 262 (1975).

406. Siebert, W., Full, R., Edwin, J., Kinberger, K., and Krüger, C., *J. Organometal. Chem.*, *131*, 1 (1977).

407. Siebert, W., Renk, T., Kinberger, K., Bochmann, M., and Krüger, C., *Angew. Chem.*, *88*, 850 (1976); *Angew. Chem. Int. Ed. Engl.*, *15*, 779 (1976).

408. Sirotkina, E.I., Nesmeyanov, A.N., and Vol'kenau, N.A., *Izv. Akad. Nauk SSSR, Ser. Khim.*, *1969*, 1524; *Bull. Acad. Sci. USSR, Div. Chem. Sci.*, *1969*, 1413.

409. Sirotkina, E.I., Nesmeyanov, A.N., and Vol'kenau, N.A., *Izv. Akad. Nauk SSSR, Ser. Khim.*, *1969*, 1605; *Bull. Acad. Sci. USSR, Div. Chem. Sci.*, *1969*, 1488.

410. Sloan, T.E., and Wojcicki, A., *Inorg. Chem.*, *7*, 1268 (1968).

411. Smith, S.R., Krause, R.A., and Dudek, G.O., *J. Inorg. Nucl. Chem.*, *29*, 1533 (1967).

412. Springsteen, K.R.M., Greene, D.L., and McCormick, B.J., *Inorg. Chim. Acta*, *23*, 13 (1977).

413. Stanghellini, P.L., Cetini, G., Gambino, O., and Rossetti, R., *Inorg. Chim. Acta*, *3*, 651 (1969).

414. Stanley, K., Groves, D., and Baird, M.C., *J. Amer. Chem. Soc.*, *97*, 6599 (1975).

415. Stevenson, D.L., Wei, C.H., and Dahl, L.F., *J. Amer. Chem. Soc.*, *93*, 6027 (1971).

416. Strohmeier, W., Guttenberger, J.F., and Müller, F.J.,

Z. Naturforsch., *B 22*, 1091 (1967).
417. Strohmeier, W., Guttenberger, J.F., and Popp, G., *Chem. Ber.*, *98*, 2248 (1965).
418. Strouse, C.E., and Dahl, L.F., *J. Amer. Chem. Soc.*, *93*, 6032 (1971).
419. Su, S.R., and Wojcicki, A., *J. Organometal. Chem.*, *27*, 231 (1971).
420. Takahashi, K., Iwanami, M., Tsai, A., Chang, P.L., Harlow, R.L., Harris, L.E., McCaskie, J.E., Pfluger, C.E., and Dittmer, D.C., *J. Amer. Chem. Soc.*, *95*, 6113 (1973).
421. Taube, R., and Steinborn, D., *J. Organometal. Chem.*, *65*, C 9 (1974).
422. Tennent, N.H., Su, S.R., Poffenberger, C.A., and Wojcicki, A., *J. Organometal. Chem.*, *102*, C 46 (1975).
423. Teo, B.K., Hall, M.B., Fenske, R.F., and Dahl, L.F., *Inorg. Chem.*, *14*, 3103 (1975).
424. Teo, B.-K., Wudl, F., Hauser, J.J., and Kruger, A., *J. Amer. Chem. Soc.*, *99*, 4862 (1977).
425. Terzis, A., and Rivest, R., *Inorg. Chem.*, *12*, 2132 (1973).
426. Thomasson, J.E., Robinson, P.W., Ross, D.A., and Wojcicki, A., *Inorg. Chem.*, *10*, 2130 (1971).
427. Tomita, A., Hirai, H., and Makishima, S., *Inorg. Chem.*, *6*, 1746 (1967).
428. Tomita, A., Hirai, H., and Makishima, S., *Inorg. Nucl. Chem. Lett.*, *4*, 715 (1968).
429. Touchard, D., Le Bozec, H., and Dixneuf, P., *Inorg. Chim. Acta*, *33*, L 141 (1979).
430. Treichel, P.M., Wilkes, G.R., and Brauner, M., *Proc. 2nd Internat. Symp. Organometal. Chem.*, Madison, 1965, p. 120.
431. Trenkle, A., and Vahrenkamp, H., *J. Chem. Res. S*, *1977*, 97.
432. Trinh-Toan, Fehlhammer, W.P., and Dahl, L.F., *J. Amer. Chem. Soc.*, *99*, 402 (1977).
433. Trinh-Toan, Teo, B.K., Ferguson, J.A., Meyer, T.J., and Dahl, L.F., *J. Amer. Chem. Soc.*, *99*, 408 (1977).
434. Trost, B.M., and Ziman, S.D., *J. Org. Chem.*, *38*, 932 (1973).
435. Tyurin, V.D., Gubin, S.P., and Nametkin, N.S., *Proc. World Ninth Petr. Congr.*, *5*, 217 (1975).
436. Ungurenasu, C., Stiubianu, G., and Streba, E., *Synth. React. Inorg. Metal-Org. Chem.*, *3*, 211 (1973).
437. Usieli, V., Gronowitz, S., and Anderson, I., *J. Organometal. Chem.*, *165*, 357 (1979).
438. Vahrenkamp, H., *Chem. Ber.*, *103*, 3580 (1970).
439. Vahrenkamp, H., *V. Internat. Conf. Organometal. Chem.*,

Moscow, 1971, Abstracts, p. 77.
440. van Santvoort, F.A.J.J., Krabbendam, H., Roelofsen, G., and Spek, A.L., *Acta Crystallogr., B 33,* 3000 (1977).
441. van Wyk, J.A., *J. Magn. Reson., 19,* 283 (1975).
442. Vergamini, P.J., Ryan, R.R., and Kubas, G.J., *J. Amer. Chem. Soc., 98,* 1980 (1976).
443. Volz, H., and Kowarsch, H., *J. Organometal. Chem., 136,* C 27 (1977).
444. Wagner, R.E., Jacobson, R.A., Angelici, R.J., and Quick, M.H., *J. Organometal. Chem., 148,* C 35 (1978).
445. Watkins, D.D., and George, T.A., *J. Organometal. Chem., 102,* 71 (1975).
446. Weber, H.P., and Bryan, R.F., *J. Chem. Soc. A, 1967,* 182.
447. Weber, L., *J. Organometal. Chem., 122,* 69 (1977).
448. Wei, C.H., and Dahl, L.F., *Inorg. Chem., 4,* 493 (1965).
449. Wei, C.H., and Dahl, L.F., *Inorg. Chem., 4,* 1 (1965).
450. Wei, C.H., Wilkes, G.R., Treichel, P.M., and Dahl, L.F., *Inorg. Chem., 5,* 900 (1966).
451. Weiss, E., and Hübel, W., *J. Inorg. Nucl. Chem., 11,* 42 (1959).
452. Welcman, N., Rosenbuch, P., quoted in R.B. King, N. Welcman, *Inorg. Chem., 8,* 2540 (1969).
453. Werner, R.P.M., *Chem. Ber., 94,* 1207 (1961).
454. Whitesides, G.M., and Boschetto, D.J., *J. Amer. Chem. Soc., 91,* 4313 (1969).
455. Whitesides, G.M., and Boschetto, D.J., *J. Amer. Chem. Soc., 93,* 1529 (1971).
456. Williams, W.E., and Lalor, F.J., *J. Chem. Soc. Dalton Trans., 1973,* 1329.
457. Wojcicki, A., Alexander, J.J., Graziani, M., Thomasson, J.E., and Hartman, F.A., *New Aspects Chem. Metal Carbonyl Deriv., Int. Symp.* 1st, 1968, 6 C; *Chem. Abstr., 72,* 12853 e (1970).
458. Wojcicki, A., Bibler, J.P., Schneider, W., and Hartman, F.A., *Proceedings IX. Internat. Conf. Coord. Chem., St. Moritz,* 1966, p. 175.
459. Wojcicki, A., and Farona, M.F., *147th Meeting Amer. Chem. Soc.,* Philadelphia, 1964, Abstracts, p. 13-L.
460. Zimmermann, J.B., Starinshak, T.W., Uhrich, D.L., and Duffy, N.V., *Inorg. Chem., 16,* 3107 (1977).
461. Yamamoto, Y., and Wojcicki, A., *J. Chem. Soc., Chem. Commun., 1972,* 1088.
462. Yamamoto, Y., and Wojcicki, A., *Inorg. Nucl. Chem. Lett., 8,* 833 (1972).
463. Yamamoto, Y., and Wojcicki, A., *Inorg. Chem., 12,* 1779 (1973).

464. Belg. Patent 574,524 (Society of European Research Associates, S.A.); *Chem. Abstr.*, *54*, 8729 f (1960).
465. Brit. Patent 913,763 (K.W. Hübel, E.L. Weiss, Union Carbide Corp.); *Chem. Abstr.*, *59*, 11568 f (1963).
466. Brit. Patent 1,149,961 (J.D. Seddon, Imp. Chem. Ind. Ltd.); *Chem. Abstr.*, *71*, 3900 v (1969).
467. Ger. Offen. 2,258,230 (Nametkin, N.S., Gubin, S.P., Tyurin, V.D., Fedorov, V.V., Usachev, V.S., Matveev, M.S., Larionov, L.I., Elinaer, A.S., and Barashkov, R.Ya.); *Chem. Abstr.*, *79*, 116823 f (1973).
468. Ger. Patent 1,274,579 (D.T. Thompson, Imp. Chem. Ind. Ltd.); *Chem. Abstr.*, *71*, 77480 p (1969).
469. Neth. Appl. 6,601,681 (Imp. Chem. Ind. Ltd.); *Chem. Abstr.*, *66*, 11796 c (1967).
470. U.S. Patent 3,280,017 (K.W. Hübel, E.H. Braye, Union Carbide Corp.).
471. USSR Patent 491,683 (Nametkin, N.S., Gubin, S.P., Tyurin, V.D., and Fedorov, V.V.); *Chem. Abstr.*, *84*, 124289 e (1976).

Reviews

472. Abel, E.W., and Crosse, B.C., *Sulfur Containing Metal Carbonyls*, *Organometal. Chem. Rev.*, *2*, 443 (1967).
473. Butler, I.S., and Fenster, A.E., *J. Organometal. Chem.*, *66*, 161 (1974).
474. Coucouvanis, D., *The Chemistry of the Dithioacid and 1,1-Dithiolate Complexes*, *Progr. Inorg. Chem.*, *11*, 233 (1970).
475. Gray, H.B., *New Structures in Transition Metal Chemistry (dithiolene complexes)*, *Coord. Chem. Rev.*, *1*, 156 (1966).
476. Jørgensen, C.K., *Spectra and Electronic Structure of Complexes with Sulphur-Containing Ligands*, *Inorg. Chim. Acta Rev.*, *2*, 65 (1968).
477. King, R.B., *Some Applications of Metal Carbonyl Anions in the Synthesis of Unusual Organometallic Compounds*, *Accounts Chem. Res.*, *3*, 417 (1970).
478. Kitching, W., and Fong, C.W., *Insertion of Sulfur Dioxide and Sulfur Trioxide into Metal-Carbon-Bonds*, *Organometal. Chem. Rev. A*, *5*, 281 (1970).
479. Livingstone, S.E., *Metal Complexes of Ligands Containing Sulphur, Selenium or Tellurium as Donor Atoms*, *Quart. Rev.*, *19*, 386 (1965).
480. McCleverty, J.A., *Metal 1,2-Dithiolene and Related Complexes*, *Progr. Inorg. Chem.*, *10*, 49 (1968).
481. Vitzthum, G., and Lindner, E., *Sulfinato-Komplexe*, *Angew. Chem.*, *83*, 315 (1971); *Angew. Chem. Int. Ed.*

Engl., *10*, 315 (1971).

482. Rosenblum, M., *Organoiron Complexes as Potential Reagents in Organic Synthesis*, *Accounts Chem. Res.*, *7*, 122 (1974).

483. Wojcicki, A., *Sulfur Dioxide Insertion Reactions of Transition Metal Alkyls and Related Complexes*, *Accounts Chem. Res.*, *4*, 344 (1971).

484. Wojcicki, A., *Insertion Reactions of Transition Metal-Carbon σ-Bonded Compounds II. Sulfur Dioxide and Other Molecules*, *Advan. Organometal. Chem.*, *12*, 31 (1974).

Index

C

D

E

F

N

O

P

Q

R

S

V

W

X